Innovations in Natural Resource Processing

Proceedings of the Jan. D. Miller Symposium

Edited by

Courtney A. Young
Jon J. Kellar
Michael L. Free
Jaroslaw Drelich
R.P. King

Published by the
Society for Mining, Metallurgy, and Exploration, Inc.

Society for Mining, Metallurgy, and Exploration, Inc. (SME)
8307 Shaffer Parkway
Littleton, Colorado, USA 80127
(303) 973-9550 / (800) 763-3132
www.smenet.org

SME advances the worldwide mining and minerals community through information exchange and professional development. SME is the world's largest association of mining and minerals professionals.

These proceedings were originated by the Mineral and Metallurgical Processing Division of SME.

ISBN 0-87335-241-6

Library of Congress Cataloging-in-Publication Data has been applied for.

Contents

Preface

Historically, faculties have engaged in research but have focused primarily on one or two subject areas, rarely venturing outside them. However, over the years that trend has diminished due to a variety of forces including the desire of granting agencies and companies to fund only cross-disciplinary research, the advent of personal computers, and the immediate availability of information on the Internet. By comparison, faculty such as Dr. Jan D. Miller have always been innovative and cross-disciplinary and have built a network of students, researchers, and postdocs, as well as research and visiting faculty, to stretch research even further. The focus of Dr. Miller and his teams has been associated primarily with the mining industry, specifically mineral and coal processing and extractive metallurgy. These areas involve fundamentals of surface chemistry to flotation (interfacial phenomena), comminution (particle characterization and liberation), hydrometallurgy, and modeling, as well as their application to other industries such as the processing of tar sands and contaminated waters, and extending to secondary resources for environmental and recycling objectives.

The minerals and associated industries are becoming more and more dependent on innovative and cross-disciplinary approaches in order to face the challenges of increasing competition, complying with environmental standards, and remaining profitable. For example, knowledge and experience gained through fundamental and applied studies encourages increased production via improved resource recoveries. This in turn can satisfy environmental requirements by minimizing and possibly preventing contamination. At the same time, through the rigors of education, a more knowledgeable student is produced for employment by associated industries. Dr. Miller and his research team have had success in all of these areas.

Over the past three years, previous and current members of Dr. Miller's research team, primarily his former students, began discussing ways to "return the favor" and opted to conduct a symposium in Dr. Miller's honor. The 2005 SME Annual Meeting was targeted because it is being held in Salt Lake City, home to the University of Utah, where Dr. Miller has been a member of the faculty for 37 years and counting. This symposium addresses innovations in areas of Dr. Miller's research and includes sessions on process fundamentals, interfacial phenomena, hydrometallurgy, applied separations and energy resource recover, and particle characterization and liberation. Prior to the meeting, 12 co-chairs, mostly former students, were selected to help manage the symposium. In addition, various potential authors were solicited and 28 papers were received from around the world, making the event truly international. Following standard review and editorial procedures, the papers were published in these proceedings.

All of the symposium participants and attendees had an influence on or were impacted by Dr. Miller in one way or another. The conference, held February 28 through March 2, 2005, and these proceedings are a special thank you to him.

ORGANIZING AND EDITORIAL COMMITTEE

Courtney A. Young
Metallurgical and Materials Engineering
Montana Tech
Butte, Montana

Jon J. Kellar
Materials and Metallurgical Engineering
South Dakota School of Mines and Technology
Rapid City, South Dakota

Michael L. Free
Metallurgical Engineering
University of Utah
Salt Lake City, Utah

Jaroslaw Drelich
Materials Science and Engineering
Michigan Technological University
Houghton, Michigan

R.P. King
Comminution Center
Metallurgical Engineering
University of Utah
Salt Lake City, Utah

SESSION CHAIRS

Process Fundamentals

Courtney Young
Jon Kellar
Michael Free

Interfacial Phenomena

William Cross
Srinivas Veeramasuneni

Hydrometallurgy

Philip Sibrell
Jose Parga

Applied Separations and Energy Resource Recovery

Jaroslaw Drelich
Dariusz Lelinski

Particle Characterization and Liberation

R.P. King
C.L. Lin
R. Rajamani

Dr. Jan D. Miller— A Biographical Sketch

DR. JAN D. MILLER

Jan Dean Miller, born 7 April 1942, son of Harry Moyer and Mary Virginia Miller, was raised in State College, Pennsylvania. His youth in Central Pennsylvania involved many activities including participation in the Episcopal Boy's Choir and Little League baseball. Early in the 1950s, before television was commonplace, odd jobs around the neighborhood (mowing lawns, washing windows, and so forth) were done to make enough money ($.20 each week) to see the Saturday movie matinee, which included a feature-length film, newsreel, cartoon, and most importantly, that week's installment of the current serial. Each Saturday meant waiting another week to find out if the good guys would escape from some pending disaster plotted by the bad guys.

Eventually, after girls were "discovered," greater financial resources were required. Caddying at the Centre Hills Country Club and later, part-time work in the local community supplied the necessary funds. Summers spent working at the golf course provided many unique and unforgettable experiences.

High school days were filled with local football, basketball, and baseball competition. In 1960 Miller graduated from State College Senior High School and began studies at Pennsylvania State University. His parents encouraged him to study science and engineering, and Miller accepted a scholarship from the Central Pennsylvania Coal Producers' Association. Miller completed a B.S. degree in mineral processing in 1964, graduating with distinction.

College memories include Penn State football at Beaver Field and Penn State basketball and wrestling at Rec Hall. More importantly, Miller's life changed when he married Patricia Ann Rossman, his high school sweetheart, friend, and wife of 41 years. It was also during his university years that Miller became a committed Christian, a step of faith and a decision that has had a most significant impact on his life.

After graduation from Penn State, Miller was encouraged by professors Maurice C. Fuerstenau and Al Schlechten to undertake graduate studies in metallurgical engineering at the Colorado School of Mines (CSM). He began classes in the summer of 1964, having been awarded the Bethlehem Steel Fellowship under the supervision of Professor Fuerstenau. This award was one of considerable significance. Under Fuerstenau's leadership, a dynamic research group was established that provided considerable opportunity for inspirational discussion of mineral processing technology. It was a great era to examine the tenants of mineral processing/hydrometallurgy with fellow graduate students.

Further stimulus at CSM was provided by visits from renowned researchers including A.M. Gaudin, M.E. Wadsworth, and D.W. Fuerstenau.

Miller ultimately completed his M.S. degree and Ph.D. at Mines. In 1968, prior to finishing his Ph.D., Miller accepted a one-year research faculty position in metallurgical engineering at the University of Utah. The following year a regular faculty appointment was offered and Miller has been with the University of Utah ever since. In the years following Miller's appointment, Dr. John Herbst and Dr. H.Y. Sohn joined the metallurgical engineering faculty at Utah and a strong program in mineral processing/extractive metallurgy developed under the guidance of Professor Milton Wadsworth.

Currently, Miller is Department Chair and Ivor Thomas is Professor of Metallurgical Engineering at the University of Utah. Miller has received the Departmental Teaching Excellence Award on three occasions. During his career at Utah, Professor Miller has supervised the research of over 70 graduate students who have successfully defended their theses. Based on their thesis research, a number of these students have received national awards, including the Marcus Grossman Award from the Minerals, Metals and Materials Society (TMS), the Taggart Award from the Society for Mining, Metallurgy and Exploration (SME), Best Paper Competition Award–Graduate Division (SME, TMS, and ASM), and the IPMI Award. In University of Utah competition, the Garr Cutler Energy Award has been won by Miller's students on six occasions. A number of his students have been recognized with the Departmental Teaching Assistant Award, and one student received the College Award. Five Utah graduates now hold tenured university faculty positions and two others have previously held university faculty positions.

Professor Miller is well known for his numerous technical contributions in the areas of surface/colloid chemistry, particle technology, coal preparation, mineral processing, hydrometallurgy, and environmental technology, with more than 400 publications to his credit. Miller's recent research activities involve both fundamental and applied aspects of surface and colloid chemistry, the use of x-ray computer tomography in the 3-D analysis of multiphase particles and packed particle beds, and the development of air-sparged hydrocyclone technology.

Professor Miller is a member of the American Chemical Society, TMS, and SME. He has served on the SME board of directors and is a past chair of SME's Mineral Processing division. He is the recipient of numerous professional society awards, being recognized for his research contributions by both SME and TMS. In 1991 Miller was recognized by AIME with the Robert H. Richards Award. He has received the van Diest Gold Medal and the Distinguished Achievement Medal from the Colorado School of Mines, as well as the Distinguished Research Award from the University of Utah. In 1993 Miller was honored with the A.M. Gaudin Award by SME and was further distinguished the same year by being elected to the National Academy of Engineering. More recently, he was recognized with the AIME Mineral Industry Education Award in 1997, and in 2003 Miller was recognized with both the Frank Aplan Award (AIME/SME) and the Stefanko Best Paper Award (SME).

Over the years many students, faculty, and staff have contributed to these numerous achievements and a special thank you is in order to all. Professor Miller enjoys many blessings; most significant in his journey are his wife, Patricia, and daughters Pamela, Jeanette, and Virginia. The motivation, support, and generosity of these women must certainly be recognized.

GRADUATE STUDENTS SUPERVISED BY J.D. MILLER, 1968–2004

Walter Gerald Peterson–M.S. (1970)
Michael Edward Kelahan–M.S. (1971)
J. Brent Hiskey–M.S. (1971)
Randolph Edward Scheffel–M.S. (1972)
Leo William Beckstead–M.S. (1973), Ph.D. (1975)
Ronald L. Atwood–Ph.D. (1973)
Roger K. Clifford–Ph.D. (1973)
Milind V. Chaubal–Ph.D. (1973)
William K. Tolley–M.S. (1975)
Juan L. Sepulveda J.–M.S. (1975)
Alfread H. Chen–M.S. (1976)
Danny L. Bauer–M.S. (1977)
Rene Antezana G.–M.S. (1977)
Jaime E. Sepulveda J.–M.S. (1977)
Pedro C. Munoz–Ph.D. (1977)
Hilarion Q. Portillo–M.S. (1978)
John B. Ackerman–M.S. (1978)
Elcio F.S. Pereira–Ph.D. (1978)
Ronald B. Mackelprang–M.S. (1979)
Royce Jay Smith–M.S. (1980)
David J. Bunte–M.S. (1981)
Christine M. Meire–M.S. (1981)
Maurits C. Van Camp–M.S. (1981)
Manoranjan Misra–Ph.D. (1981)
Rosme Aguilar H.–M.S. (1982)
Sam Shao-Sue Chang–M.S. (1982)
Chen-Luh Lin–M.S. (1982), Ph.D. (1986)
Michael B. Mooiman–Ph.D. (1984)
Rong-Yu Wan–Ph.D. (1984)
Arturo Cortes–M.S. (1985), Ph.D. (1994)
Philip Sibrell–M.S. (1985), Ph.D. (1992)
Maria Cristina Ruiz–Ph.D. (1985)
Jian Sheng Hu–Ph.D. (1985)
Mustafa Akser–Ph.D. (1987)
Jose Parga T.–Ph.D. (1987)
Ahmed Yehia Abdel Rahman (University of Cairo, Egypt)–Ph.D. (1987)
Jesus L. Valenzuela–M.E. (1988)
Youmin Liu–M.S. (1989)
Luis Bittencourt–M.S. (1989)
Ruiren Jin–Ph.D. (1989)
Woo-Hyuk Jang–M.S. (1990), Ph.D. (1994)
Nathan Rich (Mining Eng.)–M.S. (1990)
Jitesh Bole–M.S. (1991)
S. Gopalakrishnan–Ph.D. (1991)
Jon J. Kellar–Ph.D. (1991)
Ximena Diaz–M.E. (1992)
David J. Kinneberg–Ph.D. (1992)
Qiang Yu–Ph.D. (1992)
Dariusz Lelinski–M.S. (1993), Ph.D. (2002)
Yih-kuan Yen–M.S. (1993), Ph.D. (1997)
Jaroslaw Drelich–Ph.D. (1993)
Madhava Rao Yalamanchili–Ph.D. (1993)
Avimanyu Das–Ph.D. (1994)
Michael L. Free–Ph.D. (1994)
Carlos A. Garcia–Ph.D. (1994)
Edmundo Alfaro-Delgado–M.S. (1995)
Courtney A. Young–Ph.D. (1995)
Yue Ma–M.S. (1997)
Xiansheng Nie–Ph.D. (1997)
Srinivas Veeramasuneni–Ph.D. (1997)
Gustavo A. Munoz-Rivadeneira–M.S. (1998)
Laurie Lee LaPlante–M.S. (1998)
Yongqiang Lu–Ph.D. (1998)
Jinshan Li–M.E. (1999), Ph.D. (2004)
Xuming Wang–M.S. (1999), Ph.D. (2004)
William M. Cross–Ph.D. (1999)
Jakub Nalaskowski (TUG, Poland)–Ph.D. (1999)
Maria A.D. Azevedo–Ph.D. (2000)
Mehmet Hancer–Ph.D. (2000)
Padmabhushana Reddy Desam–M.S. (2001)
Ewelina Mutkowska–M.S. (2001)
Marcin Michal Niewiadomski–M.S. (2001)
Minhua Li–M.S. (2002)
Christian Roldan–M.S. (2004)
Bartek Dabrowski–M.S. (2004)
Ronel du Plessis Kappes–Ph.D. (2004)
Keqing Fa–Ph.D. (2004)
Vamsi Paruchuri–Ph.D. (2004)
Sylwia Wisniewska–Ph.D. (2004)

RESEARCHERS, POSTDOCS AND RESEARCH FACULTY

R. Aranowski
A. Atia
K. Baba
M.W. Baker
R.P. Bokotko
K. Bukka
Miroslav Colic
A. Datta
H.H. Haung
T. Jiang
S.M. Khandrika
J.H. Kim
Diana Kotlyar
N. Liu
Z. Nikolov
I. Palencia
Z. Sadowski
Q.Y. Song
L. Szewczulak
Roberto Trindade
K.R. Upadrashta
G.E. Valadao
Francois Vos
B. Wawrzacz
G. Yamauchi
Y. Ye
D. Yin
A. Zaleska
W. Zmierczak

VISITING FACULTY

M.S. Celik
N. Chakraborti
J.C. Davidtz
Y. Hu
J. Hupka
J.S. Laskowski
T. Mankhand
D. Muir
A.V. Nguyen
O. Ozcan
Josef Pitel
J. Ralston
I. Ritchie
R.G. Robins
R.F. Sandenbergh
Roger Sperline
Michael Sukop
Ron Woods

FACULTY AND ADJUNCT FACULTY

J.G. Byrne
W. Callister
R. Chandran
S. Duyvesteyn
Z. Fang
M. Free
S. Guruswamy
D. Halbe
G. Healy
J.A. Herbst
R.P. King
F.A. Olson
C.H. Pitt
R.K. Rajamani
H.Y. Sohn
M.E. Wadsworth

DEPARTMENT AND COMMINUTION CENTER STAFF

E. Aoki
K. Argyle
D. Bailey
J. Davis
K. Haynes
D. Spurlock

Process Fundamentals

Kinetics of Enhanced Gold Dissolution: Activation by Dissolved Lead

Milton E. Wadsworth* and Ximeng Zhu*

Impurity metals can markedly influence the anodic dissolution of gold in the presence of cyanide. Several important findings related to the influence of lead are reviewed and some new observations presented. Based on the effect of pH, the scan direction, and voltage, it may be shown there are three separate voltage regimes where lead influences the anodic dissolution of gold. These are related to $Pb°/HPbO_2^-$*, Au-Pb /*$HPbO_2^-$ *and* $HPbO_2^-/PbC_2^-$ *oxidation-reduction reactions which provide low resistance paths for electron exchange. The gold dissolution rate was determined by direct solution analysis and is compared to the rate implied by electrochemical measurements.*

INTRODUCTION

It has been shown that passivation results from product layers formed during anodic oxidation of gold (Cathro and Koch, 1964; Kirk, et al., 1978; Kirk and Foulkes, 1980; Nicol, 1980; Thurgood et al.,1981; Sawaguchi et al. 1995). The rate of leaching of gold is considerably less than would be expected either by diffusion of cyanide or oxygen (Zheng, et al., 1995), and the reduced rate is attributed to the formation of a passive layer. Metals such as lead, mercury, bismuth and thallium enhance the rate of dissolution of gold (Nicol 1980). Deschenes and co-workers have demonstrated the influence of lead on the leaching of complex gold ores containing a variety of minerals. (Deschenes and Wallingford, 1995; and Deschenes et al. 2000).

Lapidus has noted the effective, though complex, influence of lead is related to its enhanced solubility (Lapidus 1995), its simplest REDOX reaction being the reduction of Pb_2^+ to Pb°. Lead modifies surface films formed during leaching (Jeffrey and Ritchie

* Department of Metallurgical Engineering, University of Utah, Salt Lake City, Utah

1996) creating reactive sites for gold dissolution. Lead, present as $PbOH_3^-$, should react by cementation forming an Au-Pb alloy which catalyzes the oxidation of the gold by refluxing the lead to solution, serving as a co-oxidant with oxygen (Musatti, Mager, and Martins 1997). Alternately, it has been proposed that cementation of metallic lead on the gold surface results in bi-metallic corrosion. The effect of lead is time-dependent and excessive lead additions passivate the electrode surface (Jeffrey and Ritchie 2000; Jeffrey and Ritchie, 2001; Bek 2001; Bek et al. 2001; Bek and Shuraeva 2002).

The catalytic effect of lead has been attributed to underpotential deposition where, in the presence of oxygen, lead forms an Au-Pb alloy directly without first forming metallic lead (Sandenbergh and Miller 2001). Using "refreshed" electrodes, freshly cut under solution, it was shown the catalytic effect occurs within seconds (Bek et al. 2001)and the cementation of metallic lead on the electrode surface was observed to be a much slower process forming a distinct anodic peak (Bek and Shuraeva 2002). Also high scan rates failed to show a metallic lead peak and the lead remained active in catalyzing gold dissolution. The catalytic effect was attributed to surface adatoms or gold-lead alloy formed by cementation by underpotential deposition, similar to the mechanism proposed by Sandenburgh and Miller (2001).

EXPERIMENTAL

A pure gold rotating disk electrode (99.99 percent from EG&G) was used to obtain the majority of data presented in this study. Equipment used and procedures for electrode preparation have been described elsewhere (Wadsworth et al. 2000). Electrochemical data were obtained by scanning polarization, cyclic voltammetry, and potentiostatic (current vs time) measurements. Tests were carried out under the following conditions, unless otherwise stated: solution composition, 500 ppmNaCN and 0.5 M Na_2SC_4 (as supporting electrolyte); pH = 10.5; electrolyte volume, 250 cm^3; electrode surface area, 0.25 cm^2, electrode rpm, 300; and, scan rate, 1 mV s^{-1}. All voltages are given versus SCE. Oxygen-free anodic curves were obtained using an argon purge. Oxygen reduction was carried out in air saturated solutions with and without cyanide and/or lead present. Lead was added to solution as $Pb(N0_3)_2$ (lead ICP/DCP standard solution, Aldrich Chemical Company, Inc.). The concentration is given as ppm and refers to the amount of lead in solution. Jandel SigmaPlot was used for data analysis and preparation of figures.

RESULTS AND DISCUSSION

A series of voltammograms were obtained with and without lead present to determine the effect of voltage limits, scan direction, cyanide concentration, electrode rpm, and pH.

The Effect of Scan Direction and Voltage Limits

Figure 1 illustrates anodic forward scans from −1.0 to +1.0 V SCE, showing results without dissolved lead (solid line) and with dissolved lead (dashed line). The main peak without lead (solid line) at ~0.18 V is no longer present. In the presence of lead, the main anodic peak appears at ~−0.5 volts. Also evident in Figure 1 is a small peak in the vicinity of −0.67 volts. This peak results from the deposition of metallic lead on the electrode surface (Bek et al. 2001). Also included in Figure 1 is a reverse scan (dotted line) which provides new information in that the upper gold oxidation peaks are missing and

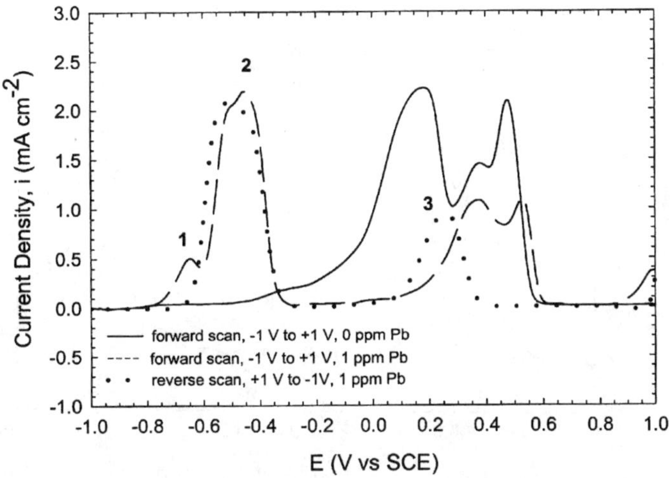

FIGURE 1 Current-Potential curves for gold dissolution in 500 ppm NaCN/0.5 M Na_2SO_4 solutions: pH 10.5; scan rate 1 mV s^{-1}; 300 rpm; Argon purge

a new peak appears at −0.25 V volts. Also apparent in the reverse scan is the absence of the low voltage peak attributed to metallic lead. The formation and nucleation of metallic lead on the surface is a slow process in agreement with Bek et al. (2001). There are three voltage peaks resulting from the presence of dissolved lead. These are shown in Figure 1 as peaks 1, 2, and 3 indicating three distinct voltage regimes in which lead-related activation and passivation is occurring. It is also important to note the higher voltage gold oxidation peaks at ~+0.35 and ~+0.5 V, Figure 1, for the forward scan remain, independent of the presence of lead. Time-dependence for the nucleation and precipitation of metallic lead is apparent when the scan rate is varied, as shown in Figure 2. Peak 1 is missing for scan rates 10 mV s^{-1} and higher.

Figures 3 and 4 illustrate the effect the voltage reversal limit has on results obtained for reverse scans. Figure 3 shows results for the four upper limits, a. 0.5 V, b. 0.6 V, c. 0.7 V and, d. 0.8 V . There is a regular replacement of the two upper gold oxidation peaks and formation of a new single peak. Figure 4 shows results when the upper voltage limits are extended to 0.9, 1.0, and, 1.1 V SCE. A separate new peak is present at ~0.25 V, the same peak as peak 3, Figure 1. At these upper voltage limits, lead reacts to passivate the electrode surface such that, in the reverse scan, the high voltage oxidation peaks for gold are missing. The peak at ~0.25 V represents a new anodic peak associated with a lead oxidation-reduction couple at higher voltage. Figure 5 illustrates results when a forward scan is started at −0.2 volts, continues to +1.0 V (dark dashed line) and is then reversed to −1.0 V. Superimposed is a typical forward scan (light solid line) from −1.0 V to +1.0 V. In the forward scan, beginning at −0.2 volts, it is apparent in this case the new anodic peak does not appear; but, the high voltage oxidation peaks for gold are present. After proceeding to +1.0 V the scan was reversed (dotted line). The new peak (peak 3) at −0.25 V and the anodic peak at ~−0.5 V (peak 2) are present. The shoulder peak at ~−0.69 V (peak 1) is missing.

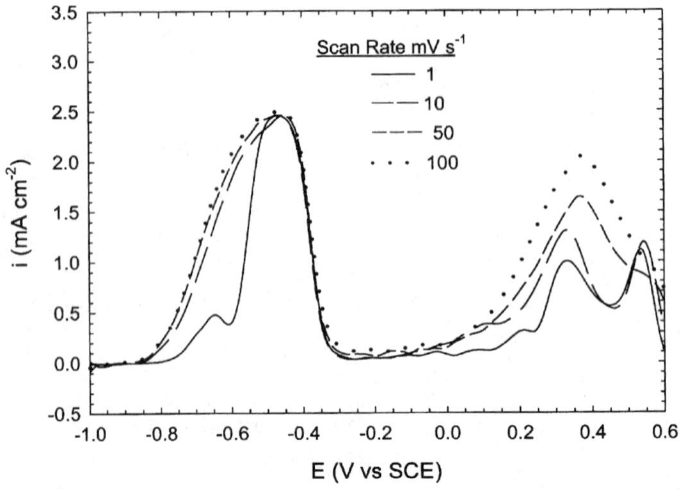

FIGURE 2 Voltammograms of gold in 500 ppm NaCN/0.5 M Na_2SO_4 solutions at various forward scan rates: 1 ppm Pb^{2+}; pH 10.6; electrode rpm 300; Argon purge

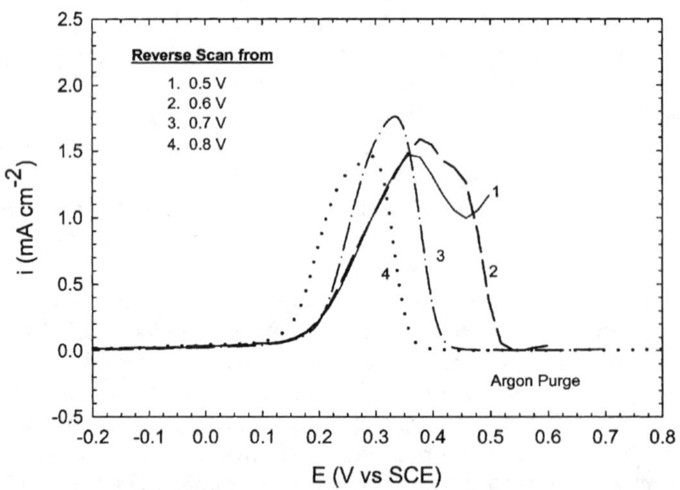

FIGURE 3 Effect of upper voltage limit on reverse scans for 500 ppm NaCN/0.5 M Na_2SO_4 solutions with 1 ppm Pb: scan rate 1 mV s^{-1}, rpm 300

To characterize the effect of lead in the low voltage region, a series ofpotentiostatic curves were obtained at −0.7 V (region of peak 1), −0.6 V (region between peak 1 and peak 2) and at −0.7 volts (region of peak 2) with and without lead present. In Figure 6, current density-time curves 1 and 2, at −0.7 V and −0.5 V illustrate the essentially inactive nature of the gold electrode in the absence of lead. Curve 3 (−0.7 V, 1 ppm Pb) illustrates current density-time results near the maximum of peak 1). The decay in current

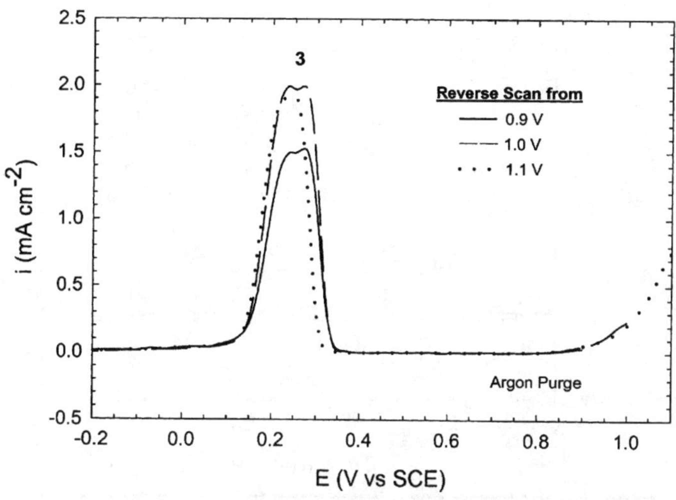

FIGURE 4 Effect of upper voltage limit on reverse scans showing a new peak (peak 3, Figure 1): 500 ppm NaCN/0.5 M Na_2SO_4, 1 ppm Pb: scan rate 1 mV s^{-1}, rpm 300

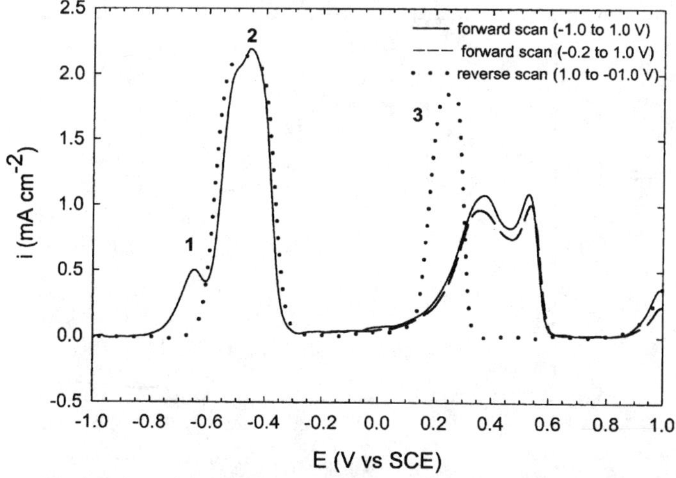

FIGURE 5 Forward and reverse voltammograms of gold showing pattern of activation and passivation in 500 ppm NaCN/0.5 M Na_2SO_4 solution: 1 ppm Pb; pH 10.6; scan rate 1 mVs $^{-1}$; 300 Argon purge

density with increasing time results from passivation caused by nucleation and crystal growth of metallic lead. Curve 4 (−0.6 V, 1 ppm Pb) is at a voltage intermediate between peaks 1 and 2. The current density is higher than that of Curve 3 but continues to decay with increasing time. The electrodes for runs 3 and 4 were slightly grey in color, characteristic of metallic lead. Curve 5 (−0.5 V, 1 ppm Pb) corresponds to the maximum of peak 2 and shows no decay in current density with time. The kinetics here are linear

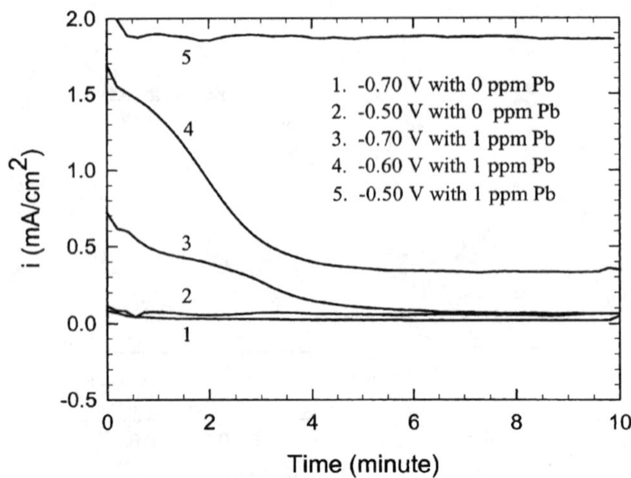

FIGURE 6 Potentostatic curves for gold electrode in 500 ppm NaCN/0.5 M Na_2SO_4 solutions with and without 1 ppm Pb: pH 10.6, rpm 300, Argon purge

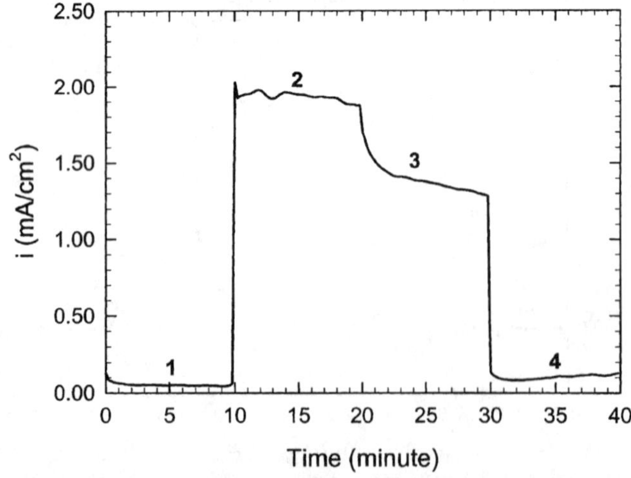

FIGURE 7 Potentiostatic (–0.5 V) current-time results for gold electrode in 500 ppm NaCN/ 0.5 M Na_2SO_4 solutions for various electrode treatments: pH 10.6, rpm 300, Argon purge

with time and reflect the true catalytic behavior of lead. The electrode appeared slightly reddish during this run.

A series of experiments were conducted to demonstrate the tenacity with which lead is held on the surface of the gold electrode, Figure 7. For the first 10 minutes, in the absence of lead, there was a very low current density, Curve 1. After 10 minutes, lead was added to the solution. The rate increased to the higher current density, Curve 2. After 10 additional minutes, the electrode was washed and transferred to a new solution

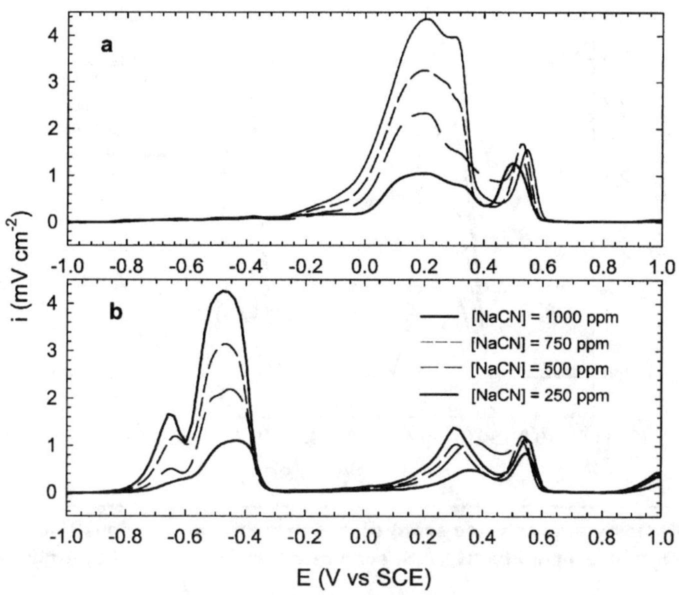

FIGURE 8 Voltammograms (forward scan) of gold in solutions of various NaCN concentrations in 0.5 M Na_2SO_4: a. 0 ppm Pb, b. 1 ppm Pb: pH 10.6, scan rate 1 mV s^{-1}, rpm 300, Argon purge

without lead. The current density, Curve 3, diminished somewhat but still remained high. At 30 minutes the electrode was re-polished and replaced in the lead-free electrolyte where it resumed its inactive behavior. These results support a mechanism in which lead bonds strongly with gold, as an alloy or as adatoms, (Musatti, Mager, and Martins 1997; Sandenbergh and Miller 2001; and, Bek et al. 2001). An important example of underpotential deposition, UDP, of Cd and Pb on silver has been well characterized by electrochemical methods (Martins et al. 1994), Clearly such a mechanism is an important component of lead-enhanced dissolution of gold.

The Effect of Cyanide Concentration and Rotation Speed

The first order effect of cyanide concentration on the rate of gold dissolution with and without lead has been well documented in many previous studies. In considering lead catalysis of gold dissolution, prior results have been presented for only the −0.25 V peak (peak 2). Included here are all three lead-enhanced anodic peaks compared to results in the absence of lead under the same conditions. Figures 8a and 8b show forward scan voltammograms (−1.0 to +1.0 V) for gold dissolution without and with lead present for four cyanide concentration: 250, 500, 700, and, 1000 ppm NaCN. Figure 9 illustrates reverse scan voltammograms (+1.0 to −1.0 V) for gold dissolution in the presence of 1 ppm Pb. Maximum current density versus cyanide concentration (Figures 8a, 8b, and 9) plots are presented in Figure 10. With the exception of the ~−0.65 V peaks, all peak current densities show similar linear dependence on cyanide concentration. These results are surprising since maximum peak current densities correspond to rates below

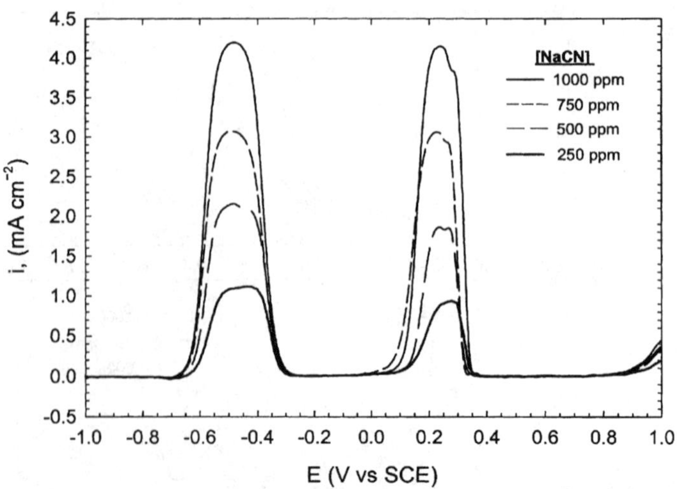

FIGURE 9 Voltammograms (reverse scan) of gold in solutions of various NaCN concentrations in 0.5 M Na_2SO_4 with 1 ppm Pb: pH 10.6, scan rate 1 mV s^{-1}, rpm 300, Argon purge

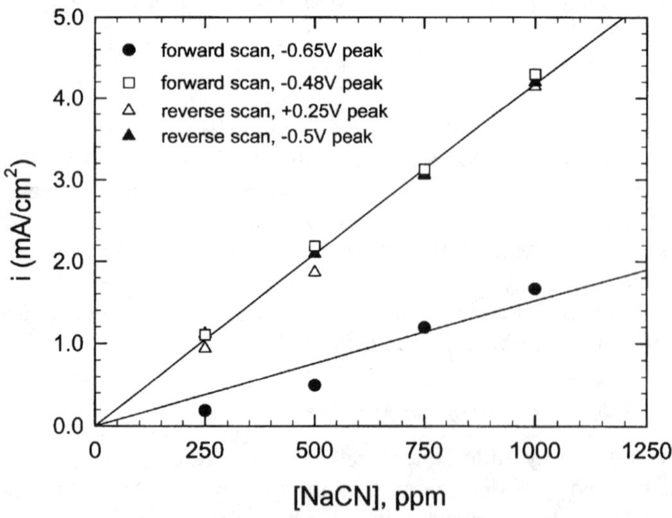

FIGURE 10 Linear relationship between current density, *i*, and NaCN concentration. Data from Figures 8 and 9.

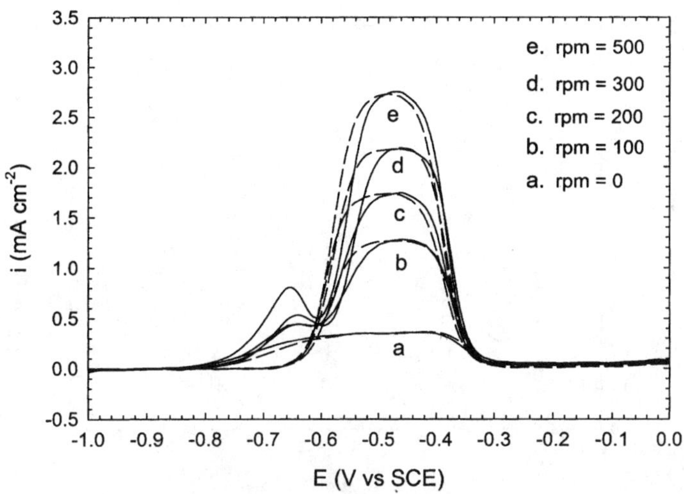

FIGURE 11 Forward (solid line) and revers (dashed line) voltammograms of gold in 500 ppm NaCN/0.5 M Na_2SO_4 solutions for different electrode rotation speeds: pH 10.6, 1 ppm Pb, scan rate 1 mVs^{-1}, Argon purge

those expected for CN^- diffusion alone. This suggests the rate is of mixed kinetic control and only reactions between gold and cyanide are controlling. The simplest model consistent with these results is one which includes diffusion of CN^- followed by activated adsorption, establishing a steady-state concentration of cyanide, $[CN^-]_s$, at the solution-electrode interface. Accordingly,

$$i = \frac{FD_{CN}}{2\delta}[CN^-] - \frac{FD_{CN}}{2\delta}[CN^-]_s \quad \text{(diffusion)} \qquad \textbf{(EQ 1)}$$

at steady-state,

$$i = F[CN^-]_s k_a \quad \text{(activated adsorption)} \qquad \textbf{(EQ 2)}$$

The equation for δ (Levich 1973)

$$\delta = 1.61\omega^{-1/2}\nu^{1/6}D_{CN}^{1/3} \qquad \textbf{(EQ 3)}$$

may be substituted in Eq. (1) and combined with Eq. (2) giving,

$$i = \frac{0.31FD_{CN}^{2/3}\omega^{1/2}\nu^{-1/6}[CN^-]}{1 + 0.31D_{CN}^{2/3}\omega^{1/2}\nu^{-1/6}k_a^{-1}} \qquad \textbf{(EQ 4)}$$

The linear dependence on $[CN^-]$ is consistent with Eq. (4) but the dependence on angular velocity is more complex. Equation (4) may also be expressed as,

$$i^{-1} = (0.31FD_{CN}^{2/3}\nu^{-1/6}[CN^-])^{-1}\omega^{-1/2} + (Fk_a[CN^-])^{-1} \qquad \textbf{(EQ 5)}$$

Figure 11 shows forward and reverse anodic peaks for four rotation speeds.

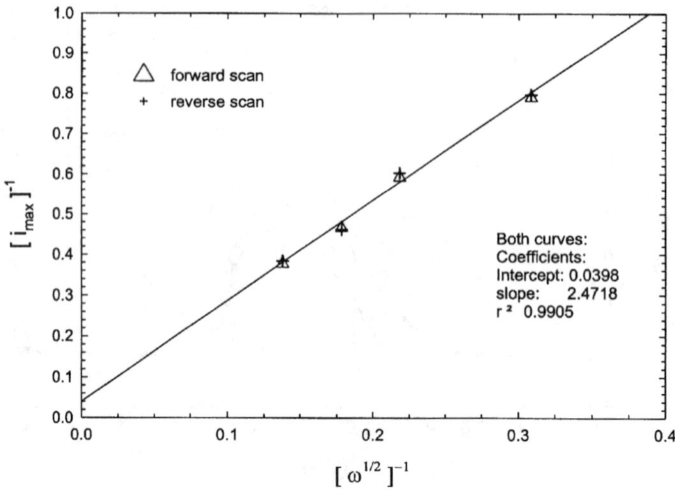

FIGURE 12 Plot showing linear correlation between $[i_{max}]^{-1}$ and $[\omega^{1/2}]^{-1}$ for the data for Figure 11 at –0.44 V

Figure 12 is a plot of i^{-1} versus $\omega^{-1/2}$ illustrating the straight-line relationship in accordance with Eq. (5). Using the value of 0.0089 $cm^2\ s^{-1}$ for the kinematic viscosity, ν, (Zheng et al., 1995) and the measured slope, the calculated value for D_{CN} is $1.48 \times 10^{-5}\ cm^2\ s^{-1}$.

Effect of pH and Thermodynamic Considerations

The influence of pH provides important information explaining the anodic dissolution of gold in the presence of lead. Figure 13 illustrates four forward scans obtained at pH values of 9.89, 10.6, 11.0, and, 11.5, showing both the low voltage as well as the high voltage anodic peaks. The increasing shift to more positive potentials with decreasing pH has been noted by other investigators (Jeffrey, Ritchie, and, LaBrooy 1996; Musatti et al., 1997; Bek and Shuraeva 2002). Figure 14 illustrates similar pH response for reverse scans for both low and high voltage anodic peaks.

These results may be compared with equilibria for the two oxidation-reduction couples for lead, $Pb°/HPbO_2^-$, and $HPbO_2^-/PbO_3^{2-}$. Lines corresponding to these reactions are shown in Figure 15. Also included in Figure 15 are measured voltages corresponding to peak maxima for the curves shown in Figures 13 and 14.

Line a of Figure 15 corresponds to the oxidation-reduction reaction (Pourbaix 1966),

$$Pb° + 2H_20 = HPbO_2^- + 3H^+ + 2e^- \quad \textbf{(EQ 6a)}$$

and for a lead concentration of 4.83×10^{-6} (1 ppm), referred to SCE,

$$E_{o\ (SCE)} = 0.303 - 0.0886\ pH \quad \textbf{(EQ 6b)}$$

The measured E_o-pH values for the 3 small peaks near –0.65 V, Figure 13, fall essentially on line a (open triangles). The anodic dissolution ofAu in the presence of cyanide,

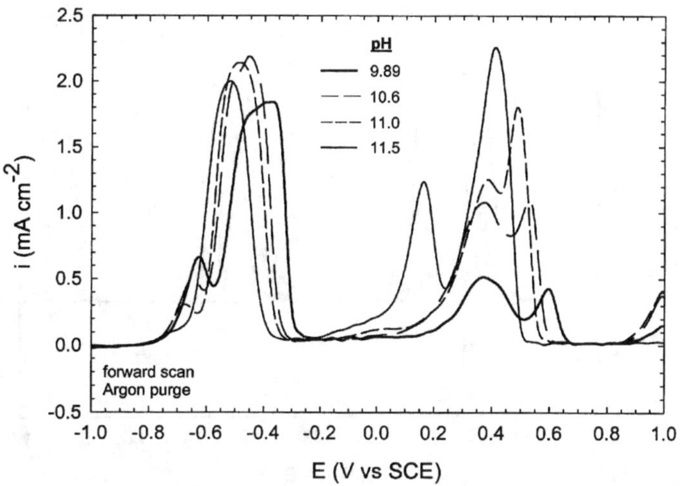

FIGURE 13 Voltammograms for gold in 500 ppm NaCN/0.5 M Na_2SO_4 solutions: various pH, 1 ppm Pb, scan rate 1 mV s^{-1}, rpm 300

$$Au + CN^- = AuCN + e^- \quad \textbf{(EQ 7)}$$

coupled with Eq. (6a) results in the cementation reaction

$$HPbO_2^- + 2Au + 2CN^- + 3H^+ = 2AuCN + Pb° + 2H_2O \quad \textbf{(EQ 8)}$$

Rapid electron exchange results in bimetallic corrosion (Jeffrey and Ritchie 2000) and gold dissolution occurs by the reaction

$$AuCN + CN^- = Au(CN)_2^- \quad \textbf{(EQ 9)}$$

rate limited by the concentration, $[CN^-]_s$, at the electrode surface.

Peak values in the -0.3 to -0.7 voltage range for the forward scan, Figure 13, and the reverse scan, Figure 14, fall on a line with an average linear regression slope of 0.0857,

$$E_{o\,(SCE)} = 0.477 - 0.0857\ pH \quad \textbf{(EQ 10)}$$

This line is shifted (ΔE_1, Figure 15) approximately +0.174 V above line a, suggesting the cementation reaction is,

$$HPbO_2^- + 3Au + 2CN^- + 3H^+ = 2AuCN + Au\text{-}Pb + 2H_2O \quad \textbf{(EQ 11)}$$

where Au-Pb is a surface alloy or surface adatom (Musatti, Mager, and Martins 1997 Sandenbergh and Miller 2001; and, Bek and Shuraeva 2002). The positive voltage shift, ΔE_1, of Figure 15, may result from the free energy of Au-Pb bonding, which for ΔE_1 = +0.174 V would be ~-8.0 kcal mol^{-1} (~33.5 kJ mol^{-1}).

Similar results were observed using "refreshed" electrodes in the 12–14 pH range and with a scan rate of 50 mV s^{-1} (Bek and Shuraeva 2002). The electrodes were held for 60 seconds at a potential 20 mV below that required for deposition of metallic lead and then scanned anodically. The $HPbO_2^-/Pb°$ and $HPbO_2^-$/Au-Pb peaks were clearly separated. Their E_o-pH plot corresponding to the $HPbO_2^-/Pb°$ exchange reaction had a slope

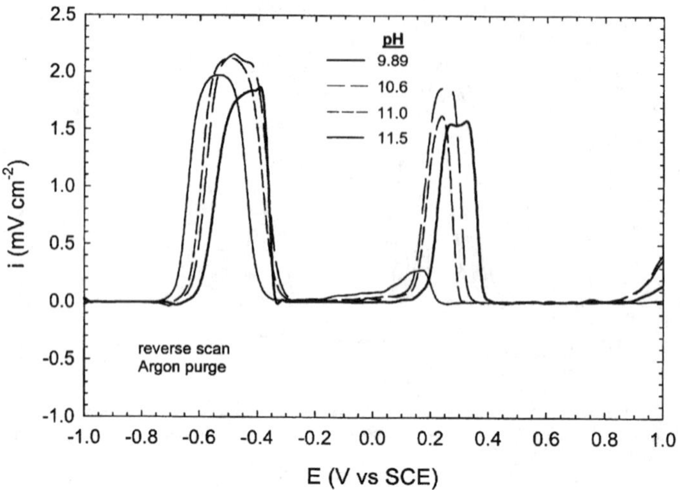

FIGURE 14 Voltammograms for gold in 500 ppm NaCN/0.5 M Na_2SO_4 solutions: various pH, 1 ppm Pb, scan rate 1 mV s^{-1}, rpm 300

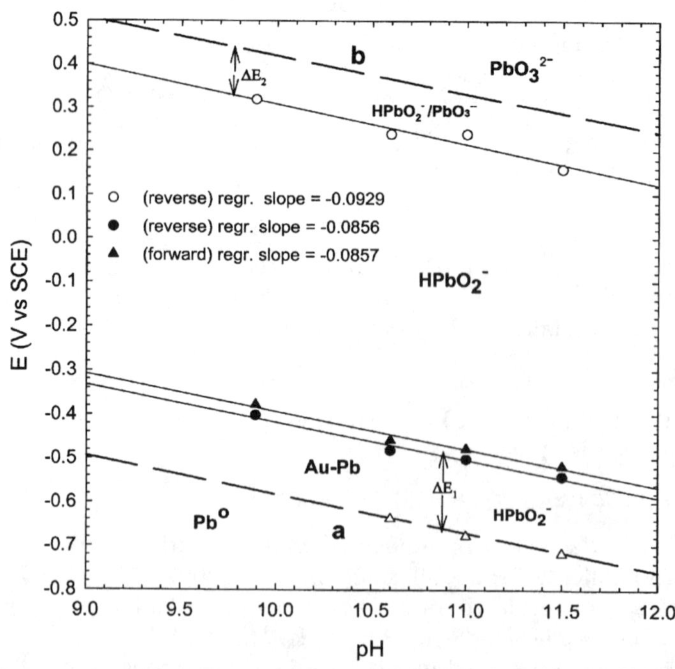

FIGURE 15 Voltage (corresponding to peak current density maxima) for voltammograms obtained at several pH values for forward (Figure 13) and reverse (Figure 14) scans for electrolytes containing 1 ppm Pb

of −0.071 and that corresponding to the $HPbO_2^-$/Au-Pb exchange reaction had a slope of −0.073. The observed voltage shift was ~+0.061 V. Results obtained under very rapid scans (Bek and Shuraeva 2002), the absence of the Pb° peak for reverse scans and the instability of lead in the presence of oxygen (Sandenberg and Miller 2001) strongly indicate Eq. 11 is the operative cementation exchange reaction for lead activated leaching of gold.

Line b, Figure 15, corresponds to the $HPbO_2^-$/Au-Pb_3^{2-} oxidation-reduction couple

$$HPbO_2^- + H_2O = PbO_3^{2-} + 3H+ + 2e^- \quad \textbf{(EQ 12a)}$$

which, for the domain limit, is

$$E_{o(SCE)} = 1.305 - 0.0886\ pH \quad \textbf{(EQ 12b)}$$

for a lead concentration of 4.83×10^{-6} M (1 ppm Pb) (Pourbaix 1966). The observed E_o-pH points for reverse scan peaks (~voltage range 0.15–0.4 V, Figure 14) are shown in Figure 15 (open circles) and have a regression slope of −0.0929, such that

$$E_{o\ (SCE)} = 1.236 - 0.0929\ pH \quad \textbf{(EQ 13)}$$

The line parallels line b (Figure 15) but is displaced ~-0.1 V, and is shown as ΔE_2 in Figure 15 The upper voltage peak (~+0.25 V, Figure 9) appears to be a doublet which may represent two single electron transfer steps.

As noted earlier, gold oxidation peaks in the 0.1 to 0.6 V range are missing for the reverse scan in the presence of lead, providing the upper voltage limit exceeds ~0.9 V. These peaks remain during the forward scan, even in the presence of lead. The oxidation of gold in this region produces $Au(OH)_3$ which passivates the gold surface. This is evidenced by the presence of oxidation peaks during the reverse scan if lead is not present. Also these oxidation peaks occur in the presence of lead if the upper voltage limit is below 0.8 V, SCE. In the high voltage range PbO_3^{2-} will form (Eq. 12a), but will compete with the reaction

$$Au(OH)_3 + 3H^+ + 3e^- = Au + 3H_2O \quad \textbf{(EQ 14)}$$

Combining the half cell reactions, Eqs. (12a) and (14), predicts reduction of $Au(OH)_3$ is slightly favored according to the reaction,

$$2\ Au(OH)_3 + 2HPbO_2^- = 2Au + 3PbO_2^- + 3H^+ + 3H_2O \quad \textbf{(EQ 15)}$$

$\Delta F°$ = −2.49 kcal (−10.4 kJ) per mol of lead, but would be more favorable if $HPbO_2^-$ and PbO_3^{2-} adsorb on the gold surface. Sustained adsorption of the hydrated lead species is required to explain the catalytic behavior of the $HPbO_2^-/PbO_3^{2-}$ couple. Assuming Langmuir adsorption the equation for the domain limit (Eq. 12b) would become

$$E_{o\ (SCE)} = 1.305 - 0.0886\ pH + 0.0295\ (\Delta F_2 - \Delta F_3) \quad \textbf{(EQ 16)}$$

where ΔF_2 and ΔF_3 are the free energy of adsorption of $HPbO_2^-$ and PbO_3^{2-}. A voltage shift (ΔE_2, Figure 15) of −0.1 V would require ΔF_2 to be more negative than ΔF_3 by only −4.6 kcal mol⁻ (−19.3 kJ mol–). On the reverse scan, the half-cell reactions Eqs 12a and 14 result in the electron exchange reaction

$$PbO_3^{2-} + 2Au + 2CN^- + 3H^+ = HPbO_2^- + 2AuCN + H_20 \quad \textbf{(EQ 17)}$$

$\Delta F°$ = −101 kcal (−423 kJ) per mol of lead.

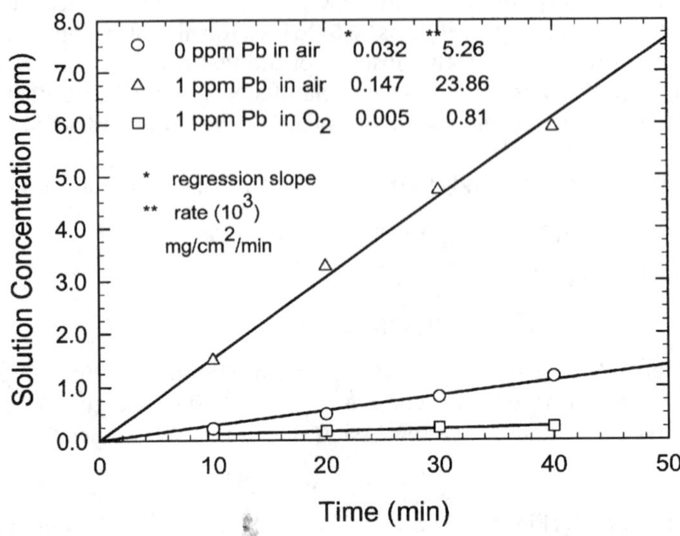

FIGURE 16 Measured gold leaching rate, with and without Pb, with air and oxygen purging in 500 ppm NaCN/0.5 M Na_2SO_4 solutions: electrode area 1.54 cm^2, pH 10.5, 25°C, rpm 300

Oxygen Reduction and the Influence of Lead

To compare rates of dissolution in the presence of oxygen, rates of gold dissolution were determined (Figure 16) by solution analysis versus time for a variety of conditions. Air sparged solutions with 1 ppm lead had a rate of 23.9 mg cm^{-2} min^{-1}, approximately 4.5 times the rate for air sparged solutions in the absence of lead, 5.3 mg cm^{-2} min^{-1}. For oxygen sparged solutions, the rate diminished, in agreement with prior findings (Jeffrey, Ritchie and LaBrooy 1996).

Figure 17 illustrates oxygen (air) reduction curves obtained for a variety of conditions. Curve 1 represents oxygen reduction on a gold electrode in the absence of Pb or NaCN. Curve 2 shows results when 1 ppm Pb and 0 NaCN are present and Curve 3 when 0 Pb and 500 ppm NaCN are present. Lead enhances oxygen reduction in the absence of cyanide while cyanide depresses oxygen reduction in the absence of lead. Figure 4 shows results when 1 ppm Pb and 500 ppm NaCN are present. The sharp rise in cathodic current with decreasing potential, shows that lead enhances oxygen reduction in the presence of cyanide, similar to Curve 2. The potential at zero current density (Curve 4) is the mixed potential E_m = −0.625 V (SCE) or −0.383 V (SHE). This result compares favorably with −0.644 (SCE) (−0.402 V (SHE)) measured for 980 ppm NaCN at pH 10.1 (Jeffrey, Ritchie and La Brooy 1996).

To approximate the true oxygen reduction curve, independent of anodic currents, anodic curves of Figure 5 (Curves 1, with metallic lead deposition and Curve 2 without metallic lead deposition) were subtracted from oxygen Curve 2, Figure 17. The results are shown in Figure 17 as Curve 5 (with metallic lead deposition) and Curve 6 (without lead metallic deposition. Curve 6, Figure 17, closely matches Curve 4, providing separate evidence that metallic lead does not precipitate during gold leaching in the presence of oxygen, as previously proposed (Sandenbergh and Miller 2001).

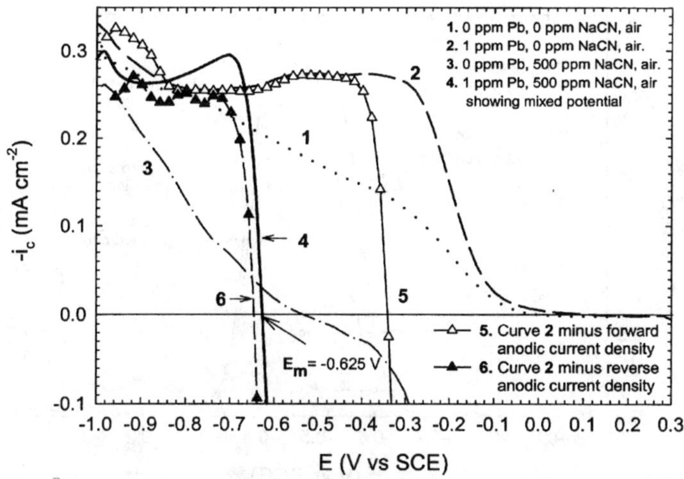

FIGURE 17 Voltammograms for oxygen reduction (air) on Au with and without additives, Pb (0 and 1 ppm) and NaCN (0 and 500 ppm) in 0.5 M Na_2SO_4 solutions: pH 10.6 scan rate 1 mV s^{-1}, rpm 300

The vertical line, Figure 18, corresponds to the measured mixed potential (-0.625 V, Figure 17). It intersects two anodic peaks at mixed currents, mc_1, and mc_2. The mc_2 current density corresponds to a rate of 61 mg cm^{-2} min^{-1} while the me, current density corresponds to a rate of 22 mg cm^{-2} min^{-1} in close agreement with the measured rate of 23 mg cm^{-2} min^{-1}, Figure 16. In actual leaching practice, in the presence of oxygen, other constraints apply depending upon the cyanide and oxygen concentrations. Firstly, the benefit of lead is limited to the threshold of the anodic peak and secondly (Jeffrey and Ritchie 2000) with increasing cyanide concentration the rate of reaction shifts from cyanide rate control to oxygen diffusion control.

CONCLUSIONS

1. Three lead REDOX reactions catalyze gold dissolution. These are Pb°/$HPbO_2^-$, Au-Pb/$HPbO_2^-$, and $HPbO_2^-$/PbO_3^{2-}, where Au-Pb is a surface alloy or adatom. These REDOX reactions provide low resistance paths for electron exchange.
2. At maximum anodic currents both Au-Pb/$HPbO_2^-$ and $HPbO_2^-$/PbO_3^{2-} exchange reactions give similar results with mixed kinetic gold-cyanide interactions.
3. In gold leaching practice the Au-Pb/$HPbO_2^-$ is the operative electron exchange reaction.
4. Lead catalysis in the presence of oxygen is limited to low threshold anodic currents, such as mc_1 of Figure 18.

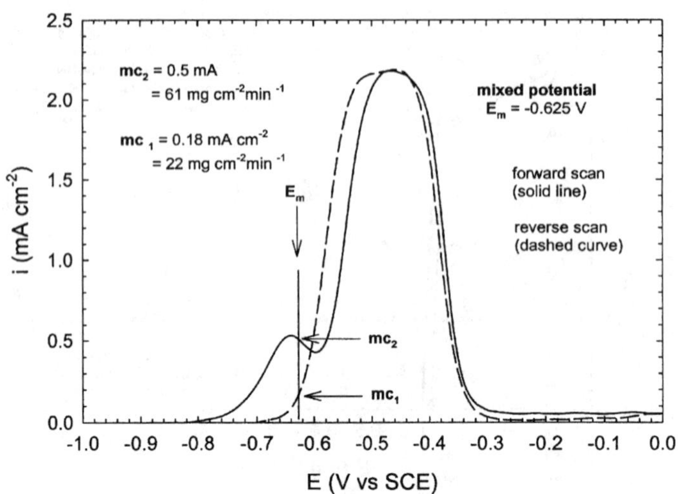

FIGURE 18 Anodic gold peaks showing mixed currents when metallic lead (solid curve) is present, mc_2, and when an Au-Pb phase (long dashes) is present, mc_1: 500 ppm NaCN/0.5 M Na_2SO_4 solutions, pH 10.6, 1 ppm Pb, scan rate 1 mV s^{-1}, electrode rpm 300

REFERENCES

Bek, R.Y., G.V. Kosolapov, L.I. Shuraeva, S.N. Ovchiimikova, P.A. Laskarzhevskii, and, A.A. Vais, 2001, Anodic Dissolution of Gold in Alkali-Cyanide Solutions: Effect of Lead Ions, Russian J.of Electrochemistry, 37 (3), 244–249, Translated from Elektrokhimiya, 37 (3), 281-286.

Bek, R.Y., 2001, Kinetics of Anodic Dissolution of Gold in Alkali-Cyanide Solutions: Effect of Lead Ions, Russian J. of Electrochemistry, 37 (4) 387–393, Translated from Elektrokhimiya, 37 (4), 448–456.

Bek, R.Y., and, L.I. Shuraeva, 2002, Anodic Dissolution of Gold in Cyanide Solutions Containing Lead Ions: Effect of Solution pH, Russian J. of Electrochemistry, 38 (5) 526–530, Translated from Elektrokhimiya, 38 (5), 589–593.

Cathro, K. J., and, D. F.A., Koch, 1964, The Anodic Dissolution of Gold in Cyanide Solutions, J. Electrochem. Soc., 111, 1416–1420.

Deschenes, G., and, G. Wallingford, 1995, Effect of Oxygen and Lead Nitrate on the Cyanidation of a Sulphide Bearing Ore, Minerals Engineering, 8, No. 8, 923–931.

Deschenes, G., R. Lastra, J.R. Brown, S. Jin, 0. May, and, E. Ghali, 2000, Effect of Lead Nitrate on Cyanidation of Gold Ores: Progress on the Study of the Mechanisms , Minerals Engineering, 13, No. 12, 1263–1279.

Jeffrey, M.I., I.M. Ritchie, and, S.R. LaBrooy, 1996, The Effect of Lead on the Electrochemistry of Gold: Myth or Magic, Electrochemical Proceedings, 96-6, 284–295.

Jeffrey, M.I., and, I.M. Ritchie, 2000, The Leaching of Gold in Cyanide Solutions in the Presence of Impurities I. The Effect of Lead, J. Electrochem. Soc, 147 (9), 3257.

Jeffrey, M.I., and, I.M. Ritchie, 2001, The Leaching and Electrochemistry of Gold in High Purity Cyanide Solutions, J. Electrochem. Soc., 148 (4), D29–D36.

Kirk, D.W., F.R. Foulkes, and, W.F. Graydon,,1978, A Study of Anodic Dissolution of Gold in Aqueous Alkaline Cyanide. J. Electrochem. Soc., 125, 1436–1443.

Kirk, D.W., and, F.R. Foulkes, 1980, Anodic Dissolution of Gold in Aqueous Alkaline Cyanide Solutions at Low Overpotentials, J. Electrochem. Soc., 127, 1993–1997.

Lapidus, G., 1995, Unsteady-State Model for Gold Cyanidation on a Rotating Disk, Hydrometallurgy, 39, 251–263.

Levich, V.G., 1973, Physiochemical Hydrodynamics, Prentice Hall, New York, N.Y.

Mac Arthur, D.M., 1972 , A Study of Gold Reduction and Oxidation in Aqueous Solutions. J. Electrochem. Soc., 672–677.

Martins, M.E., A. Hemandez-Creus, R.C. Salvarezza, and, A.J. Arvia, 1994, Dynamics of Surface Alloying Promoted by Cadmium Underpotential Deposition-Anodic Stripping Cycles on Irregular Silver Deposits, J. Ellectroanal. Chem., 375, 141–155.

Musatti, D., J. Mager, and, G.P. Martins, 1997, D.B. Dreisinger, ed., Electrochemical Aspects of the Dissolution of Gold in Cyanide Electrolytes Containing Lead, in Aqueous Electrotechnologies: Progress in Theory and Practice, The Minerals, Metals & Materials Society, 248–265.

Nicol, M.J., 1980, The Anodic Behaviour of Gold. Gold Bulletin, Part II-Oxidation in Alkaline Solutions, 13, 105–111.

Pourbaix, M., 1966, Atlas ofElectrochemical Equilibria in Aqueous Solutions, Pergamon Press, New York, N.Y.

Sawaguchi, S., T. Yamada, and, Y. Okinaka,, 1995, Electrochemical Scanning Tunneling Microscopy and Ultrahigh-Vacuum Investigation of Gold Cyanide Adiayers on Au(III) Formed in Aqueous Solution, J. Phys. Chem, 99, 14149–14155.

Sandenbergh, R.F., and, J.D. Miller, 2001, Catalysis of the Leaching of Gold in Cyanide Solutions by Lead, Bismuth, and Thallium, Minerals Engineering, 14 (11), 1379–1386.

Thurgood, C.P., D.W. Kirk, F. R. Foulkes,, and, W.F. Graydon, 1981, Activation Energies of Anodic Gold Reactions in Aqueous Alkaline Solutions, J. Electrochem. Soc., 128,1670–1685.

Tshilombo, A.F., and, R.F. Sandenbergh 2001, An Electrochemical Study of the Effect of Lead and Sulphide Ions on the Dissolution Rate of Gold in Alkaline Cyanide Solutions, Hydrometallurgy, 60, 55–67.

Wadsworth, M. E., X. Zhu, J. S. Thompson, and, C. J. Pereira, 2000, Gold Dissolution and Activation in Cyanide Solution: Kinetics and Mechanism, Hydrometallurgy, 57, 1–11.

Zheng, J., I.M. Ritchie, S.R. La Brooy, and, P. Singh, 1995, Study of Gold Leaching in Oxygenated Solutions Containing Cyanide-Copper-Ammonia Using a Rotating Quartz Crystal Microbalance, Hydrometallurgy, 99, 277–292.

Metal Deposition and the Electrocrystallization Process

J. Brent Hiskey*

INTRODUCTION

Metal electrodeposition processes are of fundamental technological importance to a number of fields. Electrorefining, electrowinning, electroplating, and electrogalvanizing have all been the subject of intense study for over a century. More recently, electrochemical deposition has played a critical role in advances in microelectronics and lately has been an enabling technology for the production of certain nanostructured materials.

Electrometallurgists have most frequently been concerned with the structure and morphology of massive deposits. For example, in metal recovery steps such as electrorefining and electrowinning, deposits demonstrating adherent coarse-grained characteristics are most desirable. In the case of electroplating, a bright fine-grained, tenacious and strongly adherent deposit is required. The fabrication of microelectronic devices and the production of nanoscale structures such as nanoparticles, nanowires, and nanorods requires a keen understanding of early stage deposition effects.

The initial stages of metal deposition, usually referred to as *electrocrystallization*, are fundamentally linked to nucleation and growth phenomena. Early stage deposits are typically very ordered. Whereas, bulk metal deposition structures are much more random. In both cases, the discharge of a metal ion at an electrified interface is represented by simple cathodic reduction as follows:

$$Me(H_2O)_x^{n+} + ne = Me + xH_2O \quad \textbf{(EQ 1)}$$

$Me(H_2O)_x^{n+}$ is a hydrated metal ion residing in the bulk solution and Me is the metal atom occupying a crystallographic lattice position. A similar reduction reaction can be

* University of Arizona, Tuscon, Arizona

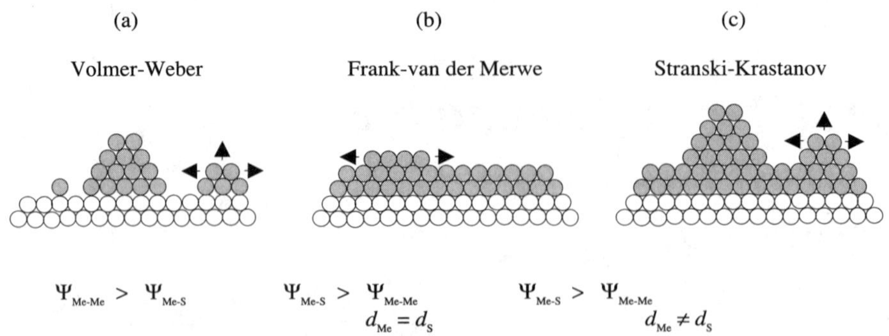

FIGURE 1 Schematics showing different growth mechanisms for deposition metal

written for a complexed metal ion, as would be the case for the deposition of Au from a cyanide bath. What appears to be a straightforward reaction is in fact a complex series of mass and charge transfer steps. Before a metal ion can be successfully incorporated into a normal lattice position, it must expel its waters of hydration or the ligands in its coordination sphere and receive its electrons. Bockris and Reddy (1970) have provided a very clear treatment of the dehydration of a hydrated ion as it moves progressively from *planar surface* to *step edge* to *kink* to *edge vacancy* and finally to a *hole site* upon a crystal-lattice plane.

Epitaxial growth is important with respect to thin-film formation and initial stages of deposition. Under these conditions crystallographic growth patterns have been found to follow three distinct mechanisms depending on the adatom-substrate interaction energy and the lattice misfit. Schematic representations of the three growth modes are depicted in Figure 1 (Staikov et al. 1994; Lorenz and Staikov 1995).

The Volmer-Weber (3-D island growth) mechanism shown in Figure 1a is favored when the Me-Me interaction energy is greater than the Me-S interaction energy, irrespective of lattice fit. On the other hand, the Frank-van der Merwe (layer-by-layer) mechanism drawn in Figure 1b is expected when the Me-S interaction energy is greater than the Me-Me energy and for a relatively low misfit and deformation of the growth layer. A thin deposit allows for a certain degree of crystallographic misfit; however, above a critical thickness deformation energy stored in the strained layer will promote the formation of dislocations and the spontaneous appearance of growth island. This situation is referred to as the Stranski-Krastanov mechanism (layer-by-layer followed by island growth) and is depicted schematically in Figure 1c.

ELECTROCHEMICAL NUCLEATION

Early stages of electrochemical metal deposition normally involve two- or three-dimensional nucleation and growth process as discussed above. The rate of nuclei formation is strongly dependent on the overpotential. Scharifker and Hills (1983) have provided a convenient method of analyzing the nucleation mechanism. Using potentiostatic transients, two limiting cases have been identified: *instantaneous nucleation,* where all nuclei are formed immediately after the potential step; and *progressive nucleation*, where nuclei are formed continuously during the entire deposition process. The equation describing the current transient for instantaneous nucleation is as follows:

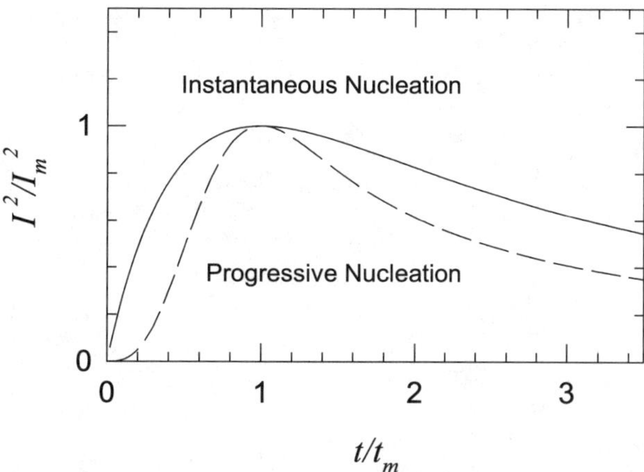

FIGURE 2 Theoretical dimensionless current-time transient plots for instantaneous and progressive nucleation

$$I_{inst} = \frac{zFD^{1/2}C}{\pi^{1/2}t^{1/2}}[1 - \exp(-N\pi kDt)] \quad \textbf{(EQ 2)}$$

where z is the charge of the depositing species, F is the Faraday constant, D is the diffusion coefficient, C is the bulk concentration, N is the total number of nuclei, and t is the time. The constant k is equal to $(8\pi C\ M/\rho)^{1/2}$ with M and ρ the molecular weight and density of the deposited material, respectively. On the other hand, progressive nucleation yields an expression of the following form:

$$I_{prog} = \frac{zFD^{1/2}C}{\pi^{1/2}t^{1/2}}[1 - \exp(-0.5AN_{\infty}\pi k'Dt^2)] \quad \textbf{(EQ 3)}$$

where A is the steady state nucleation rate constant per site and N the number density of active sites. The constant k' is equal to 4/3 k. Analysis of the current maxima in the potentiostatic current density transients allows rendering equations (2) and (3) into dimensionless form as follows:

$$\text{Instantaneous} \quad (I/I_m)^2 = 1.9542\{1 - \exp[-1.2564(t/t_m)]\}^2(t/t_m)^{-1} \quad \textbf{(EQ 4)}$$

$$\text{Progressive} \quad (I/I_m)^2 = 1.2254\{1 - \exp[-2.3367(t/t_m)]\}^2(t/t_m)^{-1} \quad \textbf{(EQ 5)}$$

Idealized plots of $(I/I_m)^2$ versus (t/t_m) can be generated showing the respective behavior of instantaneous and progressive nucleation. These reduced-variable plots of the transients are depicted in Figure 2. Experimental data can be represented in this form and compared directly with the theoretical predictions for each limiting nucleation mechanism.

The effects of overpotential, metal ion concentration and additives are reviewed with respect to their influence on the nucleation mechanism.

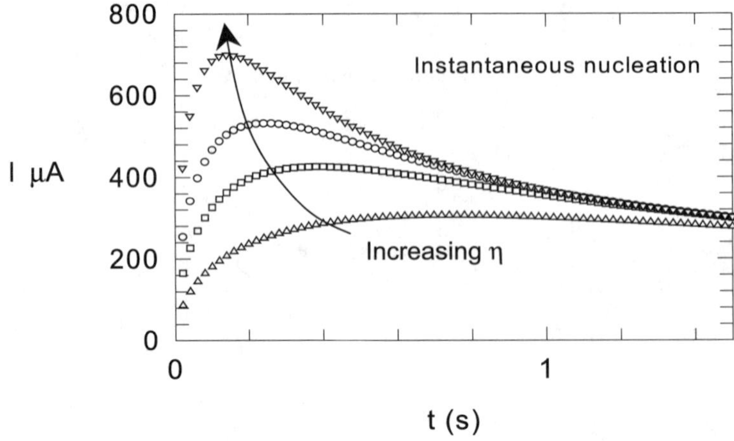

FIGURE 3 Calculated potentiostatic transients for the deposition of Pb on single crystal ZnO with increasing overpotential (η)

EFFECT OF OVERPOTENTIAL

Scharifker and Hills (1983) studied the electrochemical deposition of lead on single crystal zinc oxide. The experiments were carried out using a solution containing 50 m*M* $Pb(NO_3)_2$ in 1 *M* KNO_3. Analysis of the potentiostatic transients and plots of the nondimensional I^2/I_m^2 versus t/t_m values as expressed in Figure 2 revealed that lead deposition closely follows that predicted for instantaneous nucleation. It was observed that I_m increases with increasing overpotential, while t_m decreases correspondingly. Examination of these trends clearly shows that the total number of individual nuclei (*N*) increases with increasing overpotential. Calculated potentiostatic transients for the process are shown in Figure 3.

Alvarez and Salinas (2004) examined the initial stages of zinc electrocrystallization on highly oriented pyrolytic graphite HOPG (0001) surfaces using 0.1 *M* $ZnSO_4$ and 0.5 *M* Na_2SO_4 solutions as a function of overpotential. Analysis of the potentiostatic transient data plotted accordingly to the reduced variable plots clearly showed an instantaneous nucleation mechanism for this system. Based on this information, the total number of nuclei (*N*) was calculated as a function of overpotential. This data is plotted in Figure 4, along with that reported for Pb deposition on ZnO. Clearly, the rate of Pb nuclei formation on ZnO is orders of magnitude greater that that for Zn nuclei formation on HOPG. Furthermore, Zn nucleation rate is not as sensitive to overpotential as that for Pb.

Correia et al. (1994) modified the relationships for instantaneous and progressive nucleation when electrocrystallization occurs on ultramicroelectrodes (UME). The expression for the current transient for instantaneous nucleation at a UME is described as follows:

$$I_{inst} = \left(4zFDCr + \frac{8zFD^{1/2}Cr^2}{\pi^{3/2}t^{1/2}}\right)[1 - \exp(-N\pi kDt)] \qquad \textbf{(EQ 6)}$$

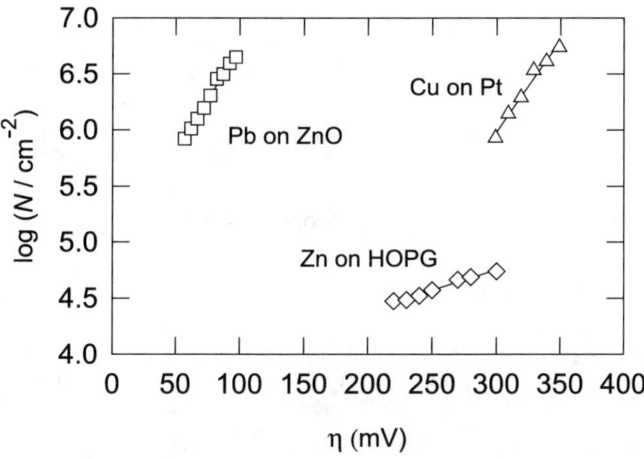

FIGURE 4 Overpotential dependence of total number of nuclei formed for various systems

where r is the radius of the UME and the other terms have the meaning previously defined. Progressive nucleation is described by:

$$I_{prog} = \left(4zFDCr + \frac{8zFD^{1/2}Cr^2}{\pi^{3/2}t^{1/2}}\right)[1 - \exp(-0.5AN_{\infty}\pi k'Dt^2)] \qquad \textbf{(EQ 7)}$$

Potentiostatic scans were obtained for copper on Pt ultramicroelectrodes by Correia, et al., (1994). Copper deposition was studied for applied potential ranging from −0.22 to −0.35 V (SCE) for 0.1 M $CuSO_4$ in 0.5M H_2SO_4. Under the experimental conditions employed, Cu electrocrystallization proceeded according to the instantaneous nucleation mechanism. The formation of Cu nuclei was found to be strongly dependent on applied potential. Total number of Cu nuclei calculated for this system ranged from 8.8×10^5 to 54×10^5 cm^{-2}. The plot of copper nucleation as a function of overpotential is depicted in Figure 4 along with the data previously discussed. It is apparent that copper nucleation parallels that of zinc under the condition employed and that it is also orders of magnitude greater than that for Zn on HOPG.

Lorenz and Staikov (1995) presented results for the electrocrystallization of Ag+ on HOPG (0001) surface from 10^{-2} M $AgClO_4$ in 1 M $HClO_4$ solutions at 298°K. An analysis of the current transients suggests that deposition clearly obeys a progressive nucleation type mechanism for this system. The deposit morphology observed during the initial stages of deposition corresponds to the "Volmer-Weber" (3-D island) growth mechanism. Using in situ scanning tunneling microscopy, Potzschke et al. (1995) in a companion study confirmed the diffusion controlled 3-D growth model. In Eq. (3), the AN_{∞} term represents the nucleation rate. Values of AN_{∞} were estimated by a fitting procedure for the data of Lorenz and Staikov (1995). Figure 5 shows the variation of AN_{∞} with overpotential.

The slope of the plot shown in Figure 5 can be used to determine the size of the critical nucleus as outlined by Milchev and Tsakova (1990).

$$\frac{d\log AN_{\infty}}{d\eta} = (n^* + \alpha)\frac{zF}{2.303RT} \qquad \textbf{(EQ 8)}$$

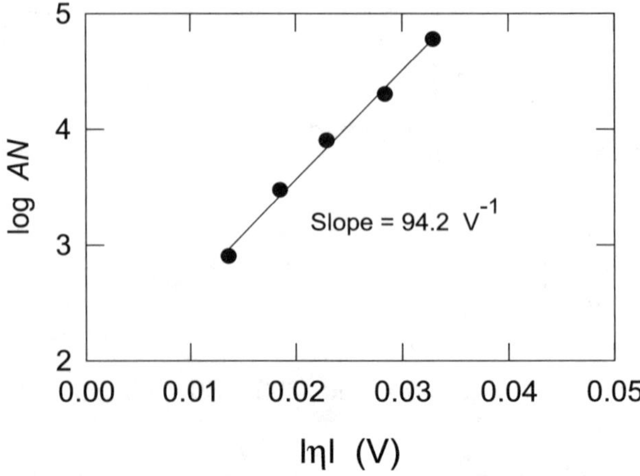

FIGURE 5 Overpotential dependence of nucleation rate for Ag^+ deposition on a HOPG (0001) surface

where n^* is the critical nucleus and α is the transfer coefficient (i.e., assumed to be 0.5 in this illustration). The value obtained for n^* was ~5 which suggest 5 Ag atoms were required to form the critical nucleus during electrocrystallization under these conditions. The in situ STM images obtained by (Potzschke et al. 1995) revealed disc shaped clusters measuring approximately 10 nm in diameter and 0.5 nm in height. Analysis of their experimental results in the 10 mV to 35 mV η| range, suggested a critical nucleus containing 4 atoms in the Ag cluster.

EFFECT OF CONCENTRATION

Copper electrocrystallization was examined by Barin et al. (2000) at different copper concentrations using ultramicroelectrodes. A distinct mechanistic change was observed when the $[Cu^{2+}]$ increased from 0.01 to 0.10 *M*. Copper deposition changed from progressive nucleation to instantaneous nucleation with increasing concentration. Initial stage deposits were studied *ex situ* by an atomic force microscope to obtain details regarding the morphological features of the respective deposits. For progressive nucleation at 0.01 *M* $[Cu^{2+}]$, AFM images revealed the presence of hemispherical nuclei of varying diameters. It was found that only one Cu atom was required to form the critical nucleus for copper electrocrystallization at the conditions employed. By contrast, instantaneous nucleation for the more concentrated 0.10 *M* solution resulted in copper nuclei uniformly covering the surface.

The nucleation of gold on a glassy carbon electrode was examined by Schmidt, Donten, and Osteryoung (1997) as a function of gold concentration. Current–time transients were measured for 5, 11, 50 m*M* [Au] in 6 *M* LiCl solutions. Under these conditions, electrodeposition proceeded according to reduction of the auric chloride anion ($AuCl_4^- + 3e \rightarrow Au^o + 4Cl^-$). The reaction has a standard potential of 1.001 V (NHE).

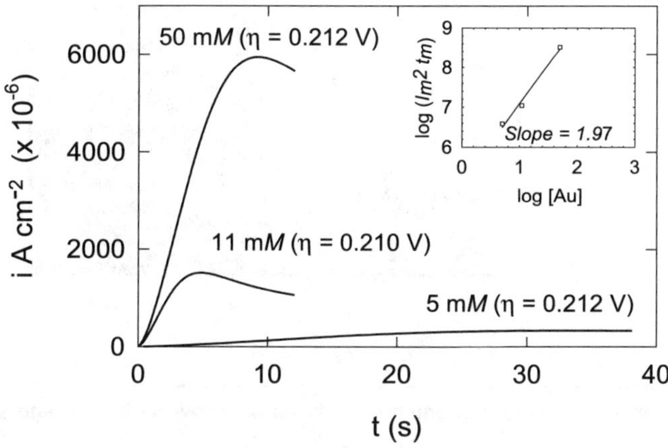

FIGURE 6 Potentiostatic transients for gold deposition on glassy carbon

Analysis of the experimental results reveals a characteristic 3-electron diffusion controlled process. Transients for 11 m*M* [Au] were plotted according to the reduced variable plot. The experimental data falls between the theoretical curves for instantaneous and progressive nucleation. In the early stages ($t < t_m$), deposition appears to agree most closely with the progressive nucleation mechanism. Whereas, at $t > t_m$ the process corresponds more nearly to that of instantaneous nucleation. Figure 6 depicts the calculated potentiostatic transients for 5 m*M* (η = 0.212 V), 11 m*M* (= 0.210 V), and 50 m*M* (η = 0.212 V).

These results show that the nucleation rate is slow for this system. It is clear that the AN_∞ decreases when the concentration increases from 11 to 50 m*M* [Au]. The authors suggest that at low gold concentrations, nucleation competes with the adsorption of $AuCl_x$ for active sites. The insert in Figure 6 shows a log I^2t_m versus log [Au]. The slope of this line is 1.97 which compares favorably with the theoretical value of 2.

Early stage nucleation and growth processes were examined by Miranda-Hernandez and Gonzalez (2004) for the electrocrystallization of silver on vitreous carbon. The supporting electrolyte contained 1.6 *M* NH_3 and 1 *M* KNO_3 at pH 11 and the electroactive species was the $Ag(NH_3)^{2+}$ ion. Concentrations of argento diamine ranging from 10^{-4} to 10^{-1} *M* were systematically investigated at various potentials. Figure 7 shows an idealized plot of the results from this investigation. Nucleation and deposit type are plotted for various $Ag(NH_3)^{2+}$ concentrations and various overpotentials.

At low concentrations of the electroactive species ($\leq 10^{-3}$ *M*), the current transients reveal two maxima which are associated with progressive 2D growth at the submonolayer level. With increasing overpotential the 10^{-3} *M* system produces layers in excess of a monolayer (3D growth) and instantaneous nucleation. In this region, nucleation is found to move from progressive nucleation of Ag on VC to instantaneous nucleation of Ag on Ag. With increasing concentration ($\geq 10^{-2}$ *M*) and at low overpotentials, 3D growth is observed and the mechanism moves from progressive to instantaneous with increasing concentration. This phenomena is generally observed for the copper, gold and silver systems discussed here.

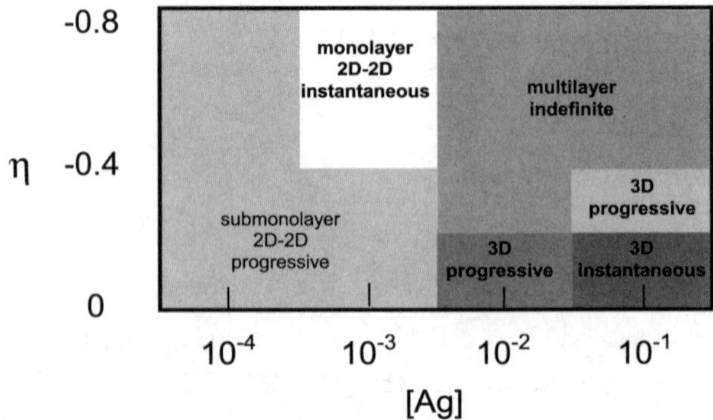

FIGURE 7 Idealized nucleation mechanism as a function of overpotential and silver concentration

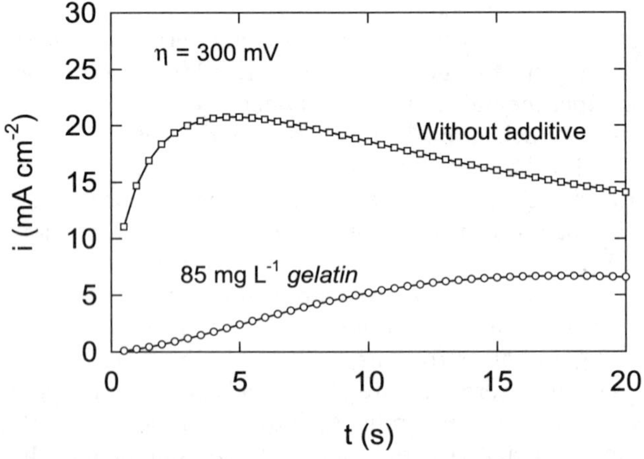

FIGURE 8 Potentiostatic transients showing the effect of gelatin on the nucleation of Zn on HOPG at 300 mV

EFFECT OF ADDITIVES

In the study of Alvarez and Salinas (2004) cited above, gelatin was examined as an additive in the deposition of zinc on HOPG. The effect of gelatin is shown in the current transients obtained at an overpotential of 300 mV and plotted in Figure 8.

Without any additive and at an overpotential of 300 mV, the deposition of Zn follows an instantaneous nucleation mechanism and reveals a current density maxima at approximately 4 seconds. In the presence of gelatin, blockage of surface sites is clearly evident. The rate of nucleation is slow and crystal growth is retarded. With gelatin in the electrolyte deposition tends toward progressive nucleation at an overpotential of 300 mV. However, with increasing overpotential nucleation rate increases and the process seems to approach instantaneous behavior.

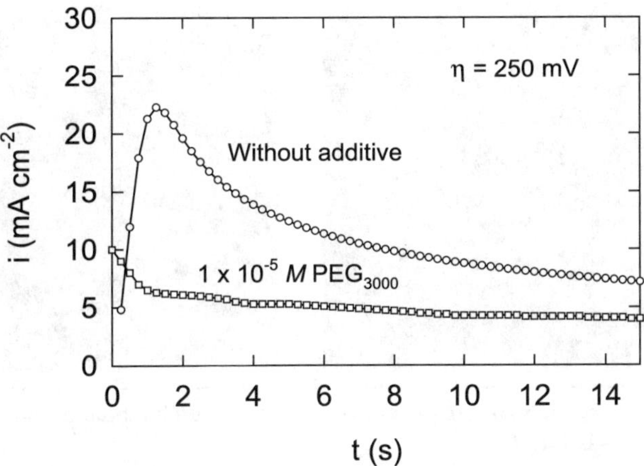

FIGURE 9 Potentiostatic transients showing the effect of PEG3000 on the nucleation of Cu on Pt at 250 mV

Polyethylene glycols are recognized for their ability to serve as a brightening agent in copper electroplating. Kapocius, Karpaviciene, and Stepanavicius (2002) examined the effect of polyethylene glycol of molecular weight 3000 (PEG_{3000}) on the initial stages of copper electrocrystallization on platinum in the underpotential deposition (UPD) region. In the absence of any additive, potentiostatic transients for 0.1 *M* $CuSO_4$ and 0.5 *M* H_2SO_4 revealed the characteristic pattern for an electrochemical nucleation process involving diffusion limited 3-D growth of nuclei. The deposition of copper under these conditions agreed closely with that predicted by the progressive nucleation model. As previously discussed, copper deposition at overpotential (OPD) conditions using a solution containing 0.1 *M* $CuSO_4$ and 0.5 *M* H_2SO_4 results in nucleation following the instantaneous mechanism. The addition of 1×10^{-5} *M* PEG_{3000} at 250 mV shows a marked suppression of copper deposition as shown in Figure 9. PEG_{3000} is an effective blocking agent at this concentration. It was observed that increasing the potential to approximately 300–320 mV restores the nucleation rate to that for the additive free case.

GOLD DEPOSITION IN NANOPORES

As deposition proceeds, epitaxy frequently breaks down resulting in the formation of random structures. In most metallurgical operations substrates are typically loaded with surface imperfections and defects. This also accounts for the random growth of polycrystalline deposits that are normally observed. Nonetheless, control of deposit morphology is a critical factor in most aspects of electrodeposition. Growth geometries (centers) typically seen in massive deposits include hemispherical, pyramidal, and ramified (dendritic) type structures.

Mantell (1960) conveniently classified massive crystalline deposits according to three categories:

1. All the initial nuclei continue to grow throughout the deposit forming isolated crystals (symmetrical and acicular).

a)

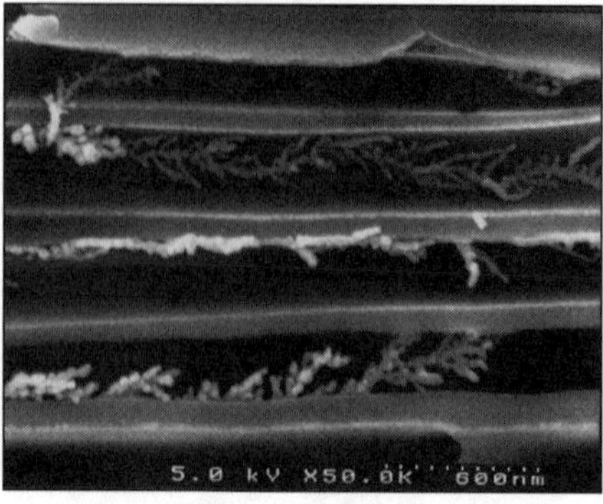

b)

FIGURE 10 SEM micrographs of cross sectional area of alumina membrane after depostion at $[Au]_o$ = 30 ppm, η = –100 mV

2. Only part of the initial nuclei or crystals continue to grow (conical and twinned).
3. No crystals continue to grow for any extended period (powdery and dendritic).

Increasing the current density and the addition of additives (i.e., colloids) favors the formation of type 3 deposits. Whereas, increasing the concentration of the electroactive species has a tendency of promoting type 1 structures.

Research is being conducted with the aim of preparing a new class of materials using templates/membranes with nanopores filled with nano-scale gold wires by electrodeposition. These so-called metamaterials have significant electromagnetic properties not encountered in nature like negative effective permittivity and/or negative permeability. Deposition of gold in polycarbonate and alumina membranes have been investigated as part of this study. A cathode surface is formed on one side of the membrane by depositing PVD (physical vapor deposition) layers consisting of Au/Cu/Au. Electrodeposition is then carried out in an electrolyte containing different concentrations of tetrachloroaurate ($AuCl_4^-$) in 0.1 M NaCl. The alumina membrane employed in these experiments is 60 µM thick and has 300 nm diameter pores.

At low gold concentration (1.52×10^{-4} M $AuCl_4^-$), electrodeposition near the PVD layers is characterized by the structure shown in Figure 10a. Isolated crystals ranging from approximately 50 nm to 150 mm.

As shown in Figure 10b, dendrites form in the middle section of the pore. This is consistent with type 6 morphological change associated with decreasing metal ion concentration. To obtain a material with the required properties it is necessary to suppress formation of dendrites. This can be achieved by increasing overall metal ion concentration.

At high gold concentration (1.02×10^{-1} M $AuCl^-$), the bottom of the pore near the PVD layer reveals a gold structure very similar to that shown in Figure 10a. However, solid gold wires were found to form in the middle region of the pore and continue toward the solution side of the membrane. The SEM micrograph shown in Figure 11 reveals several solid gold wires partially detached from the pore channel. Solid gold wires have not been observed in the deepest regions of the pore where depletion of the electroactive species favors an increase in the rate of nuclei formation.

FIGURE 11 SEM micrograph of cross sectional area of alumina membrane after deposition at [Au]o = 20,000 ppm, η = –100 mV

CONCLUSIONS

Potentiostatic transients can be used to analyze electrochemical nucleation of metals on various substrates. Electrocrystallization kinetics typically involve either an instantaneous or a progressive nucleation mechanism. Overpotential, metal ion concentration, and additives were all shown to have a pronounced effect on nucleation mechanism. These parameters control the total number of nuclei formed and the nucleation rate for the respective mechanisms.

Gold deposition was carried in an alumina membrane containing an array of nanopores. The morphology and structure of the deposit depended on the concentration of the gold in the electrolyte. Low gold concentrations tend to promote a fine dendritic type structure. Whereas, high gold concentrations yield solid gold wires.

REFERENCES

Alvarez, A.E. and D.R. Salinas. 2004. Nucleation and growth of Zn on HOPG in the presence of gelatin as additive. Journal of Electroanalytical Chemistry. 566. pp. 393–400.

Barin, C.S., A.N. Correia, S. A.S. Machado, and L.A. Avaca. 2000. The effect of concentration on the electrocrystallization mechanism for copper on platinum ultramicroelectrodes. J. Braz. Chem. Soc. 11:2. pp. 175–181.

Bockris, J. O'M. and A.K.N. Reddy. 1970. *Modern Electrochemistry–An Introduction to an Interdisciplinary Area*. New York: Plenum Press

Correia, A.N., S.A. S. Machado, and L.A. Avaca. 1994. Electrocrystallization of Cu and Hg on Pt Ultramicroelectrodes. J. Braz. Chem. Soc. 5:3. pp. 173–177.

Kapocius, V., V. Karpaviciene, and A. Stepanavicius. 2002. Electrochemical processes at a platinum electrode in CuSO4 solutions in the UPD region and at initial electrocrystallization stages: Effect of polyethylene glycol. Russian Journal of Electrochemistry. 38:3. pp. 274–279.

Lorenz, W.J. and G. Staikov. 1995. 2D and 3D thin film formation and growth mechanisms in metal electrocrystallization–an atomistic view by in situ STM. Surface Science. 335. pp. 32–43.

Mantell, C.L. 1960. Electrochemical Engineering. New York: McGraw-Hill Book Company, Inc.

Milchev, A. and V. Tsakova. 1990. Theory of progressive nucleation and growth accounting for the ohmic drop in the electrolyte. I. Journal of Applied Electrochemistry. 20. pp. 301–306.

Miranda-Hernandez, M. and I. Gonzalez. 2004. Effect of potencial on the early stages of nucleation and growth during silver electrocrystallization in ammonium medium on vitreous carbon. J. Electrochem. Soc. 151:3. pp. C220–C228.

Potzschke, R.T., C.A. Gervasi, S. Vinzelberg, G., Staikov, and W.J. Lorenz. 1995. Nanoscale studies of Ag electrodeposition on HOPG (0001). Electrochimica Acta. 40:10. pp. 1469–1474.

Scharifker, B. and G. Hills. 1983. Theoretical and experimental studies of multiple nucleation. Electrochimica Acta. 28:7. pp. 879–889.

Schmidt, U., M. Donten, and J. G. Osteryoung. 1997. Gold electrocrystallization on carbon and highly oriented pyrolytic graphite from concentrated solutions of LiCl. J. Electrochem. Soc. 144:6. pp. 2013–2021.

Staikov, G., K. Juttner, W.J. Lorenz, and E. Budevski. 1994. Metal deposition in the nanometer range. Electrochimica Acta. 39:8/9. pp. 1019–1029.

Chelating Agents as Flotation Collectors

Maurice C. Fuerstenau*

INTRODUCTION

The term, chelate, from the Greek for 'claw,' has its origin in the pincer-like representation of more than one atom of a single ligand bonded to the same metal atom, forming a ring-like structure. Depending on how many attachments exist on the sphere of coordination around the metal ion, ligands are referred to as bidentate, tridentate, tetradentate, pentadentate and hexadentate

According to Calvin and Martell (1952) and Somasundaran and Nagaraj (1984), chelating agents must contain suitable functional groups, and their location within the ligand molecule must enable ring formation that is sterically possible. Typical structures that meet these requirements have been given by Fuerstenau and Herrera-Urbina (1988) and are listed in Figure 1.

As noted, the functional groups of these reagents contain electron donor atoms such as oxygen, nitrogen and sulfur. Basic groups, such as amino ($-NH_2$), are those that contain an atom with a lone pair of electrons. Acid groups are those, such as carboxylic ($-COOH$) which lose a hydrogen ion and coordinate with the metal atom.

Examples of each of these types chelating reagents are given in Tables 1–4.

The specificity of chelating agents for various metal ions has long been known and was utilized early in the search for new flotation collectors. In 1927, Vivian used ammonium nitrosophenylhydroxylamine as a collector for cassiterite. Holman (1929) tested dimethylglyoxime as a collector for nickel oxide. He also suggested taurin as a collector for oxidized lead ores. DeWitt and von Batchelder (1939) suggested the use of oximes as flotation collectors. Gutzeit (1946) conducted an extensive investigation on the use of chelating agents as collectors and as depressants for gangue minerals. The depressants are those reagents that form soluble chelates with metal ions that could function as

* Department of Metallurgical and Materials Engineering, University of Nevada, Reno, Nevada

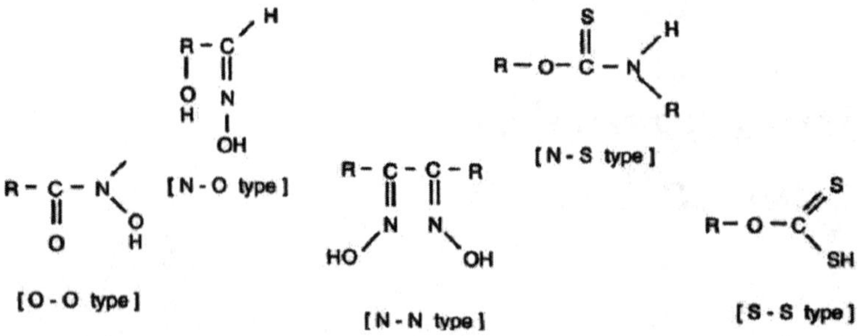

FIGURE 1 Typical structures of chelating agents (Fuerstenau et al. 2000)

TABLE 1 Molecular structures, stability constants of N-O type chelating reagents and their Cu(II) complexes (Fuerstenau et al. 2000)

Chelating Reagent	log Ka	log K_1	log K_2	References	Molecular Structure
8-Hydroxyquinoline	Ka_1 = 9.63 Ka_2 = 4.95	12.20	11.20	Perrin (1979); Sillen and Martell (1964)	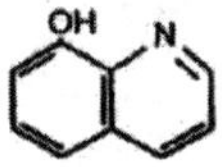
Salicylaldoxime	11.72	12.64	11.17	Sillen and Martell (1971)	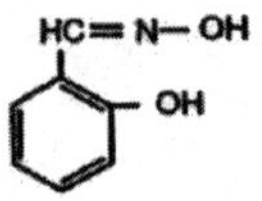
Dinitrosoresorcinol	Ka_1 = 4.73 Ka_2 = 8.10	6.10	4.42	Zayan et al. (1973) Issa and Maghrabi (1973)	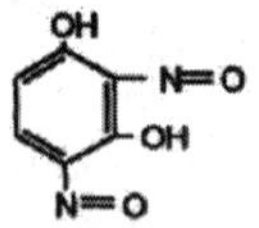
Phenylthiohydantoic Acid	NA	NA	NA		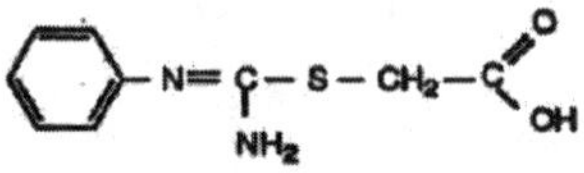
α-Benzoinoxime	12.00	NA	NA	Cheng et al. (1982)	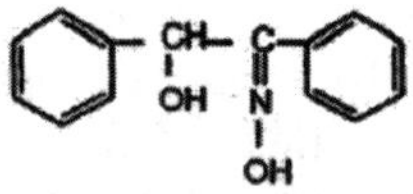

activators for gangue minerals. These metal ions are calcium, magnesium, iron, aluminum, copper and lead. Some compounds suggested by Gutzeit (1946) were aminoethanol, lactic acid, and quinolinic acid. Ludt and DeWitt (1949) reported the flotation of copper silicate with the dye, butyl, hexyl, octyl malachite green.

In the 1960s, my group at the Colorado School of Mines published papers showing the efficacy of hydroxamates as flotation collectors for copper and iron minerals (Peterson et al., 1965, Fuerstenau et al., 1967, Fuerstenau and Peterson 1969, Fuerstenau et al.,1970). These papers generated considerable interest worldwide in the use of these reagents in froth flotation. Bogdanov et al. (1973) studied the use of hydroxamic acids as collectors for wolframite, cassiterite and pyrochlore.

TABLE 2 Molecular structures, stability constants of O-O type chelating reagents and their Cu(II) complexes (Fuerstenau et al. 2000)

Chelating Reagent	log Ka	log K_1	log K_2	References	Molecular Structure
Cupferron	4.16	NA	NA	Sillen and Martell (1964)	
Neocupferron	4.10	NA	NA	Cheng et al. (1982)	
2-Nitroso-1–Napthol	7.46	7.50	6.30	Perrin (1979)	
Acetoacetanalide	10.34	6.20	5.30	Perrin (1979)	
Pentane-2,4-Dione	8.88	8.29	6.70	Perrin (1979)	
Potassium Octyl Hydroxamate	9.69	NA	NA	Ryaboi et al. (1980)	

TABLE 3 Molecular structures, stability constants of S-N type chelating reagents and their Cu(II) complexes (Fuerstenau et al. 2000)

Chelating Reagent	log Ka	log K_1	log K_2	References	Molecular Structure
Rhodanine	5.18	NA	NA	Sillen and Martell (1971)	
Dithizone	Ka_1 = 5.60 Ka_2 = 4.45	9.35	B_2= 19.18	Perrin (1979) Sillen and Martell (1964)	
1-Phenylthiosemi-carbazide	NA	7.73	NA	Koshkin and Dobryshin (1970)	
p-Dimethylamino benzilidone rhodanine	Ka_1 = 6.70 Ka_2 = 1.40	6.08	NA	Sandell and Onishi (1978) Sillen and Martell (1964)	
Diphenylthio-carbazide	NA	NA	NA		
Phenylthiourea	1.10	NA	NA	Maslowska et al. (1974)	
Dithiooxamide	10.89	NA	30.22	Burger (1973)	

TABLE 4 Molecular structures, stability constants of S-S and N-N type chelating reagents and their Cu(II) complexes (Fuerstenau et al. 2000)

Chelating Reagent	log Ka	log K_1	log K_2	References	Molecular Structure
Potassium Ethyl Xanthate	1.70	NA	13.70	Kratochvil and Nepras (1972) Sparrow et al. (1977)	$C_2H_5-O-C(=S)SK$
Sodium Diethyl Dithiocarbamate	3.35	14.9	13.9	Sillen and Martell (1971)	$(C_2H_5)_2N-C(=S)SNa$
Dimethylglyoxime	NA	NA	NA		$H_3C-C(=N-OH)-C(=N-OH)-CH_3$
Diethylenetriamine	$Ka_1 = 4.46$ $Ka_2 = 9.17$ $Ka_3 = 9.96$	15.9	NA	Calvin and Martell (1952)	$NH_2-CH_2-CH_2-NH-CH_2-CH_2-NH_2$

Palmer et al. (1975) presented the flotation of manganese silicate with this collector, while Raghavan and Fuerstenau (1975) studied the adsorption of octyl hydroxamate on ferric oxide. Evrard and DeCuyper (1975) studied the flotation characteristics of complex hydrated cobalt oxides and found that the floatability of these minerals with octyl hydroxamate was strongly affected by small amounts of contained copper. Nagaraj and Somasundaran (1979, 1981) presented extensive studies on the use of hydroxy oximes as a flotation collector or copper silicate and oxide minerals. In 1983 Pradip and Fuerstenau presented adsorption studies of hydroxamate on barite, calcite and bastnaesite, and Fuerstenau and Pradip (1984) published a general paper on hydroxamate collectors. A paper on applications of chelates in mineral processing was published by Pradip in 1988.

Rinelli and Marabini studied the use of chelating agents as flotation collectors extensively. In 1973 they presented the use of 8-hydroxyquinoline as a collector for oxides of lead and zinc. The addition of a neutral oil was found to be necessary for favorable response. These authors (Marabini and Rinelli, 1973) also studied the flotation of pitchblende with cupferron. Flotation occurred only under acidic conditions in the presence of fuel oil. Rinelli et al. (1976) studied the use of salicylaldehyde as a collector for cassiterite. Barbaro et al. (1997) presented the flotation response of chrysocolla with aminothiophenol as collector.

Excluding the S–O structured chelating agents, such as dithiophosphate, xanthate, and dithiocarbamates, the only chelating agent that has found commercial application in flotation is hydroxamate. In this view most of this paper is devoted to this collector. The response of various minerals to flotation when it is used as collector and mechanisms involved in adsorption are presented.

HYDROXAMATE

A very important property of hydroxamic acid is its weak acid characteristic. Given the dissociation reaction in 0.1 M ionic strength (Ryaboi et al.1980):

$$HXm_{(aq)} \rightarrow H^+ + Xm^- \qquad K = 2.82 \times 10^{-10}$$

where Xm^- is octyl hydroxamate ion. The pH at which the activity of octyl hydroxamate ion and undissociated hydroxamic acid molecule is pH 9.55. As will be noted, most flotation occurs below pH 9.55 with octyl hydroxamate which is a strong indication that the undissociated molecule of hydroxamic acid is the active collector species in this system.

Hydroxamic acid exists in two tautomeric forms in equilibrium (Chaterjee 1978, Pradip and Fuerstenau 1983):

1) Hydroxyamide, R—C(=O)—N(H)—OH and 2) Hydroxyoxime, R—C(OH)=N—OH

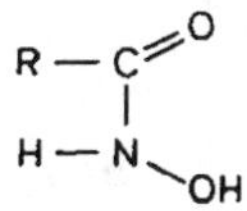

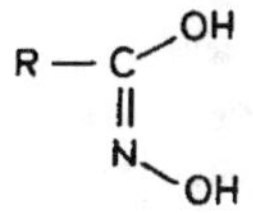

The tautomer that functions as a chelating agent is the hydroxyamide. Chelates are formed by replacement of the hydrogen by a metal ion together with a dative bond with carbonyl oxygen to form the closed ring structure.

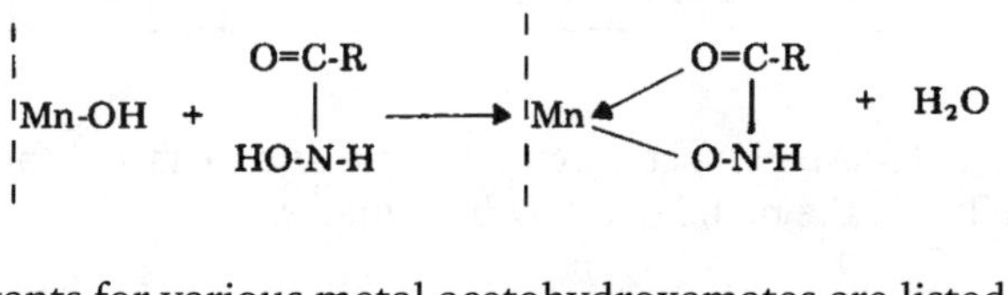

Mn-OH + O=C-R / HO-N-H ⟶ Mn(←O=C-R / O-N-H) + H_2O

Stability constants for various metal acetohydroxamates are listed in Table 5.

Corresponding equilibrium constants for octyl hydroxamate are not available, but it is expected that their values would not be significantly different from those of acetohydroxamate. The weakest complexes are formed with the alkaline earth metals. As would be expected, complexes with trivalent metal ions are considerably more stable than those formed with divalent metal ions. The strongest complex listed is that of Fe^{3+}.

The equilibrium constants are for formation reactions, such as,

$$Fe^{3+} + Xm^- \rightarrow FeXm^{2+} \qquad K = 2.63 \times 10^{11}$$

$$FeXm^{2+} + Xm^- \rightarrow Fe(Xm)_2^+ \qquad K = 4.79 \times 10^9$$

$$Fe(Xm)_2^+ + Xm^- \rightarrow Fe(Xm)_{3(aq)} \qquad K = 1.70 \times 10^7$$

My group and I at the Colorado School of Mines had the privilege of presenting hydroxamate as a flotation collector to the industrial public in 1965. The first reagent that we produced was dodecyl hydroxamate. Professor Miller played a significant role in this development. We could not get chrysocolla to float with this reagent due to obvious solubility problems, so we heated the solution. I can still remember the thrill of seeing the the chrysocolla float under these conditions. In view of the solubility problems with dodecyl hydroxamate, we started producing hydroxamate with shorter hydrocarbon

TABLE 5 Stability constants for metal acetohydroxamates at 20°C and I = 0.1 ($NaNO_3$) (Schwarzenbach and Schwarzenbach 1963)

Cation	log K_1	log K_2	log K_3
Ca^{2+}	2.4		
Mn^{2+}	4.0	2.9	
Cd^{2+}	4.5	3.3	
Fe^{2+}	4.8	3.7	
Co^{2+}	5.1	3.8	
Ni^{2+}	5.3	4.0	
Zn^{2+}	5.4	4.2	
Pb^{2+}	6.7	4.0	
Cu^{2+}	7.9		
La^{3+}	5.16	4.17	2.55
Ce^{3+}	5.45	4.34	3.0
Sm^{3+}	5.96	4.77	3.68
Gd^{3+}	6.10	4.76	3.07
Dy^{3+}	6.52	5.39	4.04
Yb^{3+}	6.61	5.59	4.29
Al^{3+}	7.95	7.34	6.18
Fe^{3+}	11.42	9.68	7.23

chains. We ultimately focused on octyl hydroxamate due to its satisfactory water solubility and the insolubility of the metal octyl hydroxamates.

Chrysocolla

Flotation of chrysocolla in a microflotation cell with various concentrations of octyl hydroxamate is shown in Figure 2. It can be noted that maximal flotation is attained at about pH 6 and that complete flotation is obtained with an addition of 3.3×10^{-4} mol/L at room temperature. The very significant improvement in response at elevated temperature is also apparent.

Palmer (1972) conducted flotation tests with octyl, nonyl and decyl hydroxamates. The results obtained with decyl hydroxamate are shown in Figure 3. It can be noted that two maxima are observed with the lower collector addition, one at about pH 6 and the other at about pH 9.5. Similar results were noted for octyl and nonyl hydroxamates.

After contact with a solution of octyl hydroxamate, the chrysocolla changes from its distinctive blue green color to apple green with just a few minutes of conditioning. Copper salts of monohydroxamic acids are green in color and very insoluble in water (Yale 1943). Such a visible change in color indicates that multilayers of cupric hydroxamate are present on chrysocolla under flotation conditions. This fact was confirmed by Palmer et al., (1975) who showed with infra-red analysis the presence of bulk precipitated copper hydroxamate on the chrysocolla surface after contact with octyl hydroxamate. See Figure 4. These spectra show clearly that the copper octyl hydroxamate on the chrysocolla surface is the same as that of bulk precipitated copper octyl hydroxamate. Similar results were reported by Fuerstenau et al. (1988).

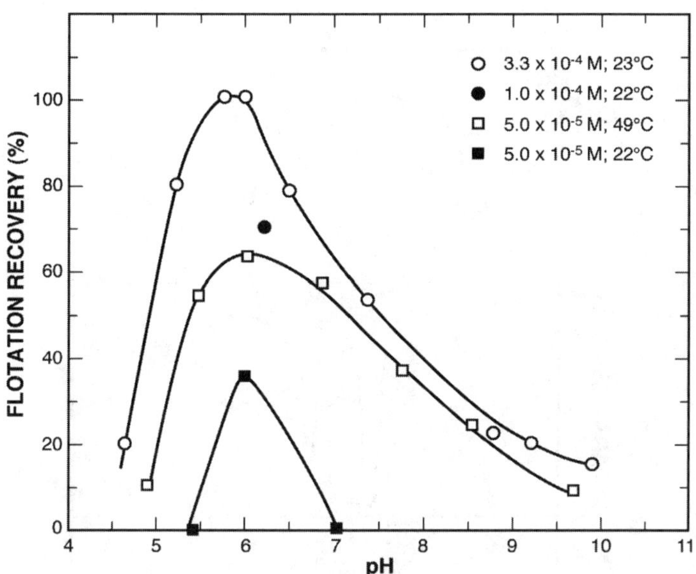

FIGURE 2 Flotation recovery of chrysocolla as a function of pH with octyl hydroxamate (Peterson et al. 1965)

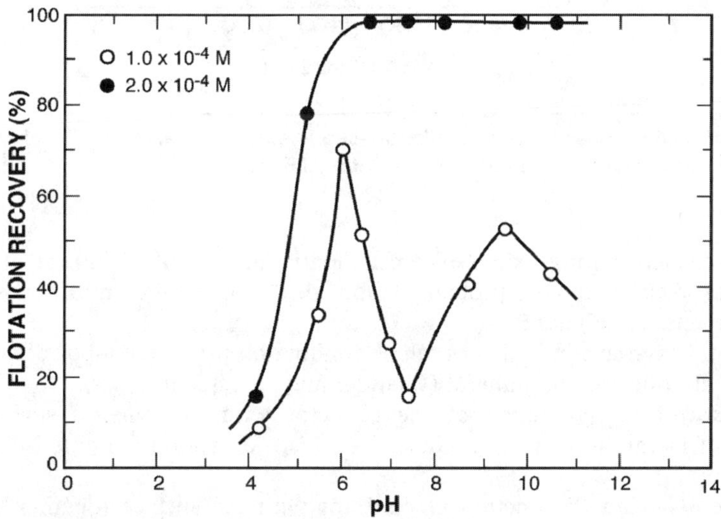

FIGURE 3 Flotation recovery of chrysocolla as a function of pH and decyl hydroxamate concentration (Palmer 1972)

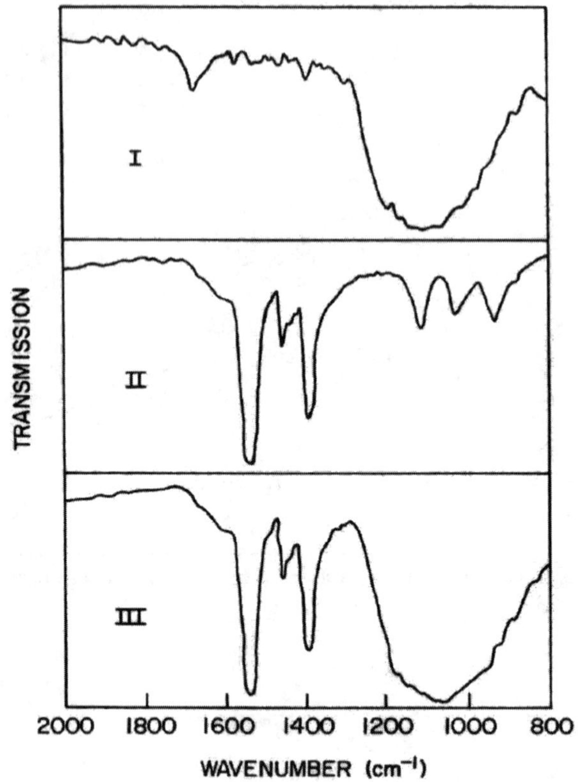

FIGURE 4 Infrared spectra: I—chrysocolla, II—cupric octyl hydroxamate, III—chrysocolla after contact with octyl hydroxamate (Palmer et al. 1975)

That a copper bearing oxide/silicate is floated around pH 6 indicates that hydrolyzed species of cupric ion are probably involved. This may be seen from the speciation diagram presented in Figure 5.

In flotation systems in which covalent bonding occurs, three types of behavior can occur between collector and mineral (Chander and Fuerstenau 1975).

Chemisorption. Interaction of the collector with the mineral surface without movement of metal atoms from their lattice sites. Adsorption is limited to monolayer coverage only.

Surface Reaction. Interaction of collector with the surface together with movement of metal atoms from their lattice sites. Multilayers of reaction product may form and adsorb.

Bulk Precipitation. Interaction of metal ions and collector away from the surface in the bulk solution.

Hydroxamate can interact chemically with surface ions that remain in their lattice sites (chemisorption) or with metal ions that have moved from their lattice sites adjacent

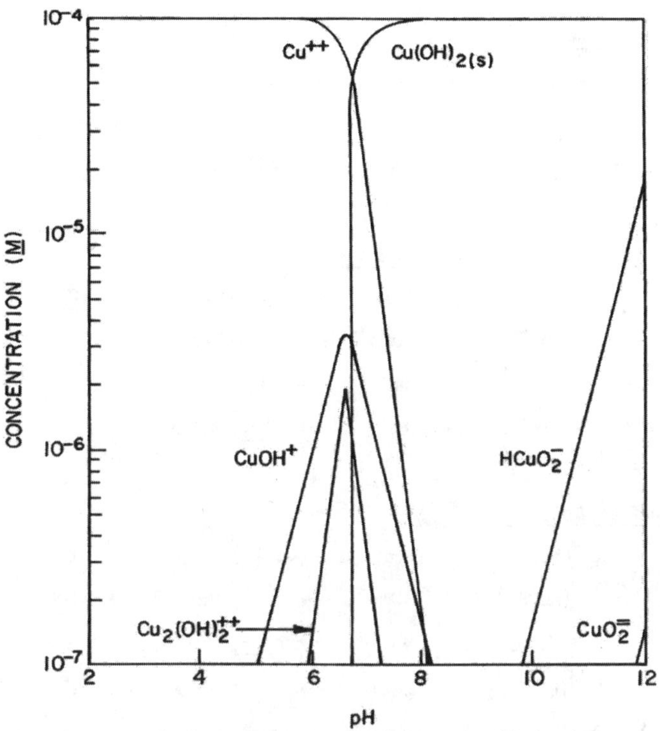

FIGURE 5 Speciation diagram for cupric ion (Palmer et al. 1975)

to the surface (surface reaction). Fuerstenau and Pradip (1984) have provided an excellent representation of these phenomena in Figure 6.

With chemisorption, adsorption is limited to monolayer coverage only. In the case of surface reaction, however, as copper ions emerge from their lattice sites, copper hydroxamate can form adjacent to the surface. The formation of multilayers of copper hydroxamate can then occur as more copper is dissolved from the mineral to form copper hydroxamate which could then adsorb on the hydrocarbon chains of the previously adsorbed species by hydrophobic bonding.

When the pH of the system is in the range in which metal ions comprising the crystal lattice hydrolyze, hydrolysis of these ions to hydroxy complexes will occur as they emerge from the lattice. The speciation diagram for Cu^{2+} (Figure 5) shows that hydrolysis of Cu^{2+} to $CuOH^+$ and $Cu(OH)_2$ occurs at about pH 6. This is the pH at which maximal flotation response of chrysocolla is obtained which indicates strongly that $CuOH^+$ or $Cu(OH_2$ is actively involved in the flotation mechanism. The formation of the hydroxy species will tend to result in greater mineral dissolution. Further, the interaction of $CuOH^+$ with the neutral hydroxamic acid molecule will result in the formation of water in addition to copper hydroxamate which will provide additional driving force for the reaction.

Whether surface reactions or bulk precipitation occurs depends on the rates of diffusion of metal ions and collector ions. If the rate of diffusion of metal ions away from the

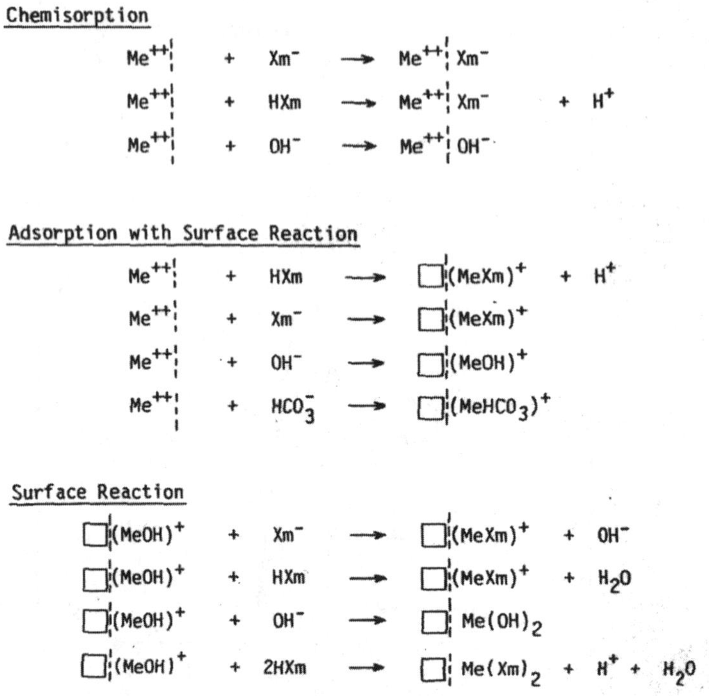

Chemisorption

$$Me^{++}| + Xm^- \rightarrow Me^{++}|Xm^-$$

$$Me^{++}| + HXm \rightarrow Me^{++}|Xm^- + H^+$$

$$Me^{++}| + OH^- \rightarrow Me^{++}|OH^-$$

Adsorption with Surface Reaction

$$Me^{++}| + HXm \rightarrow \square|(MeXm)^+ + H^+$$

$$Me^{++}| + Xm^- \rightarrow \square|(MeXm)^+$$

$$Me^{++}| + OH^- \rightarrow \square|(MeOH)^+$$

$$Me^{++}| + HCO_3^- \rightarrow \square|(MeHCO_3)^+$$

Surface Reaction

$$\square|(MeOH)^+ + Xm^- \rightarrow \square|(MeXm)^+ + OH^-$$

$$\square|(MeOH)^+ + HXm \rightarrow \square|(MeXm)^+ + H_2O$$

$$\square|(MeOH)^+ + OH^- \rightarrow \square|\ Me(OH)_2$$

$$\square|(MeOH)^+ + 2HXm \rightarrow \square|\ Me(Xm)_2 + H^+ + H_2O$$

FIGURE 6 Representation of chemsorption and surface reaction between sparingly soluble minerals and chemically bonding reagents (Fuerstenau and Pradip 1984)

surface is faster than the diffusion of collector ions through the boundary layer, bulk precipitation of metal collector will occur. On the other hand, if diffusion of collector species through the boundary layer to the mineral surface is faster than metal ion diffusion away from the surface, reaction of metal ion and collector species can occur at the surface and in the boundary layer.

Good flotation of chrysocolla is also possible at about pH 9.5. This is the pH at which hydroxamate ion and hydroxamic acid molecule are present at equal activity. Co-adsorption of the two hydroxamate species is apparently occurring under these conditions. Should sufficient Cu^{2+} be available, precipitation and adsorption of cupric hydroxamate could also occur.

Hematite

Ferric iron and hydroxamate also form very insoluble complexes, and iron-bearing minerals respond well to flotation with this collector. The response of hematite to flotation with 2×10^{-4} mol/L octyl hydroxamate is shown in Figure 7. Optimal pH for flotation is about 9. Further, it can be noted that conditioning time is an important parameter in this system also. With an additional conditioning time of seven minutes, flotation recovery

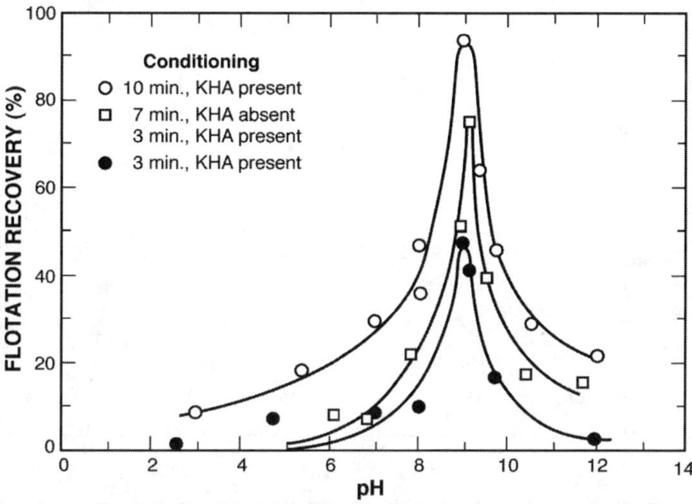

FIGURE 7 Flotation recovery of hematite as a function of pH and conditioning time with 2×10^{-4} M octyl hydroxamate at 25°C (Fuerstenau et al. 1970)

was doubled. The enhanced recovery obtained with longer conditioning times is probably related to the slight dissolution of hematite and subsequent surface reactions.

When the hydroxamate concentration is increased, but the pH range in which flotation is obtained is increased substantially (Figure 8). The fact that flotation is achieved at pH 9 and below indicates that the hydroxamic acid molecule is probably the collector in this system.

In the region of flotation, hydrolysis of ferric ion will have occurred. See Figure 9. Similarly to chrysocolla, the hydrolyzed species of the metal ion comprising the mineral obviously assume an important role in this system.

Infrared studies have shown the presence of ferric hydroxamate on the hematite surface after contact with octyl hydroxamate (Palmer et al. 1975). Similar to chrysocolla, multilayers of ferric hydroxamate were visibly apparent on the hematite surface after contact with hydroxamate. A color change on the hematite surface was noted, but the color change was not as dramatic as that with chrysocolla.

Adsorption studies by Raghavan and Fuerstenau (1975) also illustrate the optimal pH for adsorption to be about pH 9 and also the formation and adsorption of multilayers of octyl hydroxamate (Figure 10). Multilayer coverage can result from co-adsorption of hydroxamic acid molecule and hydroxamate ion or by ferric hydroxamate formation/adsorption. Raghavan and Fuerstenau (1975) noted a decrease in pH upon hydroxamate adsorption which indicates hydroxamic acid molecule is adsorbing. Since Fuerstenau et al. 1967) observed ferric hydroxamate on the hematite surface also, it seems likely that all three species are involved in flotation at pH 9.

Due to the very insoluble nature of ferric hydroxamate and the relatively-short chain length of octyl hydroxamate, flotation of very finely-divided hematite is possible. By way of example, a finely-disseminated iron ore which contained 45 % Fe in the form of red

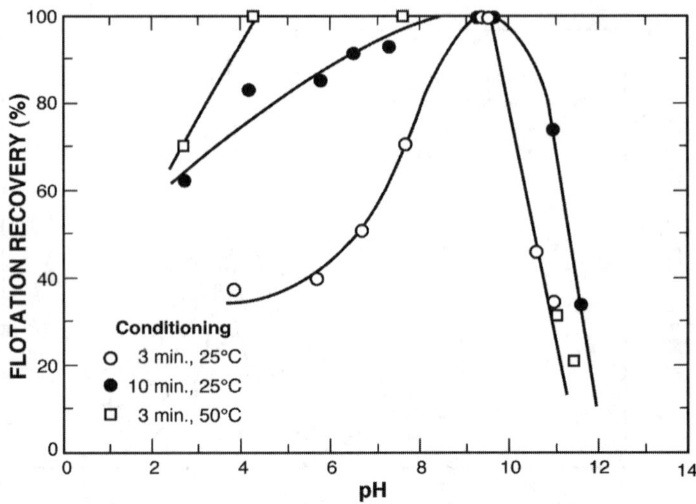

FIGURE 8 Flotation recovery of hematite as a function of pH, conditioning time and temperature with 5×10^{-4} M octyl hydroxamate (Fuerstenau et al. 1970)

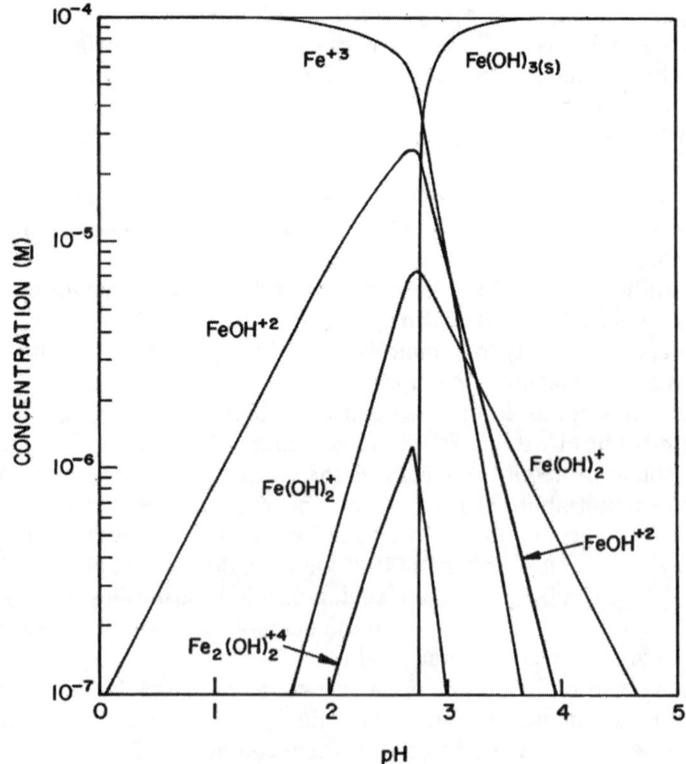

FIGURE 9 Speciation diagram for ferric ion (Palmer et al. 1975)

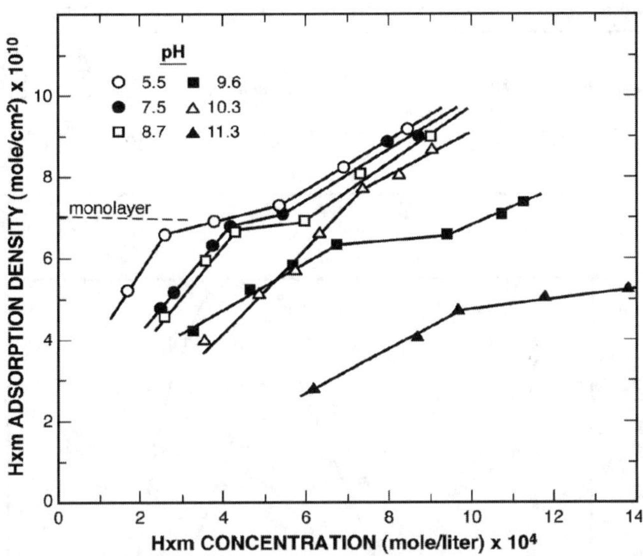

FIGURE 10 Adsorption density of octyl hydroxamate on iron oxide as a function of pH (Raghavan and Fuerstenau 1975)

hematite was ground for 50 minutes at 60 % solids to effect liberation (Fuerstenau et al. 1970. The distribution of the product was as follows:

Size (μ)	Cumulative % Finer
42	99
32	98
23	96
15	70
11	53

In the flotation experiments, the ground ore was not deslimed prior to flotation. The ore was conditioned for 10 minutes in a Fagergren flotation cell prior to collector addition and for 3 minutes after collector addition. Additions of 0.2, 0.4 and 0.5 lb/ton octyl hydroxamate resulted in recoveries of 24, 86 and 87 %, respectively, and concentrate grades of 53, 64 and 68 %, respectively.

Yoon and Hilderbrand (1986) have taken advantage of the specificity of octyl hydroxamate for ferric iron in developing technology for brightening kaolin clay by removing anatase (with iron impurity) and iron oxide minerals. They examined a number of ores from Georgia; the products obtained from one ore are listed in Table 6.

This ore contained 1.42 %TiO_2. With an addition of 1.5 lb/ton octyl hydroxamate, a concentrate containing 86.4 % of the original amount of clay with a TiO_2 content of 0.16 % was obtained.

This technology is being used commercially to brighten clays used for milk cartons. It is necessary to oxidize the hydroxamate prior to use as a product for holding food stuffs which is done with hydrogen peroxide. The value added to the clay with this technology is large, i.e., \$90–\$130 per ton depending on brightness.

TABLE 6 Potassium octyl hydroxamate flotation of a middle Georgia clay

Collector Addition (lb/ton)	TiO_2 Content in Product (%)	Clay Yield (wt %)	Brightness of Clay Products ($lb/ton\ Na_2S_2O_4$)			
			0	3	6	10
1.5	0.16	86.0	92.4	93.0	93.0	93.0
Feed	1.42	100.0				

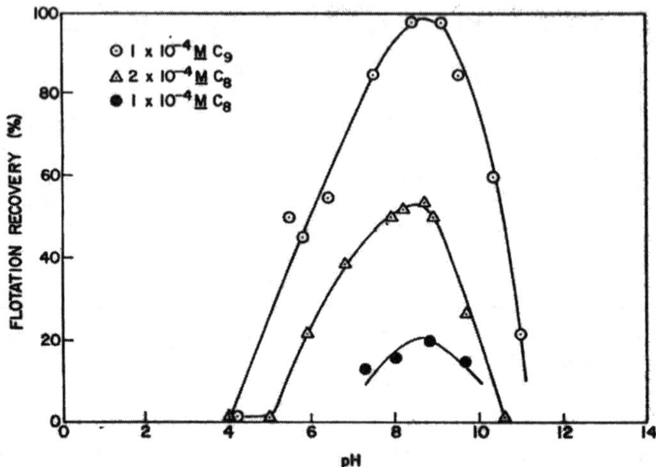

FIGURE 11 Flotation recovery of rhodonite as a function of pH with octyl and nonyl hydroxamate (Palmer et al. 1975)

Rhodonite, Hubnerite, Pyrolusite. The flotation characteristics of various manganese minerals, rhodonite, hubnerite and pyrolusite, with hydroxamate have also been studied. Palmer et al. (1975) floated rhodonite ($MnSiO_3$) with octyl and nonyl hydroxamate (Figure 11). Maximal flotation response is achieved around pH 8.5 (Figure 10). Bogdanov et al. (1973) observed very similar results with hubnerite ($MnWO_4$). Natarajan and Fuerstenau (1983) floated pyrolusite (MnO_2) with various additions of octyl hydroxamate and observed similar flotation phenomena.

Electrophoretic experiments conducted with rhodonite in the absence and presence of manganous salts show a pzc at about pH 3 for rhodonite which is typical for silicates. In the presence of an addition of 1×10^{-4}M Mn^{2+}, the pH range in which the zeta potential changes sign is the same in which flotation is achieved (Fuerstenau and Rice 1968). See Figure 12. The reversal in sign of the zeta potential with metal ion addition occurs in the pH range in which hydrolysis of the Mn2+ occurs. A speciation diagram for Mn^{2+} is presented in Figure 13.

Infrared analysis of both the rhodonite (Palmer et al. 1975) and pyrolusite (Natarajan and Fuerstenau 1983) systems showed the two major absorption bands of manganese hydroxamate on their surfaces after contact with octyl hydroxamate. These facts show that surface reactions occur in these systems and that flotation is enhanced under these conditions. Natarajan and Fuerstenau (1983) observed good flotation of

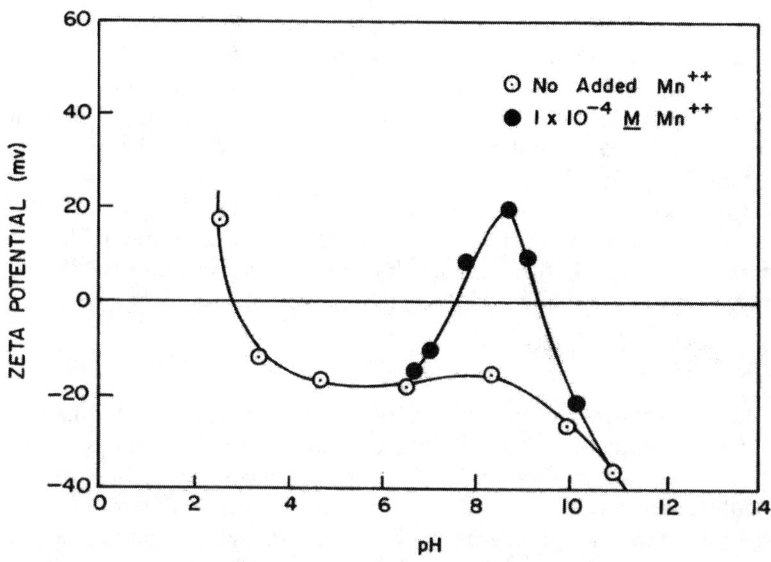

FIGURE 12 Zeta potential of rhodonite as a function of pH in the absence and presence of Mn^{2+} (Fuerstenau and Rice 1968)

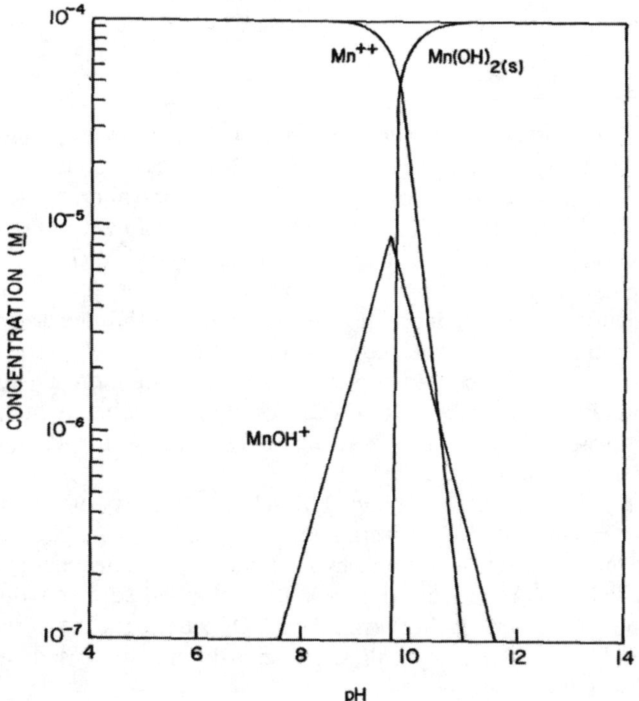

FIGURE 13 Speciation diagram for manganous ion (Fuerstenau and Rice 1968)

pyrolusite at monolayer coverage. As the hydroxamate concentration was increased, these authors noted that monolayer coverage was exceeded. This could be due to the adsorption of hydroxamate on chemisorbed hydroxamate or to the adsorption of manganous hydroxamate formed from surface reaction. A brown colloidal solution was observed indicating that manganous hydroxamate was present. Bogdanov et al. (1973) observed flotation of hubnerite when hydroxamate was chemisorbed with only 20–25 % of monolayer coverage.

In their infrared study Nataranjan and Fuerstenau (1983) observed the disappearance of the surface hydroxyl band when pyrolusite was conditioned with hydroxamate. This phenomenon coupled with the fact that flotation occurs in a pH region in which Mn^{2+} hydrolyzes, suggests that surface reaction may occur as follows.

Effect of Temperature

Elevating the temperature at which flotation is undertaken results in enhanced flotation response of chrysocolla, hematite and pyrolusite. Figure 14 shows that multilayer adsorption is obtained at hydroxamate equilibrium concentrations less than 1×10^{-4} M with flotation temperature of 40°C. This phenomenon might be due to the reaction between hydroxamate and the metal ions, copper, iron and manganese, being endothermic. In fact, utilizing the Clausius-Clapeyron equation, Raghavan and Fuerstenau (1975) calculated a heat of adsorption of +10 to +14 kcal/mol for the reaction of ferric oxide and octyl hydroxamate. Another very plausible explanation for the increased floatability of these minerals with temperature is the greater mineral dissolution that will occur with increased temperature. Under these conditions more hydrolyzed species would be available. This explanation is in accord with the enhanced flotation response of hematite with conditioning time.

Semi-salt Minerals

Flotation studies on semi-salt minerals have been conducted by Bogdanov et al. (1973) and Fuerstenau et al. (1982). Bogdanov et al. (1973) observed good flotation of fluorite from about pH 7–10 with recovery dropping off sharply above and below these values. Calcite flotation with various additions of hydroxamate is shown in Figure 15. Surprisingly, good flotation is achieved at around pH 3 with 3×10^{-4} M collector but decreases steadily with increasing pH. An explanation of this phenomenon might be that as the pH is decreased, calcite solubility is increased substantially. With increased Ca^{2+}, precipitation of calcium hydroxamate is expected to occur. If this is formed in the boundary layer of the mineral surface, enhanced adsorption of the precipitate on the surface may preclude dissolution of the mineral. Lesser Ca^{2+} ion will be available as the pH is increased resulting in lesser amounts of calcium hydroxamate available for adsorption and could result in decreased flotation response.

At pH 9.5 flotation is again enhanced which is probably due to co-adsorption of hydroxamic acid molecule and hydroxamate ion. Above this pH, hydroxamic acid molecule concentration will decrease, while hydroxamate ion concentration will increase. Repulsion of hydroxamate ion from the negatively-charged surface would result in the drop in flotation recovery under these conditions. Of course, the stability of calcium carbonate compounds and complexes will also increase with increased pH which would result in depression of the calcite.

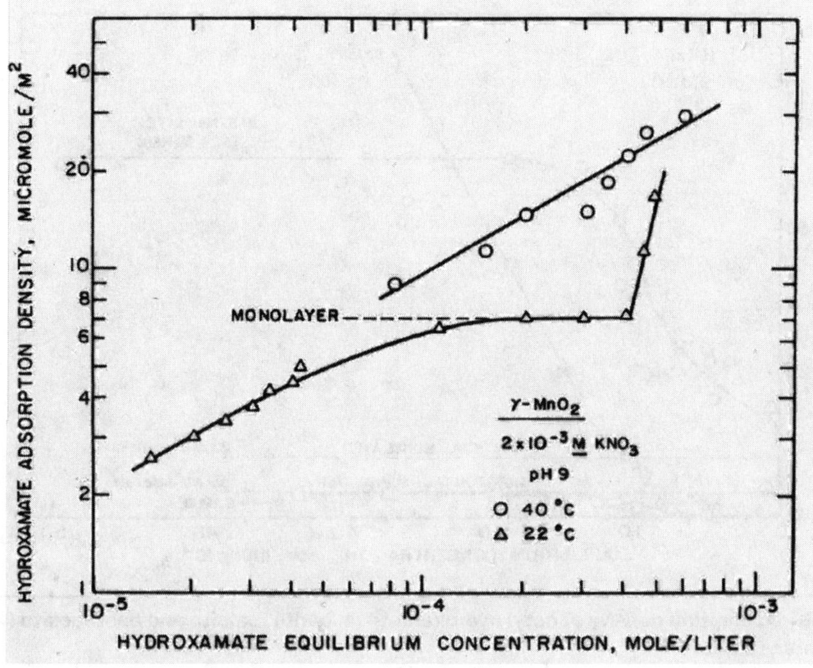

FIGURE 14 Adsorption density of octyl hydroxamate on pyrolusite at two temperatures (Nataranjan and Fuerstenau 1983)

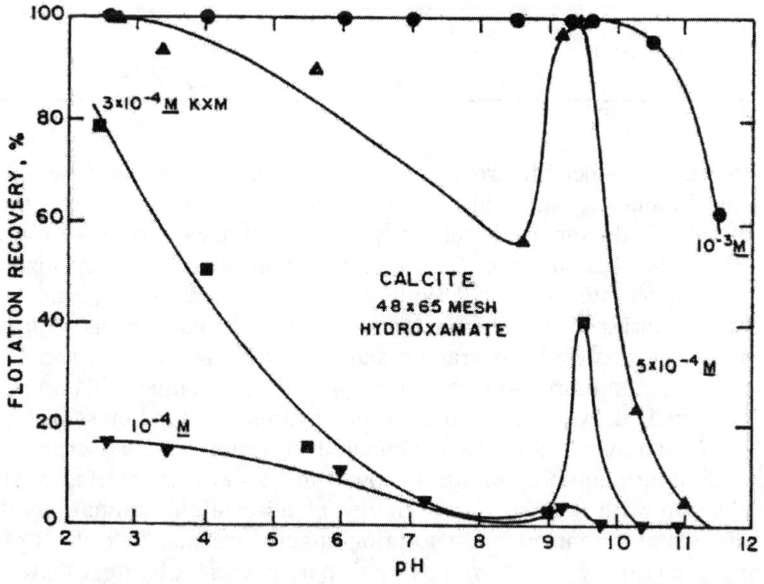

FIGURE 15 Flotation recovery of calcite as a function of pH and octyl hydroxamate concentration (Fuerstenau et al. 1982)

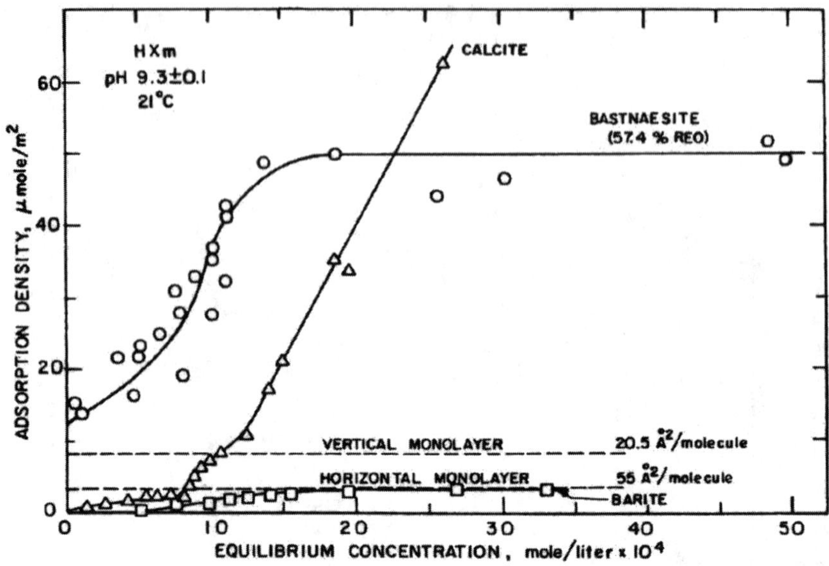

FIGURE 16 Adsorption density of octyl hydroxamate on barite, calcite and bastnaesite (Pradip and Fuerstenau 1983)

TABLE 7 Thermodynamic properties of adsorption of potassium octyl hydroxamate

Mineral	ΔH^o_{ads} (kJ mol^{-1})	ΔS^o_{ads} (J/mol^{-1} K^{-1})
Barite	20	157
Bastnaesite	187	830
Calcite	45	250

Adsorption studies of octyl hydroxamate on barite, calcite, and baestnesite have also been conducted by Fuerstenau et al. (1982). As shown in Figure 16, the interaction between barite and hydroxamate is relatively weak. Monolayer coverage only (chains lying flat on the surface) is obtained. Calcite on the other hand, shows stronger interaction with hydroxamate. Interestingly, discontinuities are seen in the adsorption isotherm. One of the discontinuities occurs when horizontal monolayer coverage is achieved and a second when vertical monolayer coverage is achieved with higher hydroxamate concentrations. Above the second discontinuity, multilayers of hydroxamic acid and hydroxamate ion or of calcium hydroxamate on chemisorbed hydroxamate are present.

Pradip and Fuerstenau (1985) have estimated free energies of adsorption in these systems using the Stern equation of adsorption at mineral/water interfaces. These values in combination with standard free energy of adsorption, standard enthalpy of adsorption, and standard entropy of adsorption enabled calculation of heat of adsorption and changes in entropy. The thermodynamic parameters for barite, calcite and bastnaesite are listed in Table 7.

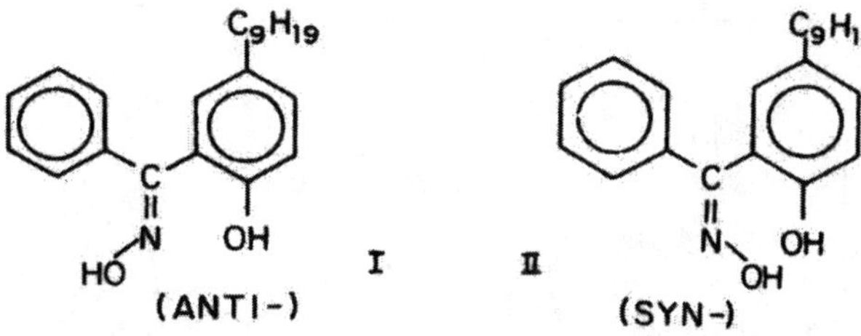

FIGURE 17 Schematic of LIX65N (2-hydroxy-5-nonyl-bezophenone oxime) (Nagaraj and Somasundaran 1979)

What is striking about these calculations is that the heat of adsorption is so much greater for bastnaesite than for barite or calcite. This is obviously related to the rare earth element content of this mineral. This phenomenon is reflected in the greater selectivity of bastnaesite over calcite at elevated temperature of flotation. Another unusual property of this system is the positive increase in entropy with adsorption. Collector adsorption is generally thought to be an ordering process with resultant decrease in entropy. In this case hydroxamate adsorption may cause the disruption of ordered water molecules at the solution/mineral interface.

HYDROXYOXIMES

A number of hydroxyamines have been investigated as collectors by Nagaraj and Somasundaran (1981). Two of these reagents are 2-hydroxy-5-nonyl-benzophenone 8oxime (LIX 65N in solvent extraction) and salicylaldoxime. There are two isomers of this benzophenone. They are shown in Figure 17. Only the anti-isomer functions as a collector. The hydroxyl on the syn-isomer blocks this species from chelating cupric ion.

Flotation of chrysocolla and cuprite (Cu_2O) with benzophenone shows similarity in response with chrysocolla floated with octyl hydroxamate. Although the mineral is Cu_2O, upon dissolution the Cu^+ emitted to solution will oxidize to Cu^{2+}. Two regions in optimal response can be noted (Figure 18). Mechanisms of adsorption of these collectors can be assumed to be the same as that involved in octyl hydroxamate flotation. That is, the undissociated oxamic acid molecule is the active species at pH values below 9.5. Bonding of the oxime to the surface of the mineral can occur through chelation with a surface metal ion or adsorbed cupric hydroxy complex in the pH 6 region. Multilayers of copper oxime can form by surface reaction. At around pH 9.5, it is surmised that mutual adsorption of undissociated acid molecule and oxime anion or the adsorption of multilayers of copper oxime are occurring.

Salicylaldoxime

Nagaraj and Somasundaran (1981) investigated the utility of a number of sparingly water soluble hydroxyoximes as collectors for chrysocolla and tenorite. One of these oximes is salicylaldoxime,

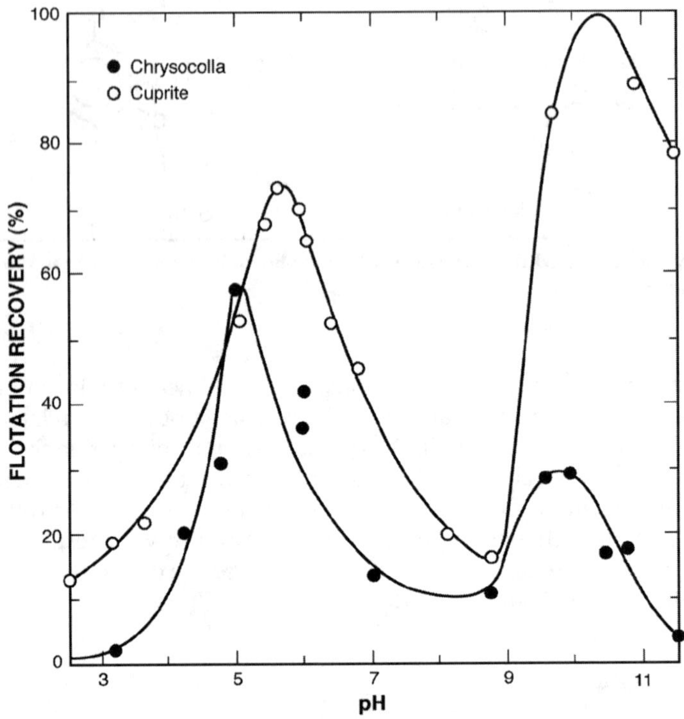

FIGURE 18 Flotation recovery of chrysocolla and cuprite with benzophenone (LIX 65) as a function of pH (Nagaraj and Somasundaran 1979)

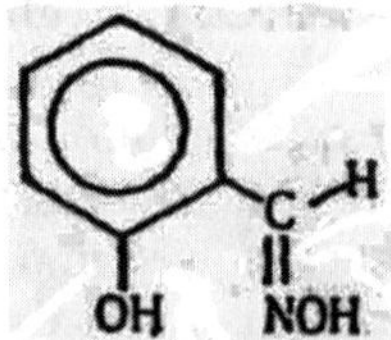

In all cases the oximes were found to be partitioned between the mineral surface and the copper in the bulk aqueous phase in the form of bulk copper chelate. The solubility of the mineral was found to largely control the extent of partitioning of the chelating agent and, hence, flotation. When the surface chelate was noted to increase significantly, flotation of the mineral was also noted to increase significantly. Flotation of tenorite is presented as a function of salicylaldoxime concentration in Figure 19. Flotation is noted to increase dramatically with salicylaldoxime concentrations greater than 5×10^{-4} mol/L. In all of the experiments, the authors noted the presence of dispersed copper salicylaldoxime. They also observed two maxima in flotation, one at about pH 2 and the other at about pH 9. At pH 2 flotation may be due to surface reaction between cupric ion and oxamic acid molecule. The enhanced flotation observed at about pH 5 to 6 can probably be attributed to cupric ion hydrolysis and the formation of cupric salicylaldoxime by

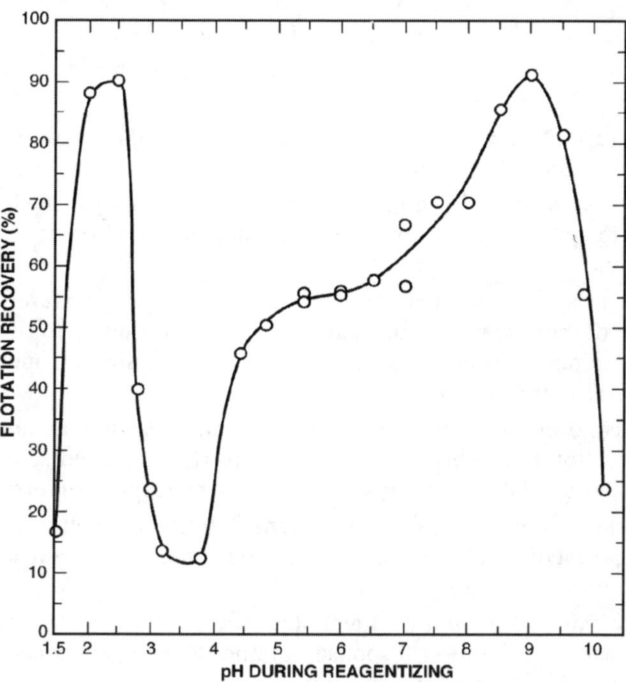

FIGURE 19 Flotation recovery of tenorite with salicylaldoxime as a function of pH (Nagaraj and Somasundaran 1981)

surface reaction. The response at around pH 9.5 appears to be due to the co-adsorption of salicylaldoxime acid molecule and anion.

SUMMARY

In summary,

1. The specificity of chelating agents for metal ions was recognized early and research as possible flotation collectors was conducted in the 1920s.
2. A large number of chelating agents have been examined over the years, but excluding the S–O compounds, such as xanthate, dithiophosphate and dithiocarbamate, the only chelating agent that is used commercially as a collector is hydroxamate.
3. Metal hydroxamates possess considerable stability which makes the use of hydroxamate particularly effective.
4. Hydroxamic acids are weak acids (pK~9). Flotation is possible in the pH range in which the neutral hydroxamic acid molecule is present.
5. Hydroxamic acid molecule chemisorbs on mineral surfaces.
6. Good flotation in some systems is attained with monolayer coverage of hydroxamate and even with fractions of monolayer coverage in other systems.

7. Hydroxamate ion co-adsorbs with hydroxamic acid molecule at ph 9 forming multilayers of collector which results in excellent flotation.
8. Hydroylsis of metal ions that have left their lattice sites in the mineral is crucial for flotation in some of the flotation systems.
9. Surface reaction between metal ions/hydroxy complexes of ions that have left their lattice sites are very important in these systems.
10. Flotation of chrysocolla and rhodonite, hubnerite and pyrolusite occurs in a pH region in which their contained metal ions hydrolyze to monohydroxy complexes.
11. Adsorption and flotation is enhanced at elevated temperature. This can be due to the endothermic nature of hydroxamate adsorption or to the increased solubility of mineral resulting in more metal ion available for metal hydroxamate formation by surface reaction.
12. Longer conditioning time with insoluble minerals, such as hematite results in enhanced flotation. This is probably due to the increased solubility of mineral and greater availability of ferric ion for ferric hydroxamate formation.
13. Due to the solubility of calcite, excellent flotation of calcite is obtained in relatively acid medium due apparently to the formation and adsorption of calcium hydroxamate on its surface.
14. Hydroxy oximes function well as flotation collectors. In the case of chrysocolla and cuprite, the flotation response obtained with these collectors is similar to that obtained with octyl hydroxamate.

REFERENCES

Barbaro, M., R. Herrera-Urbina, C. Cozza, D.W. Fuerstenau, and A.M. Marabini. 1997. Flotation of Oxidized Minerals of Copper Using a New Synthetic Chelating Reagent as Collector. *Int. J. Mineral Proc.* 50:275.

Bogdanov, O.S., Y.I. Yeropkin, T.E. Koltujova, N.P. Khobotoa, and N.E. Shtchuina. 1973. Hydroxamic Acids as Collectors in the Flotation of Wolframite, Cassiterite and Pyrochlore. *Proceedings 10th International Mineral Processing Congress.* Jones, M.J. (ed.). 553.

Burger, K. 1973. Organic Reagents in Metal Analysis. Int. Series of Monographs in *Anal. Chem.* Vol 54. Pergamon. 266 pp.

Chander, S. and D.W. Fuerstenau. 1975. Electrochemical Reaction Control of Contact Angles on Copper and Synthetic Chalcocite in Aqueous Potassium Diethyldithio-phosphate Solution. *Int. J. Mineral Proc.*, 2:333.

Chaterjee, B. 1978. Donor Properties of Hydroxamic Acids. *Coordination Chem. Rev.*, 26:281.

Cheng, K.L., K. Ueno, and T. Imamura. 1982. *CRC Handbook of Organic Analytical Reagents.* CRC Press, 534 pp.

DeWitt, C. C., and F. von Batchelder. 1939. Normal Oximes as Flotation Compounds. *J. Am. Chem. Soc.*, 61:1247.

Evrard, L., and J. DeCuyper. 1975. Flotation of Copper-Cobalt Ores with Alkylhydroxamates. *Proceedings 11th International Mineral Processing Congress,* 655.

Fuerstenau, D.W., Pradip, L.A. Ryan, and S. Raghavan. 1982. An Alternate Reagent Scheme for the Flotation of Mountain Pass Rare-Earth Ore. *Proceedings of XIVth International Mineral Processing Congress,* IV-6.2.

Fuerstenau, D.W., and Pradip. 1984. Mineral Flotation with Hydroxamate Collectors. *Reagents in Minerals Industry.* Jones, M.J. and R. Oblatt (eds.). IMM, London, 161.

Fuerstenau, D.W., and R. Herrera-Urbina. 1988. Flotation Reagents. Chander, S. (ed.). *Advances in Coal and Mineral Processing Using Flotation.* AIME, p 3.

Fuerstenau, D.W., R. Herrera-Urbina, and J. Laskowski. 1988. Surface Properties and Flotation Behavior of Chrysocolla in the Presence of Potassium Octylhydroxamate. *Proceedings of 2nd Latin American Congress on Froth Flotation.* Concepcion. Developments in Mineral Processing 9, Elsevier, Amsterdam.

Fuerstenau, D.W., R. Herrera-Urbina, and D.W. McGlashan. 2000. Studies on the Applicability of Chelating Agents as Universal Collectors for Copper Minerals. *Int. J. Mineral Proc.*, 58:15.

Fuerstenau, M.C., J.D. Miller, and G. Gutierrez-Blanco. 1967. Selective Flotation of Iron Oxide. *Trans. AIME*, 238:200.

Fuerstenau, M.C., and D.A. Rice. 1968. Selective Flotation of Pyrolusite. *Trans. AIME*, 241:453.

Fuerstenau, M.C. and H.D. Peterson. 1969. Chrysocolla and Hematite Flotation with Hydroxamate. U.S. Patent 3,438,494.

Fuerstenau, M.C., R.W. Harper, and J.D. Miller. 1970. Hydroxamate vs. Fatty Acid Flotation of Iron Oxide. *Trans. AIME*, 247:69.

Gutzeit, G. 1946, Chelate-forming Organic Compounds as Flotation Reagents. *Trans. AIME*, 169:272.

Holman, B.W. 1929. Flotation Reagents. *Trans. IMM*, 39:588.

Issa, R.M., and J.Y. Maghrabi. 1973. *S. Phys. Chem.* 253:353.

Koshin, N.V., and K.D. Dobryshin. 1970. *Prom. Teknol. Sekts.* p. 118.

Kratpohvil, V., and M. Nepras. 1972. *Coll. Czech. Chem. Commun.* 37:1533.

Ludt., R.W., and C.C. DeWitt. 1949. The Flotation of Copper Silicate from Silica. *Trans. AIME*, 184:49.

Martell, A.E., and M. Calvin. 1952. *Chemistry of Metal Chelate Compounds.* Prentice-Hall, New York, 613 pp.

Marabini, A.M., and G. Rinelli. 1973. Flotation of Pitchblende with a Chelating Agent and Fuel Oil. *Trans. IMM*, London, 82:C225.

Maslowska, J, and R. Soloniewicz. 1974. *Rocz. Chem.* 48:1391.

Nagaraj, D.R., and P. Somasundaran. 1979. Commercial Chelating Extractants as Collectors: Flotation of Copper Minerals Using 'LIX' Reagents. *Trans. AIME*, 266:1892.

Nagaraj, D.R., and P. Somasundaran. 1981. Chelating Agents as Collectors in Flotation: Oximes-Copper Minerals Systems. *Trans. AIME*, 270:1351.

Natarajan, R., and D.W. Fuerstenau. 1983. Adsorption and Flotation Behavior of Manganese Dioxide in the Presence of Octyl Hydroxamate. *Int. J. Mineral Proc.*, 11:139.

Palmer, B.R. 1972. Ph.D. Dissertation. University of Utah.

Palmer, B.R., G. Gutierrez-Blanco, and M.C. Fuerstenau. 1975. Mechanisms Involved with Anionic Flotation of Oxides and Silicates. Part I. *Trans. AIME*, 258:257.

Perrin, D.D. 1979. Stability Constants of Metal Ion Complexes. *2nd Edition, IUPAC* Chemical Data Series, No. 22, Part B, Organic Ligands. 1263 pp.

Peterson, H.D., M. C. Fuerstenau, R. S. Rickard, and J.D. Miller. 1965. Chrysocolla Flotation by the Formation of Insoluble Surface Chelates. *Trans. AIME*, 232:1.

Pradip and D.W. Fuerstenau. 1983. The Adsorption of Hydroxamate on Semi-soluble Minerals. Part I. Adsorption on Barite, Calcite and Bastnaesite. *Coll. and Surf.*, 8:103.

Pradip, and D.W. Fuerstenau. 1985. Adsorption of Hydroxamate Collectors on Semi-soluble Minerals. Part II. Effect of Temperature on Adsorption. *Coll. and Surf.*, 15:137.

Pradip. 1988. Applications of Chelates in Mineral Processing. *Minerals and Met. Proc.* 5:80.

Raghavan, S., and D.W. Fuerstenau. 1975. The Adsorption of Aqueous Otylhydroxamate on Ferric Oxide. *J. Coll. Int. Sci.*, 50:319.

Rinelli, G., and A.M. Marabini. 1973. Flotation of Zinc and Lead Oxide-Sulfide Ores with Chelating Agents. *Proceedings of 10th International Mineral Processing Congress*. Jones, M.J. (ed.), 493

Rinelli, G., Marabini, A.M., and V. Alesse. 1976. Flotation of Cassiterite with Salicylaldehyde as a Collector. *Flotation*. Fuerstenau, M.C. (ed.). AIME, p. 549.

Ryaboi, V.I., V.A. Shenderovich, and E.F. Stizhev. 1980. Protolytic Dissociation of Alkylhydroxamic Acids and their Derivatives. *Russian J. Phys. Chem.*, 54:730.

Sandell, E.B., and H. Onishi. 1978. Photometric Determination of Traces of Metals. Part l. *Chemical Analysis,* 4th Edition, Volume 3. Wiley and Sons, New York, 1085 pp.

Schwartzenbach, G., and K. Schwartzenbach. 1963. Hydroxamatkomplexe. Die Stabilitat der Eisen (III)–Komplexe Einfacher Hydroxamsauren und der Ferrioxamins B. *Helv. Chim. Acta,* 46:1403.

Sillen, L.G., and A.E. Martell. 1964. Stability Constants of Metal Ion Complexes. *Special Publication No. 17.* The Chemical Society, London, 754 pp.

Sillen, L.G., and A.E. Martell. 1971. Stability Constants of Metal Ion Complexes, *Supplement No. 1. Special Publication No. 25.* The Chemical Society, London.

Somasundaran, P., and D. R. Nagaraj. 1984. Chemistry and Applications of Chelating Agents in Flotation and Flocculation. *Reagents in Minerals Industry*. Jones, M. J., and R. Oblatt (eds.). IMM, London, p. 209.

Sparrow, G., Pomianoswki, A., and J. Leja. 1977. *Separation Science*. 12:87.

Vivian, A.C. 1927. Flotation of Tin Ores. *Min. Mag.,* London, 36:348.

Yale, H.L. 1943. The Hydroxamic Acids. *Chem. Rev.* p. 241. Yoon, R. H., and T. H. Hilderbrand. 1986. Purification of Kaolin Clay by Froth Flotation using Hydroxamate Collectors. U.S. Patent 4,629,556.

Yoon, R.H., and T.M. Hildebrand. 1986. Purification of Kaolin Clay by Froth Flotation Using Hydroxmate Collectorrs. US Patent Number 4,696,556.

Zayan, S.E., R.M. Issa, J.Y. Maghrabi, and M.A. El-Dessoukey. 1973. *Egypt. J. Chem.,* 16:459.

. . . .

Dissolved Gas, Very Small Bubbles, and Interparticle Interactions

David R.E. Snoswell,* Jingwu Yang,* Jinming Duan,* Daniel Fornasiero,* and John Ralston*

The colloidal stability of synthetic silica spheres with clean, methylated and dehydroxylated surfaces was studied at different concentrations of dissolved gas and KCl electrolyte at a fixed pH of 4.2. A classic stability ratio/electrolyte concentration analysis shows that hydrophobic, methylated particles undergo faster rates of aggregation with increasing concentrations of dissolved carbon dioxide. Similar data for hydrophilic particles and dehydroxylated particles show no change as a function of dissolved carbon dioxide concentration. Zeta potential data behave similarly, showing a strong influence of dissolved gas only for methylated particles. The results are interpreted in terms of DLVO theory, with the surface-to-surface interaction dominated by the presence of very small, protruding bubbles. Direct evidence for the existence of these very small gas bubbles (or "nanobubbles") was obtained using tapping mode atomic force microscopy. Small bubbles do not form on smooth, hydrophilic, or dehydroxylated silicon oxide water surfaces immersed in aqueous solutions under known levels of gas supersaturation. Randomly distributed small bubbles were observed over the whole surface of observation on methylated surfaces of controlled roughness. Bubbles formed on rough, methylated surfaces were larger and less-densely distributed than those on a smooth surface of similar hydrophobicity. The process of bubble coalescence was observed as a function of time. The macroscopic contact angle, measured with respect to the aqueous or gas phase, is very different from the microscopic contact angle detected by TMAFM and appears to be due to the influence of line tension at the pinned three-phase contact line. The latter has a value of -3×10^{-10} *N and acts to stabilize the small bubbles, flattening them and thereby reducing the Laplace pressure.*

---

* Ian Wark Research Institute, University of South Australia, Mawson Lakes, Adelaide, SA, Australia

INTRODUCTION

Interparticle forces in a colloidal suspension can often be described by classical DLVO theory; however, in systems where the particle surface is hydrophobic, an additional non-DLVO force is often introduced. This is an attractive force, with some studies linking its range and magnitude to the concentration of dissolved gas.[1–3] Different theories exist to explain the origin of this hydrophobic attraction and, in particular, it has been proposed that these attractive forces are due to the presence of very small bubbles on the particle surface.[4–6] The presence of small gas bubbles on hydrophobic surfaces has been demonstrated indirectly in the key early experiments of Harvey et al.[7–10] and observed on hydrophobic silica, glass, mica and graphite surfaces using tapping-mode atomic force microscopy.[11–13] It appears that both the physical and chemical nature of the solid surface influence the hydrophobic force, by promoting or suppressing the formation of very small surface bubbles.[14]

Bubble nucleation was explored in the early work of Harvey et al.[7–10] In this work visible bubbles were shown to grow from trapped "gas pockets" on a solid surface. Harvey et al. proposed that gas nuclei form when air is trapped in surface defects during the immersion of a dry hydrophobic surface in water. Similar observations were made by Wrobel,[15] who referred to the trapped air as "submicroscopic air-bells". Indirect evidence for the presence of surface gas nuclei was provided by Wood and Sharma,[16] who demonstrated that when particles are hydrophobised in solution, and not allowed to dry, the range of hydrophobic attraction is greatly reduced, since these "gas pockets" have no opportunity to form. Intuitively, it would seem that the amount of trapped gas on a surface should be related to the surface roughness and hydrophobicity. Indeed a theoretical explanation by Harvey et al.[7] shows how these stable gas nuclei can form given a certain combination of advancing/receding contact angles and crevice angles. Furthermore, experimental evidence for the effect of pore size and surface hydrophobicity on bubble nucleation can be seen in the work of Ryan and Hemmingsen,[17] and the work of Lubetkin[18] where defined pitted surfaces were investigated. The critical roles played by surface roughness and hydrophobicity in bubble nucleation is thereby demonstrated.

Regardless of the source of these very small surface bubbles, it is evident that the reported range of hydrophobic attraction varies greatly in surface force studies, ranging from those that match DLVO theory,[4,19] to studies that show long-range forces extending up to 250 nm.[20] A surface covered by very small bubbles will by nature be heterogeneous, displaying regions with different physical and chemical properties. Therefore, since only a small portion of the surfaces can interact in these experiments, a random distribution of surface bubbles may explain some of the variation observed in the measured ranges of hydrophobic attraction.

Before the advent of modern force-measurement techniques, traditional studies of aggregation rates and stability ratios allowed the indirect assessment of interparticle forces. For studies of heterogeneous surfaces, there are obvious statistical advantages in dealing with a large ensemble of particles, particularly when bulk properties like colloid stability are to be predicted. An additional advantage of the aggregation rate approach over studies that employ fixed surfaces, is the ability of the particles to rotate freely during collisions, a factor which may prove important for heterogeneous surfaces. However, despite these advantages, recent studies of this type on colloidal systems are very rare. This may partly be due to systems where classical DLVO theory fails to accurately predict the dependence of colloid stability on electrolyte concentration accurately.[21] Recent studies have shown however, that when the surface charge of the particles is sufficiently low, classical DLVO theory adequately predicts the stability/electrolyte dependence.[22] In the case of silica, the surface charge is both low and well-defined.[23]

Empirical, hydrophobic attractions are usually described using a single- or double-exponential term where one or two decay lengths are invoked.[14] Recently, a theoretical analysis by Mishchuk et al.[24] has shown that the presence of very small bubbles on the surface of a particle, can lead to an attraction that appears similar to the action of hydrophobic forces. The origin of the effect lies in a change in the magnitude (and even sign) of the van der Waals interactions between a particle and a macrobubble.

In this study, the first objective is to investigate the effects of surface hydrophobicity and heterogeneity, together with dissolved gas concentration on the stability of a colloidal particle suspension. On the basis of the premise that very small bubbles attached to particle surfaces influence hydrophobic attraction, conditions are varied to suppress or promote the formation of bubbles. The results are modeled using classical DLVO theory, introducing the influence of gas at the particle-solution interface.

Recently, with the advent of tapping mode atomic force microscopy (TMAFM), imaging of nano- or sub-micron bubbles formed on solid surfaces has become of considerable interest to colloid science[25–27]. This research was spurred by a desire to explain the nature of the long-range hydrophobic forces often observed between hydrophobic solid surfaces immersed in aqueous solutions[28].

The TMAFM technique can '*image*' very small bubbles on solid substrate surfaces, since it has the capacity to determine both the physical structure and physicochemical nature of soft and fragile surfaces at the molecular level, provided that these surfaces are stable during scanning.

To date the results of the TMAFM imaging of very small bubbles have yielded quite diverse results for "nanobubble" formation at solid-water interfaces. Lou et al.[25] showed that small bubbles less than 100 nm in diameter could form on both hydrophobic (highly oriented pyrolytic graphite) and hydrophilic (freshly cleaved mica) surfaces. Using silicon wafers substrates modified by octadecyltrichlorosilane (OTS) (root mean square (RMS) roughness < 0.2 nm; θ_w, water advancing contact angle of 110°), Ishida et al.[26] observed that small bubbles with a base diameter of 650 nm and a height of 40 nm were *linearly* distributed on the substrate surface of observation. Accompanying force measurements by AFM suggested the existence of a long-range hydrophobic force between surfaces with bubble domains. Recently, TMAFM imaging by Tyrrell and Attard[27] showed domains which were very different from those observed by Ishida et al. The bubble domains cover the whole hydrophobised surface of the substrate (methylated by dichlorodimethylsilane vapor; RMS roughness < 0.5 nm; $\theta_{water,advancing} = 101°$, $\theta_{water,receding} = 80°$, hysteresis of 21°). The domains had a mean height of 20–30 nm and mean diameter of 71–87 nm. Since the surface roughness of the substrate surfaces used in these two studies is rather similar, the differences in behaviour are not readily explained.

The second objective of this study is to investigate the formation of very small bubbles at solid-water interfaces through TMAFM imaging of well-defined substrate surfaces with known degrees of surface roughness, hydrophobicity, and gas supersaturation. We demonstrate that there is a significant difference between the macroscopic contact angle for these substrates and the microscopic angle when a very small bubble is formed. The origin of this difference may be due to the line tension acting at the solid-water-vapour contact line.

EXPERIMENTAL

Materials and Reagents

High purity (>99.9%) synthetic silica spheres were supplied by Geltech Inc. The silica spheres were cleaned in a caustic solution of KOH to remove any organic contamination,[23] followed by thorough rinsing in high purity water. The particles had a mean diameter of 1.1 μm [standard deviation of 0.16 μm], determined by an Accusizer 770 particle-counting system. The advancing water contact angle, determined by the sessile drop technique for a similarly treated silicon wafer with a thin native oxide layer was found to be less than five degrees. As further confirmation of cleanliness, a bubble cling test was performed on a submerged layer of particles in high-purity water using a video microscope. No particles were found to attach to the air bubble when pressed against the layer, confirming that the particles were free from hydrophobic contaminants. The surface of these particles after cleaning was found to be very smooth, with an RMS roughness of 0.16 nm as determined by atomic force microscopy (AFM) over a 100×100 nm area.

A certain quantity of these clean particles was then hydrophobised by immersion in a solution of 2% v/v trimethylchlorosilane (99+% purity) in AR-grade cyclohexane. Both the particles and the silica vessels were dried in a clean oven at 110°C prior to reaction to minimize any moisture present. After reaction, the particles were thoroughly washed in AR-grade cyclohexane to remove excess reagents and subsequently oven dried at 110°C. The advancing contact angle, determined by a sessile drop of water on a silicon wafer with a thin native oxide layer, treated under the same conditions, was 82°. A bubble cling test showed that these particles became attached to an air bubble, confirming their hydrophobicity. The surface RMS roughness, determined by AFM, was found to be 0.37 nm over a 100×100 nm area.

Another quantity of the cleaned particles was heated to 1050°C for half an hour. At this temperature, substantial dehydroxylation occurs at the surface, and the surface becomes hydrophobic. A silicon wafer with a thin native oxide layer was treated under the same conditions, producing a surface with an advancing water contact angle of 41°. The surface RMS roughness was found to be 0.17 nm over a 100 100 nm area.

High purity water with a conductivity of less than $1.0 \times 10^{-6}\ \Omega^{-1}\ cm^{-1}$, a pH of 5.6 ± 0.1 and a bubble residence time of less than 1 s was used throughout. Analytical-grade reagents supplied by BDH were used for the preparation of salt (KCl 99.5% purity) and acid (HCl). Food-grade carbon dioxide (99.95% purity) was passed though a suspension of precipitated silica to remove any oil or particulate contaminants prior to preparation of saturated solutions. High-purity (99.997 %) argon, passed through the same suspension, was also used in some preliminary experiments with macrospheres.

Oxidised Si(100) wafers (one-side polished, Philips, Eindhoven/NL) were used as substrates. The RMS roughness of the silicon wafers was less than 0.1 nm, whilst the maximum peak to trough distance was approximately 0.3 nm over an area of $1 \times 1\ \mu m^2$.

Surface Cleaning

Clean wafer substrates were prepared according to the following steps: (1) sonication in ethanol and cyclohexane for 30 seconds, then rinsing with Milli-Q water and treatment in an air plasma (Harrick Plasma Cleaner/Sterilizer PDC-32) for at least 30 s; (2) the substrates were then placed in a bath of H_2SO_4 (98%)/H_2O_2 (30%) (3:1 in volume) at

80°C for 30 min, generously rinsed with Milli-Q water and bathed in hot Milli-Q water for 30 min; (3) dried in a stream of N_2 (N_2 was passed through a filter with pore size of 50 nm before it passed into the glove bag, so as to remove any possible particles), then stored in a glass reaction vessel and finally dried in a clean oven at 120°C for 2–3 hours; (4) the substrates were sealed in a clean container prior to surface modification (see below). This process was performed prior to surface modification in order to minimise any contamination.

Surface Modification

Dehydroxylated Surfaces. Substrates were heated at 1050°C in a clean oven overnight and allowed to cool in air. At this temperature, substantial dehydroxylation occurs at the oxidised silicon wafer surface, resulting in an advancing water contact angle of about 42°.

Methylation with TMCS Vapour. The clean substrates were transferred to a glove bag filled with nitrogen, and then immediately placed inside a sealed bottle containing a small beaker filled with 5 cm^3 of pure TMCS liquid for 20 min. After methylation, the substrates were rinsed with cyclohexane again in order to remove residual silane and then dried in a clean room at room temperature.

Methylation with TMCS Solution. The substrates in the sealed glass vessel were first transferred to a glove bag from the oven (see surface cleaning) while a continuous N_2 gas supply to the bag was maintained. 40 cm^3 of 2×10^{-4} M TMCS solution was then rapidly transferred to the reaction vessel in the bag. Using this process, the reaction of TMCS with water vapour in the air is prevented. After a measured reaction time, the substrate was rinsed thoroughly with cyclohexane several times in the glove bag and then dried with high purity nitrogen in a clean room at room temperature. An exposure time of 20 min was used to produce a rough and hydrophobic surface.

The water advancing and receding contact angles of the prepared substrates were measured by using the sessile-drop method, in order to compare the surface hydrophobicity and assess the degree of hysteresis.

Coagulation Rate

To determine the rate of coagulation in the stability tests the classical method was adopted, where the change in turbidity was measured with respect to time. A custom benchtop apparatus was fabricated specifically for this purpose. This unit used a super-high-intensity (10 000 mcd) GalnN light-emitting diode supplied by Dick Smith Electronics as the light source, with a peak wavelength of 520 nm. The amount of light transmitted through a 1-cm pathlength glass cuvette was then determined by the electric output of a photoamplifier (IPL 10530DAL) supplied by Integrated Photomatrix Ltd. Ambient light was excluded by a light-proof lid. During experiments the photoamplifier voltage was recorded at 1-s intervals on a computer via a 16 bit A/D card, which resulted in a resolution of approximately 0.02 mg/dm^{-3} for the silica spheres supplied by Geltech Inc.

Throughout the experiments a Radiometer PHM85 Precision pH Meter was used in conjunction with a Radiometer pHC3006–9 combined pH electrode. The meter was calibrated daily using Orion pH buffers of pH 4.00 and 7.01, which are certified as traceable to NIST standard reference material. pH readings were reproducible to within ±0.03. The probe was maintained by periodic filling with Radiometer KCl/Ag solution (Part number S21M004). The temperature was maintained at 20±1°C.

TMAFM Imaging

Atomic force microscopy (AFM) measurements of surface roughness were performed using a Nanoscope model III manufactured by Digital Instruments using tapping mode.

Unlike AFM contact mode imaging, in TMAFM, the cantilever is vibrated at or near its resonant frequency; the tip gently taps the surface and then lifts off vertically so that it can protect both the sample and the tip from 'damaging' during scanning. In TMAFM scanning, topography and phase shift images can be collected at the same time. It is known that the phase image in TMAFM is more sensitive for detecting variations in material properties such as adhesion and viscoelasticity. This enables TMAFM to image very small and similar soft matter such as very small bubbles on solid surfaces.

TMAFM imaging for very small bubble observation was performed with a Nanoscope III (Digital Instruments), using a tapping mode fluid cell supplied by the manufacturer. The TMAFM imaging in air was performed only when dry surface feature characterization was of interest. The resonance frequency and spring constant of ultrasharp, non-contact silicon cantilevers (Silicon-MDT Ltd. Company) in air were 265–400 kHz and 20–75 N/m, respectively. Imaging in aqueous solutions was performed using narrow, thin Si_3N_4 cantilevers (0.34 N/m, DI Inc) at frequencies in the range of 6.5–9.3 kHz. The cantilever surfaces were rinsed with ethanol and then cleaned in the plasma reactor for 30 seconds each time before use. All the experiments were performed at $22 \pm 2°C$ in a Class 100 clean room.

Two kinds of images were obtained simultaneously: height and phase images. The tapping strength can be controlled by the amplitude of the free oscillation (A_0) and the set-point amplitude (A_{sp}) of the cantilever. Normally, a suitable driving amplitude was selected to give a 0.5 V RMS amplitude. It was found in this study that the selection of A_{sp} is critical to obtaining a high quality image. However, the optimal A_{sp} is not a constant value, but varies according to the substrate. Once the optimal A_{sp} was selected for an individual sample, it remained stable for several hours. Typically, the phase image was dominated by the mechanical properties of the substrate surfaces if a higher RMS amplitude and lower A_{sp} were applied. A lower A_{sp} reflects a stronger interaction between the tip and substrate, thus a better indication of the surface features may be obtained. Hence, the A_{sp} must be selected so that it is small enough to 'sense' the domains on substrate surface, yet not too low so as to damage the fragile surface features.

Phase imaging can be used as 'compositional mapping' since it reflects differences in energy dissipation due to variations in adhesive, mechanical and viscoelastic properties over heterogeneous surfaces, depending on tapping strength. If very small bubbles are detected on the silicon wafer surface, a significantly-high phase shift (>25°) should be seen between a bubble and bare parts of a substrate surface. A weak repulsive force between the hydrophilic tip and a gas bubble is essential for the TMAFM bubble imaging.

All images were obtained using the "E" scanner with a scan rate of 1 Hz. For each separate experiment, at least three separate images were obtained. The images were found to be both stable and reproducible.

Dissolved CO_2 Control

CO_2 solutions were used in this coagulation and imaging study for the purposes of bubble formation and observation, because: (i) CO_2 is the most soluble "common" gas in water, so that a high supersaturation can be obtained easily without application of very high pressure; (ii) the CO_2 concentration can be easily and accurately determined simply by measuring the solution pH. In all of the experiments, the Milli Q water (at pH 5.6) was supersaturated with CO_2 in a sealed pressure vessel so that the pH of the CO_2 water solution was 4.2, corresponding to a concentration of 0.01 M of gaseous, molecular CO_2.

This dissolved CO_2 in Milli-Q water solution was used for the purpose of surface bubble nucleation. It was first injected into the AFM fluid cell (with the substrate to be scanned inside), then the cell was flushed with Milli-Q water for several times before TMAFM imaging. This flush with Milli-Q water was designed to yield a high level of gas supersaturation with respect to CO_2 in the water layer adjacent to the prepared substrate surfaces, which is favorable for bubble nucleation. Using this method, very small bubbles that formed at the solid-water interface were imaged reproducibly.

RESULTS AND DISCUSSION

Stability Ratio as a Function of Dissolved Gas and Hydrophobicity

The stability ratio (W) as a function of salt concentration (C) is shown in Figure 1. The most important features of the data in Figure 1 are the variation between the curves for the particles with hydrophilic and hydrophobic surfaces and the response of these systems to increased concentrations of dissolved gas. In the case of the hydrophilic and dehydroxylated particles, it can be clearly seen that there is no detectable difference between the stability of a suspension under normal conditions compared with that of an identical system with an increased concentration of dissolved gas. Smooth hydrophilic surfaces will wet completely on immersion in water and are not expected to nucleate very small surface bubbles under supersaturated conditions to the extent that there is any detectable impact on colloid stability. In the case of smooth dehydroxylated but hydrophobic particles where the advancing contact angle is 41°, dissolved gas does not influence the dispersion stability. This is in accord with prior colloid probe determinations of the interaction force between dehydroxylated silica surfaces where the "jump distances" in the presence of gas-saturated (air, argon, and carbon dioxide) solutions were generally found to be less than a few nanometers. Similar conclusions have been reached for smooth hydrophobic ($\theta_{ad} < 38°$) alumina surfaces.

Conversely, there is a clear link between the level of dissolved gas and stability of the suspensions where the particles have a methylated surface. First, there is a marked decrease in the gradient of the stability curve (Figure 1) as the dissolved gas concentration increases. There is an increased attraction between the particles compared with the clean and dehyroxylated cases. This behaviour is in accord with our earlier in situ FTIR study of the aggregation behaviour of smooth, (a) dehydroxylated and (b) methylated 0.5-μm-diameter Geltech silica spheres at 10^{-3} M KNO_3 in the presence of carbon dioxide.[29] A second feature of note in Figure 1 is the displacement of the stability curves for the methylated silica spheres to lower critical coagulation concentration (ccc) values compared with the clean and dehydroxylated cases. We also remark that the gradient of the log W versus log C plot for methylated particles under degassed conditions matches the gradient for the clean, hydrophilic particles.

To establish whether or not the presence of dissolved gas influences the electrical double-layer characteristics of the particle surfaces, the zeta potential of the silica particles (clean, dehydroxylated and methylated) was determined at pH 4.2 and 10^{-3} M KCl at various dissolved gas concentrations. The average zeta potential data, representing three separate experiments, are shown in Table 1.

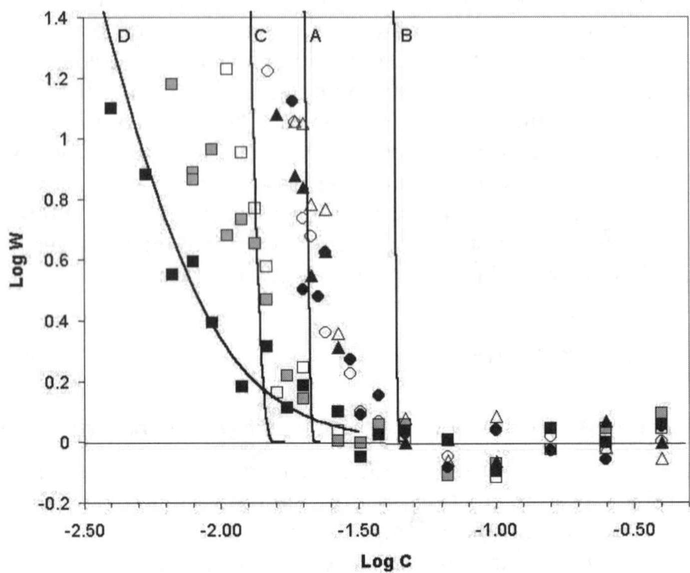

FIGURE 1 Stability ratio vs KCl concentration for Geltech silica spheres at pH 4.2 with a clean hydrophilic surface under normal conditions (○), clean hydrophilic surface / dissolved CO_2 concentration 10^{-2} M (•), methylated hydrophobic surface / normal conditions (■), methylated hydrophobic surface / dissolved CO_2 concentration 10^{-2} M (■) and methylated hydrophobic surface / degassed (□), dehydroxylated surface / normal conditions (△), dehydroxylated surface / dissolved CO_2 concentration 10^{-2} M (▲). Solid lines, A, B, C and D indicate calculated stability curves. (Reprinted with permission Journal of Physical Chemistry B, 107, No. 13, 2003. Copyright 2003 American Chemical Society.)

TABLE 1 Average zeta potential of Geltech silica spheres at pH 4.2 and KCl concentration of 10^{-3} M as a function of dissolved gas concentration (Reprinted with permission Journal of Physical Chemistry B, 107, No. 13, 2003. Copyright 2003 American Chemical Society.)

Particle Type / Conditions	Average zeta potential ± 95% confidence limits (mV)
Clean / Degassed	–29.9 ± 1.2
Clean / Normal	–31.4 ± 1.0
Clean / Gassed, CO_2 10^{-2} M	–30.0 ± 1.3
Dehydroxylated / Degassed	–21.4 ± 1.1
Dehydroxylated / Normal	–16.7 ± 0.7
Dehydroxylated / Gassed, CO_2 10^{-2} M	–17.6 ± 1.6
Methylated / Degassed	–36.6 ± 3.6
Methylated / Normal	–16.4 ± 4.5
Methylated / Gassed, CO_2 10^{-2} M	–22.2 ± 3.7

For the clean, hydrophilic particles, there is no influence of dissolved gas, within experimental error. In the case of dehydroxylated particles, again there is no influence of dissolved gas under normal and elevated CO_2 conditions, with a slight increase only in magnitude when the system is degassed. The reason for the significant difference in zeta potential between these two silica samples lies in the dehydroxylation process. The latter removes surface charging sites, where charge can develop, as silanol groups are converted to siloxane bridges[30]. In the case of the methylated surfaces, there is a marked influence of dissolved gas concentration on zeta potential when the suspensions are degassed.

We have observed similar behaviour in a related study, where nitrogen-saturated solutions of aqueous electrolyte solution, at elevated pressures, were forced through methylated silica capillaries in streaming-potential investigations. In this study, the marked (up to 50%) change in the magnitude of the zeta potential was ascribed to the presence of very small bubbles attached to the rough (on the nanometre scale), heterogeneous (patchy) methylated silica surfaces. These bubbles could be stripped from the capillary surfaces under conditions of very high shear, or prevented from forming by rendering the surface hydrophilic with a nonionic surfactant.

COLLOID STABILITY ANALYSIS

Classically, the forces experienced between two like particles in suspension can be described as the sum of the attractive van der Waals forces and the repulsive electrostatic forces. In terms of interaction energies this can be expressed as a function of H, the shortest distance between the particle surfaces.

$$V(H) = V_R(H) + V_A(H) \quad \textbf{(EQ 1)}$$

There are various equations describing the electrostatic repulsion as a function of distance. The constant-charge model for identical particles shown in Equation 2 was chosen since it was best able to model the experimental data. AFM colloid probe studies using similar silica spheres have also found that the constant charge model gives the best fit to experimental data. The approximations used for deriving eq 2 restrict its use to low surface potentials. Figure demonstrates the low zeta potentials of the silica in these experiments.

$$V_R(H) = -\frac{64\pi a n k_B T}{\kappa^2}\gamma^2 \ln(1 - \exp\{-\kappa H\}) \quad \textbf{(EQ 2)}$$

where

$$\gamma = \tanh\frac{ze\Psi_d}{4k_B T} \quad \textbf{(EQ 3)}$$

In regard to Equations 2 and 3, k_B is the Boltzmann constant, T is the temperature in Kelvin, z is the valency of the ions (assuming a symmetric electrolyte), e is the charge of one electron, Ψ_d is the potential at the outer Helmholtz plane,*a* signifies the radius of the particle, n is the bulk number density of ions and κ is the Debye-Huckel parameter with units of reciprocal length. It is generally assumed that Ψ_d and zeta potential are similar.

When considering the attractive van der Waals forces at short distances (H<<a), the interaction energy between identical particles can be adequately described by[31]

$$V_A = -\frac{Aa}{12H} \quad \textbf{(EQ 4)}$$

where A refers to the Hamaker constant for two identical particles immersed in a bathing medium.

Using these expressions, it is therefore possible to calculate the theoretical stability ratio W, at any given salt concentration where s = R/a and R denotes the distance between the centers of the approaching spheres.

$$W = 2\int_2^\infty \exp\left\{\frac{V(s)}{k_B T}\right\}\frac{ds}{s^2} \quad \textbf{(EQ 5)}$$

This stability ratio was later redefined by McGown and Parfitt[32] taking into consideration the fact that the attractive van der Waals force still operates during rapid coagulation.

$$W = \frac{\int_2^\infty \exp\frac{\left\{\frac{V(s)}{k_B T}\right\}}{s^2}ds}{\int_2^\infty \exp\frac{\left\{\frac{V_A(s)}{k_B T}\right\}}{s^2}ds} \quad \textbf{(EQ 6)}$$

This equation can be further refined by including the factor β, proposed by Honig et al.[33] to correct for the influence of hydrodynamic interactions, which were shown to be of importance in the work of Spielman.[34]

$$W = \frac{\int_2^\infty \beta(s)\exp\frac{\left\{\frac{V(s)}{k_B T}\right\}}{s^2}ds}{\int_2^\infty \beta(s)\exp\frac{\left\{\frac{V_A(s)}{k_B T}\right\}}{s^2}ds} \quad \textbf{(EQ 7)}$$

where

$$\beta = \frac{6s^2 - 11s}{6s^2 - 20s + 16} \quad \textbf{(EQ 8)}$$

Since a theoretical plot of the stability ratio vs salt concentration can be obtained, a comparison of this theoretical plot in the slow regime below the ccc with the experimental version can be performed. By using both the Hamaker constant and surface potential as fitting parameters, the calculated stability curve can be matched to the experimental results. This has been performed recently using these equations by Puertas et al.[35] in a study of the stability of latex colloids. Although the accuracy of the Hamaker constant

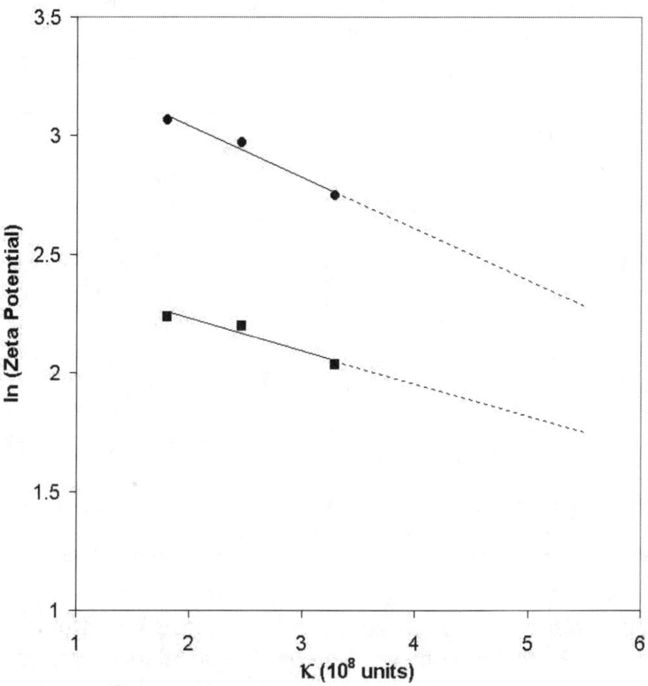

FIGURE 2 Zeta potential data expressed in a plot of ln[zeta potential] vs the Debye-Huckel parameter, κ, at pH 4.2 using KCl electrolyte under normal conditions. Filled circles represent clean hydrophilic particles, filled squares represent methylated particles. Dashed lines indicate the extrapolated region. (Reprinted with permission Journal of Physical Chemistry B, 107, No. 13, 2003. Copyright 2003 American Chemical Society.)

and surface potential obtained by this dual-parameter technique can be debated, the shape of the calculated curve does correspond very closely to the experimental curve in this particular study. For the stability curves presented in this present paper however, the particle surface potentials have been obtained from experimental results.

Using electrophoresis techniques, zeta potentials at specific salt concentrations may readily be obtained through experiments performed at KCl concentrations up to 10^{-2} M. At higher salt concentration, electrolysis becomes an issue. From a of a plot of ln (Zeta potential) versus κ, the zeta potential near the ccc can be accessed by simple extrapolation under conditions where experimental measurements of mobility are difficult. Examples of these linear extrapolations are shown in Figure 2, whilst their empirical forms are given in Table 2.

With respect to the stability curves, a sensitivity analysis performed on the theoretical stability curve reveals several important points. An increase in the Hamaker constant shifts the stability curve to lower salt concentrations, but has little influence on the gradient, whereas a decrease in particle radius decreases the gradient of the curve whilst the ccc remains relatively constant. It is also found that any variation in the Zeta potential changes both the position of the curve and, to a small degree, the gradient. This sensitivity analysis is valuable for it provides guidance in the following data interpretation.

TABLE 2 Zeta potentials and composite Hamaker constants used to model the stability ratio. For each calculation the Hamaker constant for a water/vacuum/water interaction is taken to be 4.38 × 10^{-20} J. (Reprinted with permission Journal of Physical Chemistry B, 107, No. 13, 2003. Copyright 2003 American Chemical Society.)

Condition	Potential Ψ_d (mV)	Hamaker Constant
Clean, Hydro-philic	$\Psi_d = -\exp[(2.15 \times 10^{-9} \times \kappa) + 3.47]$	8.4×10^{-20} J silica/vac/silica
Methylated Degassed	$\Psi_d = -\exp[(2.15 \times 10^{-9} \times \kappa) + 3.47]$	Particle 8.4×10^{-20} J silica/vac/silica Methylation Layer 0.3 nm thick 4.0×10^{-20} J layer/vac/layer Air Layer 3 nm thick 0 J layer/vac/layer
Methylated Gassed CO_2 10^{-2} M	$\Psi_d = -\exp[(1.39 \times 10^{-9} \times \kappa) + 2.51]$	Particle 8.4×10^{-20} J silica/vac/silica Methylation Layer 0.3 nm thick 4.0×10^{-20} J layer/vac/layer Composite Air/Water Layer 3 nm Thick 2.4×10^{-20} J layer/vac/layer

A theoretical stability curve was determined for the clean hydrophilic particles (Figure 1, Curve A). The Hamaker constant used in the calculation was adjusted to obtain the best fit with the experimental data. Values of the Zeta potential and Hamaker constant used are given in Table 2. It is notable that the agreement between the experiment and theory for the case of hydrophilic or dehydroxylated silica is acceptable in line with recent studies of the stability of colloidal dispersions of low surface charge. The presence of *very* short range hydration forces has been detected between silica surfaces, irrespective of whether the latter are hydrophilic or dehydroxylated. Under the conditions of this present study, they do not influence the essential coagulation behaviour, since regimes of both slow and fast aggregation are clearly evident as the salt concentration is increased.

We now address the case of methylated systems in the near-absence of dissolved gas and where gas is present. Before doing so, we remark that the evidence linking the presence of very small bubbles with surface heterogeneity (physical and/or chemical) is very strong indeed.[25,26,27] There are three reported AFM tapping-mode observations of "nanobubbles" on surfaces immersed in aqueous solutions reported to date, involving silane-treated silica or glass surfaces treated with silanes, cleaved mica and pyrolitic graphite. In only one of these studies was any attempt made to quantify the surface heterogeneity through the rudimentary determination of contact angle hysteresis (substantial, at 21°, compared with fluoropolymer surfaces). In view of the zeta-potential data in Table 1, and the observed heterogeneous nature of our methylated surfaces, the presence of very small surface bubbles would appear to influence the observed stability behaviour. We have already demonstrated that various surface distributions of very small bubbles can alter the van der Waals interaction when a particle and a macrobubble interact. We proceed, knowing that we are dealing with a statistically large number of particles and therefore use average values for methylation and gas "layer thicknesses". In the case of the methylated particles, the methylation layer thickness was taken to be 0.2 nm, representing the difference between the RMS roughness data for the clean and

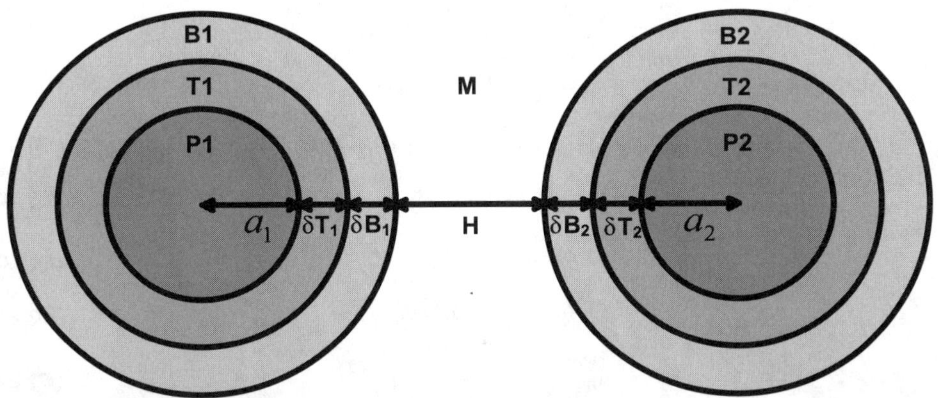

FIGURE 3 Arrangement of two identical spheres with a hydrophobic layer and air layer defining the nomenclature used in equations 10–28. [Refer to Table 1 and text for definition of "normal" and "degassed". Refer to Table 2 for the values of parameters used in the calculation of curves A to D] (Reprinted with permission Journal of Physical Chemistry B, 107, No. 13, 2003. Copyright 2003 American Chemical Society.)

methylated silica spheres. This value is consistent with the known chemistry of silanation processes for silica surfaces.

For the case of the hydrophobic methylated particles under degassed conditions (Figure 1, Curve C), it was first assumed that the observed shift in the stability curve was due to the methylation layer altering the Hamaker constant of the surface, rather than to the presence of very small surface bubbles. To allow for this and any "air layer" present in later cases, a more complicated model incorporating two adsorbed layers was used for the calculation of Hamaker constant. Using the generic approach of Usui and Barouch[36] as a basis, the following equation was obtained for the case of two interacting spheres of radius a_1 and a_2 with two adsorbed layers on each particle. Figure 3 describes the nomenclature used in equations 12–30.

$$V_A = -\frac{1}{6}\left(\frac{a_1 a_2}{a_1 + a_2}\right) \times \left[\frac{A_1}{D1} + \frac{A_2}{D_2} + \frac{A_3}{D_3} + \frac{A_4}{D_4} + \frac{A_5}{D_5} + \frac{A_6}{D_6} + \frac{A_7}{D_7} + \frac{A_8}{D_8} + \frac{A_9}{D_9}\right] \quad \textbf{(EQ 9)}$$

where

$$A_1 = (A_{B1}^{1/2} - A_M^{1/2})(A_{B2}^{1/2} - A_M^{1/2}) \quad \textbf{(EQ 10)}$$

$$D_1 = H \quad \textbf{(EQ 11)}$$

$$A_2 = (A_{T1}^{1/2} - A_{B1}^{1/2})(A_{B2}^{1/2} - A_M^{1/2}) \quad \textbf{(EQ 12)}$$

$$D_2 = H + \delta B_1 \quad \textbf{(EQ 13)}$$

$$A_3 = (A_{P1}^{1/2} - A_{T1}^{1/2})(A_{B2}^{1/2} - A_M^{1/2}) \quad \textbf{(EQ 14)}$$

$$D_3 = H + \delta T_1 + \delta B_1 \quad \textbf{(EQ 15)}$$

$$A_4 = (A_{B1}^{1/2} - A_M^{1/2})(A_{T2}^{1/2} - A_{B2}^{1/2}) \quad \text{(EQ 16)}$$

$$D_4 = H + \delta B_2 \quad \text{(EQ 17)}$$

$$A_5 = (A_{T1}^{1/2} - A_{B1}^{1/2})(A_{T2}^{1/2} - A_{B2}^{1/2}) \quad \text{(EQ 18)}$$

$$D_5 = H + \delta B_1 + \delta B_2 \quad \text{(EQ 19)}$$

$$A_6 = (A_{P1}^{1/2} - A_{T1}^{1/2})(A_{T2}^{1/2} - A_{B2}^{1/2}) \quad \text{(EQ 20)}$$

$$D_6 = H + \delta T_1 + \delta B_1 + \delta B_2 \quad \text{(EQ 21)}$$

$$A_7 = (A_{B1}^{1/2} - A_M^{1/2})(A_{P2}^{1/2} - A_{T2}^{1/2}) \quad \text{(EQ 22)}$$

$$D_7 = H + \delta B_2 + \delta T_2 \quad \text{(EQ 23)}$$

$$A_8 = (A_{P1}^{1/2} - A_{T1}^{1/2})(A_{P2}^{1/2} - A_{T2}^{1/2}) \quad \text{(EQ 24)}$$

$$D_8 = H + \delta B_1 + \delta B_2 + \delta T_2 \quad \text{(EQ 25)}$$

$$A_9 = (A_{P1}^{1/2} - A_{T1}^{1/2})(A_{P2}^{1/2} - A_{T2}^{1/2}) \quad \text{(EQ 26)}$$

$$D_9 = H + \delta T_1 + \delta B_1 + \delta T_2 + \delta B_2 \quad \text{(EQ 27)}$$

Using these equations, a single 0.3 nm organic layer (T1 and T2) with Hamaker constant of 4.0×10^{-20} J (with respect to vacuum[36]) was chosen to simulate the TMCS coating, since this value lies within the range reported for similar hydrocarbons. The calculated stability curve for these conditions however does *not* explain the experimental observations, rather the predicted displacement is to the right as a result of the lower van der Waals attraction (Figure 1, Curve B).

Even under degassed conditions, some very small gas pockets are likely to remain trapped at surface sites. Although these sites do not constitute a uniform layer over the surface of the particle, to simplify the present calculation and allow their potential effect on the van der Waals forces to be determined, a second, outer air "layer" (B1 and B2) was introduced into the calculations. If this layer is 3 nm thick, then the corresponding calculated stability curve matches the experimental curve quite closely (Figure 1, Curve C).

We note that the surface potential used in our model is obtained for an ensemble of particles. The description of the surface charge distribution on a surface decorated, to some degree, with very small bubbles is not tractable at present, and remains as a future problem. The recent approach of Zembala and Adamczyk[37] for particles on surfaces is very encouraging, however. The data in Table 1 and Figure 1 indicate that if the concentration of surface bubbles increases, the net effect is to decrease the magnitude of the zeta and therefore the surface potential. Although the exact potential of the gas-water interface remains elusive, studies indicate that it is likely to be rather small under the experimental conditions of this investigation. The observed change in the potential of the surface covered by very small bubbles is due to the underlying solid surface being screened by these bubbles.

Given that these surface bubbles increase their coverage of the methylated surface under normal and gassed conditions, it is helpful to consider the various possible surface

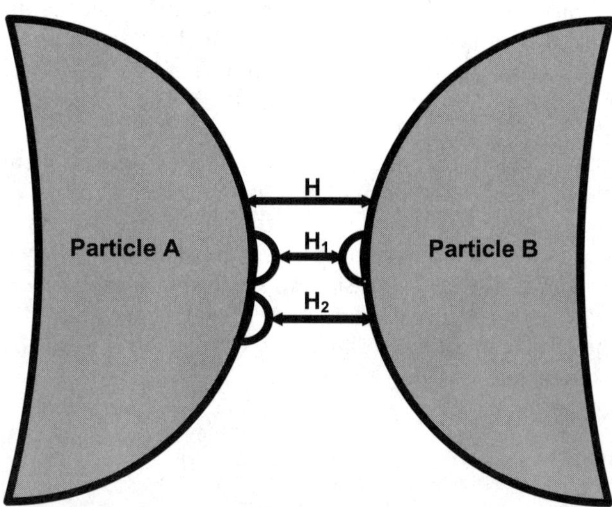

FIGURE 4 Scale diagram representing two 1 μm diameter particles with 100 nm diameter surface bubbles demonstrating the three separate interactions that can occur. In terms of the van der Waals forces these are: strongly attractive bubble-water-bubble interactions occuring over distance H; strongly repulsive silica-water-bubble interactions occuring over distance H_1; and attractive silica-water-silica interactions occuring over distance H_2. (Reprinted with permission Journal of Physical Chemistry B, 107, No. 13, 2003. Copyright 2003 American Chemical Society.)

interactions that may occur before attempting to describe the stability curves for these remaining cases. With regard to Figure 3 it is obvious that three distinct interactions can occur. Significantly, the bubble-water-silica van der Waals interaction is repulsive, whereas the bubble-water-bubble and silica-water-silica interactions are attractive, as shown in Figure 4.

In reality the total interaction energy between the particles will reflect a combination of these forces. The distribution of bubble coverage and diameters over the surface will further complicate any calculations. From the simplified interaction analysis, as the size and coverage of surface bubbles increase, one would expect that the bubble-water-bubble interactions would predominate. The AFM studies below have shown that very small bubbles on methylated silica surfaces immersed in aqueous solutions with similar electrolyte and dissolved gas levels, have radii between 20 and 60 nm and a surface coverage of approximately one-half. As a consequence, the effective radius of interaction between the particles will be that of the protruding surface bubbles rather than that of the particles themselves (see Figure 4).

Acknowledging the problems associated with the precise determination of the Hamaker constant in this situation, we have used the following procedure to calculate Curve D in Figure 1. The effective interaction radius was taken as 40 nm on the basis of extant AFM evidence. Since the surface coverage was not known exactly, it was not possible to calculate a composite Hamaker constant with acceptable reliability by following the procedure we have used previously. Hence, we have used the Hamaker constant as an effective fitting parameter while still retaining all of the other features used to calculate Curve C in Figure 1. In practice, this means that the outermost layer is taken as a composite air/water region with a thickness of 3 nm. (see Table 2). After the fitting exercise, the effective Hamaker constant for the layer B_1 (= B_2 for identical particles, Figure 3) is obtained.

Most significantly, the stability curve calculated for interacting surfaces decorated with protruding bubbles with a radius of 40 nm fits the gradient of the experimental data very closely (Figure 1, Curve D). The change in the effective interaction radius caused by the protruding bubbles explains the change in the gradient of the experimental stability curves for the methylated particles at dissolved gas levels are increased (Figure 1). Furthermore, sensitivity analysis reveals that these changes in the gradient, which is the main feature of the data, cannot be explained by changes in the effective Hamaker constant because the latter does not alter the gradient.

TMAFM IMAGING

Imaging of Substrate Surfaces Immersed in CO_2 Saturated Aqueous Solutions

The various wafer substrate surfaces were examined by TMAFM when immersed in dissolved CO_2 Milli-Q water solution (CO_2: 0.01 M). The results are given below.

Imaging of Clean Hydrophilic Si-Wafer Surfaces

A TMAFM image of hydrophilic Si-wafer substrates immersed in the CO_2 solution is shown in Figure 5.

A smooth substrate surface is evident and is very similar to that obtained in air.

Note that the surface roughness and morphology shown in Figure 5 are slightly different from those shown in air.[44] However, these minor differences are not due to any physical change of the substrate surface itself; rather, they are likely due to the effect of the medium on TMAFM imaging. Hydrodynamic and/or viscous effects can influence the vibration of the cantilever in water and, in addition, modification of the interfacial forces between the cantilever and substrate surface in water may cause a slight difference in the TMAFM images obtained in air and aqueous solutions for the same substrates.

Dehydroxylated Si-Wafer Surface

A TMAFM image for the dehydroxylated Si-wafer surface in CO_2-saturated solution is shown in Figure 6.

As in the case of hydrophilic substrates, the image obtained showed smooth surface features.

Slight morphological variations in comparison to those obtained in air were observed. There was no evidence for bubble formation on the dehydroxylated wafer surface (with an intermediate hydrophobicity, $\theta_w = 42°$). This observation[29] is consistent with the earlier observation that a clean, dehydroxylated silicon-wafer surface exhibits no sign of long-range attraction or long-range jump distances in the AFM force versus distance measurements between dehydroxylated silicon surfaces. Moreover, we do not

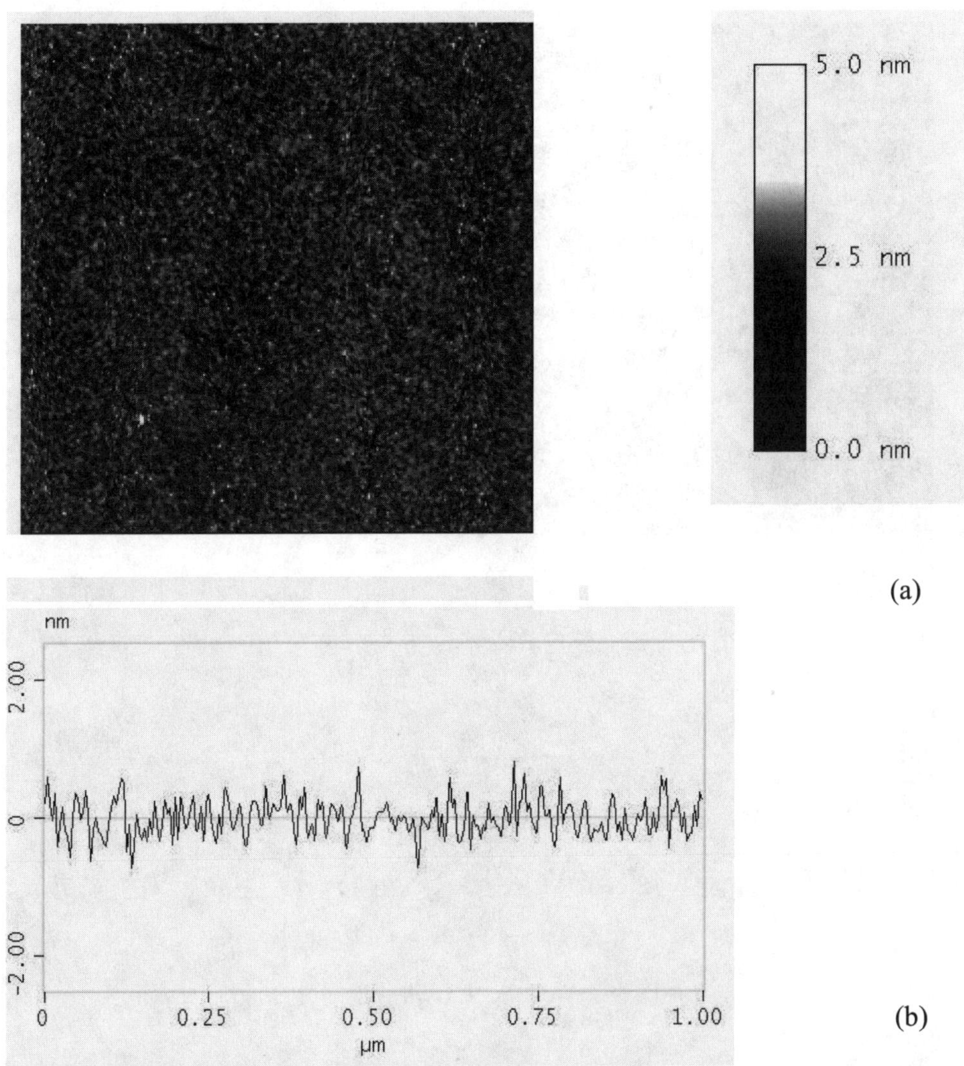

FIGURE 5 AFM tapping mode image of (a) clean Si surface and (b) height variation of cross section in saturated CO_2 solution: 1 µm × 1 µm, height = 5 nm (Reprinted with permission Journal of Physical Chemistry B, 107, No. 25, 2003. Copyright 2003 American Chemical Society.)

detect any influence of CO_2-saturated solutions on the coagulation stability of dehydroxylated silica particle dispersions.

TMCS Vapor Prepared Si-Wafer Surface

TMAFM height images for TMCS vapor prepared Si-wafer surfaces are shown in Figure 7. In contrast to the smooth surface features obtained in air for the same kind of substrate surface, clear domains cover quite randomly the whole surface of observation.

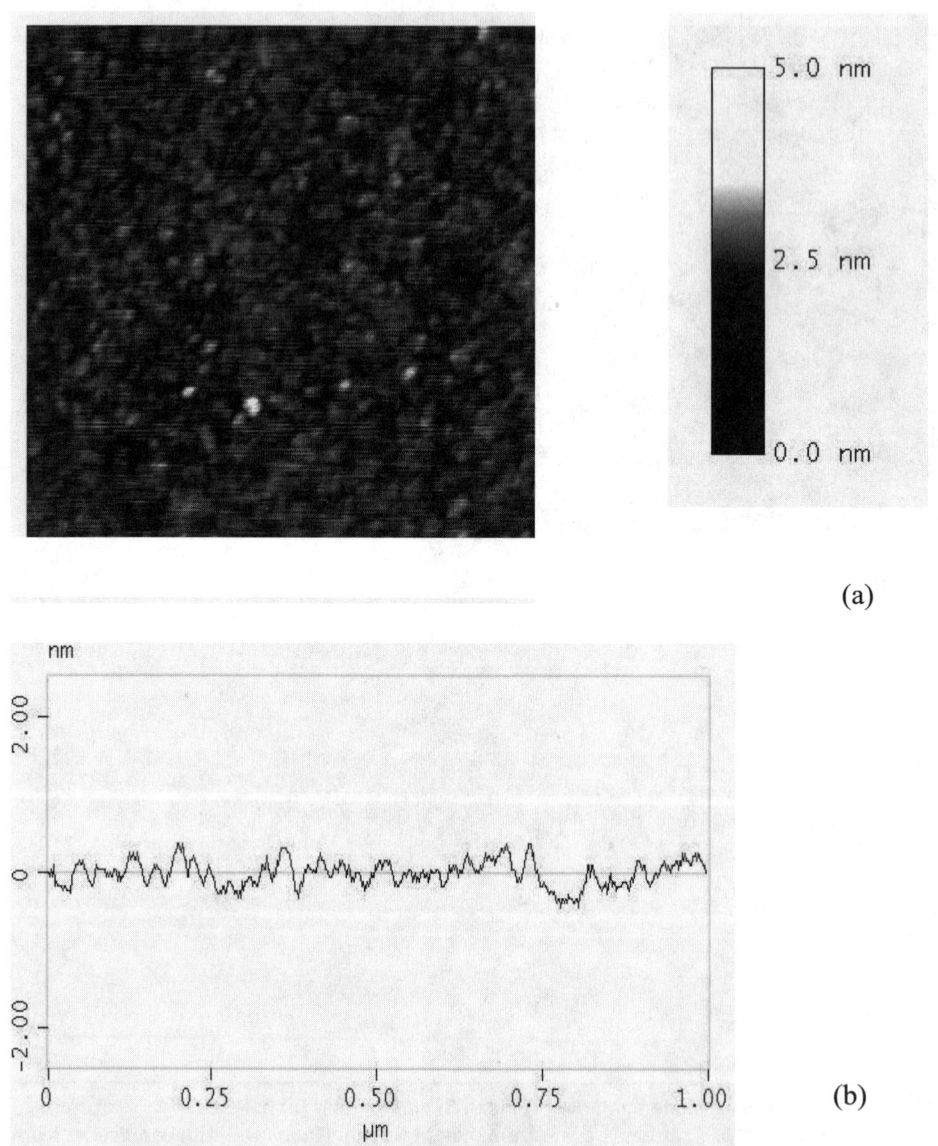

FIGURE 6 AFM tapping mode image of (a) dehydroxylated Si surface and (b) height variation of cross section in saturated CO_2 solution: 1 µm × 1 µm, height = 5 n (Reprinted with permission Journal of Physical Chemistry B, 107, No. 25, 2003. Copyright 2003 American Chemical Society.)

If, for now, we assume that these domains are bubbles, the base radius and height of these bubbles are in the range between 50 and 400 nm and 20 and 80 nm, respectively. The size range of these bubbles and the manner of their distribution are similar to those observed by other researchers. However, the bubbles shown in the image in Figure 7 are much more densely populated and much more evenly distributed over the whole surface of observation compared with those observed by Ishida et al.[26]

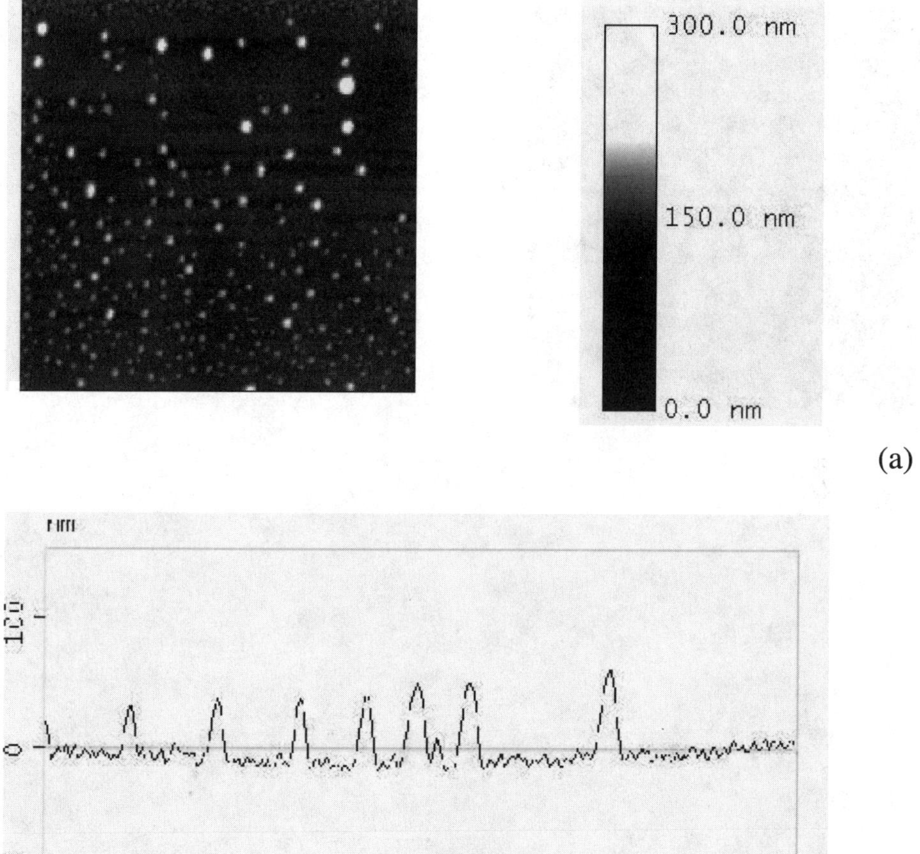

FIGURE 7 AFM tapping mode image of (a) Si surface modified with TMCS vapor for 20 min and (b) height variation of cross section in saturated CO_2 solution: 10 µm × 10 µm, height = 300 nm (Reprinted with permission Journal of Physical Chemistry B, 107, No. 25, 2003. Copyright 2003 American Chemical Society.)

TMCS/Cyclohexane Solution Prepared Si-Wafer Surface

The TMAFM image for the "rough" surface obtained by TMCS/cyclohexane solution methylation is shown in Figure 8. In comparison with the bubbles for the smooth surface (see Figure 7), one major difference is that bubbles on the rough surface are much less densely populated and the bubble sizes are relatively larger than those for the smooth surface. The bubble base radius and height are around in the range of 150–400 nm and 60–200 nm respectively.

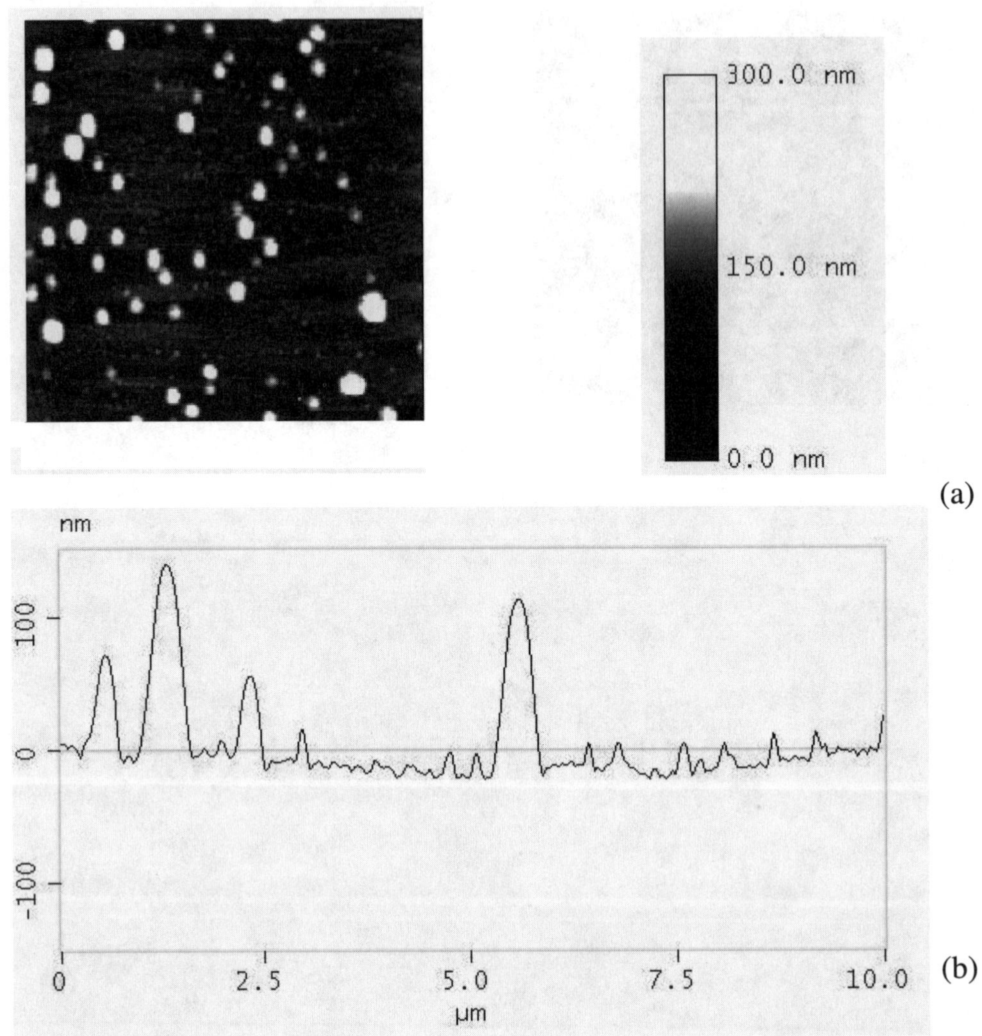

FIGURE 8 AFM tapping mode image of (a) Si surface modified with TMCS solution for 20 min and (b) height variation of cross section in saturated CO_2 solution: 10 µm × 10 µm, height = 300 nm (Reprinted with permission Journal of Physical Chemistry B, 107, No. 25, 2003. Copyright 2003 American Chemical Society.)

Note that a distinct surface pattern is reflected in the baselines in the cross section profile, evidenced by small individual peaks. These different features of bubble size and distribution reflect the effects of surface roughness on bubble formation. Note that the same degree of phase shift was observed over the domains as in the case of the TMCS vapor prepared substrate.

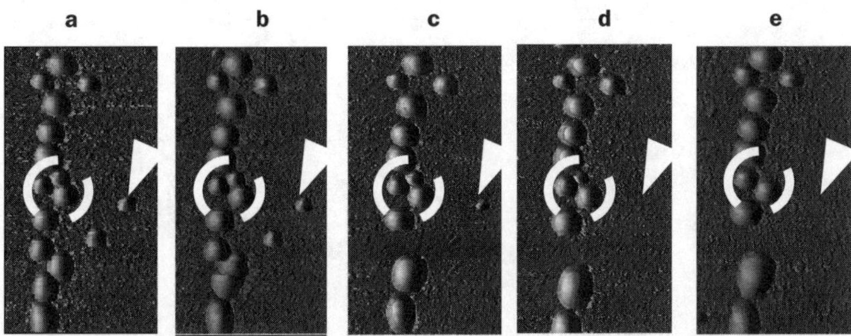

FIGURE 9 TMAFM bubble coalescence images: (a) made at 20 hours after CO_2 saturated water was injected into the fluid cell. (b), (c), (d) and (e) made at time intervals of 20 minutes after (a). Scan area: 10 μm × 5 μm, height = 300 nm. (Reprinted with permission Journal of Physical Chemistry B, 107, No. 25, 2003. Copyright 2003 American Chemical Society.)

Images of Bubble Coalescence

Consecutive imaging by the TMAFM at time intervals over a long time scale was performed to provide further evidence as to the character of the domains observed by the TMAFM imaging. The substrates used were the TMCS/cyclohexane solution prepared Si-wafer (θ_w = 88°, rms roughness = 2.7 nm). The early stage of the TMAFM imaging process was the same as for normal imaging (Figure 8). However, after the completion of the first TMAFM image, the fluid cell was flushed gently using a 10 cm^3 syringe with Milli-Q water every 20 min, and the images were followed by TMAFM scanning. This whole process lasted for more than 20 h. The first image was similar to the bubble image shown in Figure 8; five sets of bubble images thereafter are presented in Figure 9.

First, one can see in this series of images that, after long-time TMAFM scanning, the bubbles that initially distribute evenly over the whole surface "line up" around a vertical line in the observation area. This phenomenon seems to be evident in TMAFM images reported elsewhere, albeit indistinctly. It can be reasonably assumed that such a pattern of behaviour may be attributed to some fluid-bubble perturbation by the cantilever during the extended horizontal movement or scanning in the fluid cell. This perturbation may also contribute to bubble coalescence.

However, the single most important observation from these images is the process of bubble coalescence and the disappearance of an individual bubble from the solid-solution interface surface, behaviour that is characteristic of gas bubbles in aqueous solutions. For example, the smaller bubble (in the broken line circle) becomes even smaller from image a to image d in Figure 9, and finally "dissolves" or merges with the neighbouring bubble so that the latter becomes a larger bubble. An individual small bubble (identified by arrow) becomes increasingly smaller from image a to image c in Figure 9 and finally disappears from image d. The observed phenomenon, namely, that a smaller bubble becomes smaller or dissolves while the neighbouring bigger bubble becomes larger, is quite similar to "Ostwald ripening". It is the first time that such a very small bubble ripening or coalescence process has been observed by TMAFM imaging.

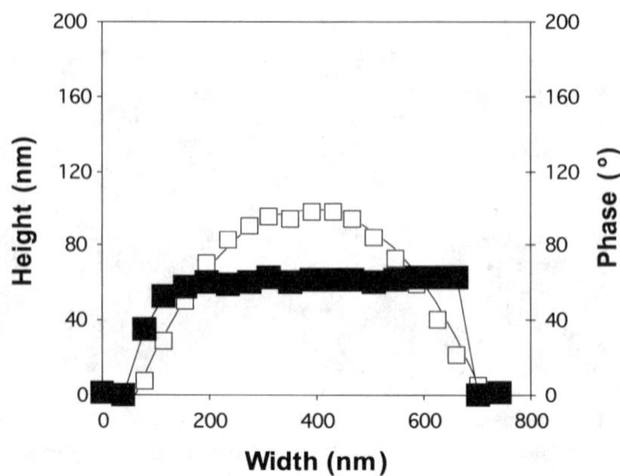

FIGURE 10 Cross section of a bubble height and phase image profile created from TMAFM image file (TMCS vapour treatment Si-wafer, supersaturated CO_2 solution). Empty square line: height image; solid square line: phase image. (Reprinted with permission Journal of Physical Chemistry B, 107, No. 25, 2003. Copyright 2003 American Chemical Society.)

Nature of Domains

Regarding the nature of the domains, several pieces of evidence identify them as gas bubbles: (1). The first important evidence is the large phase shift (approximately 50°) observed over the domains during phase imaging (See Figure 10).

This large phase shift indicates that the domains are soft and different in nature from the hydrophobized solid surfaces. A typical view of the phase shift across a bubble, generated from TMAFM data files, is shown in Figure 10. From this figure, one can see that, although the cross section of the bubble can be fitted to a spherical cap curve (note the different scales for the base and the height), a constant phase value of around 50° across virtually the whole distance of the selected bubble was seen. (2) Because TMCS, a small molecule (MW = 108.64), is attached through –Si-O-Si- bonds with the Si-wafer surface, it is very unlikely that these adsorbed TMCS molecules after vapor treatment of the solid surface, will accumulate into large domains on the substrate surface when it is immersed in the fluid cell. (3) The evidence for bubble coalescence or Ostwald ripening (see Figure 9) also supports very strongly that the domains are gas bubbles. These domains would only coalesce or dissolve if they are indeed gas bubbles; domains of polymeric material would not be expected to coalesce nor change in size over the time frame of observation. Furthermore, we note that in very strongly degassed solution the structures observed did not change with time. Their mechanical properties are different from bubbles. The force versus distance behaviour is indicative of the presence of bubbles on the solid surface.

The force measurements for a colloid probe approaching and then separating from the bubble-covered, solution-methylated Si wafer surface were obtained by using a methylated silica colloid probe (radius = 8 μm)(Figure 11).

We have described the procedures elsewhere. The long-range jump distance (>70nm) for the approaching probe, coupled with the high adhesion force during retreat, is characteristic of the presence of gas bubbles on these surfaces. The stepwise character of the force versus distance curve is consistent with bubble bridging for this rough surface.

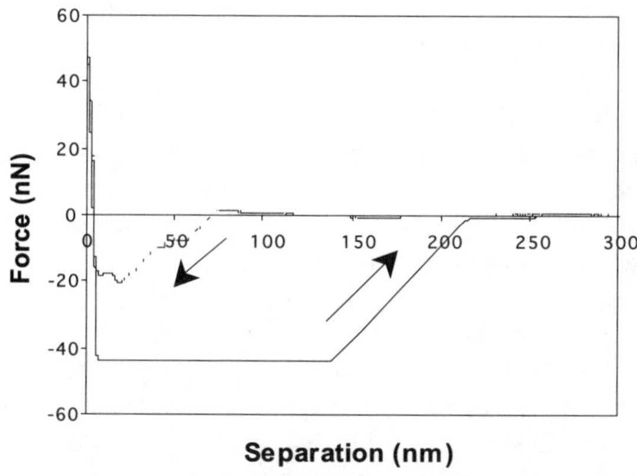

FIGURE 11 Force-separation curves measured upon a hydrophobised silica colloid probe approaching (broken line) and retreating from (solid line) a small bubble formed on a methylated Si-wafer surface (Reprinted with permission Journal of Physical Chemistry B, 107, No. 25, 2003. Copyright 2003 American Chemical Society.)

TABLE 3 Surface roughness and water contact angle for clean and hydrophobised Si wafers (air contact angles are given in brackets—there was no detectable difference between air and carbon dioxide) (Reprinted with permission Journal of Physical Chemistry B, 107, No. 25, 2003. Copyright 2003 American Chemical Society.)

Substrate Surfaces	Advancing contact angle	Receding contact angle	RMS roughness (nm)	Peak-to-valley distance (nm)
Clean Si-wafer	Spreading	—	0.09	0.3
Dehydroxylated (1050°C)	42 (146)	34 (138)	0.2	0.5
Exposure to TMCS vapour (20 min)	74 (113)	67 (106)	0.1	0.4
Exposure to TMCS-cyclohexane solution (20 min)	88 (113)	67 (92)	2.7	12.3

Bubble Formation on Solid-Water Interfaces

TMAFM images show that there is no nanobubble formation on smooth *hydrophilic* surfaces. This is in agreement with earlier observations in the literature but in contrast to a recent report that nanobubbles are present on freshly cleaved (hydrophilic) mica surfaces.

For dehydroxylated surfaces, there was no nanobubble formation. Bubble formation does not occur on this smooth surface with intermediate hydrophobicity ($\theta_W = 42°$) and a small degree of hysteresis (see Table 3).

However, when the surface hydrophobicity increased, to the level of the TMCS vapour methylated Si wafer substrate ($\theta_{Wadv} = 74°$), with approximately the same physical roughness and degree of hysteresis as that of dehydroxylated surfaces, bubble formation occurs. This evidence is important for two reasons. In the first instance, it demonstrates that the degree of hydrophobicity is important for bubble formation. This agrees with early theoretical analyses, which showed that quite apart from the geometry and size of surface sites, hydrophobicity is essential for bubble formation on solid surfaces. Secondly, considering that the physical roughness or 'defects' are also only of nano-scale size, the observation of bubble formation on such smooth hydrophobic surfaces suggests that small defects in the nanometre range are large enough to act as physical sites for bubble growth on solid surfaces. There was no detectable bubble formation in the bulk solution and thus sites on the solid surface favour bubble formation. The small degree of hysteresis evident for these vapour treated silicon wafers is sufficient to pin the three phase contact line at the surface heterogeneity.

There are distinct differences in bubble size and distribution observed for bubble images obtained for the TMCS vapour treated surface and the TMCS in cyclohexane solution prepared surface. The differences in bubble formation are due to the differences in surface roughness and hysteresis of the two kinds of substrates. The TMCS cyclohexane treated surface has a much higher surface roughness and larger hysteresis (see Table 3) compared with the TMCS vapour treated surfaces. We speculate that there may well be a larger number of surface sites on the former surface that stimulate bubble formation, perhaps leading to bubble growth and coalescence between adjacent, closely spaced sites. At this point we cannot identify these sites or their distribution with any more certainty.

Contact Angles and Line Tension

The water contact angles (or air contact angles) of the small bubbles on the methylated substrate surfaces are much greater (smaller) than the contact angles measured by the sessile drop methods (Table 3). The contact angles of the small bubbles can be determined from the cross-section profile of each bubble. This may be extracted from the data file of a bubble examined by TMAFM (such as the one presented in Figure 10). For Figure 10, the advancing air contact angles of bubbles were found statistically to lie in the range from 23–30°. Similar observations were made by Ishida et al.[26] in their study. The radius of curvature of the AFM tips used falls between 5 to 20 nm. Following detailed work reported elsewhere on fluid droplets, our very small bubble images are not distorted under our experimental conditions.

The difference between these macroscopic and microscopic contact angles may be linked to the influence of line tension. The Young angle is modified by a line tension term when the droplet or bubble size is small, viz

$$\cos\theta = \cos\theta_Y - \frac{\tau}{\gamma_{lv}R}$$

where θ is the actual contact angle that the very small bubble forms with the substrate; θ_Y is the Young contact angle; γ_{lv} is the liquid-vapour surface tension; $1/R$ is the local curvature of the bubble base on solid surface and τ is the line tension. It should be pointed out that Eq. 1 reflects only line tension effects, but does not consider the surface roughness and texture effects on the actual macroscopic or microscopic contact angle[38].

Using Eq. 1, and examining some tens of bubbles, the dependence of local, microscopic advancing air contact angles on the corresponding curvatures of the very small bubbles (examined from above) is shown in Figure 12. The line tension of the bubbles, calculated from the gradient of the regression line is -3×10^{-10} N. This value is similar in magnitude to the reported values of line tension for small droplets[39,40], very close to the

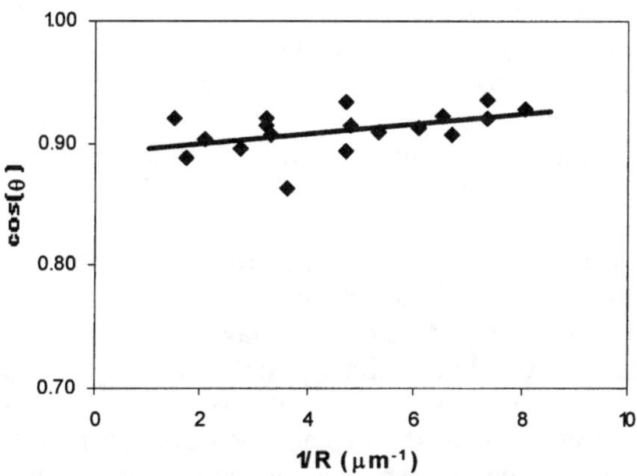

FIGURE 12 The cosine of contact angles of the nano-bubbles are plotted against the corresponding local curvatures based on the nano-bubble images by TMAFM. The value of line tension calculated from the slope is approximately -2.9×10^{-10} N. (Reprinted with permission Journal of Physical Chemistry B, 107, No. 25, 2003. Copyright 2003 American Chemical Society.)

values determined by Scheludko et al.[41] and within the range anticipated theoretically[42]. The negative line tension of the three phase contact line acts to flatten the bubbles, reducing the Laplace pressure and thus stabilizing the very small bubbles formed at these solid-water interfaces. We note that the large bubble contact angle is not recovered as R goes to infinity in Figure 12, behaviour noticed elsewhere[39,40]. We cannot be certain as to the reason at this juncture, but suspect that it may reflect differences in local surface texture for small bubbles compared with large ones with, perhaps, additional mechanisms contributing as well.

CONCLUSION

The debate concerning the origin of large attractive forces, observed between hydrophobic surfaces, has continued for many years. In our experiments, we have found that in the case of very smooth, uniform hydrophobic (dehydroxylated) and hydrophilic surfaces, classical DLVO theory can explain the interactions between these surfaces. Surface heterogeneity, both chemical and physical, and hydrophobicity, promote the formation of gas bubbles. These very small surface bubbles are predicted to change both the van der Waals and electrostatic forces acting between the particles. The methylated silica particles used in this study are hydrophobic and display both chemical and physical heterogeneity. The data presented here show a clear link between the level of dissolved gas present in solution and the related colloid stability of the particles. In contrast, the colloidal stability of dehydroxylated and hydrophilic silica particles is unaffected by dissolved gas levels. The colloidal interactions have been described using an appropriate modification of DLVO theory. The analysis shows that surface-to-surface interactions between particles are dominated by the presence of very small, protruding surface bubbles.

The formation of very small gas bubbles (so-called "nanobubbles") at structured solid-water interfaces has been studied using the tapping mode atomic force microscopy (TMAFM) imaging technique. Silicon oxide wafer surfaces were prepared with different degrees of nanometre scale surface roughness and hydrophobicity. Small bubbles do not form on smooth, hydrophilic or dehydroxylated silicon oxide wafer surfaces, immersed in aqueous solutions under known levels of gas supersaturation. Randomly distributed small bubbles were observed over the whole surface of observation on methylated surfaces of controlled roughness. Bubbles formed on rough, methylated surfaces were larger and less-densely distributed than on a smooth surface of similar hydrophobicity. The existence of these very small gas bubbles on the surface was further demonstrated by the observation of bubble coalescence with time. The macroscopic contact angle, measured with respect to the aqueous or gas phase, is very different from the microscopic contact angle detected by TMAFM, and may be due to the influence of line tension at the three-phase contact line. Surface heterogeneities act to pin the contact line. The line tension has a value of -3×10^{-10} N and acts to stabilize the small bubbles, flattening them and thereby reducing the Laplace pressure. This Laplace pressure ranges from several to upwards of ten atmospheres for the very small bubbles encountered in this study[43,44].

ACKNOWLEDGEMENT

Financial support from the Australian Research Council Special Research Centre Scheme is gratefully acknowledged. Fruitful discussions with Nataliya Mishchuk are warmly acknowledged.

REFERENCES

1. Considine, R.F.; Hayes, R.A.; Horn, R. G. *Langmuir* 1999, 15, 1657.
2. Meagher, L.; Craig, V.S.J. *Langmuir* 1994, 10, 2736.
3. Zhou, Z.A.; Zhenghe, X.; Finch, J. A. *J. Colloid Interface Sci.* 1996, 179, 311.
4. Mahnke, J.; Stearnes, J.; Hayes, R.A.; Fornasiero, D.; Ralston, J. *Phys. Chem. Chem. Phys.* 1999, 1, 2793.
5. Parker, J.L.; Claesson, P.M.; Attard, P. *J. Phys. Chem.* 1994, 98, 8468.
6. Yakubov, G.E.; Butt, H.-J.; Vinogradova, O.I. *J. Phys. Chem. B* 2000, 104, 3407.
7. Harvey, E.N.; Barnes, D.K.; McElroy, W.D.; Whiteley, A.H.; Pease, D.C.; Cooper, K.W. *J. Cell. Comp. Physiol.* 1944, 24, 1.
8. Harvey, E.N.; Whiteley, A.H.; McElroy, W.D.; Pease, D.C.; Barnes, D.K. *Journal of Cell Comp. Physiology* 1944, 24, 23.
9. Harvey, E.N.; Cooper, K. W.; Whiteley, A.H. *J. A. C. S.* 1946, 68, 2119.
10. Harvey, E.N.; McElroy, W. D.; Whiteley, A.H. *J. Appl. Phys.* 1947, 18, 162.
11. Ishida, N.; Inoue, T.; Miyahara, M.; Higashitani, K. *Langmuir* 2000, 16, 6377.
12. Lou, S.T.; Ouyang, Z. Q.; Zhang, Y.; Li, X. J.; Hu, J.; Li, M. Q.; Yang, F.J. *J. Vac. Sci. Technol. B* 2000, 18, 2573.
13. Tyrrell, J.W.G.; Attard, P. *Langmuir* 2002, 16, 6377.
14. Ralston, J.; Fornasiero, D.; Mishchuk, N. *Colloids and Surfaces A* 2001, 192, 39.
15. Wrobel, S. *Mine & Quarry Engineering* 1952, 313.
16. Wood, J.; Sharma, R. *Langmuir* 1995, 11, 4797.
17. Ryan, W.L.; Hemmingsen, E.A. *J. Colloid Interface Sci.* 1998, 197, 101.
18. Lubetkin, S.D. *J. Applied Electrochemistry* 1989, 19, 668.
19. Karaman, M.E.; Antelmi, D.A.; Pashley, R.M. *Colloids and Surfaces A* 2001, 182, 285.
20. Christenson, H.K.; Claesson, P.M. *Adv. Colloid Interface Sci.* 2001, 91, 391.
21. R.J. Hunter *Foundations of Colloid Science*; Clarendon Press: Oxford, 1987; Chapter 7.

22. Behrens, S.H.; Christl, D.I.; Emmerzael, R.; Schurtenberger, P.; Borkovec, M. *Langmuir* 2000, 16, 2566.
23. Iler, R. K. *The chemistry of silica : solubility, polymerization, colloid and surface properties, and biochemistry*; Wiley & Sons: New York, 1979.
24. Mishchuk, N.; Ralston, J.; Fornasiero, D. *J. Phys. Chem. A* 2002, 106, 689.
25. Lou, S.-T.; Ouyang, Z.-Q.; Zhang, Y.; Li, X.-J.; Hu, J.; Li, M.-Q.; Yang, F.-J. *J. Vac. Sci. Technol., B: Microelectronics and Nanometer Structures* 2000, *18*, 2573.
26. Ishida, N.; Inoue, T.; Miyahara, M.; Higashitani, K. *Langmuir* 2000, *16*, 6377.
27. Tyrrell, J.W.G.; Attard, P. (a) *Phys. Rev. Lett.* 2001, *87*, 176104/1; (b) Langmuir, 2002, 18, 160.
28. Considine, R.F.; Hayes, R.A.; Horn, R. G. *Langmuir* 1999, *15*, 1657.
29. Gong, W.; Stearnes, J.; Fornasiero, D.; Hayes, R. A. *Phys. Chem. Chem. Phys.* 1999, 1, 2799.
30. Grieser, F.; Lamb, R.N.; Wiese, G.R.; Yates, D.E.; Cooper, R.; Healy, T. W. *Radiat. Phys. Chem.* 1984, 23, 43.
31. Derjaguin, B.V.; Churaev, N.V.; Muller, V.M. *Surface Forces*; Kitchener, J. A., Ed.; Plenum Press: New York, 1987.
32. McGown, D.N.L.; Parfitt, G.D. *J. Phys. Chem.* 1967, 71, 449.
33. Honig, E.P.; Roebersen, G. J.; Wiersema, P.H. *J. Colloid Interface Sci.* 1971, 36, 97.
34. Spielman, L.A. *J. Colloid Interface Sci.* 1970, 33, 562.
35. Puertas, A.M., Nieves, F.J., *J. Colloid Interface Sci.*, 1999, 216, 221.
36. Usui, S.; Barouch, E. *J. Colloid Interface Sci.* 1989, 137, 281.
37. Zembala, M.; Adamczyk, Z. *Langmuir* 2000, 16, 1593.
38. Drelich, J.; Miller, J.D.; Good, R. J. *J. Colloid Interface Sci.* 1996, *179*, 37.
39. Hemmingsen, E.A. *J. Appl. Phys.* 1975, *46*, 213.
40. Pompe, T.; Herminghaus, S. *Phys. Rev. Lett.* 2000, *85*, 1930.
41. Scheludko, A., Toshev, B. and Bogadiev, B. *J. Chem. Soc. Far. Trans.*, 72, 2815 1976.
42. Buff, F.P. and Saltsburg, H.T., *J.Phys.Chem.*, 26, 23 1957.
43. D.R.E. Snoswell, J. Duan, D. Fornasiero and J. Ralston, *J. Physc.Chem.B.*, 107, No. 13, 2986–2994 (2003).
44. J. Yang, J. Duan, D. Fornasiero and J. Ralston, *Journal of Physical Chemistry B.*, 107, 25, 6139–6147 (2003).

Interfacial Phenomena

. . . .

Dynamic Adsorption of Surfactants at the Gas-Liquid Interface

Anh V. Nguyen,* Chi M. Phan,* Geoffrey M. Evans,* and Graeme J. Jameson*

Surfactants are used to control the interfacial properties of the gas-liquid and liquid-solid interfaces in flotation. This paper investigates the dynamic adsorption of surfactants at the gas-liquid interface. The dynamic adsorption process was modeled by considering the surfactant mass transfer to the gas-liquid interface by diffusion and the Langmuir and Frumkin adsorption isotherms. The numerical computation was applied to solve the non-linear governing equations. The dynamic adsorption was measured in terms of the dynamic surface tension using the pendant drop method. The theoretical models were compared against the experimental data obtained with non-ionic (Dowfroth 250) and ionic (sodium dodecylbenzene sulfate, SDBS) surfactants. The theoretical results agreed well with the experimental data for SDBS, but not for Dowfrother 250, especially for the initial stage. Reasons for the variation in results are given in the paper.

INTRODUCTION

Numerous inorganic and organic reagents (surface active agents or surfactants) are employed in flotation to improve the selectivity and recovery of valuable minerals. These surfactants strongly adsorb at both the air-water and solid-water interfaces, change the interfacial properties, and ultimately control the interaction between particles and bubbles in the capture process. The adsorption of surfactants at the solid-liquid interface has intensively been investigated for decades (Fuerstenau et al., 1985; Parekh and Miller,

* Discipline Chemical of Engineering, School of Engineering, University of Newcastle, Callaghan, New South Wales, Australia

1999). The focus of this paper is on the dynamic adsorption of surfactants at the gas-liquid interface.

The adsorption of surfactants at the gas-liquid interface is commonly studied by means of the dynamic surface tension measurements. The surface tension is reduced due to the surfactant adsorption. Therefore, monitoring the dynamic surface tension reduction can reveal the adsorption kinetics of the surfactant. The dynamic surface tension at the interface is also determined by the rate that surfactant molecules diffuse and adsorb onto the interface. Various adsorption models have been developed to analyze and interpret the measured dynamic surface tension data (Dukhin et al., 1995). The available models can be divided into two groups. In the first group of models, the mass transfer of surfactant molecules to the interface by diffusion is assumed to be the most decisive step of the dynamic adsorption process. The actual adsorption of molecules from the subsurface layer onto the interface takes place simultaneously. In these models, the diffusion process from the bulk to subsurface is the rate-controlling step, and the timescale of adsorption from the subsurface to the interface is very fast. These models are usually referred to as the diffusion-controlled models for dynamic adsorption (Chang and Franses, 1995). In the second group of models, the actual adsorption from the subsurface layer is not simultaneous due to the adsorption energetic barrier. The energetic barrier may be due to increased surface pressure or a decrease in 'vacant sites' available for adsorption. There may also be steric restraints on the molecule in the proximity of the interface, whereby the molecule may have to be in the correct orientation to adsorb. The energetic barrier will cause the molecule to back diffuse into the bulk rather than adsorbing, thereby increasing the timescale of the dynamic surface tension decay. The adsorption is, therefore, not instant. These models are referred to as the kinetics-controlled adsorption models (Chang and Franses, 1995).

The purpose of this study is to present a theoretical and numerical computation framework for describing the dynamic adsorption of surfactants at the gas-liquid interface. The suitability of the models is investigated by measuring the dynamic surface tension as a function of time and comparing the results with model predictions. The experimental data were obtained with sodium dodecylbenzene sulfate (SDBS) and Dowfroth 250 using pendant drop tensiometry.

THEORETICAL BACKGROUND AND DEVELOPMENT

The dynamic adsorption of surfactants at the gas-liquid interface is determined by two processes, including,

1. Diffusion of the surfactant molecules from the bulk solution to the subsurface layer at the interface and
2. Actual adsorption of the surfactant molecules from the subsurface layer onto the interface.

The two processes are illustrated in Figure 1. The modeling will be developed based on the two characteristic processes.

Surfactant Diffusion

The molecular diffusion of surfactants is well described by the diffusion equation (Fick's law)

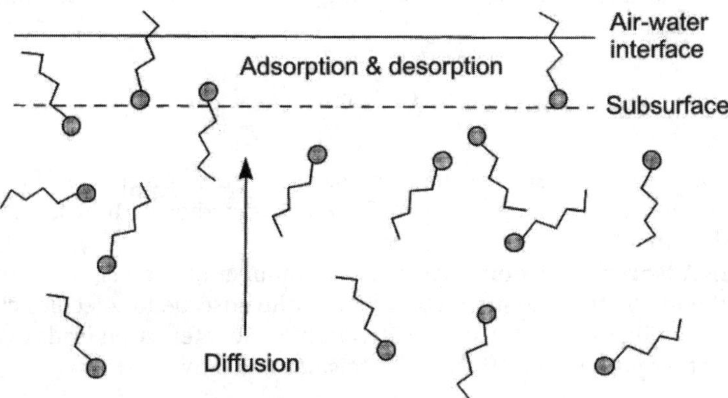

FIGURE 1 Diffusion and adsorption of surfactant molecules at the gas-liquid interface

$$\frac{\partial c}{\partial t} = D\nabla^2 c \quad \textbf{(EQ 1)}$$

where c is the surfactant concentration, D is the diffusivity of surfactant molecules, t is the reference time and ∇^2 is the Laplace operator.

The partial differential Eq. 1 was first solved for the surfactant adsorption by Ward and Tordai (Ward and Tordai, 1946). In terms of the dynamic surface excess, $\Gamma(t)$, of the adsorbed surfactant molecules, the celebrated Ward and Tordai expression gives

$$\Gamma(t) = \sqrt{\frac{D}{\pi}}\left[2c_b\sqrt{t} - \int_0^t \frac{\phi(t)}{\sqrt{t-\tau}}d\tau\right] \quad \textbf{(EQ 2)}$$

where c_b is the surfactant concentration in the bulk solution, $\phi(t)$ is the surfactant concentration at the subsurface layer, and τ is the integration variable.

Adsorption of Surfactants from the Subsurface Layer

Langmuir Adsorption Theory. The adsorption of surfactant molecules from the subsurface layer is usually described by the Langmuir adsorption isotherm by (Adamson and Gast, 1997)

$$\phi(t)K = \frac{\Gamma(t)}{\Gamma_m - \Gamma(t)} \quad \textbf{(EQ 3)}$$

where K is the adsorption constant and Γ_m is the maximum of Γ. The dynamic surface excess can be described in terms of the experimentally measured dynamic surface tension, σ, using the Gibbs adsorption equation (Adamson and Gast, 1997)

$$d\sigma = -RT\sum \Gamma d\ln c \quad \textbf{(EQ 4)}$$

where R is the universal gas constant and T is the absolute temperature. The summation is carried out for all adsorbed surfactant species.

If the concentration c in Eq. is replaced by the subsurface concentration, $\phi(t)$, in Eq. 3, one obtains

$$\sigma(t) = \sigma_0 + nRT\Gamma_m \ln\left\{1 - \frac{\Gamma(t)}{\Gamma_m}\right\} \quad \textbf{(EQ 5)}$$

where σ_0 is the surface tension of the clean surface (pure solvent), and $n = 1$ for non-ionic and 2 for ionic surfactant, respectively. Eq. 5, together with Eqs. 3 and 2, can be used to validate the dynamic adsorption theory.

Frumkin Adsorption Theory. There exist a number of extensions for the Langmuir adsorption theory. Firstly, the interaction among the adsorbed molecules can affect the Langmuir adsorption isotherm. The consideration of the interaction leads to the Frumkin adsorption isotherm (Lyklema, 1991), which is described by

$$\phi(t)K = \frac{\Gamma(t)}{\Gamma_m - \Gamma(t)} \exp\left[-2\beta\frac{\Gamma(t)}{\Gamma_m}\right] \quad \textbf{(EQ 6)}$$

where β is a parameter characterizing the lateral interaction among adsorbed surfactant molecules. The corresponding surface tension isotherm determined from the Gibbs adsorption equation gives

$$\sigma(t) = \sigma_o + nRT\Gamma_m\left[\ln\left(\left\{1 - \frac{\Gamma(t)}{\Gamma_m}\right\} + \beta\left\{\frac{\Gamma(t)}{\Gamma_m}\right\}^2\right)\right] \quad \textbf{(EQ 7)}$$

Equation 7, together with Eqs. 6 and 2, can also be used to validate the dynamic adsorption theory, in particular, to examine the non-ideal behavior and lateral interaction of the adsorbed molecules.

Rate-Controlled Adsorption. Both Eqs. 3 and 6 imply that the adsorption from the subsurface layer is considered instantaneous and the surface concentration is always in equilibrium with the subsurface concentration. Therefore, the dynamic adsorption is controlled by the surfactant diffusion from the bulk solution to the interface and is often referred to as the diffusion-controlled dynamic adsorption.

In practice, the adsorption from the subsurface layer need not be instantaneous and may be controlled by the adsorption kinetics. Since the adsorption is not instantaneous, the local equilibrium is no longer applied and the dynamic adsorption is known as the kinetics-controlled adsorption. The generalized adsorption kinetics gives (Miller and Kretzschmar, 1980; Dukhin et al., 1995)

$$\frac{d\Gamma(t)}{dt} = k_a\phi(t)\{\Gamma_m - \Gamma(t)\} - k_d\Gamma(t)\exp\left\{-2\beta\frac{\Gamma(t)}{\Gamma_m}\right\} \quad \textbf{(EQ 8)}$$

where k_a and k_d are the adsorption and desorption rate constants, respectively. It is noted that the adsorption constant is equal to $K = k_a/k_d$. If $k_a \to \infty$ (the adsorption is instantaneous), the kinetics-controlled model reduces to the diffusion-controlled model.

Numerical Computation Methodology

Numerical computation has to be applied to solve the governing Eqs. 2–8 for the dynamic adsorption process. Firstly, the time domain $(0,t)$ can be partitioned into i small intervals by $\Delta t_j = t_{j-1}$, where $j = 1,2,3,...i$. Eq. 2 is then discretized by applying the trapezoidal rule approximation to the integral, leading to the following prediction for the dynamic surface excess, $g(t_i)$, at time t_i by:

$$g(t_i) = A - \phi(t_i)B \quad \text{(EQ 9)}$$

where

$$g(t) = \frac{\Gamma(t)}{\Gamma_m} \quad \text{(EQ 10)}$$

$$A = \frac{\sqrt{D/\pi}}{\Gamma_m}\left\{2c_b\sqrt{t_i} + \sum_{j=1}^{i-1}[\phi(t_j) + \phi(t_{j-1})](\sqrt{t_i - t_j} - \sqrt{t_i - t_{j-1}} - \phi(t_{i-1})\sqrt{t_i - t_{j-1}})\right\} \quad \text{(EQ 11)}$$

$$B = \frac{\sqrt{D/\pi}}{\Gamma_m}\sqrt{t_i - t_{i-1}} \quad \text{(EQ 12)}$$

The initial conditions are determined based on the subsurface concentration and the dynamic surface excess are zero at time $t = 0$, i.e.,

$$\phi(0) = 0 \quad \text{(EQ 13)}$$

$$\Gamma(0) = 0 \quad \text{(EQ 14)}$$

For the diffusion-controlled model, Eq. 9 can be used in conjunction with Eq. 3 or 6 to numerically solve for the transient subsurface concentration and the dynamic adsorption excess.

Langmuir Adsorption Isotherm. If the Langmuir adsorption is considered, substituting $\phi(t_i)$from Eq. 3 into Eq. 9 gives

$$[g(t_i)]^2 - g(t_i)\left\{A + \frac{B}{K} + 1\right\} + A = 0 \quad \text{(EQ 15)}$$

This quadratic equation can be analytically solved for $g(t_i)$ which can be inserted into Eq. 5 to calculate the dynamic surface tension.

Frumkin Adsorption Isotherm. If the Frumkin adsorption is considered, substituting $\phi(t_i)$ from Eq. 6 into Eq. 9 gives

$$g(t_i) = A - \left\{\frac{B}{K}\right\}\frac{g(t_i)}{1 - g(t_i)}\exp[-2\beta g(t_i)] \quad \text{(EQ 16)}$$

This non-linear algebraic equation can be solved for the dynamic surface excess, $g(t_i)$, employing the bisection method. The dynamic surface excess, $g(t_i)$, is bounded by the condition $0 \le g(t_i) < 1$, which can be used to determine the initial values for iteration in the bisection method.

Rate-controlled adsorption. For the kinetics-controlled model, Eq. 8 can be further discretized by applying the implicit scheme, giving

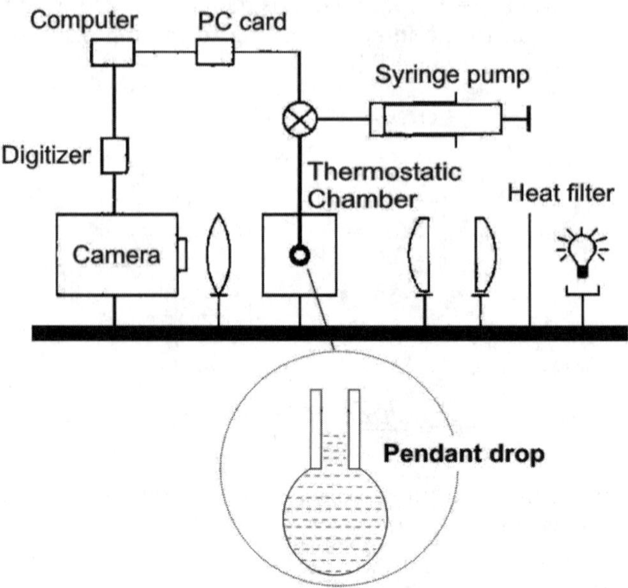

FIGURE 2 Schematic of the apparatus and the video image digitization equipment

$$\frac{g(t_i)-g(t_{i-1})}{k_a\Delta t_i} = \phi(t_i)[1-g(t_i)]-\frac{g(t_i)}{K}\exp[-2\beta(g(t_i))] \quad \textbf{(EQ 17)}$$

Equations and can be simultaneously solved for the dynamic surface excess, $g(t_i)$, by applying the initial conditions described by Eqs. 17 and 9. For instance, solving Eq. 9 for the subsurface concentration and then substituting into Eq. 17 yields

$$\frac{g(t_i)-g(t_{i-1})}{k_a\Delta t_i} = \frac{A-g(t_i)}{B}[1-g(t_i)]-\frac{g(t_i)}{K}\exp[-2\beta(g(t_i))] \quad \textbf{(EQ 18)}$$

Equation 18 can be solved for the dynamic surface excess, $g(t_i)$, employing the bisection method.

EXPERIMENTAL

Apparatus

The dynamic adsorption was experimentally determined by measuring the dynamic surface tension versus concentration using pendant drop tensiometry (Adamson and Gast, 1997) with a fully computer-controlled apparatus (System OCA 20, DataPhysics, Germany). The schematic of the experimental setup is shown in Figure 2.

The setup was used to create a pendant drop, then video imaged the drop silhouette, and before digitizing and processing the image. The image forming and recording system consisted of a constant-intensity light source with a heat filter, a lens system for producing a collimated beam, an optically transparent cell enclosed in a thermostatically controlled chamber, and a CCD video camera fitted with long-working distance objective lenses. The camera was connected to a digitizer device and a computer for image/data storage. The drop-forming system consisted of a stainless steel needle, which is connected to the computer-control system via a PC card, and a gastight micro syringe fitted with a motorized syringe pump driven by the computer.

Materials

The surfactants used to this study were sodium dodecylbenzene sulfate (SDBS) (Aldrich, Germany) and Dowfroth 250 (Dow Chemical, USA). The Dowfroth 250 surfactant was used as supplied. A Millipore Milli-Q water purification system was used to rinse all components and to make all solutions. This system provides water with a surface tension of about 72 mN/m. All glassware and components used in the experiments were soaked for several hours in 6% (by volume) solutions of Deconex 15E (Borer, Switzerland), rinsed in the Milli-Q water, and dried in a cabinet to protect them from dust. All surfactant solutions were freshly prepared and used.

Experimental Procedure

The experimental protocol was as follows: A liquid drop was formed at the end of the needle by the use of the micro syringe and the syringe pump. Typically, a drop with volume of approximately 11 L was formed within a time less than 0.5 seconds. The drop was placed in the path of a collimated light beam. The light beam cast a droplet shape onto the CCD camera. The camera was set to strobe, thereby obtaining the drop shape as a function of time. Images of the bubbles were taken at the frequency of 50 Hz. The drop shapes were digitized and analyzed offline.

The shape of drop is governed by the Young-Laplace equation, which relates the pressure drop across a curved interface by

$$\sigma\left(\frac{1}{R_1} + \frac{1}{R_2}\right) = \Delta P \qquad \textbf{(EQ 19)}$$

where R_1 and R_2 are the two principal radii of the interface, and ΔP is the pressure difference across the interface. Since the pendant drops have the rotational symmetry, the two principal radii can be described in terms of the profile coordinates of drop images (Hartland and Hartley, 1976). The pressure difference ΔP in Eq. is balanced by the hydrostatic pressure on the drop surface, which is a function of the drop profile coordinates.

The Young-Laplace equation can be described as a second order differential equation in terms of the drop profile coordinates and was numerically solved. The digitized drop profiles were fitted numerically with the solution of the Young-Laplace equation to obtain the surface tension value. The numerical solution and fitting were carried out for each drop image and the surface tension corresponding to each drop image was found.

The needle and syringe were cleaned using 6% by volume Deconex 15E solution (from Borer, Switzerland) and then rinsed with distilled water. Drops were enclosed in the sealed thermostatically controlled chamber to eliminate air turbulence effects. A small amount of solution was also left in the bottom of the chamber for 2 hours before the experiment commenced to reduce the effect of evaporation on the drop. Prior to each measurement using surfactant solutions, a distilled water surface tension measurement was undertaken and compared with standard values to ensure no

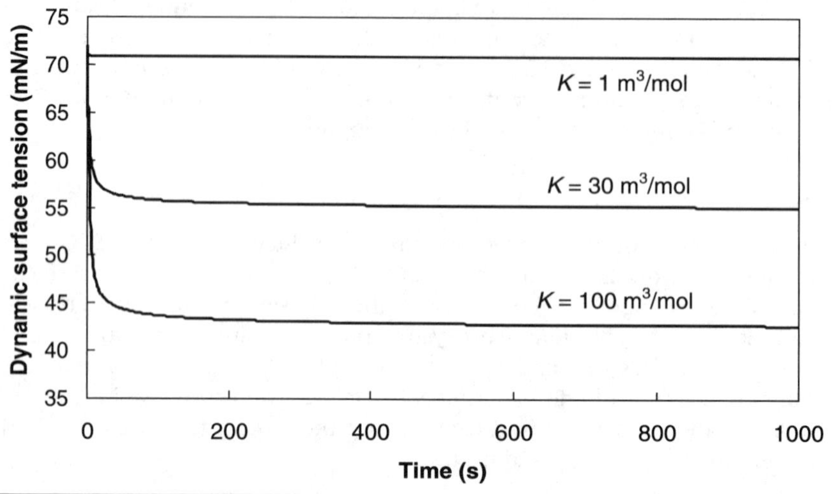

FIGURE 3 Effect of adsorption constant *K* on dynamic surface tension as predicted by the diffusion-controlled model with $D = 5 \times 10^{-10}$ m²/s , $\Gamma_m = 5 \times 10^{-6}$ mol/m², $\beta = 0$, $n = 1$, $\sigma_o = 72$ mN/m, $c_b = 1 \times 10^{-4}$ mol/L, and RT = 2,475 J/mol

contamination was present in the system. The system was calibrated using the outer diameter of the needle.

RESULTS AND DISCUSSION

Parametric Study of Dynamic Adsorption Models

The numerical computations were carried out to study the effect of the model parameters on dynamic adsorption and dynamic surface tension. The investigated parameters included the (equilibrium) adsorption constant K, the surfactant diffusivity D, the lateral interaction between adsorbed molecules and the adsorption kinetics constants (k_a and k_d). Typical results are shown in Figures 3 to 6.

The maximum adsorption density, Γ_m, is a theoretical limit which is important but cannot normally be reached because of the constraint of the micelleization and the solubility. Typical values for Γ_m are of the order of 1×10^{-6}. For close-packed insoluble monolayers of single-chain hydrocarbon surfactants, the cross-sectional area is about 20 Å² per molecule, and corresponding (highest) value for Γ_m is $\sim 1 \times 10^{-5}$. The value of 5×10^{-6} used in our parametric study gives the corresponding cross-sectional area of ~33 Å² per molecule.

The numerical computations (Figure 3) show that the equilibrium adsorption constant significantly determines the decrease in surface tension due to adsorption of surfactants. In order to produce a significant decrease in surface tension, the magnitude of the adsorption constant has to be larger than ~10 m³/mol.

It is also noted that while the ultimate decrease in surface tension is determined by the adsorption constant, K, the rate of the decrease very much depends on the surfactant diffusivity as shown in Figure 4. However, in the case of rate-controlled adsorption, the

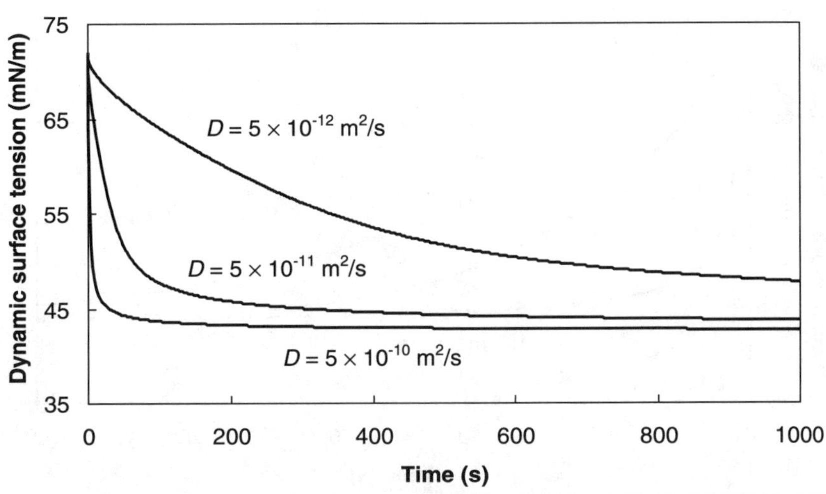

FIGURE 4 Effect of surfactant diffusivity *D* on dynamic surface tension as predicted by the diffusion-controlled model with K = 100 m³/mol, $\Gamma_m = 5 \times 10^{-6}$ mol/m², β = 0, n = 1, σ_o = 72 mN/m, $c_b = 1 \times 10^{-4}$ mol/L, and RT = 2,475 J/mol

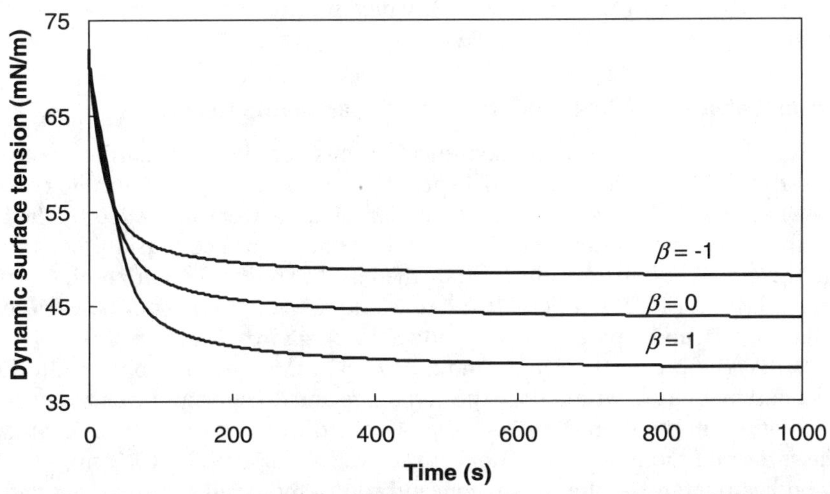

FIGURE 5 Effect of lateral interaction (β) between adsorbed surfactants on dynamic surface tension as predicted by the diffusion-controlled model with K = 100 m³/mol, $D = 5 \times 10^{-11}$ m²/s, $\Gamma_m = 5 \times 10^{-6}$ mol/m², n = 1, σ_o = 72 mN/m, $c_b = 1 \times 10^{-4}$ mol/L, and RT = 2,475 J/mol

rate of the decrease also significantly depends on the adsorption rate constants, K_a, as can be seen from Figure 6.

Effect of the lateral interaction between adsorbed surfactant molecules on dynamic surface tension is shown in Figure 5. The interaction is usually attractive ($\beta > 0$) and enhances the decrease in dynamic surface tension. The lateral interaction can also be

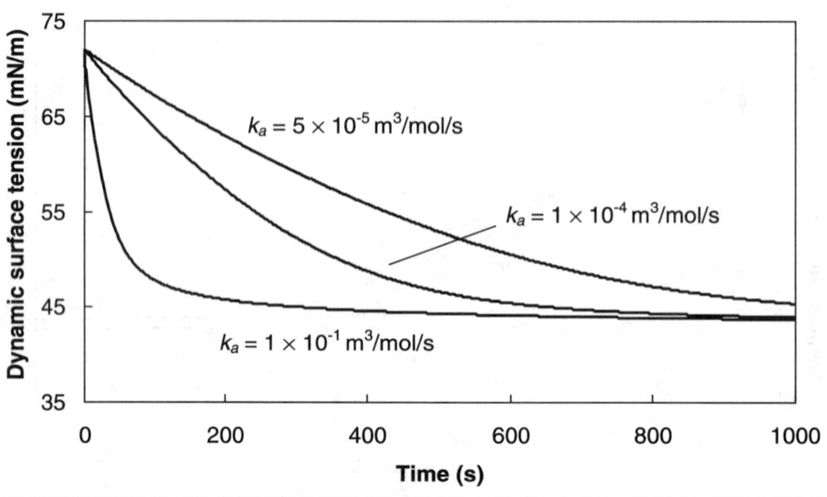

FIGURE 6 Effect of adsorption kinetics (k_a) on dynamic surface tension as predicted by the kinetics-controlled model with $D = 5 \times 10^{-11}$ m²/s, $K = 100$ m³/mol, $\Gamma_m = 5 \times 10^{-6}$ mol/m², $\beta = 0$, $n = 1$, $\sigma_o = 72$ mN/m, $c_b = 1 \times 10^{-4}$ mol/L, and $RT = 2{,}475$ J/mol

repulsive ($\beta < 0$) (Karakashev et al., 2004), which retards the decrease in dynamic surface tension.

Comparison between Model Predictions and Experimental Results

Typical experimental results and theoretical results for dynamic surface tension for SDBS and Dowfroth solutions with concentrations below the critical micelle concentration (CMC) are shown in Figure 7. The model predictions were improved by the best fit between the experimental data and Eq. for the dynamic surface tension. For the SDBS solution, the model parameters obtained by the best fit are $K = 6.45$ m³/mol, $\Gamma_m = 6.24 \times 10^{-6}$ m³/mol, $\beta = 1.146$, $D = 4.66 \times 10^{-10}$ m²/s and $K_a = 6.735$ m³/mol/s. For the Dowfroth solution, the model parameters include $K = 202.42$ m³/mol, $\Gamma_m = 2.00 \times 10^{-6}$ mol/m², $\beta = -0.137$, $D = 2.19 \times 10^{-12}$ m²/s and $K_a = 74.47$ m³/mol/s. The obtained values for the model parameters are within the expected range and do not significantly deviate from other estimations in the literature. For example, the diffusivity (at 22.3°C) estimated by using the surfactant molar volumes (Reid et al., 1986) is about 4.3×10^{-10} m²/s for SDBS.

For both surfactants at the shown concentrations, the small variation of β highlights the insignificance of the lateral interaction between adsorbed molecules, which implies that the Langmuir adsorption isotherm is appropriate.

In contrast to SDBS, the model predictions do not fit the Dowfroth 250 experimental data very well, especially for the first 10^5 ms and high concentrations (not shown in this paper). The adsorption-desorption kinetics, which depends on molecular structure and orientation at the interface, differentiates the behavior of SDBS and Dowfroth 250. As an ionic surfactant, SDBS molecules can be oriented quickly in favor of adsorption at the air-liquid interface, resulting in faster adsorption kinetics. Therefore, the overall process is controlled by the diffusion step only. On the other hand, the adsorption process of Dowfroth 250 at the interface is much slower after the initially stage and hence

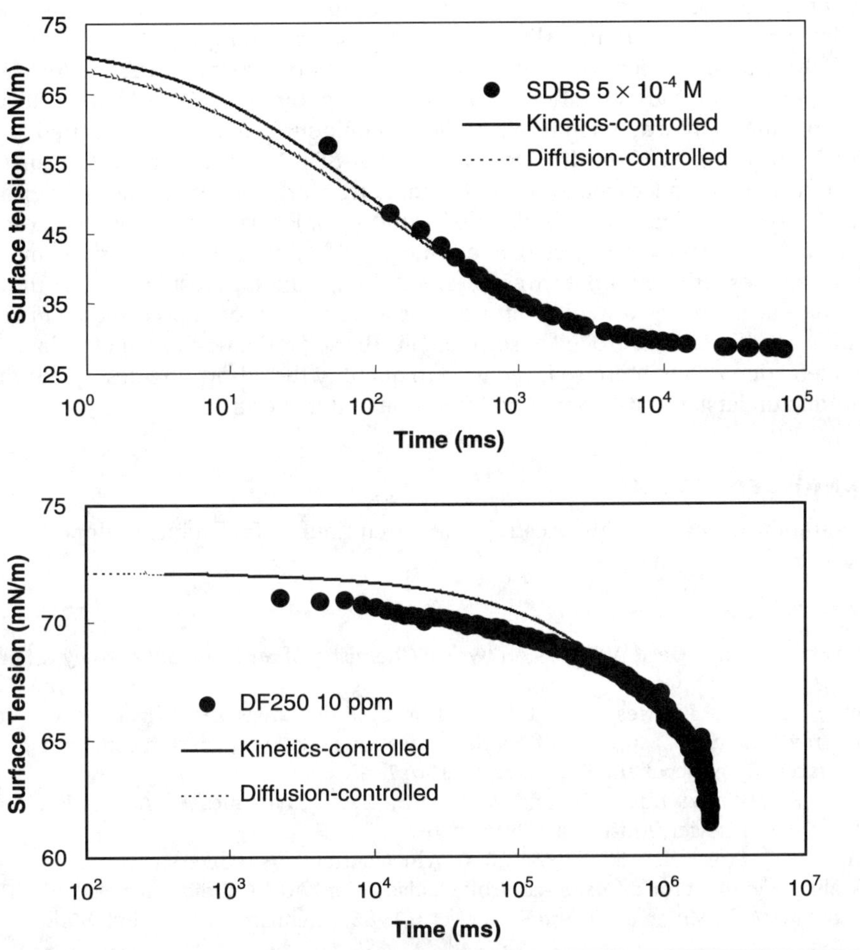

FIGURE 7 Comparison of model predictions (lines) with experimental results (points) for dynamic surface tension for 5×10^{-4} M SDBS and 10 ppm Dowfroth 250 solutions

the overall adsorption is controlled by both diffusion and adsorption-desorption processes. Further study of Dowfroth 250 behavior at the interface (e.g., the influence of orientation, the intermolecular interactions) is required to develop a successful model. It is anticipated that the molecular structure and properties are needed to explain the evident differences in adsorption dynamics of different surfactants (Ferri and Stebe, 2000). Nevertheless, these results indicate that the dynamic adsorption of Dowfroth 250 is controlled by the adsorption kinetics.

CONCLUSIONS

The dynamic adsorption of surfactants at the gas-liquid interface was theoretically and experimentally investigated. The model development incorporated the surfactant mass transfer to the gas-liquid interface by diffusion and the Langmuir or Frumkin adsorption isotherms. The obtained differential-integral governing equations were successfully solved by numerical computation. The parametric study of the model equations highlighted the physical significance of the model parameters. The dynamic adsorption was experimentally studied in terms of the dynamic surface tension measured as a function of time using pendant drop tensiometry for SDBS and Dowfroth 250 surfactants. The diffusion-controlled mass transfer and Langmuir adsorption model satisfactorily described the experimental data obtained with SDBS. For Dowfroth 250, the adsorption-desorption rate was evidently slow and a measurable difference between the model predictions and experimental data was observed during the initial stage and in the case of high concentrations. As a weakly diffusive surfactant, Dowfroth 250 could be modeled by the diffusion-kinetics-controlled adsorption. Further investigation into molecular structure and interaction at the interface is required to quantity the adsorption-desorption rate constants and understand the kinetics of adsorption at the interface.

ACKNOWLEDGEMENT

The authors acknowledge the Australia Research Council for financial support.

REFERENCES

Adamson, A.W. and Gast, A.P., 1997. Physical Chemistry of Surfaces. John Wiley & Sons, Inc., New York, 784 pp. pp.

Chang, C.H. and Franses, E.I., 1995. Adsorption dynamics of surfactants at air/water interface: a critical review of mathematical models, data and mechanisms. Colloids Surfaces A: Physicochem. Eng. Aspects, 100: 1–45.

Dukhin, S.S., Kretzschmar, G. and Miller, R., 1995. Dynamics of Adsorption at Liquid Interfaces. Elsevier, Amsterdam, 604 pp.

Ferri, J.K. and Stebe, K.J., 85: 61–97, 2000. Which surfactants reduce surface tension faster? A scaling argument for diffusion-controlled adsorption. Adv. Colloid Inter. Sci., 85: 61–97.

Fuerstenau, M.C., Miller, J., D. and Kuhn, M.C., 1985. Chemistry of Flotation. SME, New York, 176 pp.

Hartland, S. and Hartley, R.W., 1976. Axisymmetric Fluid-Liquid Interfaces: Tables Giving the Shape of Sessile and Pendant Drops and External Menisci, with Examples of Their Use. Elsevier, Amsterdam, 782 pp.

Karakashev, S.I., Manev, E.D. and Nguyen, A.V., 2004. Interpretation of negative values of the interaction parameter in the adsorption equation through the effects of surface layer heterogeneity. Adv. Colloid Inter. Sci. (in press).

Lyklema, J., 1991. Fundamentals of interface and colloid science I. Academic Press, San Diego, 673 pp.

Miller, R. and Kretzschmar, G., 1980. Numerical solution for a mixed model of diffusion kinetics-controlled adsorption. Colloid Polymer Sci., 258: 85–87.

Parekh, B.K. and Miller, J.D. (Editors), 1999. Advances in Flotation Technology. SME, Littleton, 463 pp.

Reid, R.C., Prausnitz, J.M. and Polling, B.E., 1986. The Properties of Gases and Liquids. McGraw-Hill Book Company, New York.

Ward, A.F.H. and Tordai, L., 1946. Time-dependence of boundary tensions of solutions. I. The role of diffusion in time effects. The Journal of Chemical Physics, 14(7): 453–461.

Developments in Trithiocarbonate (TTC) Chemistry and Industrial Applications

John C. Davidtz*

INTRODUCTION

The effectiveness of collector type on sulfide mineral flotation is generally determined by the amount and type of functional groups of collector molecules that interact with surface and water at the mineral/collector/water/gas interface. The final interfacial state when bubble contact takes place is preceded by chemical and physical reactions that involve these gas/liquid/solid interactions. These interactions differ not only for different minerals, solution and gas composition, but also for different collector molecules.

The purpose of this paper is primarily to summarize the developments and status of short (C2-C6) and long (C12) chain, aliphatic collectors that contain three sulfur atoms in their polar function. These are trithiocarbonates (TTCs). For purposes of comparison, similarities in molecules containing O, N and S are compared. Mostly short chain ionic collectors molecules of dithiocarbonates (DTCs) and TTCs were compared. The studies, largely bench scale flotation and plant testing, were carried out on ores containing Cu, Ni, Au, and PGM's. Studies of long chain TTC collectors are reported on Ni, PGM and auriferous pyrite ores containing uranium.

Under starvation dosage conditions, surface activity coefficients of short chain collector molecules and interacting water were calculated. Based on the UNIFAC group contribution method of activity coefficient calculations, the Gibbs Excess Free Energy

* Department of Materials Science and Metallurgical Engineering, University of Pretoria, Pretoria, South Africa

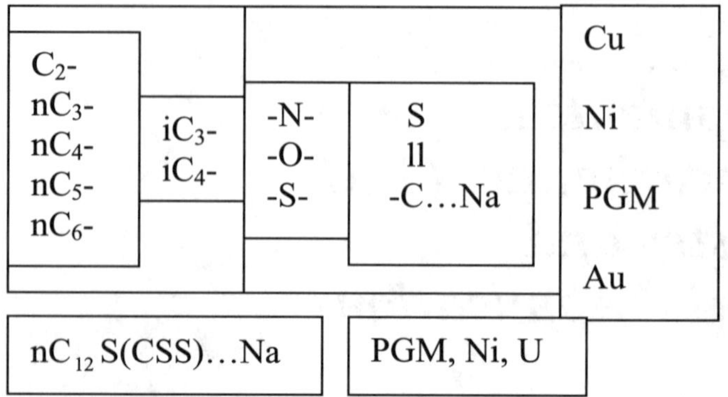

FIGURE 1 Schematic summary of ionic collectors and valuables floated. Associated minerals are Cu (chalcopyrite, bornite); PGM (chalcopyrite, pentlandite, pyrrhotite, pyrite); Au and U (pyrite); and Ni (pyrrhotite and pentlandite).

helped to answer some of the physical questions as to why some collectors were more effective in producing surface hydrophobicity.

Finally a brief summary of the TTC collector chemistry is given. This clearly shows how the TTCs differ from other thiocarbonates.

An outline of chemicals and ores investigated is shown in Figure 1.

FLOTATION OF COPPER ORE

Ionic DTCs and TTCs

Data from triplicate bench flotation tests in a Denver flotation cell with one kilogram copper ore 40 percent -75γ obtained from Phalborwa Mining Co. were used to generate time-recovery data. Dow 250 was used as frother and collector dosage was 120 micro moles pre kilogram ore. Seven concentrates were taken at intervals of 0.5, 1, 2, 3, 5, 7 and 10 minutes respectively. Similar to Ackerman et al. 1984, cumulative recovery after 30 min (R) and a mean rate (K) for the first 30 seconds were used in calculations (Slabbert, 1985).

The results reported in and Figures 2 and 3 show a steady increase in rate and recovery for normal alkanes up to C_5. Also, at the equi-molar dosage of 120 micro moles per kilogram ore the TTCs showed higher initial rates as well as final recoveries. The upward trend to C_5 follows the predicted solubilities of normal alkanes (Krescheck, 1975). Furthermore, with branched isomers, the TTCs were more effective than DTCs.

Figure 4 summarizes the rates versus recoveries for Mercaptans DTCs and TTCs and the data falls into three families. The different members correspond to the number of sulfur atoms per collector molecule.

Grade-Recovery Relationships on Copper Ore

Grade recovery relationships for all collectors followed the same pattern with iso-butyl given as example (Figure 5).

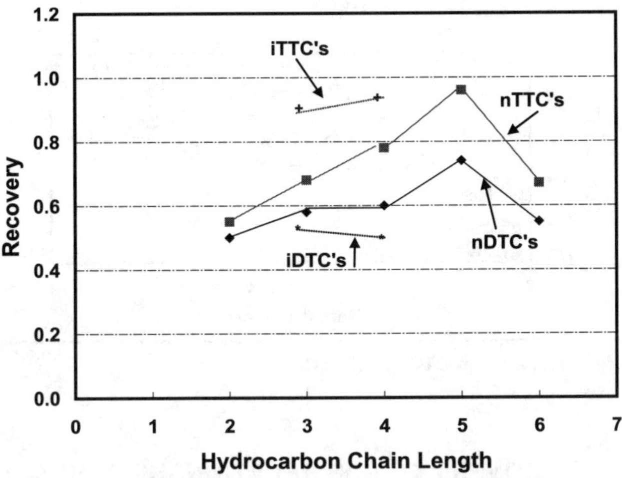

FIGURE 2 **Copper recovery after 30 minutes versus chain length**

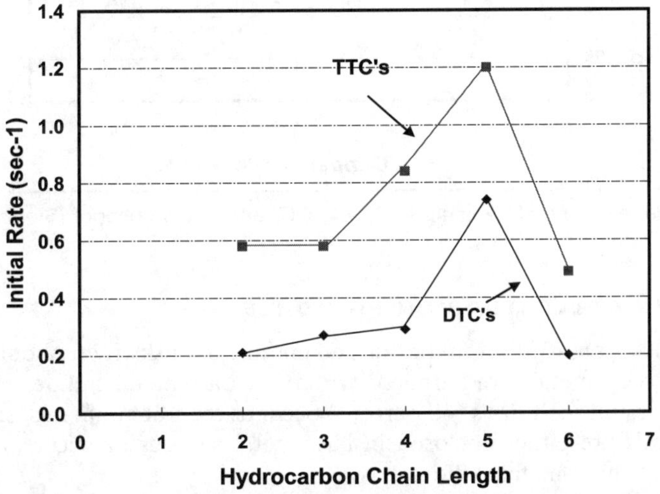

FIGURE 3 **Initial rate vs chain length on copper flotation**

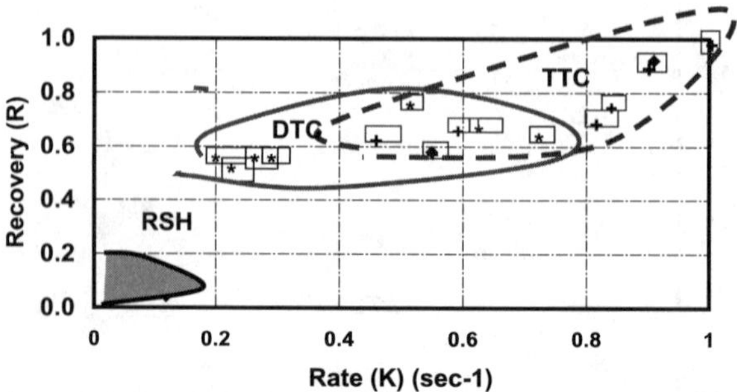

FIGURE 4 R vs K for Mercaptans DTCs and TTCs

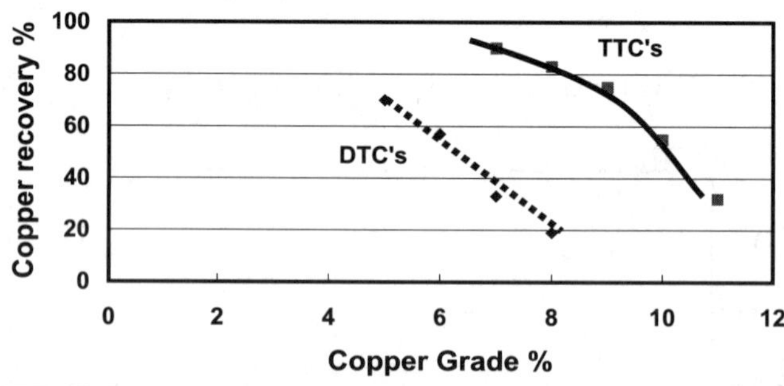

FIGURE 5 Grade recovery relationship for iso-C_4 DTC and TTC on copper (Slabbert 1985)

SHORT CHAIN COLLECTORS ON PGM ORES

Most of the studies on PGM's were performed on the Merensky reef. About 37 percent of the platinum group metals are harbored within the base-metal sulfides mainly in solid solutions. The rest of the PGM's (63 percent) occur as separate minerals, such as cooperite. The minerals are either enclosed in base-metal sulfides or occur along the grain boundaries of the base-metal sulfides.

The main minerals constituting the base-metal sulfides are: pentlandite, $(Fe,Ni)_9S_8$; pyrrhotite, (FeS); chalcopyrite, $CuFeS_2$ and pyrite, FeS_2. A small amount of troiilite (FeS) is also known to be present. The average grain size of the base-metal sulfides vary between 5 and 450 microns, whereas the grain size of the PGM minerals vary between 1 and 44 microns. The PGM minerals are braggite, (Pt,Pd,Ni)S; cooperite, PtS; electrum, AuAg; gold; kotulskite, $Pd(Te,Bi)_{1-2}$; laurite, RuS_2; merenskite, $(Pt,Pd)(Te,Bi)_2$; moncheite, $(Pt,Pd)(Te,Bi)_2$; Pt-Fe alloy; sperrylite, $PtAS_2$ and stibiopalladinite Pd_3Sb.

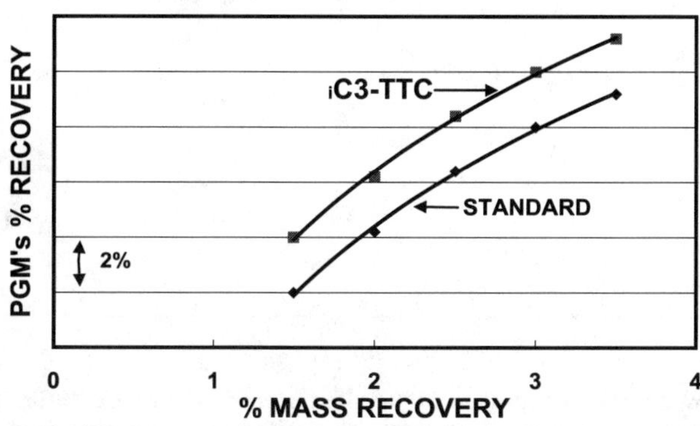

FIGURE 6 PGM recovery vs mass recovery with iC_3 TTC

TABLE 1 Results of MF1 flotation tests (Merensky)

	Standard	iC_3-TTC
PGM Rate (min^{-1})	74.3	93.7
PGM Recovery (%)	89.2	90.7
Cu Rate (min^{-1})	89.7	103.4
Cu Recovery (%)	87.4	89.4
Ni Rate (min^{-1})	45.7	74.3
Ni Recovery (%)	82.4	84.4
Cr_2O_3 Rate (min^{-1})	0.3	0.3
Cr_2O_3 Recovery (%)	1.5	1.4

Earliest work by Slabbert (1985) was with iC_3 TTC replacement of SiPX in the synergistic mixture of SiPX and diisobutyl dithiophosphate (DTP) at a combined dosage of 140 g/t. The PGM recovery vs. percent mass recovered is shown in Figure 6. For low and high mass recoveries iC_3TTC recovered two percent more PGM's than the standard, indicating not only recovery of PGM's increased but in addition grade was always higher. Less talc was recovered.

MF1 Test on Merensky Ore

In Table 1 are listed the average initial rates and recovery data obtained from twenty sets of floats of 4 kg samples in an 8 liter Denver cell. Sampling was continuous and concentrates were collected after 30 seconds, 2.5, 7.5, and 17.7 minutes. Analyses were done on PGM's, Ni, Cu and Cr_2O_3. For these twenty sets of data, generated under plant conditions, for both the standard and TTC the statistical analysis was found to fall within a 95% confidence interval (Steyn, 1996).

TABLE 2 Results of MF2 flotation tests (Merensky)

	Standard	iC_3-TTC
PGM Rate (min^{-1})	95	95.9
PGM Recovery (%)	63.7	65.0
Cu Recovery (%)	91.5	92.5
Cu Grade (%)	1.0	1.1
Ni Grade (%)	1.8	1.9
Ni Recovery (%)	82.4	84.4
Cr_2O_3 Grade (5)	0.92	0.93
Cr_2O_3 Recovery (%)	3.59	3.97

TABLE 3 Mineral mass percentages on MF1 tails obtained with QEM*SEM

	–106 µm to + 53 µm			–53 µm to + 25 µm		
Mineral	STD	iC_3TTC	Decrease (%)	STD	iC_3TTC	Decrease (%)
Pyrrhotite	0.072	0.058	19.4	0.049	0.018	63.3
Pentlandite	0.039	0.024	38.5	0.013	0.012	7.7
Chalcopyrite	0.032	0.021	34.4	0.012	0.010	16.7
Chromite	4.521	4.199	7.1	4.321	4.521	–4.4
Feldspar	45.77	46.68	–2.0	45.54	45.15	0.8
Orthopyroxene	37.72	37.70	0.1	35.09	35.30	–0.6
Other	1.85	11.31	4.5	14.97	14.99	–0.1

MF1 Tests (Merensky)

The primary float of the MF2 tests was done in exactly the same way as MF1, except that concentrates were collected continuously for 15 minutes (Table 2). Again tests were repeated twenty times and again the statistical analysis was found to fall within a 95% confidence interval.

QEM*SEM Analysis on MF1 Tails

The analyses were carried out on two size fractions: between 106 and 53 µm as well as between 25 and 53m. The results are presented in Table 3

From Table 3 less base metal sulfides occur in the TTC tailing than in the standard. The effect is most pronounced with pyrrhotite in both size fractions. As will be seen later the calculated surface hydrophobicity based on Gibbs Excess Free Energy (G^{ex}) calculations would explain these differences.

Inert Gas and Covalent TTC Ester Collectors

Since iron sulfide minerals readily oxidize, and dithiolates (dixanthogen for example) are important in floating in the presence of air, a TTC ester was prepared that could not oxidize further. The iC_3-TTC-iC_4 ester has been found to be a most effective bulk flotation chemical on a mixed Cu, Pb, Zn, Fe sulfide ore (Coetzer and Davidtz, 1989).

TABLE 4 Comparison of a covalent TTC ester with the standard SiBX/DTP mixture

Collector	Flotation Gas	Pgm Recovery (%)
SiBX/DTP	Air	91.0
SiBX/DTP	N_2	85.8
iC_3-TTC-iC_4	Air	92.4
iC_3-TTC-iC_4	N_2	93.2

TABLE 5 Calculated activity coefficients and G^{ex} values for the binary systems iC_3DTC/water and iC_3TTC/water

Molecule	Active Functional Groups	$\gamma_{collector}$ $X_c = 0.5$	γ_{water} $X_w = 0.5$	G^{ex}(Joules) $= RT \Sigma X_i = ln\gamma_i$
iC_3DTC	2xCH_3; 1xCHO	1.58	2.497	1701
iC_3TTC	2xCH_3; 1xCH	3.079	3.039	2772

In Table 4 the results when air and nitrogen was used as flotation gas for the SiBX/DTP and covalent esters. The Merensky ore was milled in nitrogen and floated with nitrogen gas.

The explanation put forward for the above behavior was based on G^{ex} values for the twocollectors. The two factors affecting the G^{ex} values are:

- the surface coverage which is influenced by surface oxidation and
- the collector activity coefficient, which is determined by the type of functional groups on the molecule. (Davidtz 1998).

GIBBS EXCESS FREE ENERGY (G^{EX})

Surface activity coefficients of adsorbed collector and adjacent water molecules were calculated from the UNIFAC group contribution method. (Sandler, 1997). It is possible to calculate activity coefficients of interacting species in solution. The requirements are the type and amount of functional groups in the interacting molecules (Davidtz, 1999).

Provided one does not exceed a monolayer surface coverage and collector molecules do not hydrophbically interact with each other, then excellent correlations are obtained between flotation activity and G^{ex}. This method has helped explain why TTCs perform better than DTCs at equivalent molar dosages. Reasoning is that the activity coefficient contribution to G^{ex} by the oxygen group in DTCs, reduces hydrophobicity relative to the sulfur in the corresponding TTC. The increased attractive interaction between water and adsorbed DTC species renders the surface less hydrophobic than is the case with TTCs.

The relationship between G^{ex} and the mole fractions of interacting species X_i, and activity coefficients γ_i and absolute temperature T, is given by:

$$G^{ex} = RT\sum X_i \ln\gamma_i$$

Table 5 lists the calculated activity coefficients for the interacting collector and water at a surface coverage of $X_c = 0.5$. The two collector molecules chosen are $(CH_3)_2O{\cdot}CS_2{\cdot}Na$

TABLE 6 Relative Gibbs excess free energies and PGM recoveries for selected collectors (N_2)

Collector(s)	PGM Recovery (%)	$\Delta G^{ex}/RT$
iC_4DTCB	61.8	–0.469
iC_4DTCB + DTP	64.5	–0.337
Standard (iC_4DTC)	71.2	–0.18
iC_3TTC + DTP	71.5	–0.002
iC_4TTC + DTP	72.7	0.053
iC_3TTC	79.7	0.422
iC_4TTC	82.1	0.528

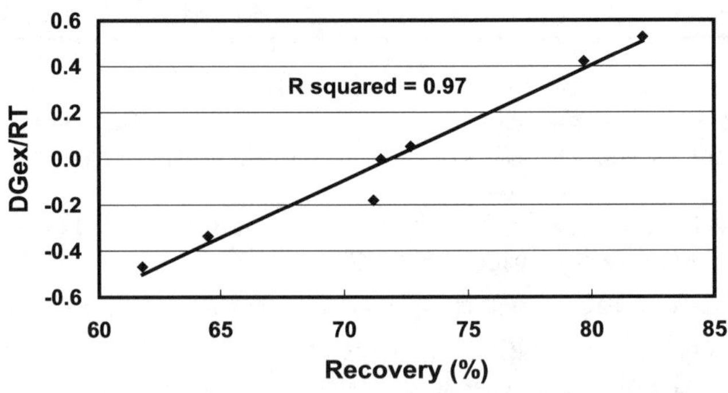

FIGURE 7 $\Delta G^{ex}/RT$ **vs PGM recovery for selected collectors and collector combinations**

(iC_3DTC) and $(CH_3)_2S{\cdot}CS_2{\cdot}Na$ (iC_3TTC). In the calculation it is assumed that all S atoms are bonded to the surface and only the O, and hydrocarbon groups interact with water.

The higher activity coefficients imply further deviations from ideality and in these cases increased repulsive interaction between water and the collector molecule. There is a 63 percent increase in G^{ex} from iC_3DTC to iC_3TTC.

Provided the mineral surface oxidation is controlled, and overdosing avoided with short chain collectors, (no hydrophobic bonding between adsorbed collector molecules) then excellent correlations are found between G^{ex} and flotation activity. (Davidtz, 1999).

In Table 6 the relative (ΔG^{ex}/RT) values for various collector systems are given. The reference state chosen is that of methyl TTC at 25°C. These comparisons were done on a Merensky ore, under nitrogen milling and flotation. (Steyn 1996).

The data from Table 6 is shown in Figure 7. It has been concluded (Davidtz 1999) that for a given particle size the only factors affecting flotation activity are the mole fraction or surface coverage (X_i) and the nature of the functional groups interacting with water. In other words the activity coefficients of interacting species at the particle /water interface. Unfortunately these calculations become less effective as the hydrophobic chains get longer. Here the hydrocarbon chains interact with one another and UNIFAC does not take this decreased interaction with water into account.

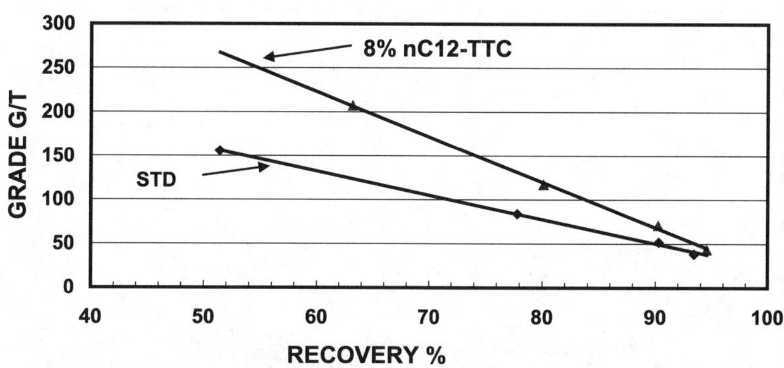

FIGURE 8 Grade recovery relationship for partial SiBX replacement by Cl2 TTC

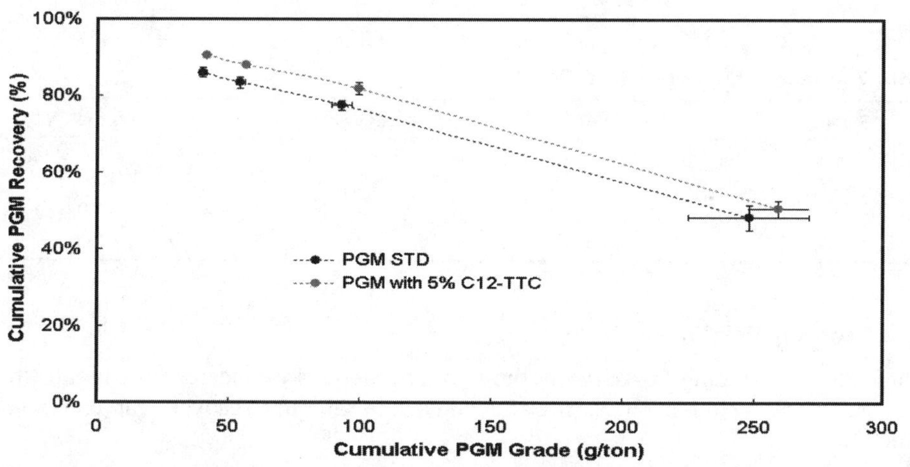

FIGURE 9 Recovery grade relationship for standard and substituted TTC

LONG CHAIN TTC COLLECTORS IN PGM RECOVERY

Grade Recovery Relationships

Perhaps the most outstanding feature of the long chain TTCs when used in conjunction with SiBX, is the improvement of grade in the initial concentrates. Typical results are shown in Figures 8 and 9. Here and eight molar percent respectively of the STD SiBX in the SiBX/DTP mixture was replaced by TTC. Similar results have been reported by Breytenbach et.al. (2003).

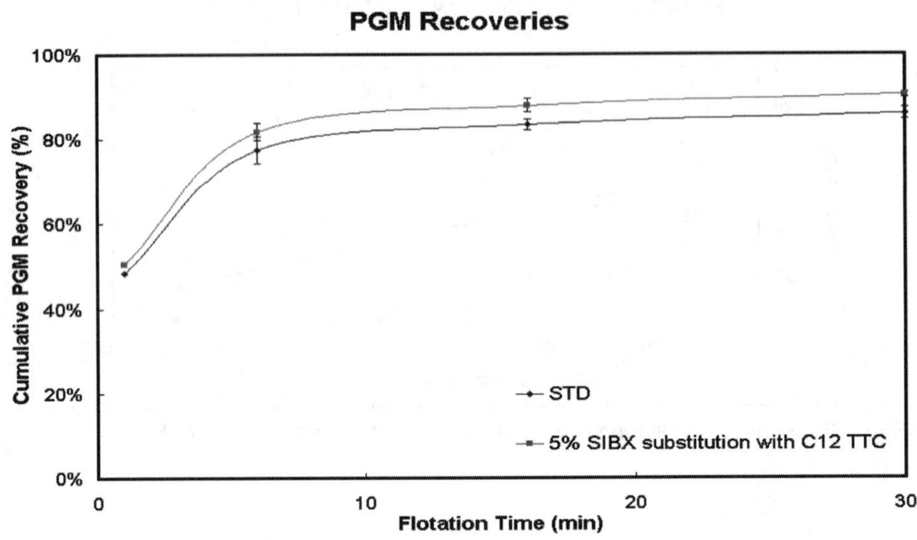

FIGURE 10 Recovery time for partial SiBX replacement by TTC on PGMs

TABLE 7 Uranium recovery with TTCs

SiBX(g/t)	C_{12}TTC(g/t)	iC_3TTC(g/t)	U_3O_8 Rec. %
12	—	—	30.1
9	4.3	1.2	40.0
9	4.3	—	41.0
3	8.6	3.6	45.0

Time Recovery Relationships

When 5 molar percent of the SIBX in the standard Merensky collector suite is substituted with nC12 TTC, a similar synergistic effect is seen in the time/recovery curve, Figure 10. (Vos 2004).

Flotation of Uranium Bearing Ore

About 7,000t/d of largely dump material is treated at the AngloGold West float plant. Twenty grams per ton SiBX and 2 g/t C12 TTC is used as collector suite. After calcination of the concentrate to produce sulfuric acid, it is further treated to recover uranium. The uranium generally occurs as uraninite an oxide (UO_3) and is associated with lead. It is not clear why this sulfide collector recovers primarily an oxide but the uranium mineral association with lead may serve as activator. Suffice it to say that as much as a ten percent improvement in uranium recovery is common with this small amount of C_{12} TTC used.

Laboratory flotation tests on a 4.5 liter Denver flotation cell and pulp with an SG of 1.427 were done in duplicate on dump material at the AngloGold West plant. The frother was Dow 250, and a gaur depressant for pyrophyllite. The averages of duplicate tests are reported in Table 7.

TABLE 8 Summary of test data

	Ni Recovery %	Cu Recovery %	S Recover %	Fe Recovery %
1 Month Before test (1/4–28/4)	81.32	92.99	61.86	23.92
Test (28/4–3/5)	83.7	94.53	71.11	29.92
Increase %	2.38	2.54	9.25	6
After test (7/5–21/5)	81.5	93.49	58.72	21.8
Decrease %	2.63	1.1	17.4	17.11

TABLE 9 Comparison of standard redox potentials for different thiocarbonate couples (solutions saturated with (thiocarbonate)$_2$, vs SHE)

Hydrocarbon Group	Monothiocarbonate MTC/(MTC)$_2$	Dithiocarbonate DTC/(DTC)$_2$	Trithiocarbonate TTC/(TTC)$_2$
Methyl	0.0200	–0.0035	-
Ethyl	0.0200	–0.0588	-
n-Propyl	–0.0022	–0.0910	-
iso-Propyl	-	–0.0917	–0.1800
n-Butyl	–0.0380	–0.1250	–0.2342
iso-Butyl	-	–0.1317	–0.2035
n-Amyl	–0.0880	–0.1530	–0.2885
sec-Amyl	-	-	–0.2158
Hexyl	–0.0120	–0.1550	–0.2158

Flotation of Nickel Ore

A six day plant trial run at the Tati nickel mine, Botswana was primarily to improve the sulfur grade, required at the smelter. Two grams per ton C12TTC was fed from the main feed tanks on the primary circuit. On line S and Fe analysis clearly showed a steady increase in the Fe grade and recovery in the first rougher bank within half an hour of initial addition. The major minerals are pentlandite, pyrrhotite and chalcopyrite. A summary of the data obtained before and after the test is shown in Table 8.

CHEMISTRY OF TTC COLLECTORS

Standard Redox Potentials for Different Thiocarbonate Couples

Most TTCs chemistry has been done on pyrite by J.D. Miller and his research group. A comparison of standard redox potentials for different thiocarbonate couples is given in Table 9 and Figure 11 (du Plessis et.al. 2000).

Electrochemically Controlled Contact Angle

Elecrochemically controlled contact angle measurements were made for pyrite for the potassium salts of n-amyl DTC and n-amyl TTC at 1×10^{-3} mol/L and pH 4.68. The results are shown in Figure 12.

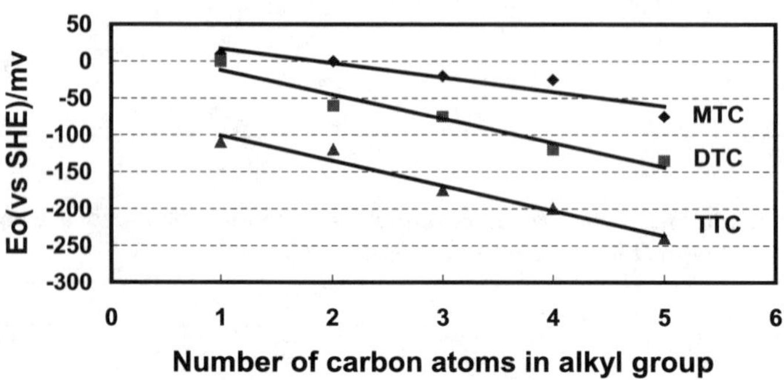

FIGURE 11 Effect of number of sulfur atoms in molecule on the standard redox potential

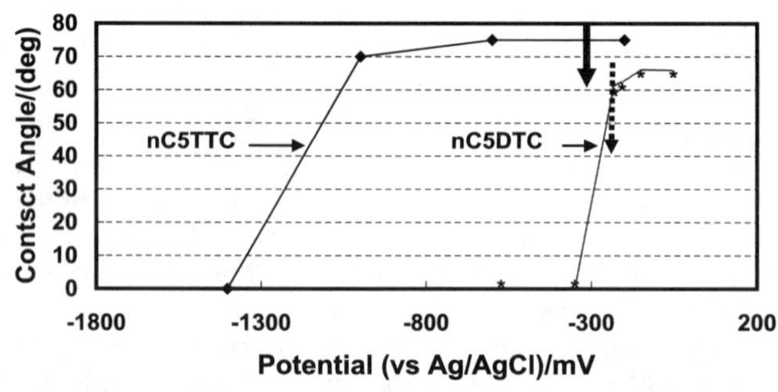

FIGURE 12 Potential (mv) vs contact angle (deg) (Arrows indicate the corresponding half-cell potential for the monomer/dimmer couple)

TTC Flotation Chemistry in the Presence of Oxygen

Results from electrochemically controlled contact angle measurements confirmed previous analysis of the DTC collector flotation of pyrite, that a hydrophobic surface state could only be created at potentials above the potential where its corresponding $(DTC)_2$ formed. However TTCs were found to be effective below the potential where the corresponding $(TTC)_2$ was formed. This was in contrast with the DTC collector, where pyrite hydrophobicity under these conditions could not be obtained (du Plessis et al. 2004). FTIR external reflection spectroscopy revealed that under oxidizing conditions, the $(DTC)_2$ and $(TTC)_2$ would form as multi-layers at the pyrite surface.

TTC Flotation in the Absence of Oxygen

Although the short chain TTCs are superior collectors to DTCs, they suffer a disabling odor that requires special handling and equipment. With the success of the N_2TEC technology, the closed system offers an alternative engineering approach where the atmosphere can be controlled.

Results on electrochemically controlled contact angle, Hallimond tube flotation, bench scale flotation and solution depletion data under nitrogen revealed that the TTC collector is a superior collector at neutral pH conditions, and shows improvement at high (pH 9.2) compared to DTC. (du Plessis et al 2004).

Analysis of UV spectroscopic data as well as FTIR external reflection spectra, suggest that the adsorption of a mercaptan at the pyrite surface which leads to the hydrophobic pyrite surface state being created at low potentials (in air and nitrogen) when the TTC is used as collector. (du Plessis, 2003). It would appear this active mercaptan has to go adsorb via adsopbed TTC species, as neither mercaptide nor mercaptan were found to be as effective as the TTC. Slabbert 1985 and van Rensburg 1987.

ACKNOWLEDGEMENTS

Special appreciation is extended to Impala Platinum and AngloGold for supporting this research. Also, Jan Miller and his group at Utah University and Pretoria University for their major involvement.

REFERENCES

Ackerman, P.K., Harris, H.G., Klimpel, R.R, and Aplan, F.F. Effect of alkyl substituents on performance of thionocarbamates as copper sulfide and pyrite collectors. Reagents in mineral industry. *Inst. Min. and Met.*, 1984, pp. 69–78.

Breytenbach W. Vermaak, M.K.G., and Davidtz J.C. Synergistic effects among dithiocarbonates (DTCs), dithiophosphate (DTP) and trithiocarbonates (TTC) in the flotation of Merensky ores. Jnl. Safri. Inst. Min. and Met. pp 667–670. Dec. 2003)

Coetzer G. and Davidtz J.C. Sulphydryl collectors in bulk and selective flotation. Part 1 Covalent trithiocarbonate derivatives. J. S. Afr. Inst. Min. Metall., vol. 89, no. 10. 1989. pp. 307–311.

Davidtz, J.C. Quantification of Flotation Activity by means of Excess Gibbs Free Energies. *Mins. Eng.* Vol 12 No. 10 pp 1147–1161, 1999.

Davidtz, J.C., 1998 Design of Molecules Specific to Base Metal Recovery in Typical PGM containing Ores. 8th Int. Platinum Symposium Geological Soc. Of S.A. and SAIMM Symposium Series S18. (1998)

du Plessis, R. Miller, J.D. and Davidtz, J.C. Preliminary examination of electrochemical and spectroscopic features of trithiocarbonates collectors for sulfide mineral flotation. Trans. Nonferrous Met. Soc. China Vol 10. Article ID: 1003- 6326(2000)S1–0012–07)

du Plessis, R., Miller, J.D., and Davidtz, J.C. 2004 “Trithiocarbonate Collectors for Sulfide Mineral Flotation-Fundamentals and Applications,” International Mineral Processing Congress, Cape Town, South Africa, to be published.}

Krescheck, (1975) In: WATER, Franks, F (Ed.), Plenium Press New York

Sandler, S.I. Chemical Engineering Thermodynamics 2nd Edition 1997.

Slabbert, W., The Role of Trithiocarbonates and Thiols on the Flotation of some Selected South African Sulfide Ores. *M.Sc. Dissertation.* Potchefstroom University South Africa. (1985).

Steyn, J.J. The role of Collector Functional Groups in the Flotation Activity of Merensky Reef Samples. *M.Sc. Dissertation. Potchefstroom University South Africa 1999.*

Van Rensburg, A.R.J. 1987. A Quantitative Evaluation of Sulfhydral Collectors in Noble Metal Recovery. M.Sc. Thesis.

Vos, F. Internal Impala Report. July 2004

Selective Separation of Carbonate Minerals by Reactive Flotation

A. El-Midany,* H. El-Shall,* R. Stana,† S. Svoronos,‡ and B. Moudgil*

Due to their similar surface properties, it is not always easy to separate carbonate minerals from other salt type minerals by froth flotation. A particular example is the difficulty of separating dolomite impurities from phosphate minerals. Such challenging fundamental and practical problems have inspired several research efforts in various laboratories around the globe to find methods for economically removing the dolomite. However, a universal and cost effective process is not in commercial use. In this paper, a new process for separation of carbonate minerals (dolomite or calcite) from other associated minerals is presented. Differential solubility of carbonates in slightly acidic solution is utilized to generate micro CO_2 bubbles at the carbonate mineral / water interface. Particle (e.g., dolomite)/ bubble aggregates float to the surface leaving phosphate concentrate with low MgO content. Interestingly, up to 6.0 mm particles can be floated by such micro bubble/particle aggregates. Different surface-active agents were tested to stabilize the bubbles. Results of these tests are presented and the mechanisms involved are discussed.

INTRODUCTION

As phosphate-mining operations continue to move south in Central Florida, there is an increase in the MgO content of the rock. This MgO increases the difficulty in processing the rock (lowers the P_2O_5 recovery) and makes it difficult to make Diammonium Phosphate (DAP) grade. In fact, much of the higher dolomite rock is simply left in the ground, as there is no economical method for removing the dolomite from the rock. At present,

* Department of Materials Science and Engineering, University of Florida, Gainesville, Florida

† Particle Engineering Research Center, University of Florida, Gainesville, Florida

‡ Chemical Engineering Department, University of Florida, Gainesville, Florida

the only method ever commercialized, heavy media separation, is sitting idle due at least partially to its relatively high operating cost and poor performance. The heavy media product typically contains 1% or above MgO and less than half the P_2O_5 tons. The high dolomite rock is now either discarded or blended off with lower MgO rock.

Over the years the Florida Institute of Phosphate Research (FIPR) and the industry have looked at a number of methods for removal of the MgO. They include selective flotation, flocculation, grinding and screening, and partial acidification (Abdel-Khalek, 2000; Amankonah and Somasundaran, 1985; Ananthapadmanabhan and Somasundaran, 1984; Anazia and Hanna, 1987; Andersen and Somasundaran, 1993; Elgillani and Abouzeid, 1993; El-Midany, 2004; El-Shall and Zhang, 2004; El-Shall et al., 1996; Houot et al., 1985; Lu and Sun, 1999; Shao et al., 1998, Somasundaran et al., 1989; Somasundaran and Zhang, 1999; Zhengxing et al., 1999; Zhizhong and Zhengxing, 1999). All flotation methods required that the rock be first ground to concentrate size material or smaller prior to the flotation. While some of these have shown promise in the laboratory, none have been commercialized to date, generally due to either a complex flow sheet, poor economics or low P_2O_5 recovery. The most promising process evaluated to date is the China Lianyungang Design and Research Institute (CLDRI) process (El-Shall and Zhang, 2004; Zhengxing et al., 1999; Zhizhong and Zhengxing, 1999). Even in the CLDRI fine flotation process, there are some concerns such as:

- Handling of fine concentrate (dewatering and transport)
- Capital cost (grinding and multiple process steps)
- Operating cost: increase in MgO percentage in the phosphate rock increases the reagent consumption. In addition, grinding prior to beneficiation will increase energy consumption as well as plant operating cost.

This study addresses a process that can be applied to separate dolomite from phosphate pebbles. The basic idea is based on a well-known reaction between carbonates and acids producing carbon dioxide gas. Dolomite, as a carbonate mineral, generates CO_2 when exposed to acidic solutions. Capturing the generated CO_2 at the dolomite particle surface into bubbles using a coating agent (surfactant or polymer) can selectively float dolomite and separate it from phosphate rock. Thus, this process is called the "Reactive Flotation" (RF) process. The RF process requires an acidic medium and a surfactant, which adheres to the dolomite surface and forms a stable membrane (film) that is acid permeable and impermeable to CO_2 gas. Preliminary testing using several surfactants and polymers suggest that the polyvinyl alcohol (PVA) forms the required membrane that leads to dolomite flotation. This paper presents the development activities as well as fundamental studies to illustrate the mechanisms involved in this process.

MATERIALS AND METHODS

Flotation

In this study, screening and central composite statistical experimental designs are used to determine the effect of various factors on the removal of MgO and its recovery. The Plackett-Burman design was used to screen out which of the seven factors being considered were the most statistically significant. The variables examined in the screening stage were acid concentration (1% and 3%), polyvinyl alcohol (PVA) concentration in the coating solution (1% and 3%), the PVA molecular weight (60,000 and 150,000), the degree of hydrolysis (88%–99%) of the PVA, particle size (−0.5 mm and 5–9 mm), and drying time of the coating (0 and 2 hrs). The levels were chosen according to the exploratory test results.

A central composite design was used to further investigate the three most significant factors (acid concentration, PVA concentration, and particle size) determined from the screening design. Five levels of each variable were tested. For each run, the products are chemically analyzed as mentioned above and the MgO percentage and MgO removal recovery were calculated for all the experimental runs. The computer program, Design expert® (SAS Institute, Inc.), was used to perform the statistical analysis for the central composite design. The data was analyzed by fitting the response variables to a second order polynomial model.

Flotation Procedures

Using 300 ml of 3% H_2SO_4 in a 400 ml beaker, particles were pre-coated by spraying 3% PVA solution (using a mist delivery nozzle) then added to the acidic solution. The experiment was left for 5 minutes. Dolomite floated to the surface, which was separated by decantation on a screen. The products (floated and unfloated fractions) were dried, weighed, digested and chemically analyzed using an Inductively Coupled Plasma (ICP) emission spectrometer (Perkin Elmer Optima 3200RL Optical Emission Spectroscopy, Norwalk, CT). The ICP was calibrated using AFPC (Association of Florida Phosphate Chemists) rock check No.22.

PVA Preparation

PVA was kindly supplied by Celanese chemicals Co., USA. Different PVA types with various ranges of molecular weights and degree of hydrolysis were used in the screening factorial design. In the rest of the study, PVA Celvol 165 (150,000 Av. MW and 99% degree of hydrolysis) was used. PVA flakes were gradually dissolved in de-ionized water at 90°C with continuous mixing by magnetic stirrer and heated at 90°C for two hours to produce different PVA concentrations (0.5, 1, 2, 3, and 4%). PVA solution of a certain concentration was treated with 25 ml of 4% boric acid solution and 2 ml of iodine solution (prepared from 1.27g of iodine and 25 g of KI/l) in that order. The resulting solution was diluted to 100 ml and kept at 25°C and its absorbance measured at 690 nm against a blank solution containing the same amount of boric acid and iodine solution. Beer's Law applies in the concentration range of 0–40 mg of PVA/l of solution. Absorbance spectra were taken using ultraviolet/visible spectrophotometer (UV-Vis spectrophotometer, Perkin-Elmer Lambda 800). All spectra were taken at room temperature.

Equilibrium Surface Tension Measurements

Equilibrium surface tension for freshly prepared solutions was measured by the Wilhelmy plate method at room temperature.

Contact Angle Measurements

The contact angle goniometer is mounted on the optical bench to examine a single liquid drop of the PVA solution resting on the smooth, planar, solid substrate. The drop is illuminated from the rear to form a silhouette, which is viewed through a telescope connected to a computer. The image of the drop appears on the computer monitor by which the brightness and contrast, focusing and amount of illumination can be adjusted. After adjusting the image, the contact angle can be measured. In this case, the contact angle is the angle included between the tangent plane to the surface of the liquid and the tangent plane to the surface of the solid, at any point along their line of contact inside the liquid drop. In other words, higher angles means spreading the liquid on the solid substrate.

Zeta Potential Measurements

In the zeta potential measurement tests, 1 g of ground mineral samples was added into a 250 ml beaker in which 100 ml 0.003 M KNO_3 solutions were added. The suspension was conditioned for 3 min during which the pH was adjusted, followed by 5 min of conditioning after adding the polymer. It was then allowed to settle for 3 min, and about 10 ml of the supernatant was transferred into a standard cuvette for zeta potential measurement using a Brookhaven ZetaPlus Zeta Potential Analyzer. Solution temperature was maintained at 25°C. Ten measurements were taken and the average was reported as the measured zeta potential.

Adsorption

These studies were carried out at 25°C using a 0.5 gm mineral sample in a 50 ml solution having the desired PVA concentration and the ionic strength fixed at 3×10^{-3} M using KNO_3. One set of experiments was performed first to determine the effect of conditioning time at a fixed pH (pH 7) and PVA concentration of 100 ppm (3.33×10^{-8} M). Adsorption isotherms were also determined for PVA/dolomite and PVA/apatite systems at pH 7 and 3×10^{-3} M KNO_3 to keep the ionic strength constant.

After the conditioning step, the suspensions were centrifuged at 3000 rpm for 5 min using a IEC-Medispin centrifuge (Model 120). Then, a volume of the clear solution was withdrawn for measurement of the PVA residual concentration.

RESULTS AND DISCUSSION

Results from the screening statistical design indicated that the main factors significantly (95% confidence level) affecting MgO % in the concentrate as well as the amount removed are PVA concentration in the coating solution, acid concentration, and particle size. These were studied further using the Central Composite Design. The results of dolomite flotation experiments are given in Figures 1 and 2 in terms of percentage of MgO removal in floated fraction and percentage of MgO in the sink (nonfloat) fractions respectively. As mentioned previously, the confidence interval of these findings is within 95%.

It can be seen from these figures, that the percentage of MgO recovery and MgO % in the concentrate are greatly and positively affected by increasing acid concentration and PVA dosage. For instance, percentage of MgO recovery increases from 30 to more than 90% and MgO % in the concentrate decreases from more than 2.1 to 0.6 respectively. However, increase in particle size negatively affects the percentage of MgO recovery and MgO % in the concentrate. With increasing particle size the percentage of MgO recovery also decreases. At the maximum acid concentration and PVA dosage, percentage of MgO recovery is as low as 75% and percentage of MgO % in the concentrate is higher than 1.1 %. It should be remembered, however, that the flotation time is five minutes, which the same for both sizes in these tests. In other words, more time is needed

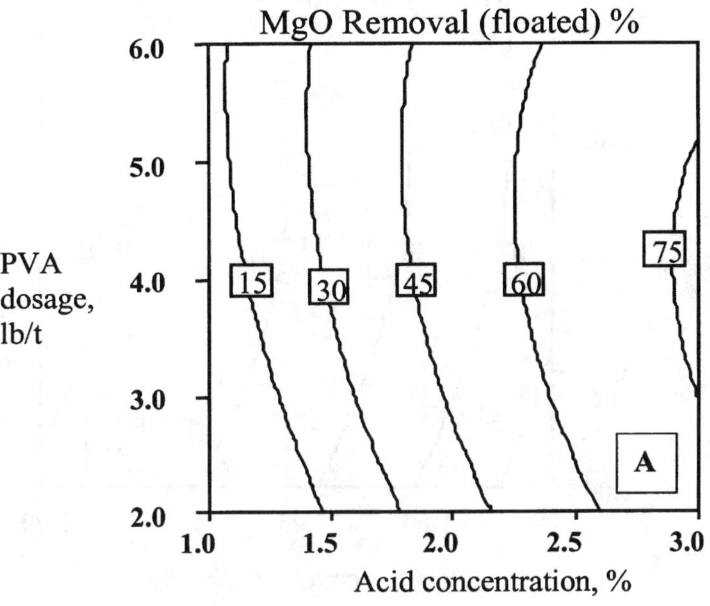

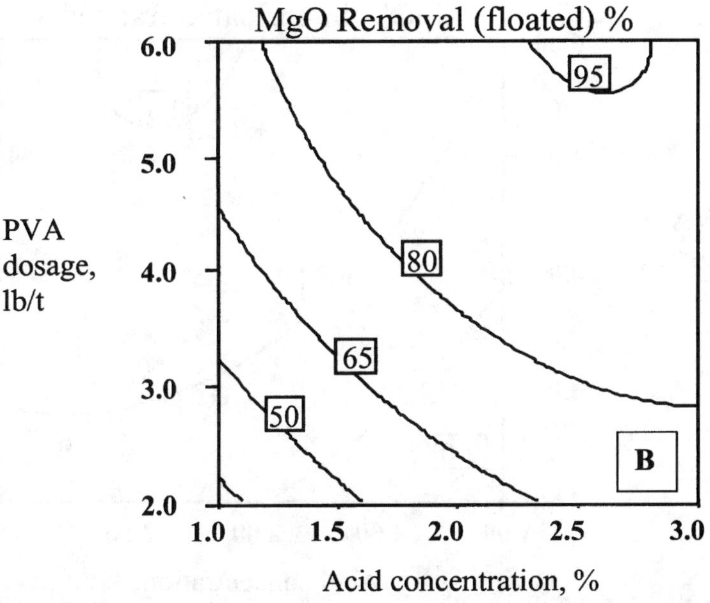

FIGURE 1 Effect of PVA dosage and acid conc. on MgO recovery % in the floated fraction at different particle sizes, a) particle size = 5–9 mm, and b) particle size = −0.5 mm

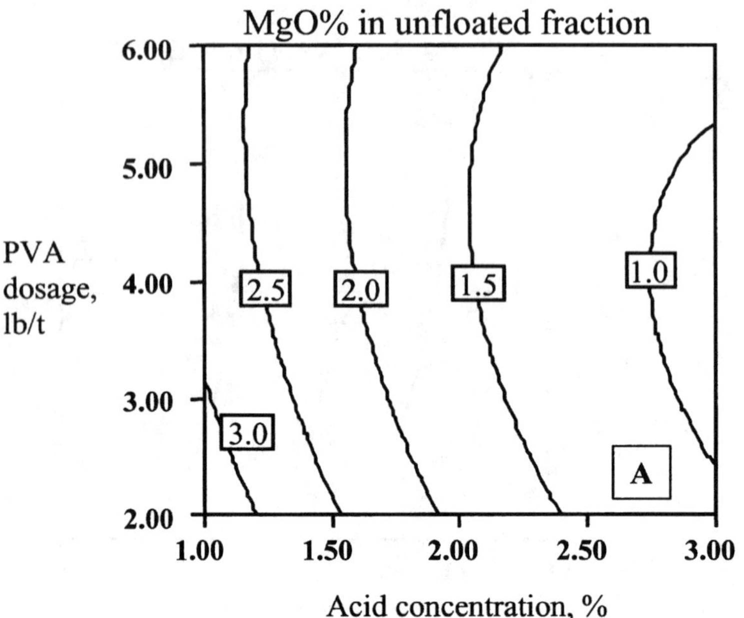

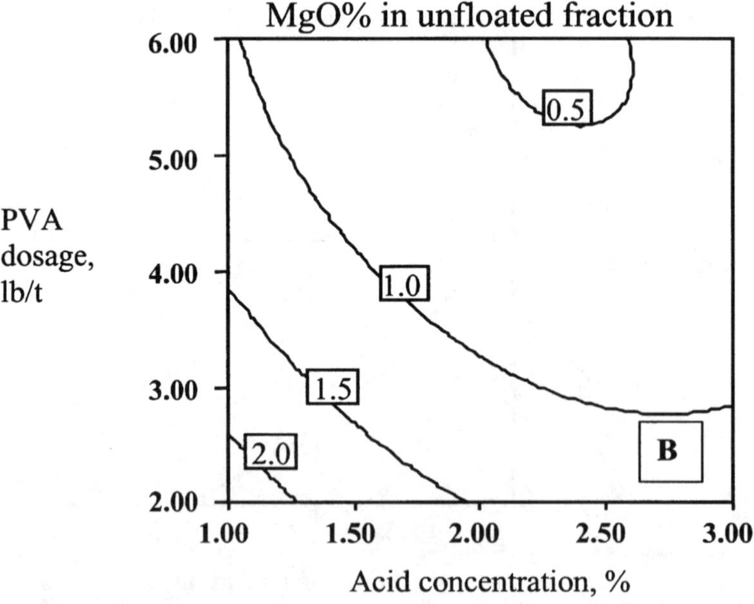

FIGURE 2 Effect of PVA dosage and acid concentration on MgO % in unfloated fraction at different particle sizes; a) particle size = 5–9 mm, and b) particle size = –0.5 mm

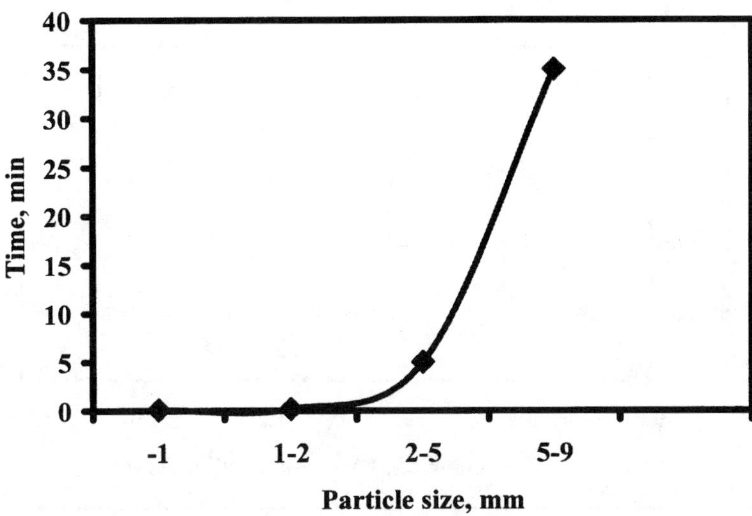

FIGURE 3 Effect of the particle size on the time needed to float the particle in 3% H_2SO_4 when the particle is pre-coated with 3% PVA

for enough CO_2 to form to make the coarse particles buoyant. This can be clear from the data presented in Figure 3 showing flotation time as a function of particle size. Recently, El-Midany (2004) has modeled the reaction rate of dolomite particles with acid. The developed model can be used to estimate the amount of gas evolved as a function of time. The amount of gas needed to induce buoyancy to particles of different sizes was estimated. Most interestingly, El-Midany (2004) has found a direct correlation between time needed for acid to react with dolomite (to produce needed amount of CO_2 gas) and the flotation data shown in Figure 3.

The data shown in the Figures 1 and 2 indicate that 2.0 kg/t (4.0 lb/t) of PVA are needed to achieve the best results. Such amount is not cost effective for the phosphate industry. Thus, further testing of PVA at different concentrations, and conditioning time, as well as method of conditioning (stirring vs. spraying), suggested that 1.0 % PVA concentration and 0.5–1.0 kg/t (1–2 lb/t) could be used to achieve concentrate of good quality (< 1.0% MgO, and high removal of MgO). Selective separation of dolomite by this reactive flotation process is obviously due to the formation and entrapment of CO_2 in a PVA membrane on dolomite surface. Such membrane has been confirmed by the adsorption of PVA on dolomite as shown in Figure 4. PVA also adsorbs, but to a lesser degree, on apatite as given in Figure 5. The S-shape of the adsorption isotherm indicates physical adsorption. This also is clear from the zeta potential data for both dolomite (Figure 6) and apatite in Figure 7. The data shows no change in the isoelectric points of these minerals, indication physical adsorption. Further confirmation of the physical adsorption mechanism is given by the FTIR data, El-Midany (2004), which suggests a hydrogen bonding between PVA molecules and isolated and hydrogen-bonded hydroxyl groups on dolomite surface. Only bonded hydroxyl groups are found on apatite samples.

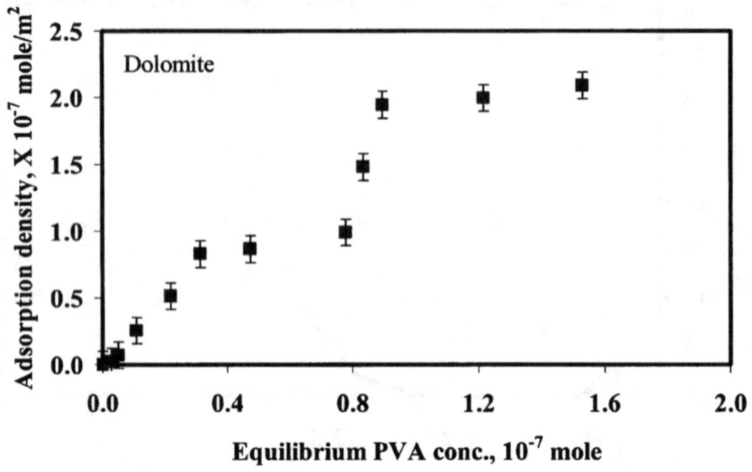

FIGURE 4 FIGURE 4. Adsorption isotherm of PVA on dolomite at 23°C

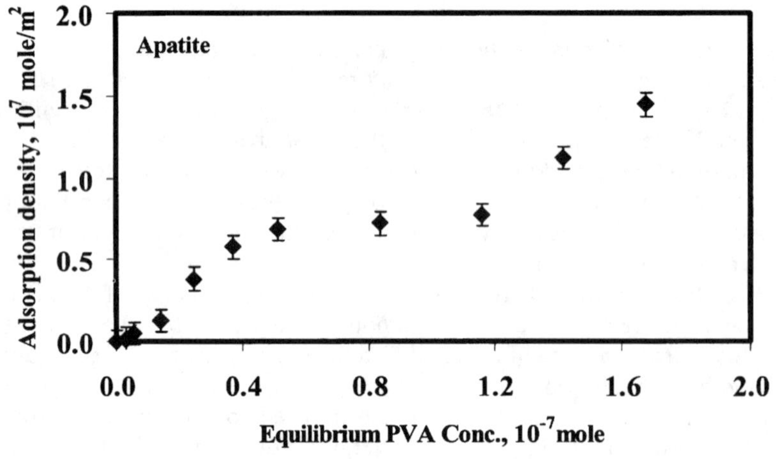

FIGURE 5 Adsorption isotherm of PVA on apatite at 23°C

It may be considered that the decrease in zeta potential in presence of PVA is not due to a decrease in charge and surface potential, but rather to a shift of the shear plane (Figure 8).

The strength of adhesion to a solid surface can be estimated from the value of the thermodynamic 'work of adhesion' or W_a,. In a simple system where a liquid L adheres to a solid S, the work of adhesion can be estimated from Eq.(1):

$$W_a = \gamma_l(1 + \cos\theta) \quad \textbf{(EQ 1)}$$

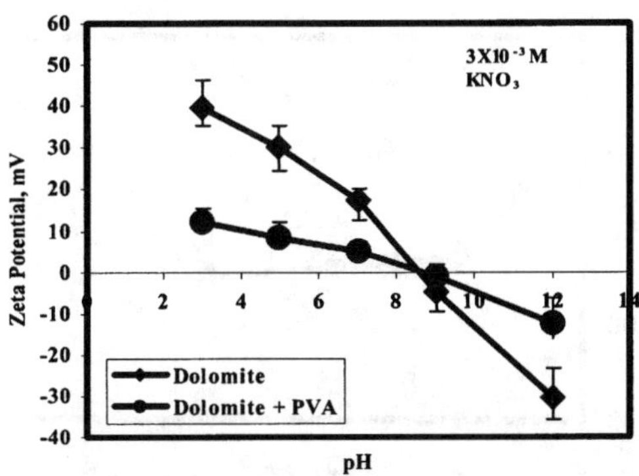

FIGURE 6 Zeta potential of dolomite in absence and presence of polymer

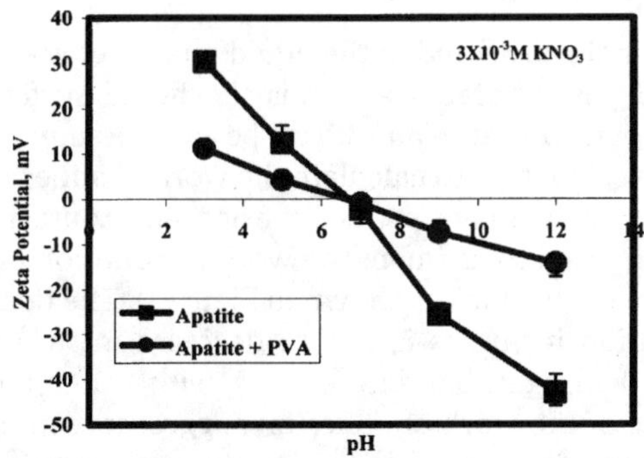

FIGURE 7 Zeta potential of dolomite in absence and presence of polymer

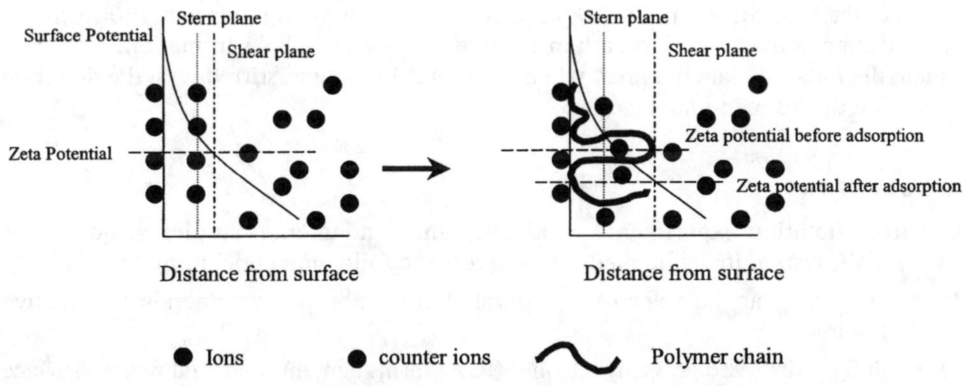

FIGURE 8 Effect of polymer on the shear plane position, without (a) polymer, (b) after polymer adsorption

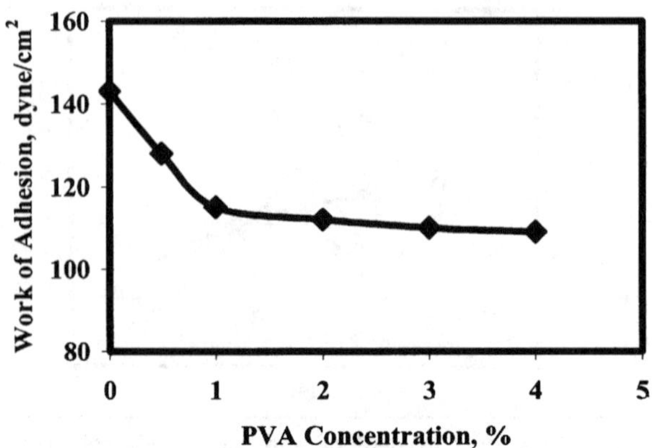

FIGURE 9 Work of adhesion as a function of PVA concentration on dolomite

where θ is the contact angle measured in the liquid phase between the tangents between solid/liquid and liquid/gas interfaces and γ_l is the liquid surface tension (liquid/gas interfacial tension). Provided that γ_l and θ can be measured experimentally, it is then possible to use the above equation to calculate the work of adhesion.

Therefore, the measurements of surface tension and contact angle were done and the data generated were used to estimate the work of adhesion of PVA to dolomite surface at different PVA concentrations as shown in Figure 9. The decrease in work of adhesion as PVA concentration increases is due to decrease in surface tension, El-Midany (2004) resulting from hydrogen bonding of PVA with water molecules. Most importantly, the lower values of the work of adhesion suggest spreading of PVA on the dolomite surface at least up to 1 % concentration. The decrease in work of adhesion at higher concentrations of PVA is not as dramatic as at 1% concentration. This is attributed to hydrogen bonding among PVA molecules leading to formation of more viscous droplets on the surface rather than a spread film. Such data together with the elasticity of the polymer film (calculated from dynamic surface tension results by El-Midany (2004) indicate that 1% PVA concentration could provide a film of enough elasticity to form the required membrane to entrap carbon dioxide leading to bubble formation. This may explain the flotation data obtained when 1% PVA solution was stirred with the dolomite before flotation as mentioned earlier.

CONCLUSIONS

Data from flotation experiments, modeling, and fundamental studies of dolomite/apatite/PVA system in acidic medium, suggest the following concluding remarks:

- Dolomite can be selectively separated from phosphate minerals by reactive flotation.
- Fully hydrolyzed polyvinyl alcohol (PVA) forms a membrane on dolomite surface entrapping carbon dioxide gas making dolomite buoyant.
- PVA adsorbs on dolomite surface via hydrogen bonding with isolated hydroxyl groups.

- Physical adsorption of PVA on dolomite is confirmed by results from FTIR characterization, zeta potential measurements, and adsorption isotherms.
- PVA reduces water surface tension and increases contact angle leading to lower work of adhesion on dolomite surface. Higher PVA concentrations than 1% do not significantly affect work of adhesion. This fundamental result can explain the flotation data showing 1% PVA concentration to be effective in flotation if PVA is stirred with the mineral to form a thin film.

REFERENCES

Abdel-Khalek, N. A., 2000, "Evaluation of flotation strategies for sedimentary phosphates with siliceous and carbonates gangues", Minerals Engineering, Vol. 13, No. 7, pp. 789–793.

Amankonah, J.O., Somasundaran, P., 1985, "Effects of dissolved mineral species on the electrokinetic behavior of calcite and apatite", Colloids and Surfaces, 15, p. 335.

Ananthapadmanabhan, K.P., Somasundaran, P., 1984, "The role of dissolved mineral species in calcite-apatite flotation", Minerals and Metallurgical Processing, 1, p. 36.

Anazia, I.J., Hanna, J., 1987, "New flotation approach for carbonate phosphate separation", Minerals and Metallurgical Processing, November, pp. 196–202.

Andersen, B., Somasundaran, P., 1993, "The role of changing surface mineralogy on the separation of phosphatic clay waste", International Journal of Mineral Processing, Vol. 38, pp. 189–203.

Elgillani, D.A., Abouzeid, A-Z.M., 1993, "Flotation of carbonates from phosphate ores in acidic media", International Journal of Mineral Processing, 38, pp. 235–256.

El-Midany, A., 2004, "Separation of Dolomite from Phosphate Rock by Reactive Flotation". Ph. D. Thesis, University of Florida.

El-Shall, H., Zhang, P., 2004, "Beneficiation technology of phosphates: challenges and solutions," Minerals and Metallurgical Processing, Vol. 21, No. 1, pp. 103–111

El-Shall, H., Zhang, P., Snow, R., 1996, "Comparative analysis of dolomite-francolite flotation techniques", Minerals and Metallurgical Processing, Vol. 8, pp. 135–140.

Houot, R., Joussement, R., Tracez, J., Brouard, R., 1985, "Selective flotation of phosphate ores having a siliceous and/or a carbonated gangue", Int. J. Miner. Process., Vol. 14, pp. 245–264.

Lu, S., Sun, K., 1999, "Developments of phosphate flotation reagents in China", In: P. Zhang, H. El-Shall, and R. Wiegel (editors) Beneficiation of Phosphates: Advances in Research and Practice, SME, USA, Chapter 2, pp. 21–26.

Shao, X., Jiang, C.L., Parekh, B.K., 1998, "Enhanced flotation separation of phosphate and dolomite using a new amphoteric collector", Minerals and Metallurgical Processing, Vol. 15, No. 2, pp. 11–14.

Somasundaran, P., Xiao, L., Viswanathan, K.V., 1989, "Interactions between oleate collectors and alizarin modifiers in francolite/dolomite systems," Mineral and Metallurgical Processing, 6, p. 100

Somasundaran, P., Zhang, L., 1999, "Role of surface chemistry of phosphate in its beneficiation", In: Zhang, P., El-Shall, H., and Wiegel, R. (eds.) Beneficiation of Phosphates: Advances in Research and Practice, SME, Co, USA, pp. 141–1154.

Zhengxing, G., Zhizhong, G., Zheng, S., 1999, "Beneficiation of Florida dolomitic phosphate pebble with a fine particle flotation process", In: Zhang, P., El-Shall, H., and Wiegel, R. (editors.) Beneficiation of Phosphates: Advances in Research and Practice, SME, pp. 155–162.

Zhizhong, G., and Zhengxing, G., 1999, "Plant practices of phosphate beneficiation in China". In: P. Zhang, H. El-Shall, and R. Wiegel (editors) Beneficiation of Phosphates: Advances in Research and Practice, SME, USA, Chapter 25, pp. 289–301.

Hydrophobic Interactions Between the Air Bubbles in Water

Liguang Wang[*] and Roe-Hoan Yoon*

The thin film balance (TFB) technique was used to measure the equilibrium film thicknesses (H_e) between two air bubbles in sodium dodecyl sulfate (SDS) and electrolyte (NaCl) solutions. The results were used to determine the magnitudes of the hydrophobic forces (K_{232}) using the extended DLVO theory. The K_{232} values determined from H_e measurement allowed us to predict disjoining pressure isotherms, which were found to be in excellent agreement with those obtained experimentally using a bike-wheel film holder.

The rate of film thinning was measured at low SDS concentrations with varying amounts of electrolyte (NaCl) added. In general, film thinning was controlled by the capillary pressure during the initial stages, while it was controlled by the surfaces forces during the later stages. At low NaCl concentrations, it was possible to predict the rate of film thinning using the Reynolds equation, provided that the driving force was corrected for the contributions from the hydrophobic force, which in turn were determined from the K_{232} values obtained from the H_e values measured experimentally. At high NaCl concentrations, where it was difficult to measure the equilibrium film thicknesses, the contributions from the hydrophobic forces were determined by fitting the film thinning data to the Reynolds equation. The K_{232} values obtained using both the equilibrium film thickness and the film thinning measurements showed that hydrophobic force deceased with increasing surfactant and salt concentrations. It was found that the hydrophobic force in foam films was very sensitive to the added electrolyte concentration.

* Center for Advanced Separation Technologies, Virginia Polytechnic Institute and State University, Blacksburg, Virginia

INTRODUCTION

Foam is a dispersion of gas bubbles in a liquid. It is generally believed that the balance between the electrostatic and van der Waals forces in foam films determines foam stability, according to the DLVO theory. It was shown, however, that the DLVO theory fails to predict the equilibrium film thicknesses at low surfactant concentrations (Tchaliovska, et al. 1994; Yoon and Aksoy 1999). These films are less stable than predicted using DLVO theory.

Craig et al. (1993) reported that the coalescence of N_2 gas bubbles in water is greatly reduced in the presence of various electrolytes. It was suggested that the coalescence is caused by the hydrophobic force at the air/water interfaces, which is dampened by the electrolyte. Deschenes et al. (1997) showed that coalescence of gas bubbles is a sensitive probe of hydrophobic forces. They produced bubbles using different gases and conducted similar experiments. They found an excellent correlation between coalescence and solubility parameter (δ^2), a measure of the hydrophobicity of solute. They found that vacuum is the most hydrophobic "solute" possible.

Recent TFB studies also showed evidence for the presence of hydrophobic force in soap films (Tchaliovska, et al. 1994; Yoon and Aksoy 1999; Wang and Yoon 2004). In these studies, it was assumed that at low surfactant concentrations, equilibrium film thicknesses (H_e) are determined as a result of the equilibrium between the electrostatic, van der Waals and hydrophobic forces. One can then determine the magnitude of the hydrophobic force from the values of H_e measured using the TFB technique and the electrostatic and van der Waals forces. Using this method, Yoon and Aksoy (1999) showed that the hydrophobic force increases with decreasing concentration of dodecylammonium hydrochloride. Similar results were also obtained with sodium dodecyl sulfate (Wang and Yoon 2004). The method of determining the contributions from hydrophobic force in this manner may have a degree of uncertainty due to the possible inaccuracies involved in determining the electrostatic force. Therefore, Angarska, et al. (2004) determined the hydrophobic forces from the critical rupture thicknesses (H_{cr}) measured at a high electrolyte concentration, where the electrostatic repulsive force can be neglected. The results obtained with the foam films stabilized by SDS showed the presence of long-range hydrophobic forces with decay length of 15.8 nm, although the pre-exponential parameters are much smaller than those measured between macroscopic solid surfaces.

In the present work, the TFB technique was used to measure H_e as a function of SDS in the presence of electrolyte (NaCl). At low electrolyte concentrations, the foam films were stable; therefore, it was possible to measure H_e and use the results to determine the hydrophobic forces as a function of SDS concentration. The hydrophobic forces determined from the H_e measurements were then used to predict the rate of film thinning using the Reynolds equation. At a high electrolyte concentration, however, the films were unstable. Therefore, the hydrophobic forces were calculated from the film thinning data using the Reynolds equation. The results are used to discuss the role of hydrophobic force in film thinning.

THEORETICAL MODEL

Surface Potentials at Air-Water Interface

In the presence of electrolyte (e.g., NaCl), the surface excess (Γ_s) for the surfactant ions (DS^-) was calculated using Gibbs adsorption equation from surface tension data. The charge density (σ_0) and the apparent surface potential (ψ_0) at the air/water interface can then be calculated using the method reported by Tchaliovska et al. (1994) and Yoon and Aksoy (1999), assuming that SDS molecules are fully dissociated at the interface.

It is unlikely, however, that the DS^- ions are fully ionized at the air/water interface due to the counterions adsorbed in Stern layer. The counterion 'binding' should cause the charge density to decrease, and the net charge density, σ is equal to $\sigma_0 + \sigma_c$, where σ_c is the charge density of counterions (Na^+) at the Stern layer. The Stern potential (ψ_s) was determined by this net charge density σ at the Stern layer, given by the following equation at 25°C (Wang and Yoon, 2004):

$$0.2078\sqrt{C_{SDS} + C_{NaCl}}\sinh(19.489\psi_s) + \psi_s - \psi_0 = 0 \quad \textbf{(EQ 1)}$$

This relation can be used to obtain the values of ψ_s from ψ_0, the molar concentration of SDS (C_{SDS}) and the molar concentration of NaCl (C_{NaCl}). The potentials ψ_0 and ψ_s are in volts.

One can also determine the surface excess of the counterions (Γ_c) as follows:

$$\Gamma_c = \Gamma_s\sqrt{\frac{8000kT\varepsilon_0\varepsilon_r(C_{sds} + C_{NaCl})}{e^2N_A}}\sinh\left(\frac{e\psi_s}{2kT}\right) \quad \textbf{(EQ 2)}$$

where ε_0 and ε_r are the permittivity of vacuum and the dielectric constant of water, respectively, e the electronic charge, N_A Avogadro's number, k the Boltzmann's constant and T the absolute temperature. This relation is useful for understanding the interaction mechanism between the surfactant and the counterions.

Determination of Hydrophobic Force Constant

During the process of film thinning, two air/liquid interfaces are brought to each other in closer ranges. The surfaces forces emanating from both interfaces interact with each other and give rise to a disjoining pressure (Π). At low surfactant concentrations, it was represented by the extended DLVO theory (Yoon and Aksoy 1999):

$$\Pi = 64(C_{sds} + C_{NaCl})RT\tanh^2\left(\frac{ze\psi_s}{4kT}\right)\exp(-\kappa H) - \frac{A_{232}}{6\pi H^3} - \frac{K_{232}}{6\pi H^3} \quad \textbf{(EQ 3)}$$

where R is the gas constant, z the valence of the electrolyte, κ^{-1} the Debye length, A_{232} the Hamaker constant, and K_{232} the hydrophobic force constant representing the hydrophobic interaction between two gas phases (or bubbles) 2 interacting with water 3 at a separation distance of H.

At equilibrium, the disjoining pressure should be equal to the capillary pressure (P_c). Thus,

$$64(C_{sds} + C_{NaCl})RT\tanh^2\left(\frac{ze\psi_s}{4kT}\right)\exp(-\kappa H_e) - \frac{A_{232}}{6\pi H_e^3} - \frac{K_{232}}{6\pi H_e^3} - \frac{2\gamma}{r_c} = 0 \quad \textbf{(EQ 4)}$$

where H_e is the equilibrium film thickness, γ is the surface tension, and r_c (=2.0 mm) is the radius of the film holder. Eq. 4 can be used to determine the values of K_{232} from the values of H_e, γ, ψ_s, and Hamaker constant ($A_{232} = 3.7 \times 10^{-20}$ J) (Israelachvili 1992).

Kinetics of Film Thinning

We studied the kinetics film thinning using the Reynolds equation (Scheludko 1967):

$$-\frac{dH}{dt} = \frac{2H^3 \Delta P}{3\mu R_f^2} \qquad \textbf{(EQ 5)}$$

where H is the film thickness, t the film drainage time, μ the bulk viscosity, R_f the film radius, and ΔP the driving force. Equation 5 is applicable for tangentially immobile films (i.e., under no slip conditions). In the present work, we used the viscosity of water (0.900 cP at 25°C) (Lide and Kehiaian 1994) for μ. We also considered that the driving force for film thinning is determined by the following relation:

$$\Delta P = P_c - \Pi \qquad \textbf{(EQ 6)}$$

In the present work, Eq. 5 was used to obtain H vs. t curve by integrating it at 0.1 nm intervals of H using the central scheme of the Euler integration method. For the integration, we used the value K_{232} determined using Eq. 4 from H_e measurements. When studying unstable films, it was not possible to determine H_e and hence K_{232}. In this case, the integration was carried out to fit the film thinning kinetics data with K_{232} as the only adjustable parameter.

EXPERIMENTS

Materials

Specially pure sodium dodecyl sulfate (SDS) was obtained from BDH and recrystallized from ethanol. Its surface tension isotherm showed no minimum near cmc, indicating that the surfactant has no significant impurities. A Nanopure water treatment unit was used to obtain double-distilled and deionized water with a conductivity of 18.2 MΩcm^{-1}. A high-purity sodium chloride (99.99%) from Alfa Aesar was used as an electrolyte.

Film Thickness and Disjoining Pressure Measurements

We used the thin film balance (TFB) technique to measure the kinetics of film thinning formed in a Scheludko cell (Scheludko 1967; Exerowa and Kruglyakov 1998). The inner radius of the film holder (r_c) was 2.0 mm. Following the recommendation by Exerowa et al. (1979), the inner wall of the film holder was scratched with a sharpening stone to improve wettability. A PC-based data acquisition system was used to record the intensities of the reflected lights, from which film thicknesses were obtained using the microinterferometric technique (Scheludko 1967). We used a rather strong surfactant, SDS, to obtain tangentially immobile film, while the film radii (R_f) were controlled to be well below 0.1 mm. These conditions were necessary to minimize hydrodynamic or thermal fluctuations so that the experimental results can be compared with the Reynolds lubrication theory. In the present work, the film radii were controlled within the 0.055–0.080 mm range. The error associated in measuring R_f was ±0.01 mm.

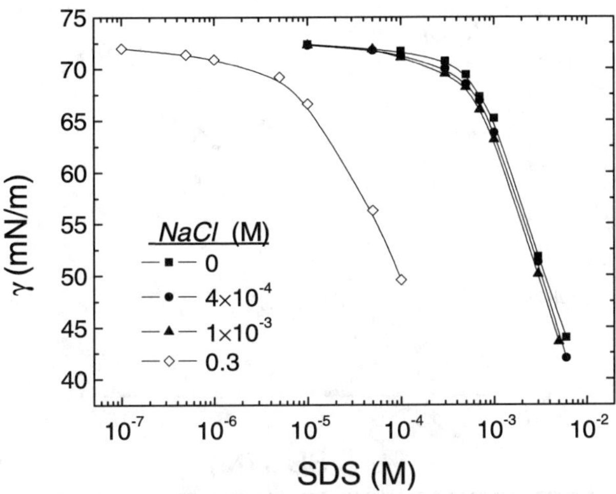

FIGURE 1 Surface tensions of SDS solutions in the presence and absence of NaCl

Scheludko cells with 1.0 and 2.0 mm radii were used to measure the disjoining pressures below 150 Pa, while a bike-wheel microcell (Cascão Pereira et al. 2001) with a hole diameter of 0.75 mm was used to measure the disjoining pressure above 150 Pa. The bike-wheel film holder was reusable, had a uniform liquid drainage inside, and exhibited an air/water entry pressure of about 10 KPa. Special care was taken to minimize pressure disturbances, as we were interested in measuring metastable films at low surfactant concentrations. A nanomover was used to drive a 20 ml gas-tight syringe and change the gas pressure in a TFB cell.

Surface Tension

The surface tension isotherms of the SDS solutions were measured using the pendant drop method both in the presence and absence of NaCl. Each surfactant solution was prepared at least one day prior to experiment and used within one week to minimize hydrolysis.

RESULTS AND DISCUSSION

Figure 1 shows the surface tension isotherms for the SDS solutions at various concentrations of NaCl. As predicted by the Gouy-Chapman theory, added salt systematically decreases surface tension by promoting surfactant adsorption since salt can decrease the free energy of forming a charged monolayer with increasing ionic strength (Persson et al. 2003).

The Stern potentials (ψ_s) at the air/water interface were calculated using Eq. 1 and are given in Figure 2. The Stern potentials (ψ_s) decreased systematically with increasing NaCl concentration, mainly due to the increase in the ionic strength.

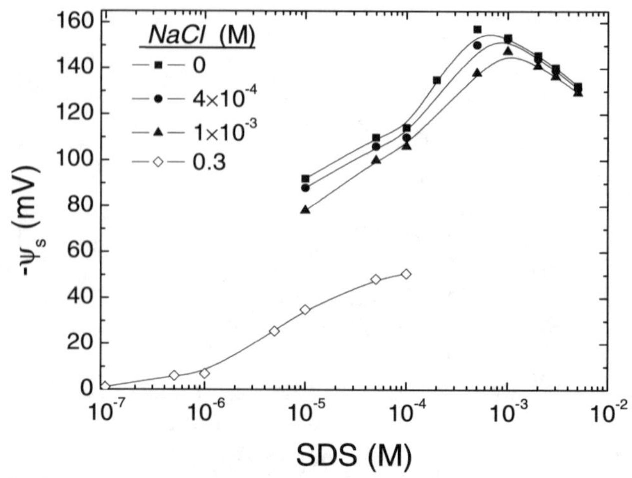

FIGURE 2 The Stern potential at the air/water interface as calculated from Eq. 1

At low concentrations of NaCl, where SDS concentration became higher than the fixed NaCl concentrations, the three curves tend to overlap with each other. This can be attributed to the fact that the ionic strengths employed in the three sets of experiments were close to each other. Likewise, the sharp decrease in ψ_s above 10^{-3} M SDS can be attributed to the increase in ionic strength (Kralchevsky et al. 1999). At 0.3 M NaCl, the increase in SDS concentration has little impact on the ionic strength; therefore, the magnitude of the Stern potential increased monotonically with increasing SDS concentration due to surfactant ion adsorption.

Figure 3 shows the equilibrium film thicknesses (H_e) measured at varying SDS concentrations both in the presence and absence of NaCl. As shown, H_e decreased sharply with increasing electrolyte concentration due to double layer compression. Note, however, that H_e increased slightly as the SDS concentration increased from 10^{-6} to 4×10^{-6} M in the absence of NaCl. Likewise, H_e increased slightly as SDS concentration increased in the presence of NaCl. These increases were observed only at low surfactant concentrations before ionic strength became an overriding factor. This phenomenon may possibly be caused by a decrease in hydrophobic force with increasing SDS concentration. The presence of an attractive hydrophobic force should cause H_e to decrease. As will be shown later, the hydrophobic force decreases with increasing SDS concentration, which should cause H_e to increase.

The data presented in Figures 2 and 3 can be used to determine the values of K_{232} using Eq. 4. The K_{232} values obtained in this manner are given in Figure 4. As shown, K_{232} decreased steadily with increasing SDS concentration. In the presence of NaCl, K_{232} decreased by orders of magnitudes. K_{232} was reduced close to or below the value of A_{232} at 0.4 mM NaCl and above 5×10^{-4} M SDS. At a higher (1 mM) NaCl concentration, the same was observed at a 10-fold lower SDS concentration (5×10^{-5} M). Thus, electrolyte addition is an effective means of reducing the hydrophobic force in soap films.

Figure 5 shows a series of disjoining pressure isotherms measured in the present work at 10^{-4} M SDS in the absence and presence of NaCl. These rupture thicknesses of the foam films containing no NaCl or 0.4 mM NaCl were larger than predicted by the

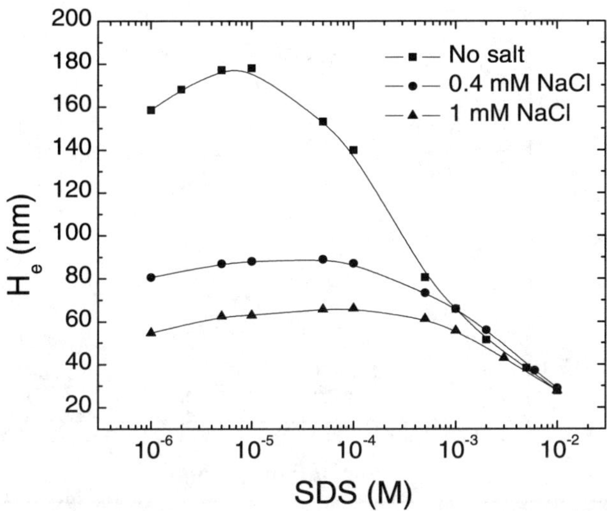

FIGURE 3 Equilibrium film thicknesses as a function of SDS concentration at 0, 0.4 and 1 mM NaCl; pH 5.7–6.0; and 25±0.1°C

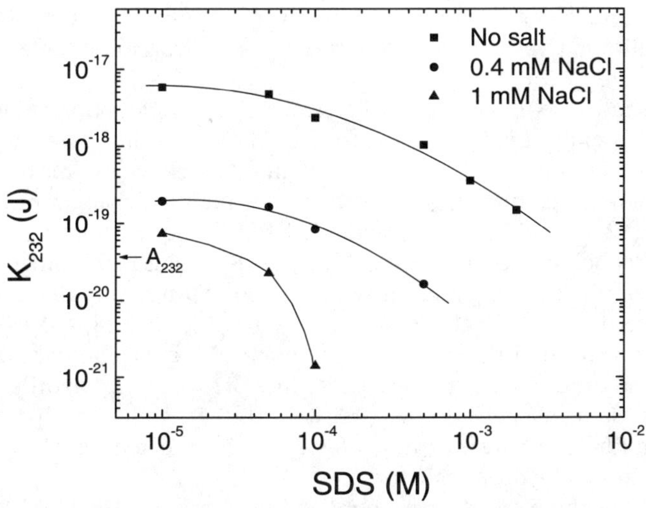

FIGURE 4 K_{232} of Eq. 4 as a function of SDS concentration at 0, 0.4 and 1 mM NaCl

DLVO theory, represented by the dotted lines. Also, the measured pressures were considerably lower than predicted by the theory, indicating the presence of hydrophobic force. The solid lines in Figure 5 represent the extended DLVO theory (Eq. 3), which includes contributions from the hydrophobic force. The extended DLVO fit was made with corresponding values of ψ_s (see Figure 2) and K_{232} (see Figure 4). The same values of ψ_s and Debye length (κ^{-1}) were used for the DLVO fit.

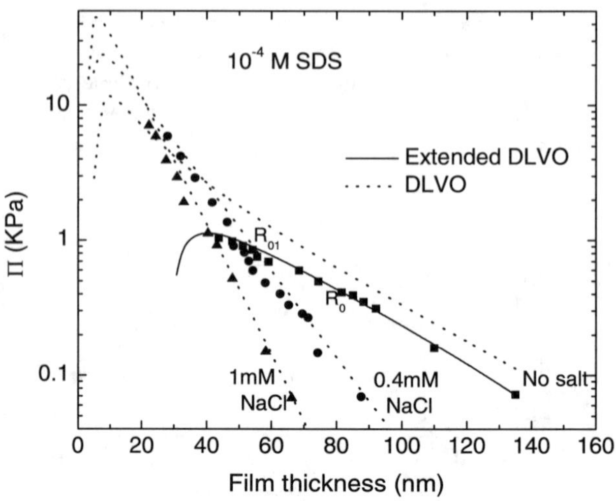

FIGURE 5 Disjoining pressure isotherms obtained in 10^{-4} M SDS solutions at different NaCl additions

The results show that as the NaCl concentration increased, the films became thinner due to double layer compression, and yet the film became more stable. The increased stability may be due to the reduction in hydrophobic force with increasing NaCl concentration.

Figure 6a shows the film thickness *(H)* versus time *(t)* plot obtained in a 10^{-5} M SDS solution in the absence of inorganic electrolyte. The radius of the flat film (R_f) that was used in the film thinning experiment was 0.08 mm. Also shown in this figure for comparison is the Reynolds equation (see Eq. 5) plotted using the driving forces *(P)* calculated with DLVO (Eq. 3 with K_{232}=0) and extended DLVO (Eq. 3) theories. At K_{232}=0 (i.e., there is no hydrophobic force in the soap film), the Reynolds equation greatly under predicted the rate of film thinning and the predicted equilibrium film thicknesses (H_e) were much larger than the experiment. At $K_{232}=5.8 \times 10^{-18}$ J, which is the value determined using the equilibrium film thickness measurement, there was an excellent fit between the predicted and experimental values. The values of K_{232} were two orders of magnitude larger than the Hamaker constant ($A_{232}=3.7 \times 10^{-20}$ J), which was probably the main reason that the film thinning was much faster than predicted without considering the contributions from the hydrophobic force.

Note here that the *H* vs. *t* curves obtained using the classical and extended DLVO theories fall on the same curve during the initial 2 seconds, when the film thickness is large and, hence, surface forces do not come into play as yet. At a large film thickness, capillary force is the overriding driving force for the film thinning process.

The next series of film thinning experiments were conducted at a higher SDS concentration (10^{-4} M) and in the presence of 4×10^{-4} M NaCl. The results are given in Figure 6b. At such high surfactant and electrolyte concentrations, there was an excellent fit between the Reynolds equation coupled with the DLVO theory (dotted line) and the experiment (circles). Also shown in this figure is the Reynolds equation plotted with *P*

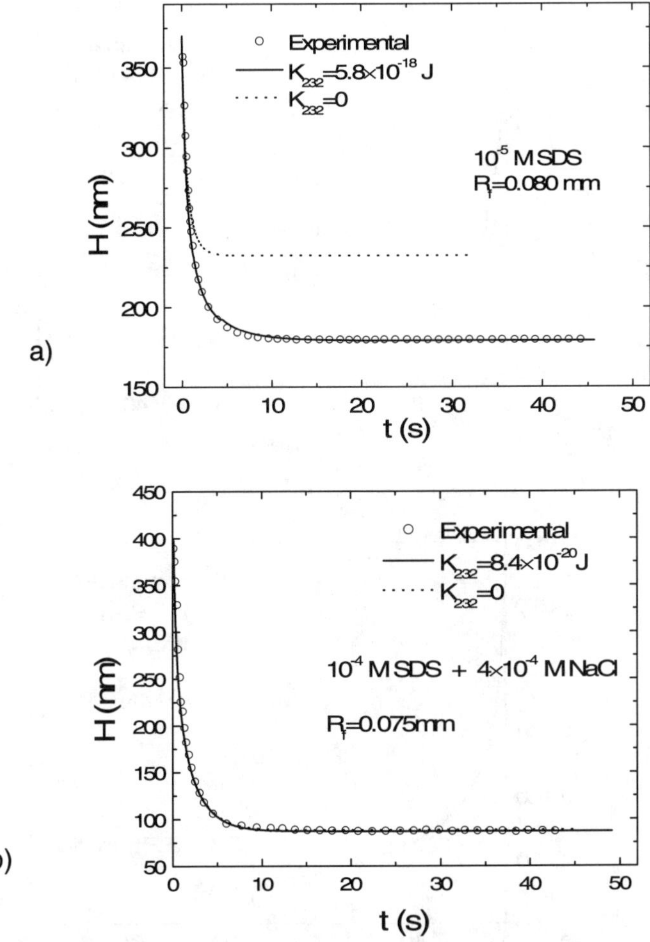

FIGURE 6 Kinetics of film thinning at a) 10^{-5} M SDS; b) 10^{-4} M SDS and 4×10^{-4} M NaCl. The solid line represents the Reynolds equation with the extended DLVO theory, while the dotted line represents the same with the DLVO theory.

corrected for the hydrophobic force with K_{232}= 8.4×10^{20} J (solid line). All three plots fall essentially on the same curve, indicating that at such high SDS and NaCl concentrations, the contribution to from the hydrophobic force to the total disjoining pressure is minimal. If it exists, the hydrophobic force is very weak, as K_{232} is only 2.3 times larger than A_{232}.

At a high electrolyte concentration (e.g., 0.3 M), electrical double layer force is effectively screened so that $\Pi_{el} \approx 0$ and the films become unstable. Figure 7 shows two film thinning curves obtained at a high electrolyte concentration. At 0.3 M NaCl and 5×10^{-7} M SDS, the film ruptures at 28.2 nm in 15.7 s. At 0.3 M NaCl and 10^{-4} M SDS, the film became more stable owing to the higher surfactant concentration, and ruptures at 23.3 nm in 30.9 s to be transformed to a Newton black film (NBF)s. The transition from

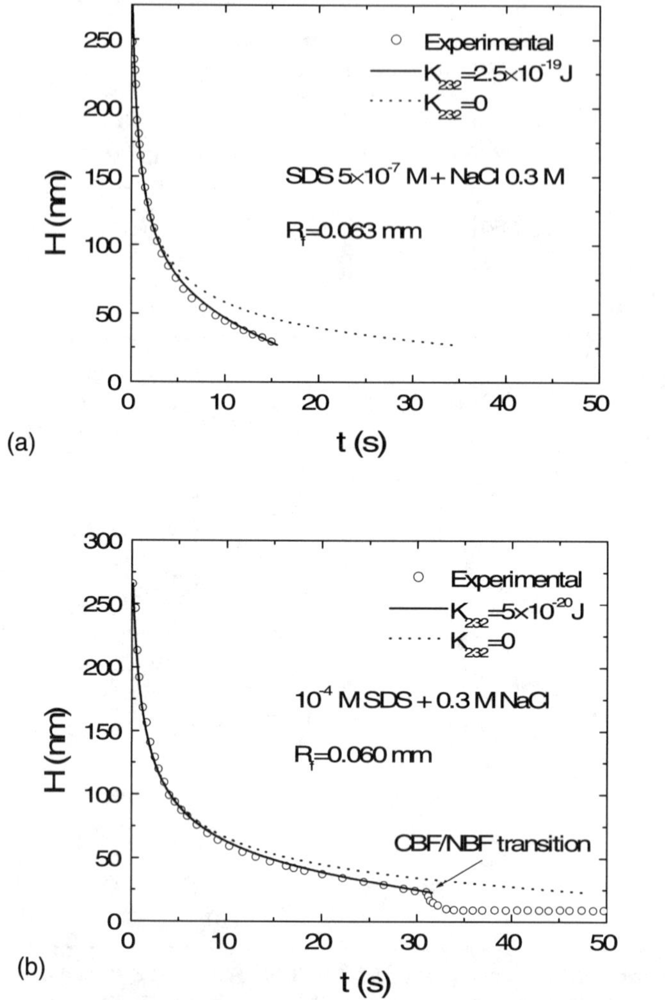

FIGURE 7 Kinetics of film thinning at a) 5 × 10^{-7} M SDS and 0.3 M NaCl; b) 1 × 10^{-4} M SDS and 0.3 M NaCl. The solid line represents the Reynolds equation with the extended DLVO theory, while the dotted line represents the Reynolds equation with the DLVO theory.

common black (CBF) film to NBF occurred above 2 × 10^{-5} M SDS at 0.3 M NaCl. The stability of NBF has been studied extensively, and a good review was given by Exerowa and Kruglyakov (1998).

In Figure 7, the dotted lines represent the Reynolds theory (Eq. 5) coupled with the DLVO theory. Obviously, there is a discrepancy between the experiment and theory. When using the extended DLVO theory (Eq. 3), the fit was much better. At 5 × 10^{-7} M SDS and 0.3 M NaCl, we used the value of K_{232}=2.5 × 10^{-19} J, which was obtained by curve fitting. At 10^{-4} M SDS and 0.3 M SDS, we used K_{232}=5.0 × 10^{-20} J to fit the film thinning data obtained before the CBF/NBF transition.

The results presented hitherto showed evidence for the hydrophobic force in soap films. It has been shown that the hydrophobic force, whose magnitude is represented by K_{232}, varies with SDS and NaCl concentrations. In the presence of 0.3 M NaCl, the values of K_{232} decreased from 3.3×10^{-19} J to 5×10^{-20} J as the surfactant concentration was increased from 10^{-7} to 10^{-4} M. These results are consistent with those reported by Angarska et al. (2004). Note here that the value of K_{232} obtained at 10^{-5} M SDS and 0.3 M NaCl was approximately 45 times lower than that obtained at the same surfactant concentration but without the salt. At 10^{-4} M SDS, the presence of 0.3 M NaCl reduced the hydrophobic force to the level of the van der Waals force (3.7×10^{-20} J). Thus, the presence of electrolyte has a profound impact on the hydrophobic forces in foam films.

It has also been shown throughout this investigation that K_{232} changes with the adsorption of DS^- and its counterions (Na^+). Therefore, the values of K_{232} are plotted in Figure 8 as a function of surface access of the DS^- and Na^+ ions. The data were plotted as a function of $(\Gamma_s + \Gamma_c)^{-1}$, which was referred to as 'inverse ionic adsorption' by Angarska et al.(2004). These investigators showed that the hydrophobic force parameter, B, increases with increasing inverse ionic adsorption. Likewise, one can see that $K_{232,}$ a measure of hydrophobicity, increases with increasing inverse ion adsorption. Figure 8 shows two sets of data plotted, one in the presence of SDS alone and the other in the presence of SDS and 0.3 M NaCl. In both cases, there are three distinct zones. In Zone I, DS^- ions form a close-packed monolayer of hydrocarbon chains at the air/water interface, in which case K_{232} decreases precipitously with increasing SDS concentration. In Zone II, the surfactant adsorption is significant but they are not well ordered. According to Eriksson and Ljunggren (1989), however, the surfactant may still adsorb as clusters as a means of minimizing free energy. In Zone III, the surfactant adsorbs individually at the air-water interface, which may be referred to as gaseous state in a sense that the surfactant solution follows the Henry's law.

The most striking future of Figure 8 is that electrolyte (NaCl) has a profound impact on K_{232}. According to the K_{232} vs. $(\Gamma_s + \Gamma_c)^{-1}$ plot, the higher the Na^+ ion adsorption as counterions is, the lower the hydrophobic force becomes. Therefore, the decrease in hydrophobic force in the presence of electrolyte may be attributed to the positive adsorption of the electrolyte at the air/water interface. It appears, however, that electrolyte can also reduce the hydrophobic force between gas bubbles under conditions of negative adsorption. It has been shown by Craig et al. (1993) that coalescence of gas bubbles in water is reduced substantially in the presence of various electrolytes. One possible explanation for this phenomenon may be that electrolytes may cause a decrease in the cohesive energy of water and, hence, the hydrophobic force. As suggested by van Oss (1994), hydrophobic interaction arise from the strong cohesive energy of water (-102 mJ/m^2). Consider two nano-size gas bubbles suspended in water. Since water molecules cannot form strong hydrogen bonds with them, the bubbles will be push away form the surrounding water molecules, which may be manifested as hydrophobic interaction (or force) between the nona-size bubbles. This explanation is similar to the mechanism for the hydrophobic effect between hydrocarbon chains (Tanford 1980; Lazaridis 2001). In the presence of electrolytes, the cohesive energy of water may be reduced, resulting in a reduced hydrophobic force.

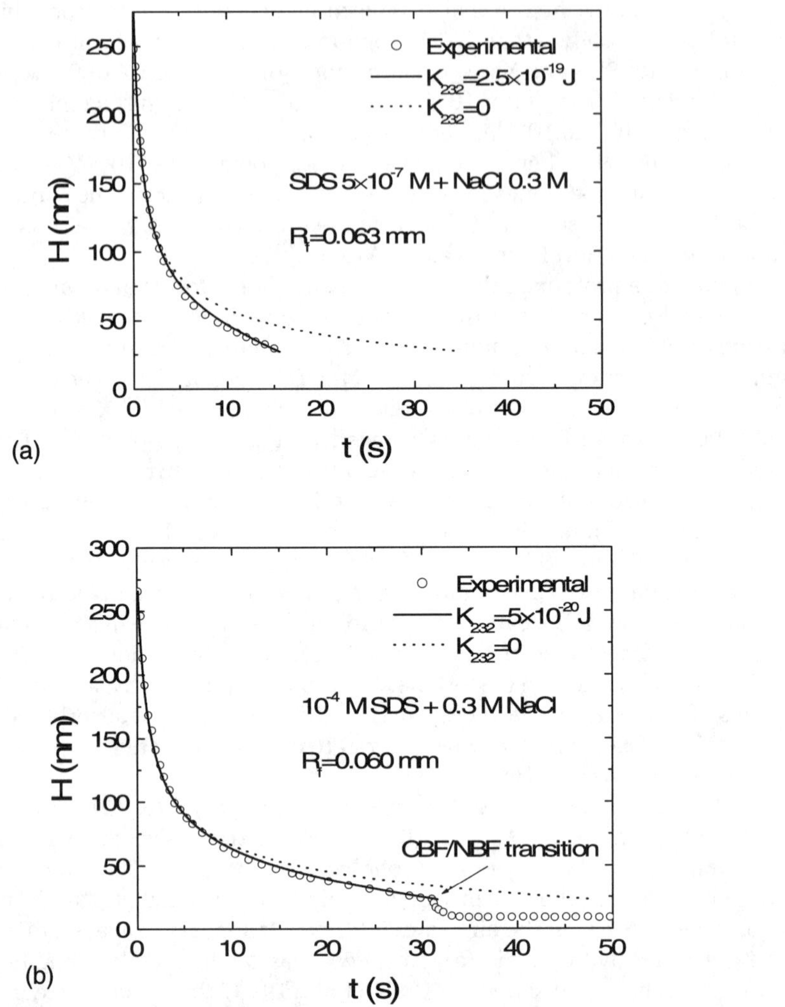

FIGURE 8 K_{232} **versus** $(\Gamma_s + \Gamma_c)^{-1}$ **as a function of SDS concentration in the absence and presence of 0.3 M NaCl**

CONCLUSIONS

A counter ion binding model was used to calculate the electrical surface potential $_s$ at the air/water interface in the presence of SDS and NaCl. The kinetics of film thinning and the equilibrium film thickness (H_e) of the horizontal foam films formed in a Schludko cell were measured using thin film balance (TFB) technique. The values of ψ_s and H_e were used to determine the magnitudes of hydrophobic force constants (K_{232}) using the extended DLVO theory. It has been found that K_{232} decreases with increasing SDS and NaCl concentrations. The K_{232} values obtained using the TFB technique allowed us to predict disjoining pressure isotherms, which were in excellent agreement with those measured using a bike-wheel film holder.

By comparing the film thinning data (*H-t*) with those predicted from the Reynolds equation, it was possible to identify the factors affecting the film thinning process. During the initial stages, the process is controlled by the capillary pressure. During the later stages, it is controlled by surface forces. At low SDS and NaCl concentrations, it was necessary to consider the contributions from the hydrophobic force. There was an excellent agreement between the theory and experiment when using the values of K_{232} determined from H_e measurement.

At a high electrolyte concentration (i.e., 0.3 M), where it was difficult to determine H_e, the values of K_{232} were determined by fitting the film thinning data to the Reynolds equation. The logarithms of the K_{232} values plotted versus inverse ionic adsorption $(\Gamma_s + \Gamma_c)^{-1}$ identified three distinct zones. The plots showed that the hydrophobic force decreases with increasing adsorption of the surfactant and its counterions, the decrease being accelerated with increasing adsorption densities.

REFERENCES

Angarska, J.K., B.S. Dimitrova, K.D. Danov, P.A. Kralchevsky, K.P. Ananthapadmanabhan, and A. Lips. 2004. Detection of the hydrophobic surface forces in foam films by measurements of the critical thickness of the film rupture. *Langmuir.* 20:1799.

Cascão Pereira, L.G., C. Johansson, H.W. Blanch, and C.J. Radke. 2001. A bike-wheel microcell for measurement of thin-film forces. *Colloids and Surfaces, A: Physicochem. Eng. Aspects.* 186:103.

Craig, V.S.J., B.W.Ninham, and R.M. Pashley. 1993. The effect of electrolytes on bubble coalescence in water. *J. Phys. Chem.* 97:10192.

Deschenes, L., A. P. Zilaro, L.J. Muller, J.T. Fourkas, and U. Mohanty. 1997. Quantitatively measure of hydrophobicity: experiment and theory. *J. Phys. Chem.* 101:5777.

Eriksson, J.C. and S. Ljunggren. 1989. A molecular theory of the surface tension of surfactant solutions. *Colloids and Surfaces.* 38:179.

Exerowa, D. and P.M. Kruglyakov. 1998. *Foam and Foam Films*. Amsterdam: Elsevier.

Exerowa, D., M. Zacharieva, R. Cohen, and D. Platikanov. 1979. Dependence of the equilibrium thickness and double layer potential of foam films on the surfactant concentration. *Colloid and Polymer Sci.* 257:1089.

Israelachvili, J. N. 1992. *Intermolecular and Surface Forces*. London: Academic press.

Kralchevsky, P.A., K.D. Danov, G. Broze, and A. Mehreteab. 1999. Thermodynamics of ionic surfactant adsorption with account for the counterion binding: effect of salts of various valency. *Langmuir.* 15:2351.

Lazaridis, T. 2001. Solvent size vs. cohesive energy as the origin of hydrophobicity. *Accounts of Chemical Research.* 34:931.

Lide, D.R. and H. V. Kehiaian. 1994. *CRC Handbook of Thermophysical and Thermochemical Data*. CRC Press, Boca Raton, FL.

Persson, C., A. Jonsson, M. Bergström, and J.C. Eriksson. 2003. Testing the Gouy–Chapman theory by means of surface tension measurements for SDS–NaCl–H_2O mixtures. *Journal of Colloid and Interface Sci.* 267:151.

Scheludko, A. 1967. Thin liquid films. *Advances in Colloid and Interface Sci.* 1:391

Sheludko, A. and D. Exerowa. 1959. Über den elektrostatischen druck in schaumfilmen aus wasseringen elektrolytlösungen. *Kolloid-Z.* 165:148.

Tanford, C. 1980. *The Hydrophobic Effect: Formation of Micelles and Biological Membanes*. 2nd ed., New York: Wiley-Interscience.

Tchaliovska, S., E. Manev, B. Radoev, J. C. Eriksson, and P.M. Claesson. 1994. Interactions in equilibrium free films of aqueous dedecylammonium chloride solutions. *J. Colloid Interface Sci.* 168:190.

Van Oss, C.J. *Interfacial Forces in Aqueous Media*. New York: Marcel Dekker, 1994.

Wang, L. and R.-H. Yoon. Hydrophobic forces in the foam films stabilized by sodium dodecyl sulfate: effect of Electrolyte. *Langmuir*. in press.

Yoon, R.-H. and B.S. Aksoy. 1999. Hydrophobic forces in thin water films stabilized by deodecylammonium chloride. *J. Colloid Interface Sci.* 211:1.

. . . .

Adsorption and Conformation of Polysaccharide Depressants on Minerals

P. Somasundaran,* Jing Wang,* and D.R. Nagaraj†

Polysaccharides have been widely used as depressants in mineral processing for decades. They constituent one of the most abundant family of biopolymers. However, only a small number of them have been used by industry so far because of a general lack of understanding of interactions between these polymers and minerals. In this work, adsorption tests, electrophoretic mobility measurements, microflotation and fluorescence spectroscopic tests were performed along with computer simulation to explore the mechanistic aspects of guar gum adsorption and conformation on talc as a function of relevant variables such as ionic strength, pH, solvent composition and polymer molecular weight. From microflotation tests, it was found that talc hydrophobicity decreased markedly due to the adsorption of guar gum even at very low concentrations and reached a plateau below maximum adsorption. Guar gum adsorption on talc was found to be not affected significantly by changes in solution conditions such as pH and ionic strength, ruling out electrostatic force as controlling factor for adsorption. Fluorescence spectroscopic studies showed no evidence of the formation of hydrophobic domains at talc-aqueous interface. Also urea, a hydrogen bond breaker, reduced the adsorption of guar on talc. All of the above results suggest that one of the main driving forces for guar adsorption on talc is hydrogen bonding in preference to electrostatic or hydrophobic force. This information should help toward designing reagents of optimum molecular architecture for efficient depression, dispersion or other such surface modification of particles.

* NSF Industry/University Cooperative Research Center for Advanced Studies in Novel Surfactants, Columbia University, New York

† Cytec Industry, 1937 West Main Street, Stamford, Connecticut

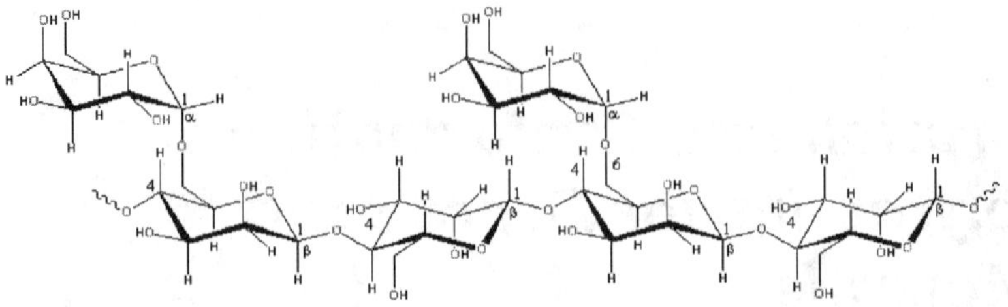

FIGURE 1 The monomeric structure of guar gum

INTRODUCTION

Talc is a layered hydrous magnesium silicate in which the layers are held together by van der Waals bonds. During grinding, two different surfaces are formed. Basal cleavage planes are formed by rupture of the van der Waals bonds. Because the resulting planes do not contain any broken Si-O and Mg-O bonds, the basal surface is neutral and hydrophobic. However, the edges of the mineral sheets contain broken Si-O and Mg-O bonds and, consequently, charged species and the edges thus exhibit hydrophilic properties.

Guar gum is a natural nonionic polysaccharide with an average molecular weight of 100,000 to 2000,000. The repeating unit of guar gum is shown in Figure 1. Each unit contains nine OH groups. These OH groups are available for hydrogen bonding with the guar gum molecule to the mineral surfaces (J.M.W.Mackenzie, 1980).

In the flotation of platinum bearing ores, guar gum is often used as a depressant for talceous gangue minerals. With a large number of studies conducted in the past, the mechanism of adsorption of guar gum on talc is well disputed. Mackenzie (1986), Pugh (1989), Healy (1974) and Rath (1997a, 1997b) proposed that the mechanisms governing the adsorption of polymers on mineral surfaces include hydrophobic interaction, hydrogen bonding, chemical and electrostatic interactions. However, the reason for the selectivity of the adsorption of depressants on minerals has not been accounted for. Steenberg et al. (1982, 1984) and Ralston et al. (1998) proposed that the adsorption of guar gum on talc occurs mainly at the basal planes via hydrophobic force. In contrast, Rath et al. (1995) proposed that the adsorption of guar occurs through hydrogen bonding on talc edges.

The objective of the present study was to clarify the mechanistic aspects of the interactions between guar gum and talc using a multipronged approach involving spectroscopic, electrostatic and adsorption measurements.

EXPERIMENTAL

Materials

Talc sample of 150–300 microns size used for the present tests was obtained from Cytec Industries. The nitrogen BET surface area measured using a Coulter Omnisorb 100 was 2.61 m^2/g after further grinding. X-ray diffraction (XRD) analysis indicated that it is a very pure sample with only a small amount of tremolite ($Ca_2Mg_5(Si_4O_{11})_2F_2$). Alumina

(AKP50, surface area: 10.9 m^2/g) obtained from Sumitomo Chemical Co. was also used in this work.

Unmodified guar gums (142C and IMP4) were obtained from Cytec Industries. GPC analysis showed that the weight average molecular weight of 142C and IMP4 were 1,450,000 and 242, 000 respectively. Their polydispersities were 1.5 and 25.9 separately. All guar gum stock solutions were prepared by quickly adding 0.045 g of gum powder into 45ml of water under vigorous stirring conditions and furthur stirring for 30 minutes. The solution was refrigerated overnight to ensure complete hydration or dissolution of guar gum and then filtered through filter paper (Whatman #4) to remove any undissolved impurities.

Experiments

Adsorption Measurements. The suspensions of talc, with ionic strength adjusted to the desired level using KCl, at 10% solid loading, were ultrasonicated for 30 minutes. It was then stirred magnetically for 2 hours after the pH was adjusted to the desired value using HCl and KOH. Guar gum stock solution was then added to the talc suspension and left overnight for conditioning. The suspensions were then centrifuged and the supernatants pipetted out for determination of the guar concentration using a Total Organic Carbon Analyzer. The adsorption density of guar gum on talc was calculated from the difference between the initial and the residual guar gum concentrations.

Colorimetric Method. A colorimetric method described by Dubois et al (1956) was used in experiments wherein other organic additives, such as urea, were used. 80% phenol and 5 ml 98% sulfuric acid were added to 2ml of supernatant obtained after centrifugation. After 4 hours of color development under warm conditions, UV absorbance was measured at a wavelength of 487.5nm. Adsorption density of guar on talc (or alumina) was calculated from the difference in absorbance.

Electrokinetic Measurements. Small amounts of talc was added to desired amounts of 10^{-3} M KNO_3 solution and ultrasonicated for 30 minutes, magnetically stirred for 2 hours and the pH adjusted using HCl and KOH. Finally, the guar gum stock solution was added and left for overnight conditioning with stirring. The zeta potential was then measured using a Zeta meter.

Fluorescence Studies. Suspensions of talc at 10% solid loading were ultrasonicated for 30 minutes and mixed for 2 hours with a magnetic stirrer. Guar gum stock solution with saturated pyrene in it or dansyl-labeled guar gum was then added to the talc suspension. The vials containing the suspensions were quickly wrapped with aluminum foil to keep the light away and left stirred overnight. Finally, the hydrophobicity of the supernatant and the solid were analyzed separately via a LS-1 fluorescence spectrophotometer (Photon Technology International).

Molecular Modeling. Molecular modeling (Macromodel) is powerful for exploring the conformation of polysaccharide in solvents. It is applied to study the conformation of this polymer.

Microflotation Test. Flotation tests were performed with a micro-flotation cell. The cell was first loaded with 10 g talc and then filled with aqueous polysaccharide solution. Frother (Betafroth 206) was used in the microflotation experiments following the polymer addition. After conditioning the mixture for 5 minutes, air was fed to the cell at a constant rate of 0.5 liter/min through the base, and the flotation tests were carried out

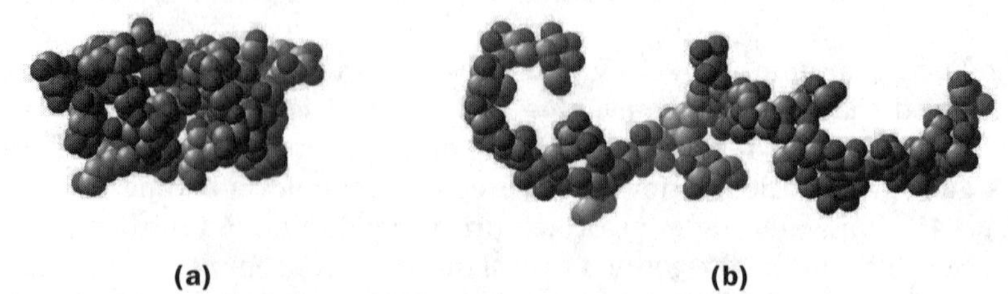

FIGURE 2 Computer simulated molecular CPK structure of guar in: (a) no solvent , (b) aqueous environment (each contains five repeat units)

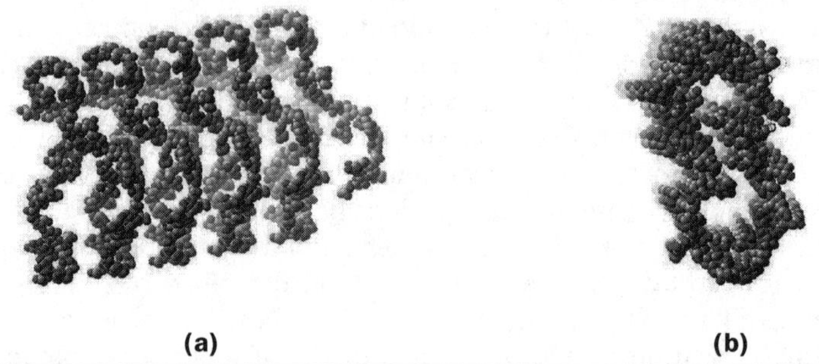

FIGURE 3 Computer simulated (a) latitudinal view and (b) longitudinal view of molecular CPK structure of CMC in aqueous environment (each contains 50 repeat units)

at room temperature for 4 minutes. Froth was removed from the head of the column using a stainless steel scrap.

RESULTS AND DISCUSSION

Molecular Modeling

Figure 2 shows the computer simulated molecular space filling structure of guar gum in both vacuum and aqueous environments. Each of them contains 5 repeat units. It is seen that polymer chains tend to be in a more stretched form of helices in water than in vacuum. The complex helical structure shown in Figure 3 is a piece of polymer chain containing 50 repeat units. It is clear that the OH groups tend to stick out of the helix. According to the simulation results, there is no change in the structure when the number of repeat units on the chain is increased further from 50 to 100.

Microflotation Tests

To study the effect of polymer adsorption on talc hydrophobicity, microflotation tests were conducted with guar gum/talc systems. The results show the talc hydrophobicity to decrease with guar adsorption at concentration below 50ppm, and stays almost constant above that. Surprisingly, molecular weight of guar gum has a very minor effect on the flotation of talc (Figure 4), which agrees with the adsorption results (Figure 6). The

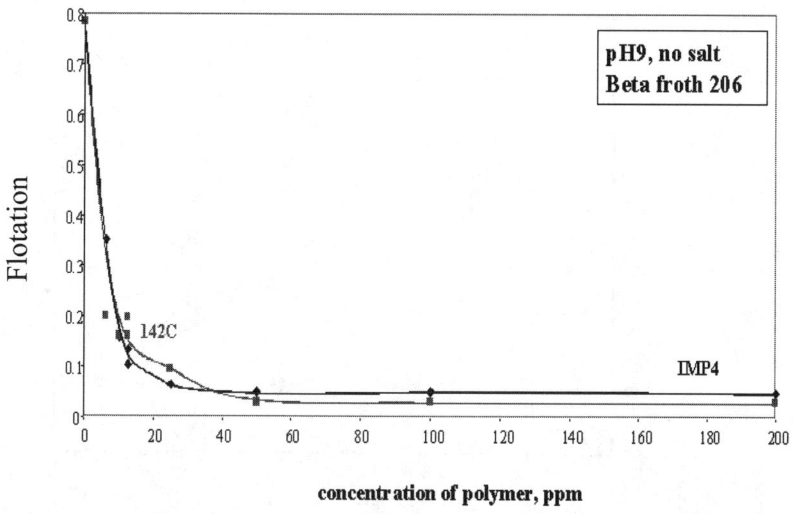

FIGURE 4 Microflotation of talc with adsorption of guar gums with different M.W.

number of polymer repeating units adsorbed on solid should be similar if guar choose flat conformation to adsorb on talc. The hydrophilicity of both surfaces should hence be almost the same.

Electrokinetic Studies

It is evident from the electrokinetic data shown in Figure 5 that the surface of talc is negatively charged and the isoelectric point is located around pH 2.5. Above pH 2.5, the electronegative character increases with increase in pH up to pH 7 and thereafter remains almost constant. Addition of different amounts of guar gum reduces, but does not reverse the negative zeta potential of talc, without any shift in the isoelectric point and does not reverse the potential. This reduction is attributed to the adsorption of guar gum on the surface of talc only to "mask" the negative charge on it. It is then possible that the electrostatic force is not the dominant driving force for the adsorption process.

Adsorption Studies

Influence of pH. Figure 6 shows the isotherms for the adsorption of guar gum on talc in 0.1M KCl solution at pH 4 and pH 9. It can be seen that the adsorption density does not vary measurably between pH 4 and 9. If electrostatic interaction plays an important role in the adsorption process, the adsorption density of guar gum on talc should have been altered since the surface charge of talc does change with change in pH from 4 to 9. Such is not the case here, and it can be concluded that electrostatic interaction is not the dominant force for the adsorption of guar gum on talc.

Influence of Ionic Strength. Adsorption isotherms of guar gum on talc at pH 9 are shown in Figure 7 at different ionic strengths. It is clear that the influence of ionic strength is also negligible, which further supports the suggestion that electrostatic force does not control the adsorption. If hydrophobic force makes a significant contribution to

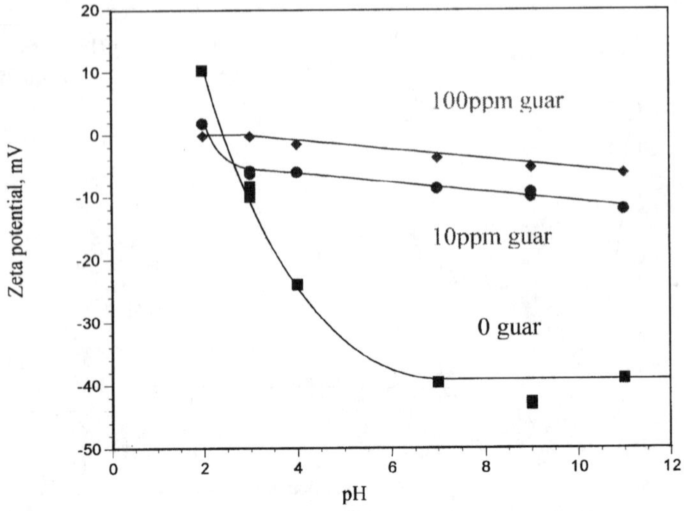

FIGURE 5 Zeta potential of talc as a function of pH in the presence and absence of guar gum, I.S. = 0.001

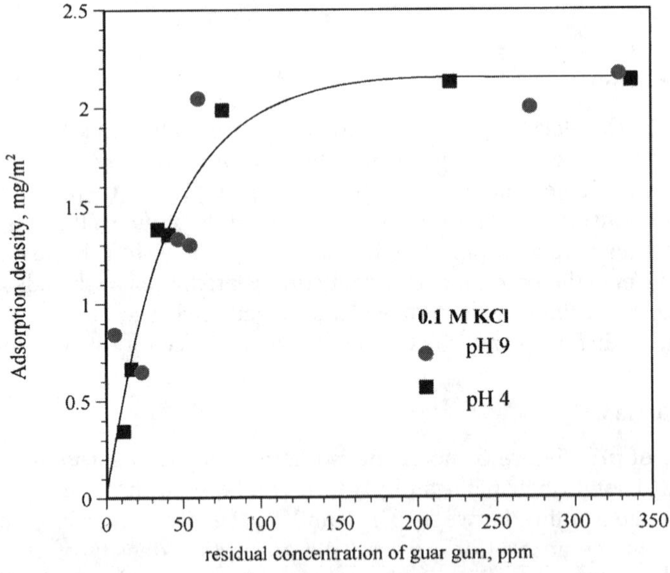

FIGURE 6 Adsorption isotherm of guar gum on talc at different pH

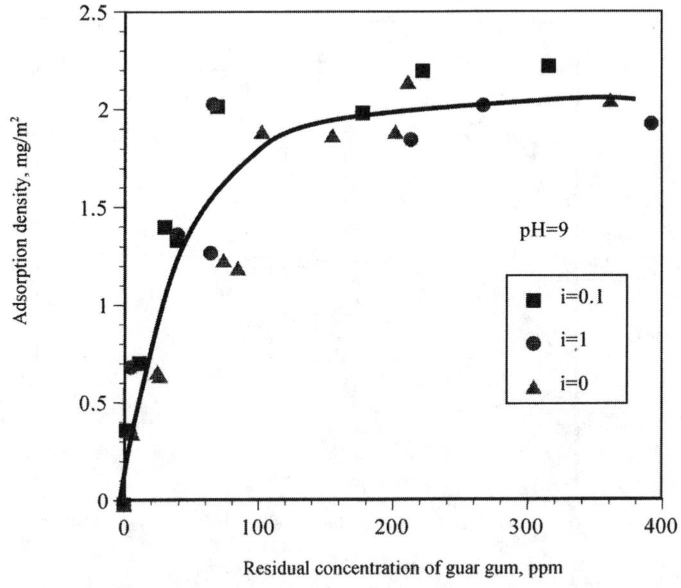

FIGURE 7 Adsorption isotherm of guar gum on talc with different ionic strength

adsorption of a polymer on a mineral, the adsorption density would increase with the addition of salt due to "salting-out" effect (Glenn Rexwinkel et al, 1999). Such an increase was not obtained here. Thus it would appear that the hydrophobic force also is not the main driving force for the adsorption of guar gum on talc.

Comparison of Adsorption on Talc and Alumina With and Without Urea. To explore the role of hydrogen bonding, adsorption tests were carried out in the presence of urea, a hydrogen bond breaker. Urea is a strong hydrogen bonding acceptor and it can affect the hydrogen bonding between the solid and the polymer in solution by preferential formation of hydrogen bonds between themselves and the polysaccharides and water. Tests were also done with alumina which is known to adsorb polysaccharides by hydrogen bonding (B.A.Jucker et al, 1997). Adsorption isotherms for guar on talc and alumina in the presence and absence of urea at pH 8.5 are shown in Figures 8 and 9 respectively. A comparison of the results presented in these figures shows that urea, a hydrogen bond breaker, reduces the adsorption of guar on talc to the same extent as that on alumina. Since as mentioned above, the driving force for the adsorption of guar gum on alumina is hydrogen bonding, it is suggested that hydrogen bonding plays a similar important role in the adsorption of guar on talc also.

Fluorescence Studies

To better understand the nature of guar gum adsorption, the microstructure of the adsorbed layer was probed using fluorescence spectroscopy. Fluorescence spectroscopy is a well-developed method (J.K.Thomas, 1984) for monitoring hydrophobic domain formation during adsorption. Fluorescence data for pyrene probe for the talc-solution interface (after guar adsorption) and for the supernatants are shown in Figure 10.

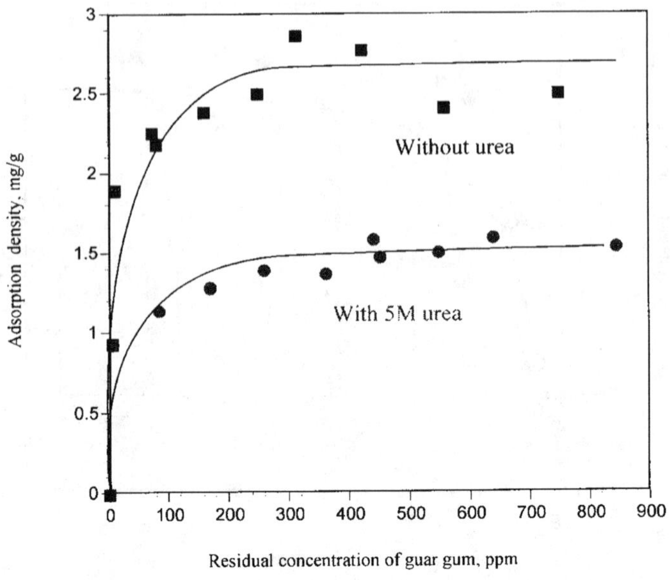

FIGURE 8 Adsorption isotherms of guar gum on talc with and without urea (I = 0)

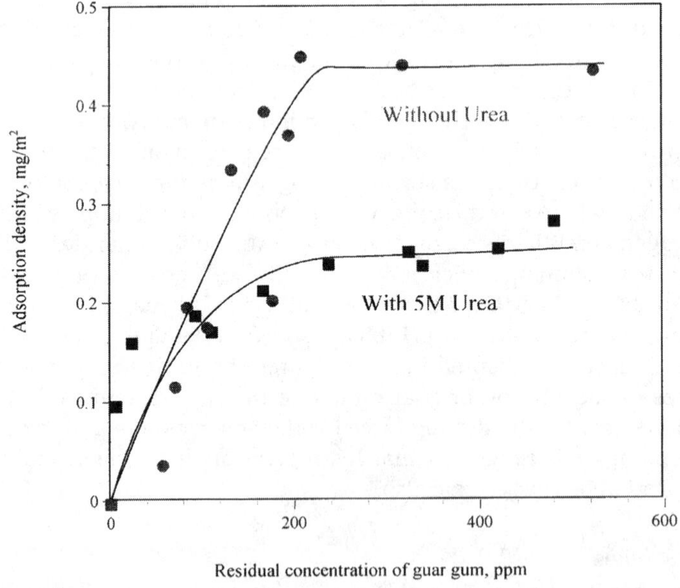

FIGURE 9 Adsorption isotherms of guar gum on alumina with and without urea (I = 0)

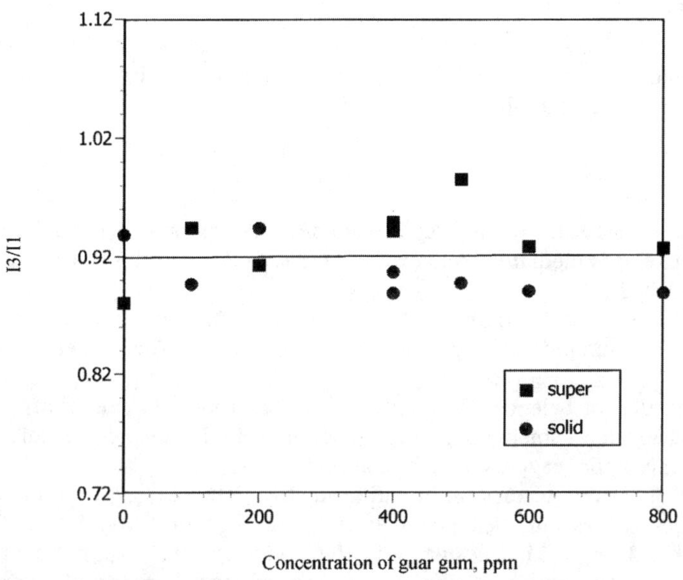

FIGURE 10 Pyrene fluorescence for guar gum and talc interface

Pyrene fluorescence showed no hydrophobic domain at the talc-aqueous interface because the polarity parameter remained almost constant at 0.92 for both the guar adsorbed talc and its supernatant with change in guar gum concentration.

CONCLUSIONS

1. Nonionic polysaccharide guar gum is an effective depressant for talc. Decrease in hydrophobicity of talc, as measured by flotation, occurred at a very low concentration of the guar gum and reached a plateau at far below the conditions corresponding to maximum polymer adsorption. Molecular weight of the guar gum showed a very minor effect on their depression capability.
2. Electrokinetic studies of adsorption of guar on talc in the pH range of 2–11 showed that guar gum decreased the negative zeta potential of talc but did not reverse it.
3. Adsorption studies also showed that the adsorption of guar gum on talc is not affected by changes in pH. These results suggest that the electrostatic force is not the dominant force for the adsorption process.
4. The adsorption of guar gum on talc was not affected by the ionic strength level. This also suggests that the electrostatic force does not play a role.
5. Pyrene fluorescence showed no hydrophobic domain formation for the adsorption of guar gum at talc-aqueous interfaces. Urea, a hydrogen bond breaker, reduces the adsorption of guar on talc to the same extent as that on alumina. This result also supports a mechanism involving hydrogen bonding rather than a hydrophobic one as the predominant mechanism.

ACKNOWLEDGEMENTS

The authors acknowledge the support of the National Science Foundation (Grant# CTS-00-89530 Dynamic Aggregation Behavior of Surfactant Mixture In Solutions at Solid/Liquid Interface) and Cytec Industries.

REFERENCES

B.A. Jucker, H. Harms, S.J. Hug and A.J.B. Zehnder, Adsorption of bacterial polysaccharides on mineral oxides is mediated by hydrogen bonds. Colloids and Surfaces B: Biointerfaces, 1997, 9, 331–343

Dubois, Michel, Gilles, K.A., Hamilton, J.K., Rebers, P.A. and Smith, Fred, Colorimetric method for determination of sugars and related substances. Anal. Chem., 1956, 28, 350–356

E. Steenberg, faculty of Science, University of Potchefstroom, Johannesburg, South Africa. PhD thesis: The depression of the natural floatability of talc: the mechanism involved in the adsorption of organic reagents of high molecular mass, 1982

E. Steenberg, P.J. Harris, Adsorption of carboxymethyl cellulose, guar gum and starch onto talc, sulphides, oxides and salt-type minerals. S. Afr. J. Chem., 1984, 37, 85–90

Glenn Rexwinkel, Bert B. M. Heesink, Wim P.M. van Swaaij, Adsorption of halogenated hydrocarbons from aqueous solutions by wetted and nonwetted hydrophobic and hydrophilic sorbents: equilibria. J. Chem. Eng. Data, 1999, 44, 1139–1145

J.K. Thomas, The chemistry of excitation at interfaces. ACS, Washington DC, 1984

J.M.W. Mackenzie, Guar-based reagents. EGMJ, 1980, October, 80–87

M. Mackenzie, in: D. Malhotra, W.F. Riggs, Chemical Reagents in the Mineral Processing Industry. Society of Mineral Engineerings, Colorado, USA, 1986, pp. 139–145

Paul Jenkins, John Ralston , The adsorption of a polysaccharide at the talc-aqueous solution interface. Colloids and Surfaces A, 1998, 139, 27–40

R.J. Pugh, Macromolecular organic depressants in sulfide flotation—a review, 1. principles, types and applications. Int. J. Miner. Process., 1989, 25, 101–130

R.K. Rath, S. Subramanian, J.S. Laskowski, in: J. S. Laskowski, G. W. Poling. Processing of Hydrophobic Minerals and Fine Coal. Canadian Institute of Mining, Metallurgy and Petroleum, Montreal, Canada, 1995, pp. 105

R.K. Rath, S.Subramanian, Study on adsorption of guar gum onto biotite mica. Minerals Engineering, 1997a, 10, 1405–1420

R.K. Rath, S.Subramanian, J.S. Laskowski, Adsorption of dextrin and guar gum onto talc. A comparative study. Langmuir, 1997b, 13, 6260–6266

T.W. Healy, Principles of adsorption of organics at solid-solution interfaces. J. Macromol. Sci. Chem. 1974, A8, 603–619

The Role of Sulfate Ions in the Flotation of Nonmetallic Minerals

Douglas W. Fuerstenau*

This paper concerns the specific adsorption of sulfate anions on nonmetallic minerals and its role in flotation. The specific surface affinity of sulfate ions in various systems has been observed in the results of electrokinetic measurements that show conditions under which sodium sulfate can reverse the sign of the zeta potential. The electrical double layer model provides a means for interpreting the results and for estimating the free energy of adsorption of sulfate ions. Because of their surface activity, sulfate ions not only can function as a depressant in the flotation of oxide minerals with physisorbed anionic collectors but also can function as an activator for flotation with physisorbed cationic collectors. Their role in depressing or enhancing the flotation of the sparingly-soluble mineral barite is also explained in terms of the double layer model.

INTRODUCTION

The adsorption of multivalent metal cations, and particularly hydrolyzed cations, has been studied in detail by researchers interested in flotation, water treatment, and fundamental colloid science. Except for research by soil scientists on the adsorption of phosphate ions, only limited research has been directed towards the specific adsorption of inorganic anions on oxide minerals. Sulfate ions have long been known to play a significant role in sulfide mineral flotation since they are a product of the surficial oxidation of most sulfide minerals. For example, the chemical exchange between carbonate and sulfate ions formed on the surface of galena under oxidizing conditions in airb and their subsequent exchange adsorption with ethyl xanthate ions were quantitatively shown 70 years ago by Taylor and Knoll (1934) at Columbia University. Interestingly, sulfate ions

* Dept. of Materials Science and Engineering, University of California, Berkeley, California

can also, but in a different way, significantly affect the flotation of nonmetallic minerals but never analyzed in terms of specific adsorption potentials. Measurement of zeta potentials showed that sulfate ions are specifically adsorbed on corundum and that this specific adsorption could lead to depression of corundum flotation when sodium dodecylsulfate was used as the collector and conversely could lead to activation if dodecylammonium acetate was used as the collector (Modi and Fuerstenau 1957, 1960). As will be discussed, other researchers subsequently observed specific adsorption of sulfate ions on rutile, ilmenite and hematite, and found that sulfate ions can function as an activator for ilmenite and hematite at low pH with dodecylammonium acetate as collector. Taggart and Arbiter (1946) published a paper concerned with the effect of barium ions and sulfate ions on contact angles on barite with aminium ion collectors and interpreted their findings in terms of a metathetic reaction. Research conducted during the last half century has shown that flotation in many systems with alkylamine and alkylsulfate salts as collectors involve adsorption in the electrical double layer at the solid/water interface. The objective of this paper is to present an analysis of the specific adsorption of sulfate ions and their role in activation and depression processes in terms of the electrical double layer model. To do this, the electrical double layer model and adsorption phenomena are first briefly reviewed.

THE ELECTRICAL DOUBLE LAYER AND ELECTROKINETIC POTENTIALS

The Electrical Double Layer

The immersion of a solid into an aqueous solution produces a charge apparently fixed at the solid surface that is exactly balanced by a diffuse region of equal but opposite charge (termed *counter ions*). This is called the *electrical double layer* (Hunter 1981; Shaw 1980). Figure 1 is a schematic representation of a simple model of the electrical double layer at a mineral/water interface, showing the charge on the solid surface and the counter ions extending out into the aqueous phase as a diffuse layer. Unless counter ions have special affinity for the surface, because of thermal motion, they extend out into the liquid as a diffuse layer. This figure also shows the drop in potential across the double layer. The closest distance of approach of (hydrated) counter ions to the surface is called the Stern plane (the layer of counter ions at a distance, δ. The total double layer potential, or *surface potential*, is ψ_o and that at the Stern plane is δ. From the Stern plane out into the bulk of the solution, the potential drops exponentially to zero. As will be discussed later, the electrokinetic or *zeta potential* ζ is the potential at the plane where shear takes place when the liquid moves relative to the solid. In the case of those ions that interact with surface sites, either chemically or by some other strong specific adsorption forces, the adsorbed ions may lie closer to the surface than the plane δ. Various parameters that quantify the electrical double layer are useful in interpreting flotation behavior, such as the magnitude of the surface charge, the point of zero charge of the mineral, interfacial potentials, thickness of the electrical double layer, specific adsorption, and ion exchange phenomena.

This paper is mainly concerned with the double layer on oxides, but also with the sparingly-soluble salt mineral, barite. In the case of salt-type minerals such as barite (barium sulfate), fluorite (calcium fluoride), argentite (silver sulfide), iodyrite (silver iodide), etc., the surface charge arises from the preference of one of the lattice ions for the solid relative to the aqueous phase. Equilibrium is attained when the electrochemical

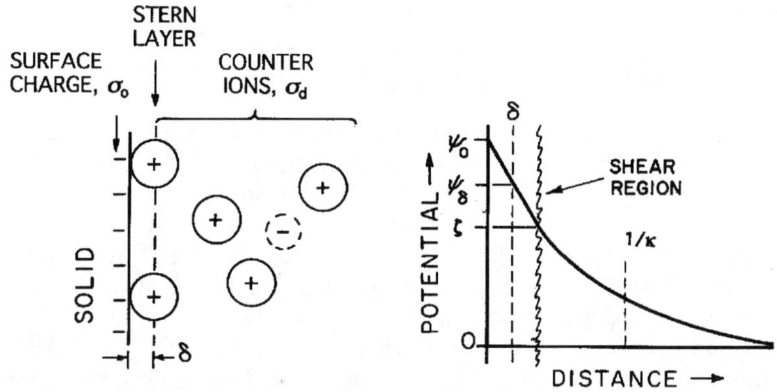

FIGURE 1 Schematic of the electrical double layer, showing the surface charge, the Stern plane which is the closest distance of approach of counter ions, and the diffuse layer of counter ions extending out into the solution. Also illustrated is the distribution of potential across the double layer, showing the surface potential, the Stern layer potential, and the zeta potential at the electrokinetic slipping or shear plane.

potential of these ions is constant throughout the system. Those particular ions which are free to pass between both phases and therefore establish the electrical double layer are called *potential-determining ions* (PD ions). The surface charge, σ_o, as well the surface potential, ψ_o, is determined by the activity, a, (or concentration, C) of PD ions in solution as described by the following equations:

$$\sigma_o = zF\left(\Gamma_{M^{+z}} - \Gamma_{M^{-z}}\right) \quad \textbf{(EQ 1)}$$

where F is the Faraday constant, Γ the adsorption density in mol per cm^2, and z the valence. For the P-D cation,

$$\psi_o = \frac{RT}{z^+F}\ln\left(a_{M^{+z}}/a_{M^{+z}(PZC)}\right) \quad \textbf{(EQ 2)}$$

where R is the gas constant and the PZC defines the condition in the aqueous solution where σ_o is zero. (An analogous equation can be written PD anions.) For a sparingly-soluble salt mineral like barite, Equations 1 and 2 are reduced to

$$\sigma_o = 2F\left(\Gamma_{Ba^{++}} - \Gamma_{SO_4^{--}}\right) \quad \textbf{(EQ 3)}$$

and

$$\psi_o = \frac{RT}{2F}\ln\left(a_{Ba^{++}}/a_{Ba^{++}(PZC)}\right) \quad \textbf{(EQ 4)}$$

For oxide and silicate minerals, hydrogen and hydroxyl ions have long been considered to be potential-determining, and Eq. 1 and 2 become:

$$\sigma_o = F\left(\Gamma_{H^+} - \Gamma_{OH^-}\right) \quad \textbf{(EQ 5)}$$

and

$$\psi_o = \frac{RT}{F}\ln\left(a_{H^{+z}}/a_{H^+(PZC)}\right) = 0.059\left[pH_{PZC} - pH\right] \text{ volts} \quad \textbf{(EQ 6)}$$

With regard to flotation behavior, the single most important parameter that describes the electrical double layer of a mineral in water is the point of zero surface charge (PZC). The importance of the PZC is that the sign of the surface charge has a major effect on the adsorption of oppositely charged ions that function as the counter ions. In contrast to the situation in which the PD ions are special for each system, any ions present in solution can function as counter ions. As has been well established, the counter ions occur in a diffuse layer that extends from the surface out into the bulk solution. For a symmetrical electrolyte with ions of valence z (where $z = z^+ = z^-$) at concentration C, the "thickness" of the diffuse double layer, is $1/\kappa$ is given by

$$\kappa = \sqrt{\frac{2\pi z^2 F^2 C}{\varepsilon\varepsilon_o RT}} = 0.329 \times 10^{10}\sqrt{Cz^2}\ m^{-1} \quad \textbf{(EQ 7)}$$

where ε is the dielectric constant of the liquid and ε_o is the permittivity constant. (For nonsymmetical electrolytes the quantity under the square root is the ionic strength.) Actually $1/\kappa$ is the center of gravity of the diffuse layer but is considered to be the thickness of the diffuse layer (Figure 1). As the ionic strength of the solution is increased, the thickness of the double layer decreases. For 1–1 valent electrolytes, $1/\kappa$ is 10 nm in 10^{-3} M and 1 nm in 10^{-1} M solutions, for example.

If counter ions are adsorbed only though electrostatic interaction with the charged surface, the added salt is called an *indifferent electrolyte*, and the diffuse layer charge, σ_d, is oppositely equal to the surface charge, σ_o. The charge in the diffuse double layer σ_d given by the Gouy-Chapman relation and as modified by Stern for a symmetrical electrolyte is:

$$\sigma_d = -\sigma_o = (8\varepsilon\varepsilon_o RTC)^{1/2}\sinh(zF\psi_\delta/2RT) \quad \textbf{(EQ 8)}$$

where ψ_δ is the potential in the Stern plane. However, some ions exhibit surface activity in addition to electrostatic attraction and adsorb strongly in the Stern plane because of such other phenomena as covalent bond formation, hydrophobic bonding, hydrogen bonding, etc. The term *specific adsorption* is used for such situations. Because these adsorbed ions may be interacting chemically with surface sites, they may lie in a plane closer to the surface than the Stern plane. This plane is generally called the inner Stern plane. Unlike so-called indifferent counter ions, these ions can adsorb in the Stern plane to a greater extent than is required to neutralize the surface charge. If we designate the charge due to specifically adsorbed ions as σ_δ, then

$$\sigma_o + \sigma_\delta + \sigma_\delta = 0 \quad \textbf{(EQ 9)}$$

The charge in the Stern plane is given by

$$\sigma_\delta = zF\Gamma_\delta \quad \textbf{(EQ 10)}$$

There are several related approaches for relating the adsorption density of specifically adsorbed ions to their concentration in the bulk solution. This can be done through a Stern-Langmuir isotherm as follows:

$$\frac{\theta}{1-\theta} = \frac{C}{55.5}\exp\left[-\frac{\Delta G^o_{ads}}{RT}\right] \qquad \text{(EQ 11)}$$

or alternatively,

$$\Gamma_\delta = \frac{\Gamma_m}{1+\frac{55.5}{C}\exp\left(\frac{\Delta G^o_{ads}}{RT}\right)} \qquad \text{(EQ 12)}$$

where ΔG^o_{ads} in the standard free energy of adsorption, $\theta = \Gamma_\delta/\Gamma_m$ is the fractional surface coverage at equilibrium molar concentration C in solution, and Γ_m is the adsorption density at monolayer coverage. For low adsorption densities, Grahame simplified Eq. 12, which will be expressed here in the form:

$$\Gamma_\delta = 2rC_o\exp(-\Delta G^o_{ads}/RT) \qquad \text{(EQ 13)}$$

where Γ_δ is the adsorption density in mol per square cm, r is the effective radius of the adsorbed ions in cm, and here C_o is the bulk concentration in mol per cubic cm to keep units consistent. Sometimes for semiquantitative purposes, the inner and outer Stern planes may be considered simply to coincide.

If ions are adsorbed only because of electrostatic interactions, then the standard free energy of adsorption is given by

$$\Delta G^o_{ads} = \Delta G^o_{elec} = zF\psi_\delta \qquad \text{(EQ 14)}$$

but when ions exhibit surface activity, such as flotation collectors, then the standard free energy of adsorption has additional terms (Fuerstenau and Healy 1972):

$$\Delta G^o_{ads} = zF\psi_\delta + \Delta G^o_{spec} \qquad \text{(EQ 15)}$$

where ΔG^o_{spec} represents the specific interaction terms. There can be one or more of several different contributions to the specific adsorption free energy:

$$\Delta G^o_{spec} = \Delta G^o_{chem} + \Delta G^o_h + \Delta G^o_{hpb} + \Delta G^o_{chem} + \Delta G^o_{solv} + \ldots \qquad \text{(EQ 16)}$$

where the individual terms represent changes in the standard free energy due to chemical bonding, hydrogen bonding, hydrophobic bonding, and solvation effects, respectively. Depending on the mechanisms involved in the interaction of a flotation collector with the mineral surface, the contributions to the change in adsorption free energy can be essentially zero or have a finite value. In this paper, we are not concerned with collectors that chemisorb (such as oleate on hematite or barite) but with physisorbed collectors (such as sodium dodecylsulfate or dodecylammonium acetate on corundum or hematite) that adsorb as indifferent ions at low concentrations but at higher concentrations begin to associate into hemimicelles causing to become finite (Modi and Fuerstenau 1960).

Ion Exchange in the Double Layer

The structure of the diffuse part of the double layer depends upon the composition of the bulk solution and any change in the composition of the bulk solution can result in ion exchange within the double layer. If two different kinds of counter ions, a and b, are present, then

$$\Gamma_a/\Gamma_b = K^*(C_a/C_b) \qquad \textbf{(EQ 17)}$$

where Γ_a and Γ_b are the adsorption densities in the double layer, C_a and C_b are their respective bulk concentrations, and K* is a constant. For indifferent counter ions such as K^+ and Na^+, K* has a value close to unity. However, ions having a strong affinity for the surface will tend to concentrate in the double layer over simple indifferent electrolytes. Where appreciable adsorption occurs in the Stern layer, the competition for sites would include specific adsorption phenomena:

$$\frac{\Gamma_a}{\Gamma_b} = \frac{r_a C_a}{r_b C_b} \exp\left[\frac{(\Delta G^o_{ads})_a}{RT} - \frac{(\Delta G^o_{ads})_b}{RT}\right] \qquad \textbf{(EQ 18)}$$

with the radius of the two kinds of ions having to be taken in account if they differ in size. If the competing ion a is an ion that adsorbs only electrostatically (such as Na^+ on quartz), and b is an ion that has specific affinity the surface (such as a flotation collector), then the exponential term would simply be $\exp[(\Delta G^o_{spec})_b/RT]$. In ion exchange, divalent ions dominate over monovalent ions because of the valence effect.

Zeta Potentials

Electrokinetic phenomena, which involve the interrelation between mechanical and electrical effects at a moving interface, have found widespread use in colloid and surface chemistry (Hunter 1981). The electrical potential involved is that at the slipping or shear plane, which is the plane at which relative motion between the two phases takes place. This potential is commonly referred to as the zeta potential, ζ (Figure 1). The problem is what potential does ζ represent in the double layer. Across the diffuse layer, the potential decays with distance according to the equation:

$$\psi_{(x)} = \psi_o \exp(-\kappa x) \qquad \textbf{(EQ 19)}$$

where x represents the distance from the surface into the solution. If x is the radius of an adsorbed ion in the Stern plane, then $\psi_{(x)} = \psi_\delta$. If x is the distance to the electrokinetic slipping plane, then $\psi_{(x)} = \zeta$. Thus, the closer the slipping plane is to the Stern plane, the closer will $\psi_{(x)}$ [the zeta potential in this case] approximate ψ_δ. In applying electrokinetic results, the assumption that $\psi_{(x)} \cong \psi_\delta$ has often been very useful. The two most common methods used for evaluating zeta potentials have been electrophoresis for fine particles and streaming potentials for coarse particles or capillaries. Electrophoresis is the process by which the application of a field causes suspended particles to move through a liquid. Streaming potential is the process in which the application of a pressure gradient causes a liquid to flow through a bed of particles or a capillary, and the moving liquid strips off ions in the diffuse layer, thereby generating an electric current which leads to an electric field. Zeta potentials can be readily calculated from electrophoretic mobilities or streaming potentials (Hunter 1981; Shaw 1980).

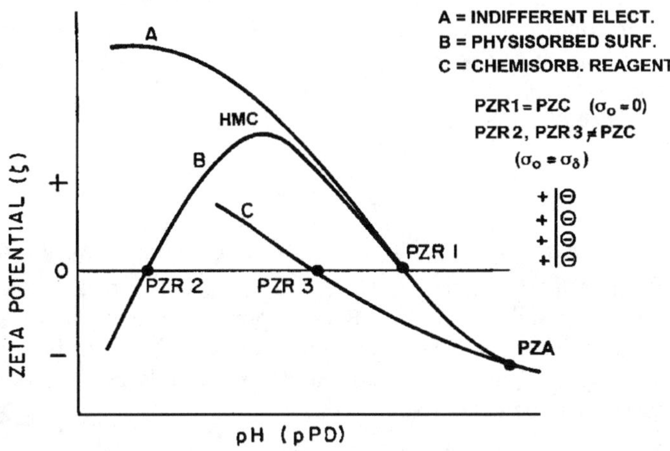

FIGURE 2 Typical plots of zeta potential versus pH for three different types of electrolytes added to the system: indifferent electrolytes, physisorbing anionic collectors, chemisorbing inorganic anions (or chemisorbing anionic collectors) showing zeta potentials reversals or isoelectric points

Any electrolyte that affects adsorption in the Stern plane can have a pronounced effect on the zeta potential. Potential-determining ions can reverse the sign of the zeta potential because they can reverse the sign of the surface charge, σ_o, whereas specifically adsorbed ions can reverse the sign of the zeta potential because the charge in the Stern plane can exceed that at the surface $|\delta| > |\sigma_o|$. Sometimes this is termed super-equivalent adsorption. Figure 2 illustrates schematically how ζ can change with pH or pPD in the presence of three different types of electrolytes of interest: an indifferent electrolyte, a physisorbed anionic collector, and a surface-active inorganic anion. The pH at which the zeta potential is brought to zero and reversed is generally termed an isoelectric point (IEP) or a point of zeta potential reversal (PZR). Since the added salt is an indifferent electrolyte, PZR 1 in Figure 2 is the PZC. PZR 2 is typical of physisorbed anionic surfactants, and results from the increase in adsorption free energy due to hemimicelle formation. It should be noted that with such surfactants, the PZC does not shift and zeta potential measurements show two IEPs, PZR 1 and PZR 2. Specifically adsorbed reagents that have a constant specific adsorption potential exhibit a single IEP, PZR 3 in Figure 2, which is at a lower pH than the PZC for anionic reagents or higher than the PZC cationic surface-active reagents. Another condition is apparent from these schematic diagrams, namely the pH (pPD) at which the ζ-vs-pH curve for the specifically adsorbed anion joins the curve for the indifferent electrolyte. In the case of a specifically adsorbed reagent, this is the PZA or point of zero adsorption, which signifies the condition where the electrical potential in the Stern plane is great enough to counteract the specific adsorption potential and electrostatically repels the surface-active anions from the surface. In the case of physisorbed alkyl surfactants, at the maximum in the ζ-vs-pH curve represents the onset of specirfic adsorption that results from hemimicelle formation and is denoted by HMC (the hemimicelle concentration in this figure).

An estimate of the magnitude of the adsorption free energy can be made in three different ways from conditions illustrated in Figure 2. At the IEP where the zeta potential is brought to zero by specific adsorption, the electrical double layer is simply a molecular condenser, of which one plate is the surface charge and the other plate is the Stern plane. The capacitance of the Stern layer, C_δ^*, is given by

$$C_\delta^* = \frac{\sigma_o}{\psi_o - \psi_\delta} \quad \text{(EQ 20)}$$

Since σ_d is zero, the total capacitance of the double layer C* is equal to the Stern layer capacitance. To evaluate ΔG^o_{spec} since $\psi_\delta = 0$, $\psi_o = 0.059\ [pH_{PZC} - pH_{IEP}]$ volt, and $-\sigma_o = \sigma_\delta = zF\Gamma_\delta$, we need an estimate of Γ_δ and C_δ^*. For this purpose, the Stern-Grahame relation, Eq. 13, can be used to determine the adsorption density in the Stern plane and a typical value of the Stern layer capacitance assumed. Thus, considering points B and C as a molecular condenser, can be evaluated:

$$\psi_o C_\delta^* = \sigma_o = -2C_o zF \exp(\Delta G^o_{spec}/RT) \quad \text{(EQ 21)}$$

Pradip generalized this approach somewhat by developing an equation in terms of ΔpH and the Stern-Langmuir equation. Finally, since the PZA gives the conditions where the negative surface charge is great enough to prevent adsorption of surface-active anions, under these conditions $\Delta G^o_{spec} = zF\,\psi_\delta$.

SPECIFIC ADSORPTION OF SULFATE IONS ON OXIDES

Hydrolyzed multivalent metal cations are strongly specifically adsorbed on virtually all subtrates and are used as activators in the soap flotation of quartz. It is the hydroxylated form of the metal ion that strongly adsorbs. Studies with multivalent anions have shown that sulfate ions and phosphate ions will specifically adsorb on oxides from aqueous solutions. Monovalent anions such as chloride, nitrate and perchlorate do not specifically adsorb on oxides. However, they can exhibit slight effects on surface charge, double layer capacitance, etc. because of their differences in ionic radii and ionic hydration.

Modi and Fuerstenau (1957) first observed specific adsorption of sulfate ions on positively charged corundum (synthetic sapphire) from determination of zeta potentials with streaming potential measurements. Their results for the effect of sodium chloride, sodium sulfate and sodium thiosulfate on the zeta potential of corundum at pH 6.5 are plotted in Figure 3. In the case of NaCl, the slight increase in ζ at low concentrations is an artifact encountered when streaming potentials are measured with plugs of particles due to overlapping double layers at low ionic strengths. Because chloride ions are not specifically adsorbed, ζ only decreases towards zero as the double layer is compressed at high NaCl concentrations. On the other hand, both sulfate and thiosulfate ions reverse the zeta potential of corundum at a concentration of 3×10^{-4} M, indicating quite strong specific adsorption and essentially equivalent to sodium dodecylsulfate at that pH. At pH 11, sodium sulfate no longer exhibits specific adsorption, and when data are plotted in terms of equivalents per liter, the ζ-log-C curve for sodium sulfate superimposes on that for NaCl. Experiments to measure as a function of pH at constant sodium sulfate concentration were never carried out. Specifically adsorbed sulfate ions (radius 0.29 nm) must

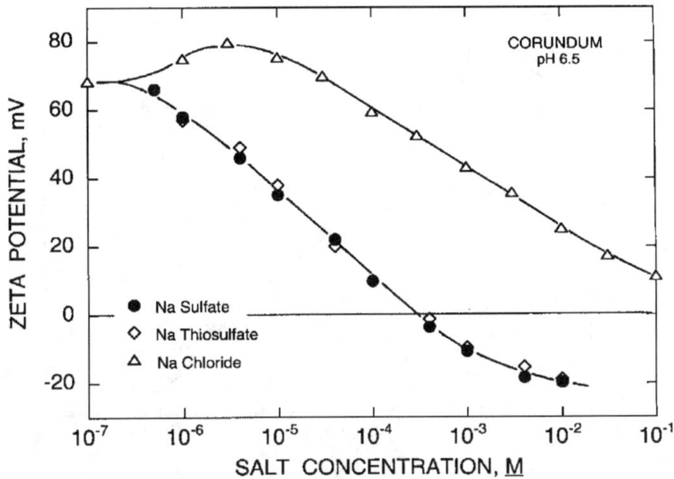

FIGURE 3 The effect of sodium chloride (an indifferent electrolyte) and sodium sulfate and sodium thiosulfate (specifically adsorbed anions) on the the zeta potential of positively charged corundum at pH 6.5 (data from Modi and Fuerstenau)

penetrate the hydration sheath at the solid-water interface, so the capacity of the Stern layer will be about 25 µfarad/cm^2. The PZC of synthetic sapphire as measured by Modi occurs at pH 9.4. With Eq. 21, the calculated magnitude of is −6.9RT.

Using streaming potential methods, Purcell and Sun (1963) measured the zeta potential of rutile as a function of pH with 0.001 M sodium sulfate and with different concentrations of sodium chloride. Their results, which are replotted in Figure 4, show that the PZC of rutile occurs at pH 6.6 as evidenced by the curves at three concentrations of NaCl all crossing at this pH. With 0.001 M Na_2SO_4 solutions, the ζ-vs-pH curve exhibits two distinct differences from the 0.001 M NaCl curve, in addition to the zeta potential being less positive in acidic solutions. First, the IEP is shifted to a pH lower than the PZC, namely 5.9. Secondly, the two curves join at pH 7, indicating that sulfate ions are no longer specifically adsorbed. From the shift in the IEP, with Eq. 21 the magnitude of is −3.6RT. As can be seen schematically in Figure 1, $\zeta < \psi\delta$, and for calculations involving the PZA, the Stern layer potential itself is needed. So here ψδ is estimated with Eqs. 8 and 19. From Eq. 19, the magnitude of ψδ is - 24 mV and with Eq. 14, is to estimated to be −2.0RT. With rutile there might be a dependence of the specific adsorption free energy on surface charge since the dielectric constant of rutile is unusually high and essentially the same as that of water, namely 80. Note that the zeta potential curve of sodium sulfate solutions joins that of sodium chloride at a pH barely above the PZC. Sun and Purcell correctly commented that if they had used a higher concentration of sodium sulfate in their experiments, ζ would have been much more negative.

Nakatsuka et al. (1970) carried out detailed electrokinetic and flotation studies on ilmentite, hematite and magnetite, and presented one plot of the effect of pH on the zeta potential of ilmenite in the presence of 0.15 mM NaCl and also 0.05 mM Na_2SO_4 with

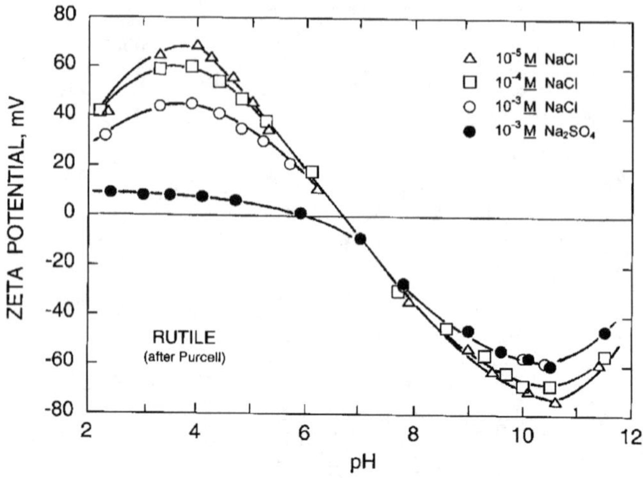

FIGURE 4 The zeta potential of rutile as a function of pH at different concentrations of sodium chloride and sodium sulfate (data from Purcell and Sun)

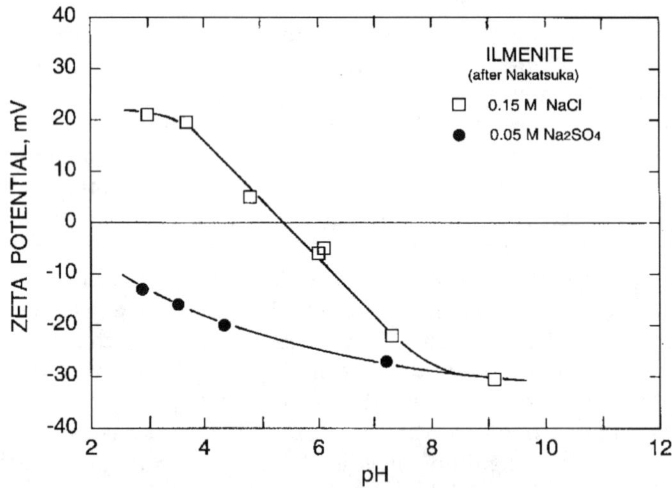

FIGURE 5 The zeta potential of ilmenite as a function of pH in aqueous solutions of sodium chloride and sodium sulfate (data from Nakatsuka et al.)

NaCl to maintain an ionic strength of 0.015 M. solutions. The results of their electrophoretic measurements are replotted in Figure 5. The NaCl results indicate that the PZC of ilmenite occurs at pH 5.6. Sulfate ions appear to strongly adsorb on ilmenite since extrapolation suggests that the IEP would not be reached until the pH was decreased below pH 1 or so. The two ζ-vs-pH curves join at about pH 8. With the potential evaluated with Eq. 19, the magnitude of is estimated to be −7.2RT.

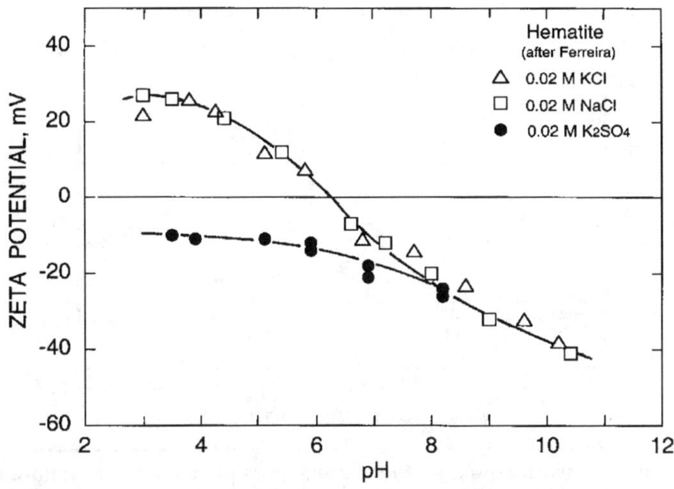

FIGURE 6 The zeta potential of hematite as a function of pH in aqueous solutions of potassium chloride, sodium chloride and potassium sulfate (data from Ferreira et al.)

Recently, Ferreira et al. (2004) published their results on the measurement of electrokinetic potentials on hematite in 0.01 M solutions of NaCl, KCl and K_2SO_4. Their electrophoretic results, replotted in Figure 6 indicate that the PZC of their hematite sample occurs at pH 7.0. The ζ-vs-pH curves for NaCl and KCl are virtually identical, but that for K_2SO_4 shows marked specific adsorption of sulfate ions. The three sets of curves join at about pH 8.5. Evaluating the potential when the two curves come together with Eq. 19, is estimated to be −5.8 RT. In 1972, Breeuwsma and Lyjklema reported on their detailed titrations to determine the effect of various cations and anions on the surface charge, σ_o, of hematite. The PZC of their synthetic hematite was pH 8.5. At lower pHs, the surface charge itself was considerably more positive in the presence of 0.001 N potassium sulfate than with 0.001 N potassium chloride, indicating specific adsorption of the sulfate ions. The two surface charge curves approach each other at pHs higher than the PZC. They did observe a shift in the PZC towards pH 9.5 in the presence of potassium sulfate, but do not consider than sulfate ions to be as strongly adsorbed as calcium ions. They showed that chloride ions are not specifically adsorbed by hematite. Zeta potentials were not measured.

These foregoing values of specific adsorption free energies can only be considered to be estimates since they are based on Nernstian behavior (59 mV change in ψ_o with per pH unit change) and the assumption that the PZC did not change appreciably by the specific adsorption process. However, these calculations show that the specific adsorption potential of sulfate ions is high.

FLOTATION DEPRESSION WITH INORGANIC IONS

In the flotation of insoluble oxides and silicates with physisorbed collectors, reducing and reversing the surface charge by changing the pH is a powerful tool to depress flotability. Examples include the flotation of corundum and hematite with sodium dodecylsulfate

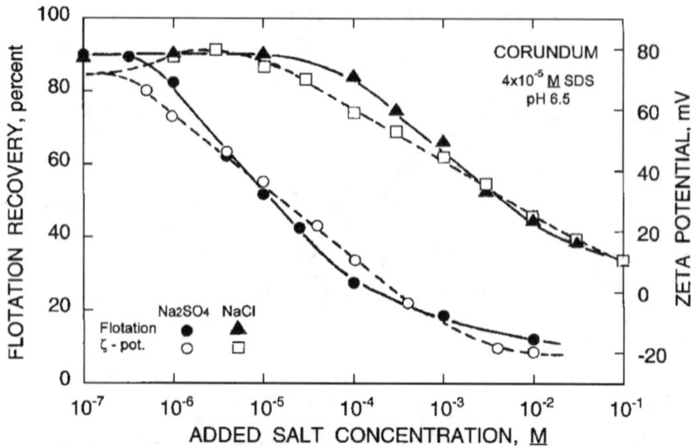

FIGURE 7 Correlation between the effect of sodium chloride and sodium sulfate in depressing the flotation of corundum with sodium dodecylsulfate as collector at pH 6.5 and on the zeta potential of corundum in the absence of collector (data from Modi and Fuerstenau)

(SDS) or dodecylammonium chloride (DAC) as collector. Since physisorbed surfactants act as flotation collectors by adsorbing as counter ions in the double layer, the addition of any inorganic salt at sufficiently high concentration should depress flotation. This might be important in the cationic flotation of nonmetallic minerals using seawater. To illustrate these phenomena, Figure 7 presents plots of the effect of sodium sulfate and sodium chloride on the zeta potential of corundum (in the absence of SDS) and also their effect on corundum flotation with 4×10^{-5} M SDS at pH 6 (Modi and Fuerstenau 1960). As seen in the previous section, sulfate ions are specifically adsorbed on corundum. The plots given in Figure 7 show excellent correlation of the effect of the two inorganic salts on the zeta potential of corundum and in depressing flotation. This figure shows that both chloride and sulfate ions inhibit the flotation of positively charged alumina but with the effect of sulfate^{-} being nearly 500 times that of chloride^{-}.

The greater effect of sulfate over chloride ions in depressing alumina flotation results from the valence effect and the specific adsorption potential of sulfate on alumina. This can be interpreted in accordance with Eq. 17. The concentration of SDS used in these experiments was 4×10^{-5} M, and hemimicelles had formed at the surface, with their adsorption free energy being −5.6 RT under these conditions (Fuerstenau and Modi 1959). From conditions leading to the PZR of corundum with sodium sulfate, the adsorption free energy of divalent sulfate ions is about −6.9 RT. The ionic radii are 0.18 nm for chloride ions and 0.29 nm for both sulfate ions and the head group of dodecylsulfate ions. For these calculations, we will consider conditions that led to the reduction of flotation recovery to 50 %. Figure 7 shows that 5.6×10^{-3} M NaCl is required to do this when the collector concentration is 4×10^{-5} M, or 140 times as much. Assuming that half the dodecylsulfate ions in the Stern plane are replaced by chloride ions, Eq. 15 estimates that the required ratio of chloride to dodecylsulfate is 170. To carry out the estimation for depression with divalent sulfate ions, we have to take into account that one divalent sulfate ion will displace two dodecylsulfate ions in the Stern plane and bring the zeta potential into the calculation. Doing so shows that the concentration of

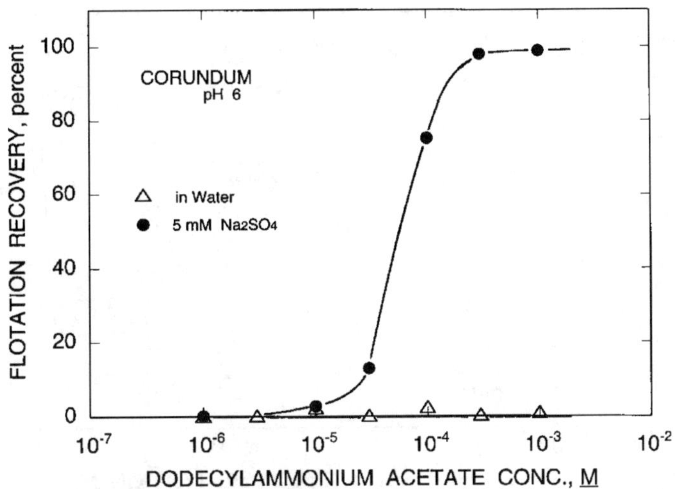

FIGURE 8 The activation of corundum with sodium sulfate for flotation with dodecylammonium acetate at pH 6.5 (data from Modi and Fuerstenau)

divalent sulfate ions necessary to reduce the recovery to 50 % is one-fourth that of the SDS in solution, or 0.9×10^{-5} M compared to the 1.3×10^{-5} M observed. Given that a rigorous calculation would require potentials at different planes and knowledge of possible hydration phenomena with adsorbed chloride, these estimates starting with electrokinetic results are interestingly close.

FLOTATION ACTIVATION WITH SULFATE IONS

We have just seen the effects of sodium sulfate in inhibiting the flotation of positively charged alumina with an anionic collector. Figure 8 illustrates how the same inorganic salt can promote flotation. At pH 6, alumina does not respond to flotation with dodecylammonium chloride as collector since these organic cations do not adsorb on a positively charged solid (Modi and Fuerstenau 1960). On the other hand, by adding sufficient divalent sulfate ions to the system, the net charge in the Stern layer exceeds that of the surface; and since the net charge is reversed, the solid now adsorbs the organic cations as counter ions. In flotation terminology, sulfate ions are said to function as an activator; they serve as a link between the surface and collector ions charged similarly. In such systems, the electrical double layer is in reality a triple layer. Although zeta potential measurements were not carried out to elucidate these conditions, chain-chain interaction could lead to stronger aminium ion adsorption such that a quaternary layer might form: surface charge, inner Stern layer of sulfate ions, outer Stern layer of aminium ions, and diffuser layer of chloride ions.

A detailed investigation to learn more about sulfate activation of corundum was carried out by the measurement of contact angles on a sapphire crystal that had been cut and carefully polished (Fuerstenau and Foreman 2004). The experiments were designed to delineate the regions of flotability of sulfate-activated corundum for flotation with high-purity dodecylammonium acetate (DAA) as the collector at pH 6 to 7. The sample

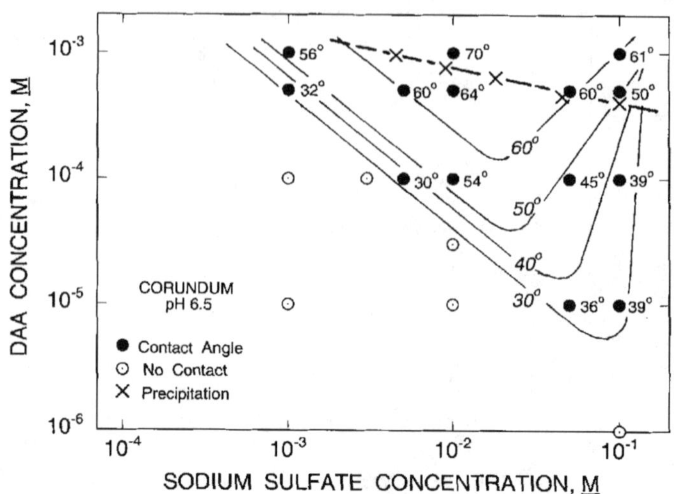

FIGURE 9 Isogonic map for bubble-particle contact for corundum at pH 6.5 as a function of sodium sulfate and dodecylammonium acetate concentration. Reagent concentrations for the precipitation of dodecylammonium sulfate are shown.

was first preconditioned with 0.01 M Na_2SO_4 for 10 minutes and checked that the contact angle was zero, and then the solution was replaced with another 0.01 M Na_2SO_4 solution containing the desired amount of DAA and then stirred again for another 10-minute period before placing and photographing a bubble on the surface of the inverted polished specimen. Two extensive series of experiments were conducted to determine the minimum time required for equilibrium to be reached. With 10^{-3} M DAA, the contact angle was found to be 53 degrees after 1, 5, 10, 25, 55 and 85 minutes of contact. Clearly equilibrium is attained very quickly in this system. Similar kinetic results were obtained with a DAA concentration of 10^{-4} M DAA. Subsequently, contact angle measurements were made at several different sodium sulfate and DAA levels. As another part of that study, the solubility of dodecylammonium sulfate was determined by titrating DDA with sodium sulfate (and repeating with the reverse order of addition). The end point was measured by light scattering. The results of these experiments are plotted as an isogonic map in Figure 9, with the concentration of DAA as the ordinate and sodium sulfate as the abscissa. The values of the measured contact angles under several different activator/collector conditions are recorded, along with regions where the contact angle is zero. The line of collector precipitation as dodecylammonium sulfate is shown. Flotation clearly occurs under conditions where there is no precipitation, but the highest contact angles do occur under conditions where the aminium sulfate has precipitated. The contact angle limits appear to be steep, that is going from no contact to significant contact occurs fairly abruptly. This is in agreement with the flotation curve given in Figure 8, which shows a sharp increase in flotation with increasing DAA concentration with 0.01 M sodium sulfate activator. At sodium sulfate concentrations above 0.1 M, the contact angles appear to decrease significantly. Under these conditions the sodium ion concentration is extremely high, and ion exchange would begin to be replace aminium ions by sodium ions at the surface.

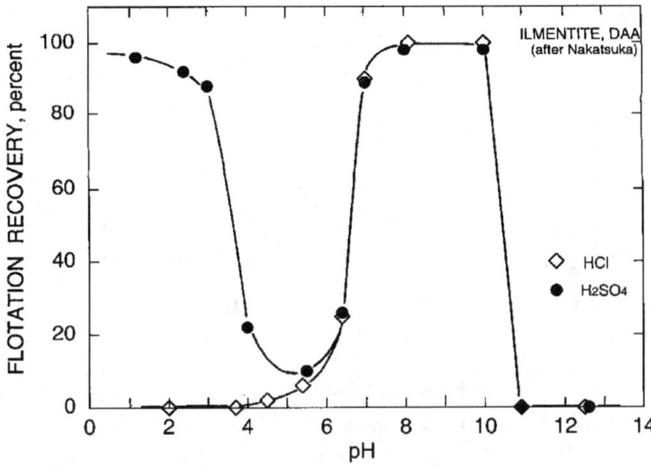

FIGURE 10 The flotation recovery of ilmenite as a function of pH regulated with sodium hydroxide, sulfuric acid and hydrochloric acid, showing activation by sulfate ions at low pH (data from Nakatsuka et al.)

In an attempt to reduce the titania content of magnetite concentrates, Nakatsuka et al. (1970) investigated the flotation of ilmenite activated by sulfate ions. They also conducted extensive flotation experiments on activation phenomena with a number of other minerals related to the magnetite concentrates. Figure 10 presents a plot of their results for the flotation of ilmenite with dodecylammonium acetate as collector as a function of pH regulated with NaOH, HCl and H_2SO_4. In the alkaline region flotation is excellent until the usual amine hydrolysis region is reached. Flotation decreases sharply as the pH is decreased below 7 when pH is regulated with either HCl or H_2SO_4, and goes to zero with continued addition of HCl. However, when the pH is regulated with H_2SO_4, ilmenite is activated by the sulfate ions and responds again to flotation as the pH is decreased below 4. They found almost indentical behavior for hematite, but magnetite was not activated by sulfate ions. They also observed that HPO_4^{2-} and PO_4^{3-} activated ilmenite for amine flotation. Their conclusions were the same as first suggested by Fuerstenau and Modi (1960) that activation occurs by superequivalent adsorption of sulfate ions in the Stern layer, and that the aminium cations then adsorb after Stern layer charge reversal.

SULFATE IONS IN THE FLOTATION OF BARITE

In 1946 Taggart and Arbiter published the results of a study of the effect of laurylamine hydrochloride (dodecylammonium chloride, DAC) on the contact angles on sparingly-soluble salt minerals, and mainly barite. First of all, barite ($BaSO_4$) is slightly soluble, with its solubility product being 1.5×10^{-9}. They found that the contact angle increased from zero with increasing DAC concentration, reaching a plateau of about 70 degrees at 2 mM collector. At a DAC concentration of 9 mM, the contact angle had decreased to 60 degrees, typical of physisorption in reverse orientation. Their interesting results

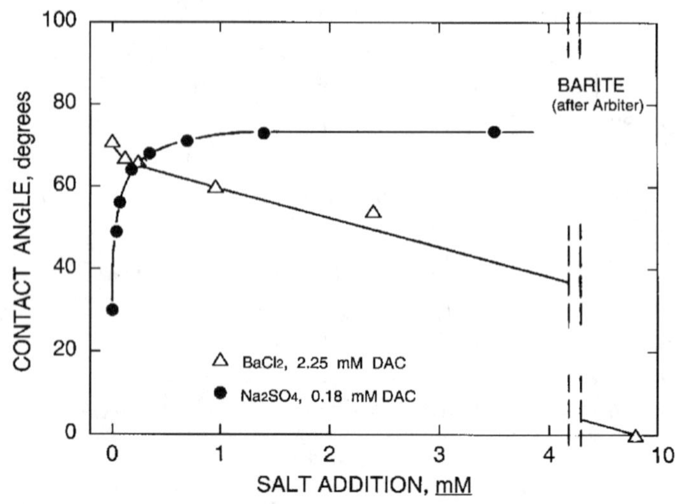

FIGURE 11 The effect of sodium sulfate and barium chloride on contact angles on barite at two different levels of dodecylammonium chloride, 39 and 500 mg per liter (data from Taggart and Arbiter)

involved investigation of the effect of added sodium sulfate and barium chloride on the contact angle. Figure 11 presents their results for the addition of sodium sulfate on the contact angle on barite in solutions containing 2.28 mM (500 mg/l) DAC and barium chloride in solutions containing 0.18 mM (39 mg/l) DAC. As can be seen in this figure, the contact angle increases sharply with the addition of sodium sulfate, reaching a plateau. On the other hand, the addition of barium chloride causes the contact angle on barite to decrease, becoming zero by the addition of less than 5 mM barium chloride. Taggart and Arbiter explained their results as a metathetic exchange reaction for the case of adding sulfate ions to the system but had no real explanation for the effect of adding barium ions, writing: "Conversely, addition of barium ion decreases the collecting effect of a given amount of amine, the probable mechanism being closure of the barium sulphate surface by driving back barium ion." They were unaware of the role of potential-determining ions in controlling interfacial phenomena at mineral-water interfaces in these sorts of systems. Somasundaran and Agar (1967) showed that dodecylammonium acetate and sodium dodecylsulfate are physically adsorbed on calcite, a similar sparingly-soluble salt mineral. In aqueous systems, barite carries a small positive charge because of a slightly greater tendency for barium potential-determining ions to be preferentially adsorbed over the sulfate potential-determining anions. The PZC of natural barite was found to occur at pBa 5.1 by Buchanan and Heymann (1948). The sample used by Taggart and Arbiter could have had a different PZC, or possibly their sample did not reach equilibrium with regard to dissolved sulfate ions and barium ions. The effect of adding barium ions is to make the surface more positively charged, thereby decreasing the adsorption of aminium ions. Conversely, adding sodium sulfate to the system causes the surface to become jincreasingly negatively charged, resulting in increased adsorption of the aminium ions. Hemimicelle formation would begin when adsorption densities were high enough. Although adsorption densities were not measured, the plateau in the contact angle probably represents monolayer coverage of the collector.

SUMMARY

Electrokinetic measurements clearly show that sulfate ions can reverse the sign of the zeta potential and are specifically adsorbed at the surface of various oxide minerals. The electrical double layer model provides not only a method for interpreting sulfate ion adsorption phenomena but also provides means for estimating their adsorption free energy. Because of the surface activity of sulfate ions, sodium sulfate can function as a depressant for the flotation of positively charged minerals with physisorbed anionic collectors. By ion exchange, both chloride and sulfate ions can depress flotation, but sulfate ions are about 500 times as effective as monovalent chloride ions because of their surface activity. This same surface activity allows sodium sulfate to be used as an activator for the flotation positively charged corundum with positively charged dodecylammonium ions as collector because of super-equivalent adsorption of sulfate ions in the Stern layer at the mineral/water interface.. At low pHs, sulfate also functions as an activator for positively charged ilmenite with DAA collector. Early work on the effect of sodium sulfate and barium chloride on contact angles on barite with dodecylammonium chloride as collector had been explained as metathesis, but those results are reinterpreted in terms of double layer phenomena.

REFERENCES

Buchanan, A.S., and E. Heymann. 1948. The electrokinetic potential of barium sulphate. *Proceedings the Royal Society, A.* 195:150.

Breeuwsma, A., and J. Lyklema. 1973. Physical and chemical adsorption of ions in the electrical double layer of hematite. *Journal of Colloid and Interface Science*. 43:437.

Ferreira, E.E., B. Klein and P.R.G. Brandao. 2004. The effect of metallic cations on the zeta potential of pure hematite. In *Particle Size Enlargement in Mineral Processing*, Proceedings of the Fifth UBC-McGill Biennial International Symposium on Mineral Processing. Edited by J.S. Laskowski. 229–241.

Fuerstenau, D.W., and H.J. Modi. 1959. Streaming potentials of corundum in aqueous organic electrolyte solutions. *Journal of Electrochemistry*. 106:336.

Fuerstenau, D.W., and T.W. Healy. 1972. Principles of flotation. In *Adsorptive Bubble Separation Techniques*. Edited by R. Lemlich. New York, Academic Press. 92–131.

Fuerstenau, D.W., and W.E. Foreman 2004. Unpublished results.

Hunter, R.J. 1981. *Zeta Potential in Colloid and Surface Chemistry*. New York: Academic Press.

Modi, H.J., and D.W. Fuerstenau. 1957. Streaming potential studies on corundum in aqueous solutions of inorganic electrolytes. *Journal of Physical Chemistry*. 61:640.

Modi, H.J., and D.W. Fuerstenau. 1960. Flotation of corundum: an electrochemical interpretation. *Transactions AIME*. 217:381.

Nakatsuka, K., I. Matsuoka and J. Shimoiizaka. 1970. On the flotation of ilmenite from magnetite sand. *Proceedings, 9 th International Mineral Processing Congress, Prague*. 251–256.

Purcell, G., and S.C. Sun. 1963. Significance of double bonds in fatty acid flotation: an electrokinetic study. *Transactions AIME*. 226:6.

Shaw, D.J. 1980. *Introduction to Colloid and Surface Chemistry*, 3 rd ed. London:Butterworths.

Somasundaran, P., and G.E. Agar. 1967. The zero point of charge of calcite. *Journal of Colloid and Interface Science*. 24:433.

Taggart, A.F., and N. Arbiter. 1946. The chemistry of collection of nonmetallic minerals by amine-type collectors. *Transactions AIME*. 169:266.

Taylor, T.C., and A.F. Knoll. 1934. Action of alkali xanthates on galena. *Transactions AIME*. 112:382.

Hydrometallurgy

FTIR Investigation of Electrochemical Flotation in the Ethyl Xanthate–Pyrrhotite System

Hu Yuehua,* Zhang Qin,* Gu Guohua*

The flotation behavior of pyrrhotite was investigated by using ethyl xanthate as a collector. Results showed that pyrrhotite had good floatability from pH 2 to pH 11, and poor floatability when pH was >12. The flotation of pyrrhotite was dependent on pulp potential on various pH values. The potential-pH range for pyrrhotite flotation was established. Cyclic voltammetry and FTIR spectroscopic analysis showed that the major adsorption product of ethyl xanthate on pyrrhotite was dixanthogen. The intensity of FTIR signals of dixanthogen adsorption on pyrrhotite and the anode current of a pyrrhotite electrode and flotation response of pyrrhotite are correlated with pulp potentials.

INTRODUCTION

Pyrrhotite ($Fe_{1-x}S$, $0<x<0.223$) is an iron sulfide mineral widely encountered in base-metal sulfide ores. Because it has changeable stoichiometries and crystal structure (i.e., x is variable), pyrrhotite can exhibit different flotation characteristics which depend upon pH as well as oxidation and pulp potentials.

Heyes and Trahar (1984) reported that, at acidic pH values and for short conditioning times, good flotation can be achieved without collector addition. Woods and his co-workers studied the reaction and the production of the surface of pyrrhotite by XPS, linear potential sweep voltammetry and chemical analysis (Buckley and Woods,1985a,b; Hamilton and Woods, 1981; Steger, 1982). Their results showed that pyrrhotite surfaces

* School of Minerals Processing and Bio-Engineering,Central South University, Changsha, China

(at fracture faces) oxidize immediately on exposure to ambient air to form an overlayer of iron(III) hydroxide (or hydrated oxide) covering an iron-deficient sulfide lattice, the metal content of which decreases with increasing exposure (i.e., oxidation) time. This air oxidation mechanism is similar to that found in XPS studies of galena (PbS), pyrite (FeS_2) and chalcopyrite ($CuFeS_2$) (Buckley and Woods, 1984a,b; 1987).

Cheng et al. (1994) reported the cathodic decomposition behavior of pyrrhotite in deoxygenated solution. They showed that pyrrhotite did not exhibit significant collectorless flotation and, when ethyl xanthate was as collector, the floatability of pyrrhotite decreased with cathodic treatments: the more negative the potentials applied and the longer the time used for polarization, the lower the recoveries became. They offered a relationship of pulp potential and flotation recovery. The electrochemical study of the surface reactions that occur during the flotation of nickeliferous pyrrhotite in the recovery of nickel and the platinum group metals was made by Buswell and Nicol (2002). Their results showed that the formation of dixanthogen on pyrrhotite surface was thermodynamically favorable in plant flotation slurries. Leppinen et al. (1989) and Leppinen (1990) studied the electrochemical reactions involving ethyl xanthate with chalcocite, chalcopyrite, pyrite and galena in situ using an electrochemical ATR cell to spectroscopically analyze the reaction products at mineral surfaces by FTIR. They found that the IR signal intensity, which is a measure of xanthate adsorption and flotation response, can be correlated with pulp potentials.

In this paper, the flotation behavior of pyrrhotite was investigated by using ethyl xanthate as a collector. The interaction mechanism for the pyrrhotite and ethyl xanthate was studied using cyclic voltammetry. Reaction products on the pyrrhotite surfaces were examined by FTIR reflection spectral analysis.

EXPERIMENTAL

Materials

Pyrrhotite of composition $Fe_{0.885}S$ (i.e., $x = 0.115$) was used in this investigation and was from Dachang number 100 ore body, Guangxi province. Mineral lumps were crushed and hand-picked. Pyrrhotite grains of several millimeters in diameter were ground in a ceramic ball mill and sieved to −0.1mm and stored under nitrogen until needed for the flotation. This sample was analyzed to be 93.86% pure pyrrhotite. Stock solutions for flotation were prepared from distilled water and pH buffering reagents: HCl was used to modify pH to 2.2; HAC and NaAC (AC = acetate) for pH 4.7; Na_2HPO_4 and KH_2PO_4 for pH 7.0; $NH_3 \cdot H_2O$ and NH_4Cl for pH 8.8; $NaCO_3$ and $NaHCO_3$ for pH 11; and NaOH for pH 12.1. The reagents used to adjust pulp potential were ammonium persulfate [$(NH_4)_2S_2O_8$] and sodium dithionite ($Na_2S_2O_4$). All reagents for tests were analytical reagent grade except for the frother butyl ether alcohol which was added at 10 mg/L concentration for the flotation tests and potassium ethyl xanthate (KEX) which was added at concentrations dependent on the tests being conducted.

Flotation Tests

Flotation tests were carried out in a microflotation cell with 25 ml of effective volume. The sample amount used for each experiment was 2.2 grams and was prepared by ultrasonic washing for five minutes to remove possible oxides on the sample followed by de-aerating and decanting. The sample was then placed into the flotation cell along with a corresponding pH buffer solution. The pulp potential was then adjusted and measured with a saturated calomel electrode, SCE (0.245 V vs. SHE) and a platinum electrode with geometric surface area of 0.28 cm^2. The SCE was checked periodically to ensure accurate measurement. Before each experiment, the platinum electrode was cleaned. Collector and frother were then added. The flotation time was 4 minutes. Flotation recovery (R) was calculated from:

$$R = \frac{m_1}{m_1 + m_2} \times 100\% \qquad \textbf{(EQ 1)}$$

where m_1 and m_2 were weights of the floated and unfloated products respectively.

Cyclic Voltammetry

Cyclic voltammetry was used to characterize redox reaction that occurs at the pyrrhotite surface. Cyclic voltammetry was conducted by using an EG&G PARC electrochemical measurement system. The setup consisted of Model 273 multiple-function potentiostat, M270 measurement software, and a PC computer. The pyrrhotite electrode was prepared from pure crystal pyrrhotite which was hand-picked from the mineral sample. The pyrrhotite electrode was polished with 600 grit silicon carbide papers and then rinsed with distilled water before each experiment. The sweep rate was fixed at 0.01 V/s (10 mV/s).

FTIR Reflection Spectral Measurement

A NEXUS 470 infrared spectrophotometer was used. The pyrrhotite sample weighing 0.7 g was suspended in 25 ml of buffer solution at the corresponding pH. After the addition of each reagent, the suspension was stirred and settled for 15 minutes each. The solution was then filtered. The treated sample was dried in the vacuum and used for FTIR reflection spectral measurements.

RESULTS AND DISCUSSIONS

Potential-pH of Pyrrhotite Flotation

The effect of pulp pH on the flotation recovery of pyrrhotite in the presence of 1×10^{-4} M KEX is shown in Figure 1. The results indicate that the pyrrhotite exhibited good floatability from pH 2 to pH 11 when using ethyl xanthate as a collector. Recoveries exceed 80% when the pH is between 4 and 8. At high pH >12, the flotation recovery declines significantly. It was also observed that the pulp potential decreased with the increasing pH.

The relationship between pyrrhotite flotation recovery and pulp potential is better presented in Figure 2. It can be shown that the flotation of pyrrhotite may be possible only at a suitable range of pulp potentials depending on the pH conditions. By defining the flotation potential with upper and lower limits at 50% recovery, the relationship of pyrrhotite flotation upper (u) and lower (l) potential limits is presented in Figure 3. It can be seen that the flotation of pyrrhotite may occur only at a range of the pulp potential Eh (l) < Eh < Eh (u) at various pH. In the neutral and weak acidic solution, there is a wider range of pulp potential for flotation. The flotation potential range decreases with

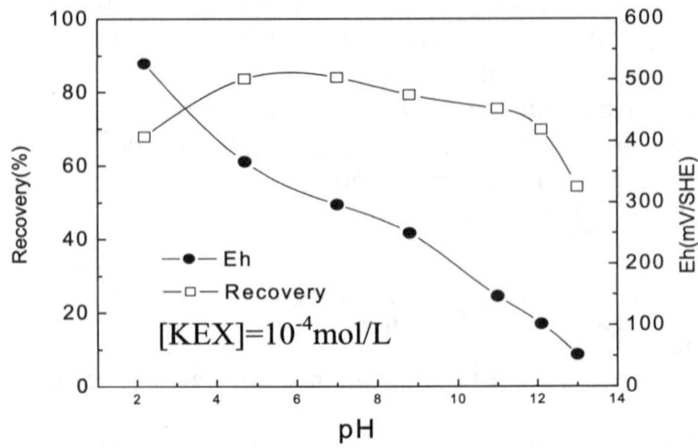

FIGURE 1 Flotation recovery of pyrrhotite as a function of pH and pulp potential

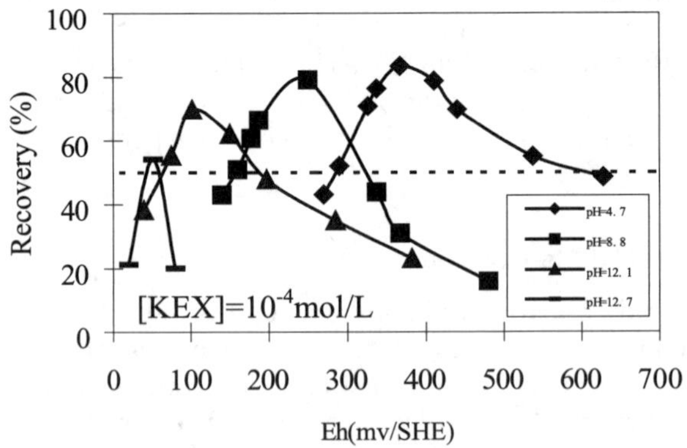

FIGURE 2 The relationship of pyrrhotite flotation recovery and pulp potential

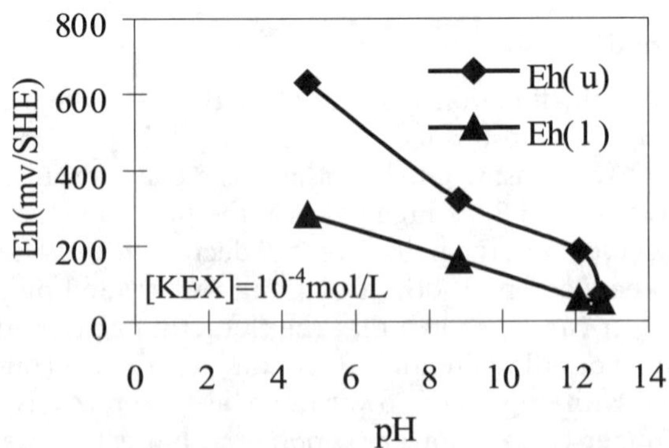

FIGURE 3 The relationship of pulp pH and pyrrhotite flotation to upper (u) and lower (l) potential limits

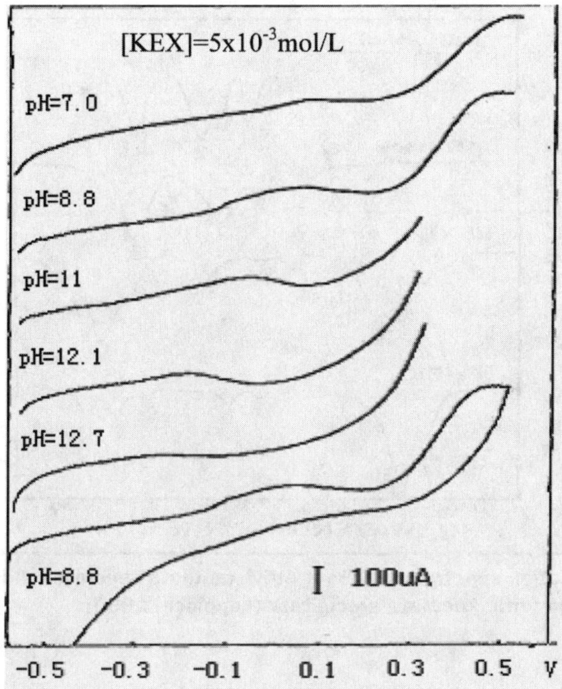

FIGURE 4 Anodic scans of cyclic voltammetry for pyrrhotite in different pH buffer solution at potential scan of 10 mV/sec

the increasing pH. For example, at pH 8.8, the optimal potential range for flotation of pyrrhotite is about 150–320 mV. The maximum flotation occurred at potential 250 mV. When the pH is high, say near 12.7; flotation of pyrrhotite falls below 50% no matter what potential is used.

Anodic Oxidation of Pyrrhotite

The anodic scan section of cyclic voltammetry for pyrrhotite at pH 7.0, 8.8, 11, 12.1 and 12.7 buffer solution with KEX are presented in Figure 4. The cyclic voltammograms curve at pH 8.8 is also presented in Figure 4. It can be seen that anodic current peak occurs at about 0.1 V. As pH increasing, the peak moves to the left. This peak may correspond to the formation of dixanthogen. The oxidation of ethyl xanthate to dixanthogen is electrochemically and thermodynamically described as shown by the following:

$$2EX^{-} \Leftrightarrow EX_{s(l)} + 2 \quad \textbf{(EQ 2)}$$

$$E° = -0.06V \quad \textbf{(EQ 3)}$$

$$Eh = -0.06 - 0.059\log[EX^{-}] \quad \textbf{(EQ 4)}$$

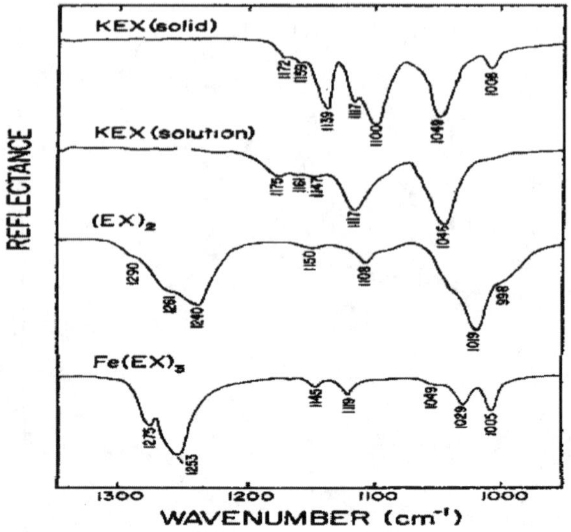

FIGURE 5 FTIR reflection spectra of different ethyl xanthate species: solid KEX, dissolved KEX, dixanthogen and ferric xanthate precipitate (Leppinen, 1990)

When the concentration of ethyl xanthate is 5×10^{-3} mol/L, *Eh = 0.08V,* which agrees reasonably well with the results in Figure 4. When hydrophobic dixanthogen is formed, the flotation of pyrrhotite becomes possible.

It can be also seen from Figure 4 that the anodic peaks move to lower potentials as the pH increases. When the pH is 12.7, the anodic peak disappears and anodic current increases rapidly. This indicates that only pyrrhotite oxidation itself takes place on the surface. In addition, no dixanthogen was formed at pH higher than 12.7 so pyrrhotite flotation can not be carried out. These phenomena may correspond to the following reactions which are written in unimolar amounts of Fe:

$$FeS_{1.13} + 7.52H_2O = Fe(OH)_3 + 1.13SO_4^{2-} + 12.04H^+ + 9.78e \qquad \textbf{(EQ 5)}$$

$$FeS_{1.13} + 4.695H_2O = Fe(OH)_3 + 0.565S_2O_3^{2-} + 6.39H^+ + 5.26e \qquad \textbf{(EQ 6)}$$

In this regard, $Fe_{0.885}S$ is equivalent to $FeS_{1.13}$.

FTIR Reflection Spectral Analysis

The FTIR reflection spectra of KEX solid, KEX in solution, diethyl dixanthogen (EX_2) and iron (III) ethyl xanthate [$Fe(EX)_3$] are shown in Figure 5. Only the wavenumbers from 950–1350 cm^{-1} are depicted since most of the important vibrations of xanthate functional groups are in this region. From Figure 5, it can be seen that the characteristic absorption bands of ethyl xanthate are the stretching vibration bands of the C-O-C at 1100–1172 cm^{-1} and C=S between 1008–1049 cm^{-1}. When diethyl dixanthogen was formed, the stretching vibration band of C=S shifted to lower wavenumbers between 998–1019 cm^{-1}, and that of C-O-C moved to higher wavenumbers between 1240–1290 cm^{-1}. For iron (III) ethyl

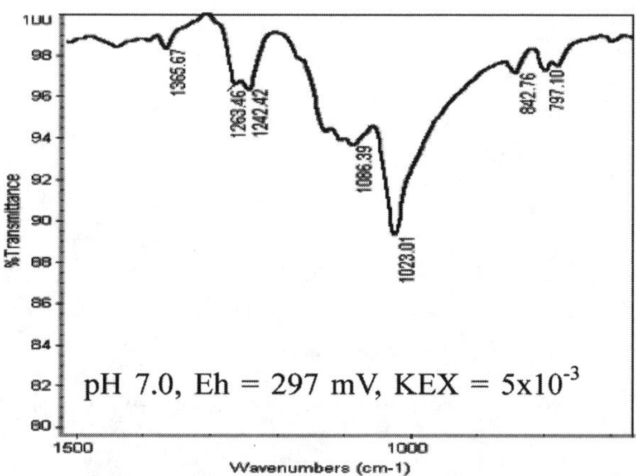

FIGURE 6 FTIR reflection spectra of 5 × 10^{-3} M ethyl xanthate adsorbed on pyrrhotite at pH 7.0 and 297 mV

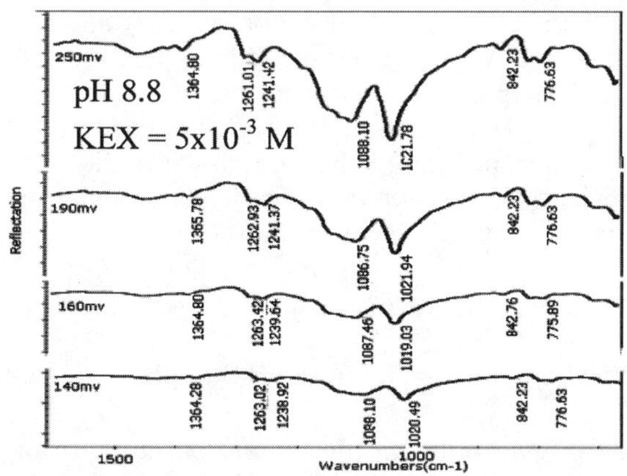

FIGURE 7 FTIR reflection spectra of 5 × 10^{-3} M ethyl xanthate adsorbed on pyrrhotite at pH 8.8 and different potentials of 250, 190, 160 and 140 mV

xanthate, the stretching vibration band of C=S shifts to lower wavenumbers between 1005–1029 cm^{-1} (Leppinen,1990).

FTIR reflection spectra of ethyl xanthate adsorption on pyrrhotite as a function of pulp potential are shown in Figures 6 and 7. It can be seen that the characteristic absorption bands of diethyl dixanthogen which at 1023cm^{-1}, 1242cm^{-1} and 1263cm^{-1} appear on surface of pyrrhotite indicating that the dominate hydrophobic species on pyrrhotite surface is dixanthogen. It follows that at pH 8.8 the ethyl xanthate adsorption on pyrrhotite is mainly of dixanthogen independent of potential in the range of 140–250 mV due to the occurrence of almost same dixanthogen characteristic bands. However, the intensity of IR signals changed at various potential values. It demonstrated that IR signals of

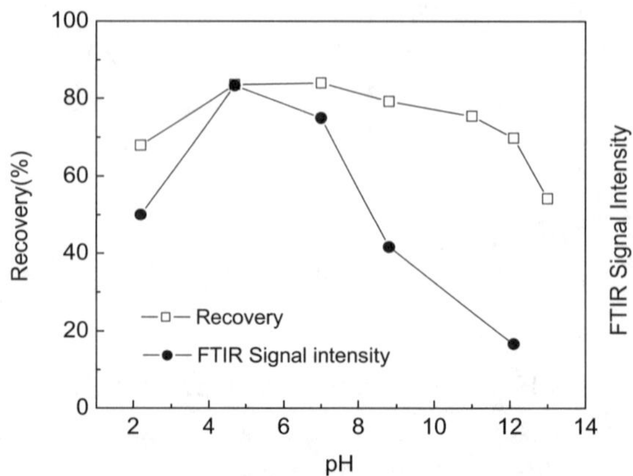

FIGURE 8 The effect of pH on FTIR spectral intensity and flotation recovery

the characteristic dixanthogen peaks and hence dixanthogen adsorption on pyrrhotite decreases with decreasing potential from 250–140 mV. This correlates well with the flotation results Figure 2 and 3. At pH 8.8, flotation recovery and dixanthogen adsorption on pyrrhotite are decreased with the decrease of pulp potential from 250–140 mV. The effect of pH on FTIR signal intensity of xanthate adsorption on pyrrhotite and pyrrhotite flotation recovery is plotted in Figure 8. The signal intensity has been measured near 1200 cm^{-1}. It can be seen from Figure 8 that IR signal intensity correspond well to flotation recovery. At weak acidic pulp solution, The IR signal intensity is the strongest and the flotation recovery of pyrrhotite is maximum.

CONCLUSIONS

1. Pyrrhotite had good floatability from pH 2 to pH 11 and poor floatability when the pH >12.
2. The flotation of pyrrhotite was dependent on pulp pH and potential. At weak acidic pH media, the potential range of pyrrhotite flotation is wider. With the increase of pH, the potential range is decreased.
3. Through potential scan of cyclic voltammograms for pyrrhotite and analysis of FTIR reflection spectra, it can be concluded that the major product of ethyl xanthate adsorption on pyrrhotite is diethyl dixanthogen.
4. Dixanthogen therefore accounts for the flotation of pyrrhotite.

ACKNOWLEDGEMENTS

The authors would like to thank China National Science Fund Key Program (NO. 50234010) and China Foundation of National Excellent Doctoral Dissertation (NO.200145) for supporting this research.

REFERENCES

Buckley, A.N. and Woods, R., 1985a, Appl. X-Ray Photoelectron Spectroscopy of Oxidized Pyrrhotite Surfaces, Exposure to Air, Appl. Surf. Sci., 22/23 :280–287.

Buckley, A.N. and Woods, R., 1985b, Appl. X-Ray Photoelectron Spectrpscopy of Oxidized Pyrrhotite Surfaces, Exposure to Aqueou solutions, Appl. Surf. Sci., .20:472–480.

Buckley, A.N. and Woods, R, 1984a, Appl. An X-Ray Photoelectron Spectroscopic Study of The Oxidation of Galena, Appl. Surf. Sci. , 17: 401–414.

Buckley, A.N. and Woods, R., 1984b, An X-Ray Photoelectron Spectroscopic Study of The Oxidation of Chalcopyrite, Aust. J. Chem., 37: 2403–2413.

Buckley, A.N. and Woods, R., 1987, The Surface Oxidation of Pyrite, Appl. Surf. Sci. , 27: 437–452.

Buswell, A.M., Nicol, M.J., 2002, Some Aspects of Applied Electrochemistry of the Flotation of Pyrrhotite, Journal of Applied Electrochemistry, 32(12): 1321–1329.

Cheng, X., Iwasaki, I., and Smith, K.A., 1994, An Electrochemical Study on Cathodic Decomposition Behavior of Pyrrhotite in Deoxygenated Solutions, Mineral & Metallurgical processing, 1(3): 160–167.

Hamilton, I.C. and Woods, R., 1981, An Investigation of Surface Oxidation of Pyrite and Pyrrhotite by Linear Potential Sweep Voltammetry, J. Electroanal Chem., 118:327–343.

Heyes, G.W. and Trahar,W.J., 1984, The Flotation of Pyrite and Pyrrhotite in the Absence of Conventional Collectors. Proceeding-The electrochemical society, 10: 219–232.

Leppinen, J.O., Bssilio, C.I. and Yoon, R.H., 1989, In-Suit FTIR Study of Ethyl Xanthate Adsorption On Sulfide Under Conditions of Controlled Potential, International Journal of Mineral Processing, 26:259–274.

Leppinen, J.O., 1990, FTIR and Flotation Investigation of The Adsorption of Ethyl Xanthate On Activation and Non-activated Sulfide Mineral, Int. J. Miner. Process., 30: 245–263.

Steger, H.F., 1982, Oxidation of sulfide minerals, VII. Effect of Temperature and Relative Humidity On the Oxidation in of Pyrrhotite, Chem. Geol. , 35: 281–295.

Wang Dianzuo and Hu Yuehua, 1988, Solution Chemistry of Flotation, Hunan Science Technology Press, Changsha, 298 (in Chinese).

Zou Tonghui et al. Analytical Chemical Handbook, 1997, Chemistry Industry Press, Beijing, 335–349 (in Chinese).

Applications of the Electrochemistry of Fine Mineral Sulfides

Michael J. Nicol* and Hajime Miki*

Relatively simple techniques have been developed which enable the electrochemistry of sulfide minerals to be studied with particles varying in size from above 100 μm to less than 1 μm. Both oxidative and reductive processes have been studied using the minerals pyrite, arsenopyrite and chalcopyrite. It has been shown that it is possible to completely oxidise or reduce the minerals during a single voltammetric sweep. The resulting voltammogram produces peaks which are characteristic of each mineral and can be used to qualitatively identify the minerals. In the case of chalcopyrite, it has been demonstrated that the charge involved in the anodic oxidation can be quantitatively related to the amount of copper dissolved. Quantitative information can be obtained using peak fitting techniques. Interesting differences in the behaviour of the minerals in sulfuric and hydrochloric acids have been observed, particularly in the case of pyrite for which a second major peak is obtained which has been attributed to the oxidation of elemental sulfur in the presence of chloride ions. The technique offers the possibility of providing mineralogical information of individual particles based on an electrochemical voltammetric fingerprint.

INTRODUCTION

Electrochemical techniques have been widely applied used in the study of leaching, precipitation and flotation of sulfide minerals. The most commonly used technique involves the fabrication of electrodes from massive samples of the relevant mineral. However, in many cases, such material is not available or alternatively, it is often desirable to study the behaviour of milled materials such as would be produced as flotation concentrates. The method most widely used for such materials involves fabrication of an electrode

* Parker Centre, Murdoch University, South Street, Murdoch, Western Australia

TABLE 1 Properties of the mineral samples

Sample	Chemical composition, %	Mineralogical composition	Size, D_{50}
$CuFeS_2$	Cu 25.5, Fe 29.5, S 33.9, Si 1.20	$CuFeS_2$	2.38 μm
FeS_2	Fe 30.8, S 35.0, Si 0.10	FeS_2, minor SiO_2	1.42 μm
FeAsS	As 42.9, Fe 36.0, S 20.2	FeAsS	1.78 μm
−1μm $CuFeS_2$ conc	Cu 26.6, Fe 32.0, S 35.8, Si 2.40	$CuFeS_2$, minor FeS_2 , SiO_2	0.70 μm
−3μm $CuFeS_2$ conc	Cu 25.5, Fe 29.0, S 30.3, Si 2.21	$CuFeS_2$, minor FeS_2 , SiO_2	2.21 μm
−10μm $CuFeS_2$ conc	Cu 25.9, Fe 26.0, S 30.0, Si 2.25	$CuFeS_2$, minor FeS_2 , SiO_2	6.25 μm
−45μm $CuFeS_2$ conc	Cu 24.7, Fe 30.2, S 34.0, Si 1.10	$CuFeS_2$, minor FeS_2 , SiO_2	22.9 μm

using carbon paste (graphite powder and a non-conducting binder such as paraffin oil) and finely milled mineral (Lázaro et al. 1995, 1997; Lu et al. 2000). An alternative technique which does not appear to have been adopted by workers in the mineral field involves direct attachment of a powdered sample to the surface of paraffin-impregnated graphite electrodes (Sholz and Meyer 1998). This generally involves attachment of the mineral particles by rubbing of the electrode over a film of the powdered mineral on a glass slide. A major drawback of these techniques is that the amount of active material in the electrode is not known and quantitative analysis is therefore not possible.

In this study, the latter technique has been improved and extended using a quantitative method for attaching microgram quantities of fine mineral particles to a graphite electrode surface. This paper summarizes the initial results obtained with this method compared with the conventional method, and suggests some applications of this technique in quantitative analysis of mineral oxidation and mineralogical identification.

MATERIALS AND METHODS

Preparation of Powder Samples

Highly purity samples of massive pyrite and arsenopyrite were crushed in a ring mill and then manually ground in a pestle and mortar. Massive chalcopyrite was ground by hand and wet milled in a stirred mill, washed with acetone and dried. A chalcopyrite concentrate was ground in a ring mill and several size fractions obtained by successive suspension in acetone in an ultrasonic bath followed by decantation for various time periods.

Each of the samples was characterized by chemical analysis, X-ray diffraction (CuK· radiation Kristalloflex-4, Siemens Inc.) and particle size (SRA 150, Microtrac Inc.). The results are shown in Table 1.

Electrochemical Measurements

Electrochemical measurements were carried out using a standard three electrode system, consisting of a working electrode, platinum counter electrode, and a saturated calomel reference electrode in a small glass cell (3.54 cm^3). The reference electrode was provided with a Luggin capillary tip. Electrolytes were prepared using analytical grade reagents and distilled-deionized water. Before each experiment, the electrolyte was purged with water-saturated nitrogen for 30 min. The electrolyte and glass cell were maintained at a temperature of 25°C by circulation of thermostatically controlled water around the glass cell. A

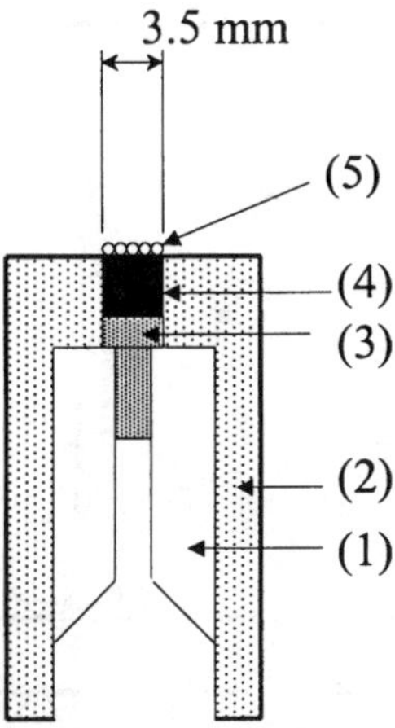

FIGURE 1 Schematic of working electrode (1) Stainless steel electrode holder, (2) Plastic coating of electrode holder, (3) Stainless steel screw, (4) Carbon paste, (5) Powder sample

potentiostat (Model 273, EG&G Instruments Inc.) was used in conjunction with corrosion measurement software (352 SoftCorr III Ver. 3.06, EG&G Instruments Inc.).

Figure 1 shows a schematic figure of the working electrode.

The cavity between the surface of the holder and the screw was filled with carbon paste (graphite powder and paraffin oil in 4:3 mixture and the surface smoothed using a spatula. The mineral powder sample was attached as follows:

> 0.05 g of the powdered sample, 5 ml each of ethanol and water was added to a 20-cm^3 glass vial, and suspended using an ultrasonic bath. A known amount of this suspension was pipetted onto the surface of the carbon paste and dried using a hot air heater. In this way a known amount (generally of the order of µg) could be deposited on the surface of the electrode.

For comparison with the conventional carbon paste method, a paste containing graphite, paraffin oil and powder $CuFeS_2$ sample was made by mixing in the ratio 4:3:0.8 and placed into the cavity of the electrode holder.

Cyclic voltammetric experiments were carried out at a scan rate of 1 mV s^{-1} starting in the anodic direction from the open circuit potential to about 1.4 V and reversed. In some cases, the electrolyte was sampled after the experiment and the metal ion concentration was measured by ICP analysis.

RESULTS

Comparison of the New Method with the Conventional Method

Figure 2 shows the voltammogram for a $CuFeS_2$ sample in 1 mol dm^{-3} H_2SO_4 in which 10 µg of the sample was studied using the new method.

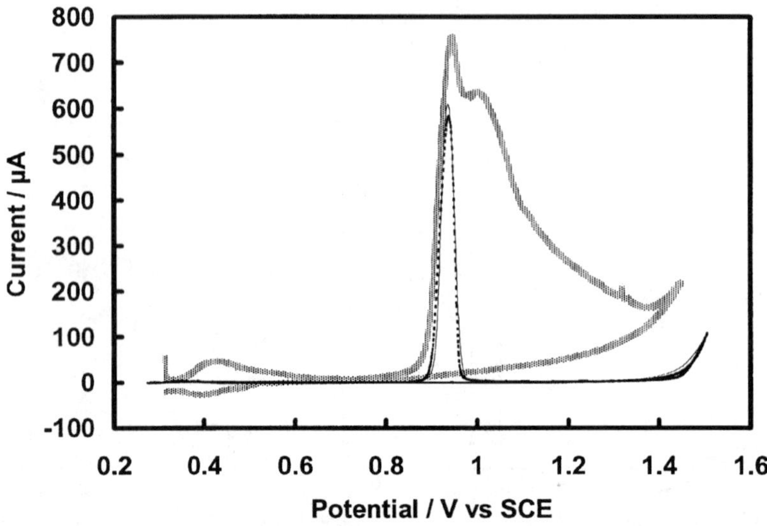

FIGURE 2 Comparison of voltammograms obtained by the new method (10 µg $CuFeS_2$; dotted and solid line) and the conventional method ($CuFeS_2$ and carbon paste mixture: thick line) in 1 mol dm^{-3} H_2SO_4

The reproducibility is good as shown by the two peaks at about 0.95 V (dotted and solid). For comparison, the voltammogram obtained using the conventional carbon paste method under the same conditions is shown as the thick solid line.

It is apparent that the new technique yields very different results in that a reproducible symmetrical sharp peak is obtained with low background currents compared to the broad peak obtained with the conventional method. This difference is largely due to the fact that complete oxidation/dissolution of the fine mineral particles occurs during the potential scan. It should be noted that the small peak normally observed with bulk chalcopyrite at about 0.45 V is not observable in the voltammogram obtained using the new method.

The Effect of Sample Mass

As indicated above the new attachment method enables the amount of material to be varied by regulating the volume of suspension used to deposit the sample onto the electrode surface. Figure 3 shows the results of voltammetric experiments in which various amounts of $CuFeS_2$ (A), FeS_2 (B) and FeAsS (C) were studied in 1 mol dm^{-3} H_2SO_4 electrolyte.

As could be expected the peak heights increase with increasing sample amount. The peak potentials for $CuFeS_2$ and FeS_2 are both at about 0.95 V although the width of the pyrite peak is considerably greater than that for chalcopyrite. In the case of arsenopyrite, it appears that there are two overlapping peaks at potentials below that for the other minerals.

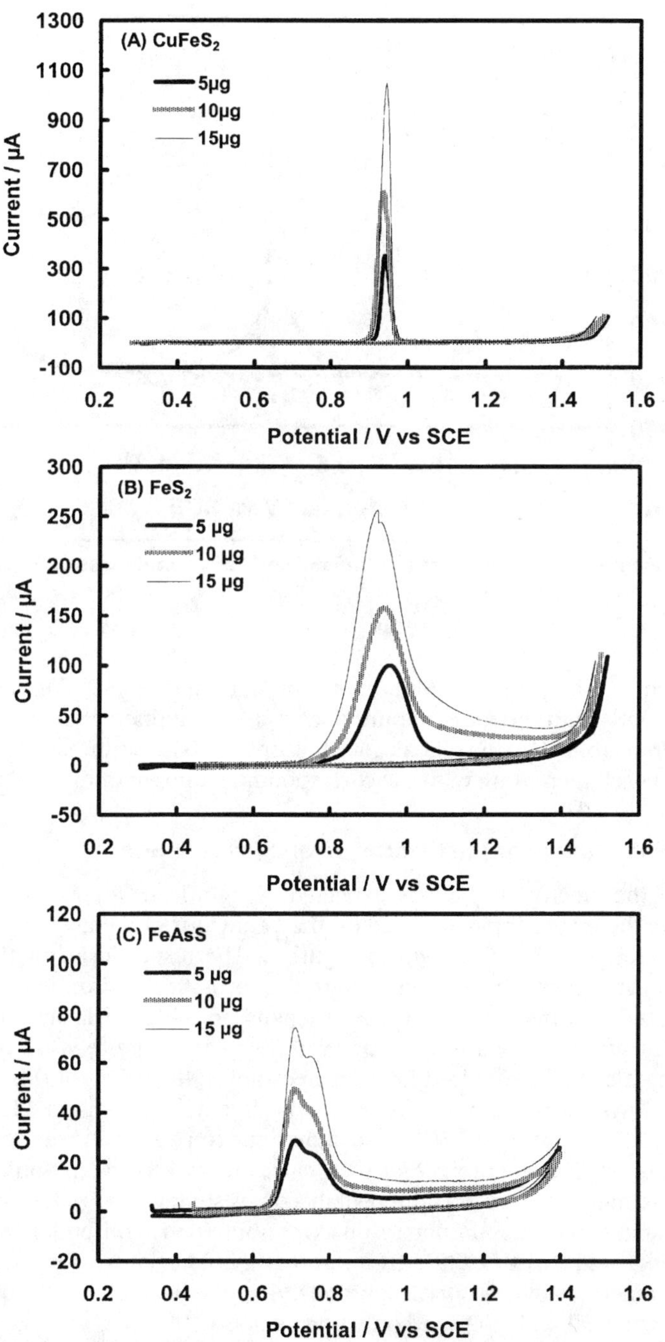

FIGURE 3 The effect of the attached amount of $CuFeS_2$ (A), FeS_2 (B) and FeAsS (C) on voltammograms in 1 mol dm^{-3} H_2SO_4

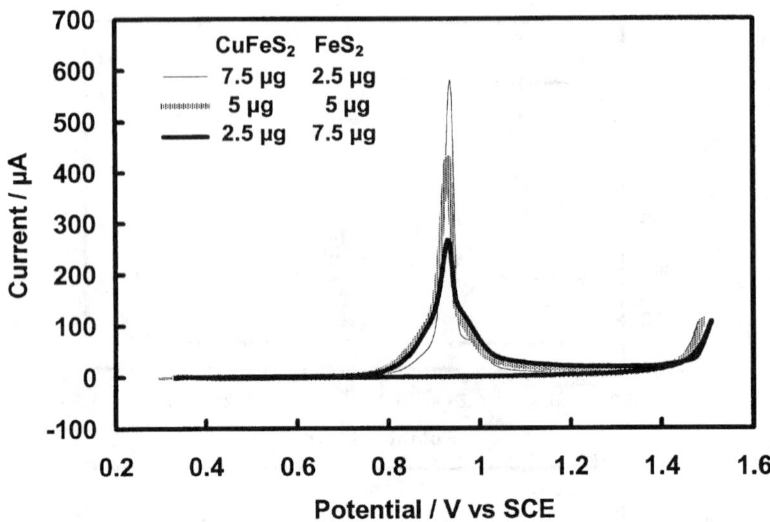

FIGURE 4 **Voltammograms for mixtures of $CuFeS_2$ and FeS_2 (total mass: 10 µg) in 1 mol dm^{-3} H_2SO_4**

Figure 4 shows the result of voltammetry in 1mol dm^{-3} H_2SO_4 for various mixtures of $CuFeS_2$ and FeS_2 with the total amount of sample maintained at 10 µg.

It is apparent that the sharp peak for chalcopyrite is superimposed on the broader FeS_2 peak with each increasing with the corresponding amount of mineral in the mixture.

The Effects of the Nature of the Electrolyte and Particle Size

The nature of the electrolyte has, as expected, a significant effect on the voltametric response. Thus the effect of particle size on the behaviour of a chalcopyrite concentrate in 1 mol dm^{-3} KCl solution is shown in Figure 5. The response in neutral solutions is more complex than in acidic solutions due to the formation of iron hydroxides and the relative magnitudes of the peaks appears to depend on the particle size. This latter phenomenon is unusual but was found to be reproducible and more work on chalcopyrite and other minerals will be required in order to establish the origin of this effect.

Figure 6 shows the result of voltametric experiments in 1 mol dm^{-3} HCl for 10 µg of $CuFeS_2$ (A), FeS_2 (B) and FeAsS (C). Also shown for comparison is that obtained in sulfuric acid solutions. It is apparent that the peak potential for oxidation of chalcopyrite shifts to slightly higher values and the peak becomes somewhat broader in hydrochloric than sulfuric acid solutions. Similar results were obtained with bulk chalcopyrite electrodes by Biegler and Swift (1979). Of interest is the fact that two peaks at 0.91 V and 1.22 V are obtained for pyrite in HCl solutions. Also the peak obtained for FeAsS in HCl is higher than the peak in H_2SO_4 and only one peak can be observed.

In order to establish the role of chloride, the effect of the chloride ion concentration on the behaviour of pyrite was investigated. Figure 7 shows the results of experiments carried out using 0.1 mol dm^{-3} HCl on the one hand and 0.1 mol dm^{-3} HCl + 0.9 mol dm^{-3} NaCl on

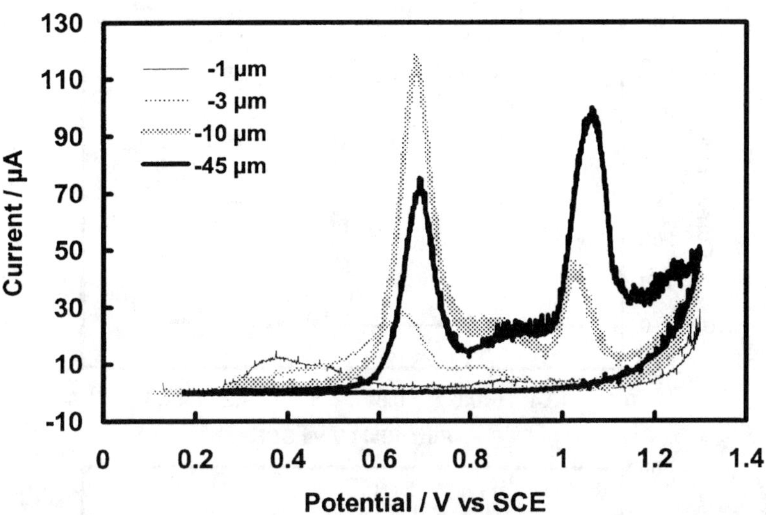

FIGURE 5 The effect of particle size on voltammograms for 25μg of chalcopyrite concentrate in 1 mol dm^{-3} KCl

the other. It is observed that the second anodic peak is only obtained in solutions of high chloride concentration.

DISCUSSION

Unlike conventional voltammetry using bulk mineral or powdered composites, the peaks obtained using the present technique appear to be symmetrical which suggests that complete oxidation of the fine mineral particles is obtained. This was confirmed by the observation that a second anodic sweep on the same sample revealed the absence of any additional oxidation. The data in the vicinity of each peak was fitted using a Gaussian function and the results are shown in Figure 8 for 10 μg of FeAsS (A) and for a mixture of 2.5 μg $CuFeS_2$ and 7.5 μg FeS_2 (B).

As shown the curve for FeAsS can be resolved into two peaks at 0.704 V and 0.762 V while the composite curve for the mixture can be deconvoluted to give a sharp peak for $CuFeS_2$ and a broad FeS_2 peak.

The charge involved in the oxidation of each mineral can be estimated from the area under each peak. If the reaction for oxidation of the mineral at the peak potential (i.e., the number of electrons involved per mole of mineral) is assumed then the amount of mineral oxidised at each peak can be simply estimated as

$$W = \frac{C \times M}{F \times n} \qquad \textbf{(EQ 1)}$$

where W is the mass(g) of mineral oxidised, C is total charge (coulomb) under the peak, M is the molecular mass(g) of the mineral, F is the Faraday (96485 C) and n is the number of electrons needed for the oxidation of 1 molecule of mineral.

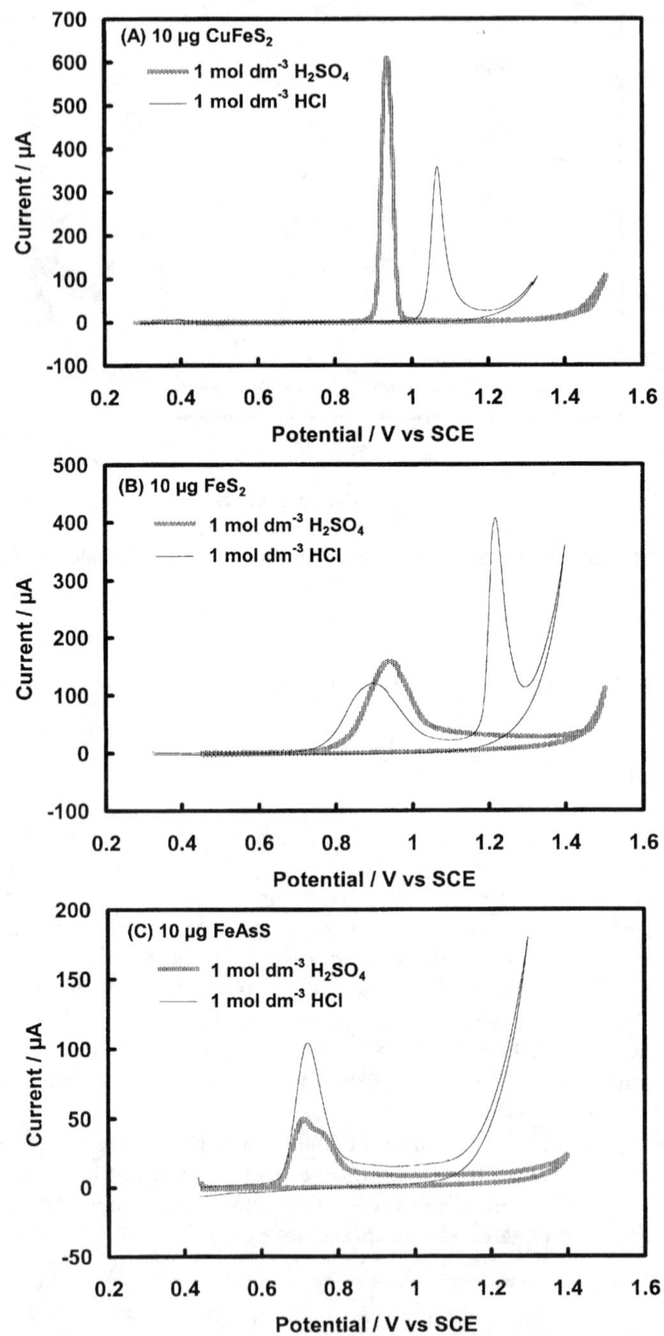

FIGURE 6 The effect of the electrolyte on voltamograms for 10 µg $CuFeS_2$ (A), FeS_2 (B) and FeAsS (C) in 1 mol dm^{-3} H_2SO_4 (thick line) and 1 mol dm^{-3} HCl (thin line)

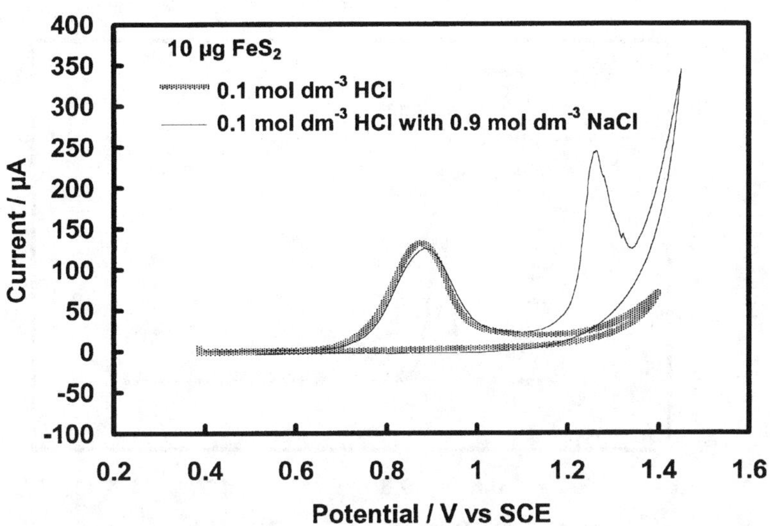

FIGURE 7 The effect of chloride ions on voltammograms for 10 µg FeS_2 in 0.1 mol dm^{-3} HCl (thick line) and 0.1 mol dm^{-3} HCl with addition of 0.9 dm^{-3} NaCl (thin line)

Thus, assuming that the reactions associated with the peaks for chalcopyrite and pyrite are given by Equations 2 and 3, the areas under the peaks in Figures 3 and 4 can be used to estimate the amount of material on the electrode surface. Note that iron(III) is the assumed product of oxidation given that the peak potentials are significantly greater than the potentials required to oxidise iron(II) to iron(III). This calculated mass is compared to that added as a suspension in Figure 9.

$$CuFeS_2 = Cu^{2+} + Fe^{3+} + 2S^0 + 5e^- \qquad \textbf{(EQ 2)}$$

$$FeS_2 = Fe^{3+} + 2S^0 + 3e^- \qquad \textbf{(EQ 3)}$$

There appears to be a good correlation between the amount of mineral added and the charge under the relevant peak. In the case of chalcopyrite, the amount calculated from the charge is about 20% less than the estimated amount added. At least part of this discrepancy can be related to the purity of the fine fraction of the mineral sample. It could possibly be expected that partial oxidation of sulfur to higher oxidation states would occur at the potentials involved in these experiments. However, this appears to be unlikely given that the charge/mole of mineral would be greater than estimated on the basis of the above equations. Thus, elemental sulfur is apparently the primary product of oxidation of these minerals.

A comparison of the amounts of $CuFeS_2$ and FeS_2 calculated from the charge under the deconvoluted peaks obtained with mixtures of these minerals (Figure 4) with the actual amounts added is shown in Figure 10.

Given the inaccuracies inherent in mixing and handling microgram quantities of the minerals, the results suggest that semi-quantitative estimates can be made of mineral composition even for cases in which the peaks overlap.

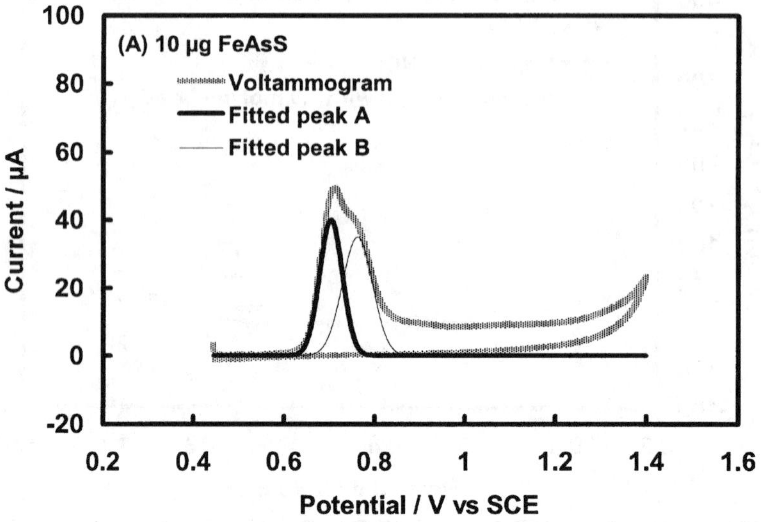

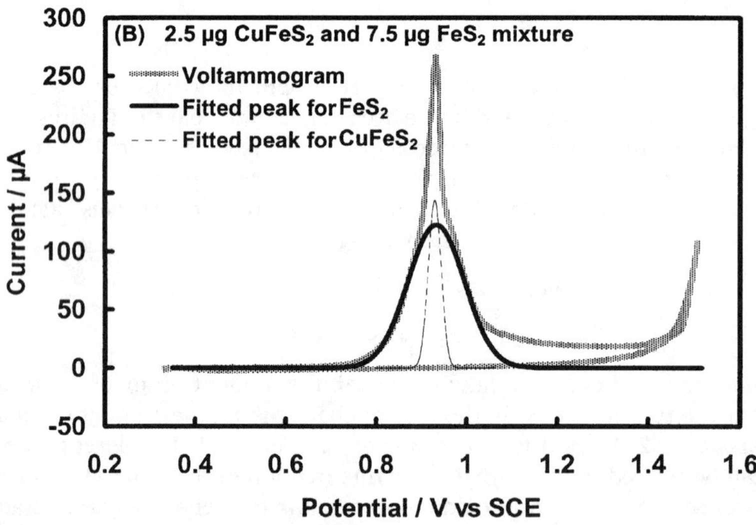

FIGURE 8 Example of peak fitting of voltammograms for 10 μg FeAsS (A) and a 2.5 μg $CuFeS_2$ + 7.5 μg FeS_2 mixture (B) in 1 mol dm^{-3} H_2SO_4

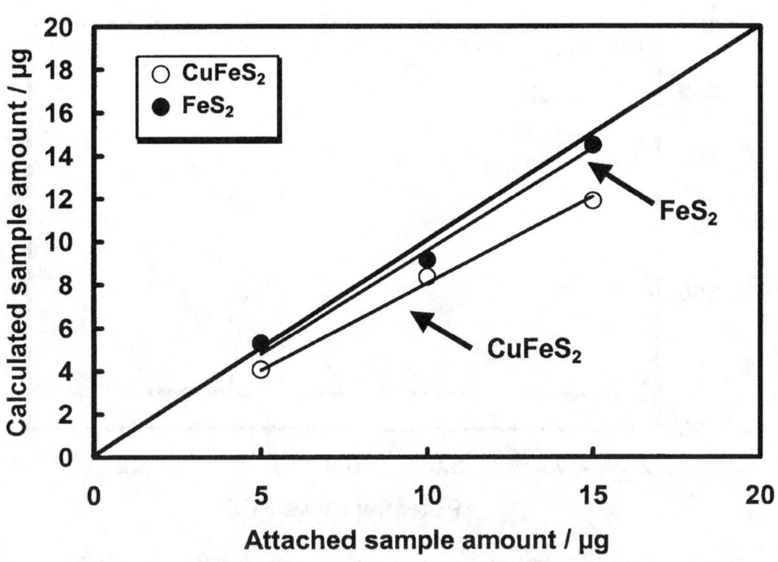

FIGURE 9 The relationship between the attached amount and that calculated from the charge for $CuFeS_2$ and FeS_2 (Data from Figure 3(A): $CuFeS_2$, and Figure 3(B): FeS_2)

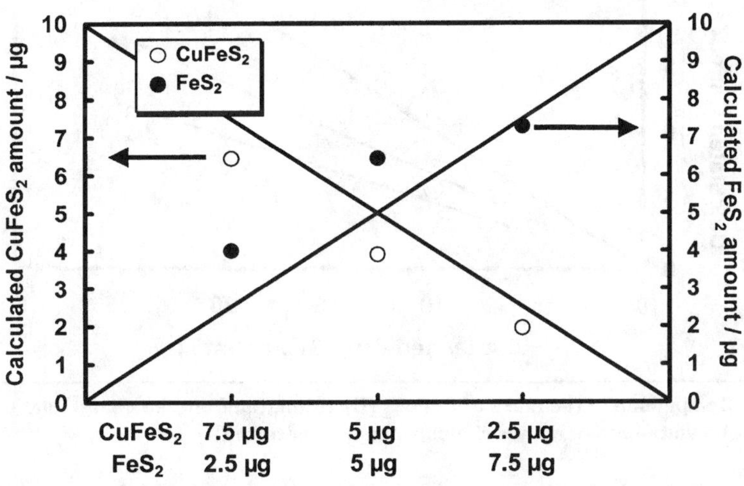

FIGURE 10 Estimated masses of $CuFeS_2$ and FeS_2 from the charge under the deconvoluted peaks in Figure 4

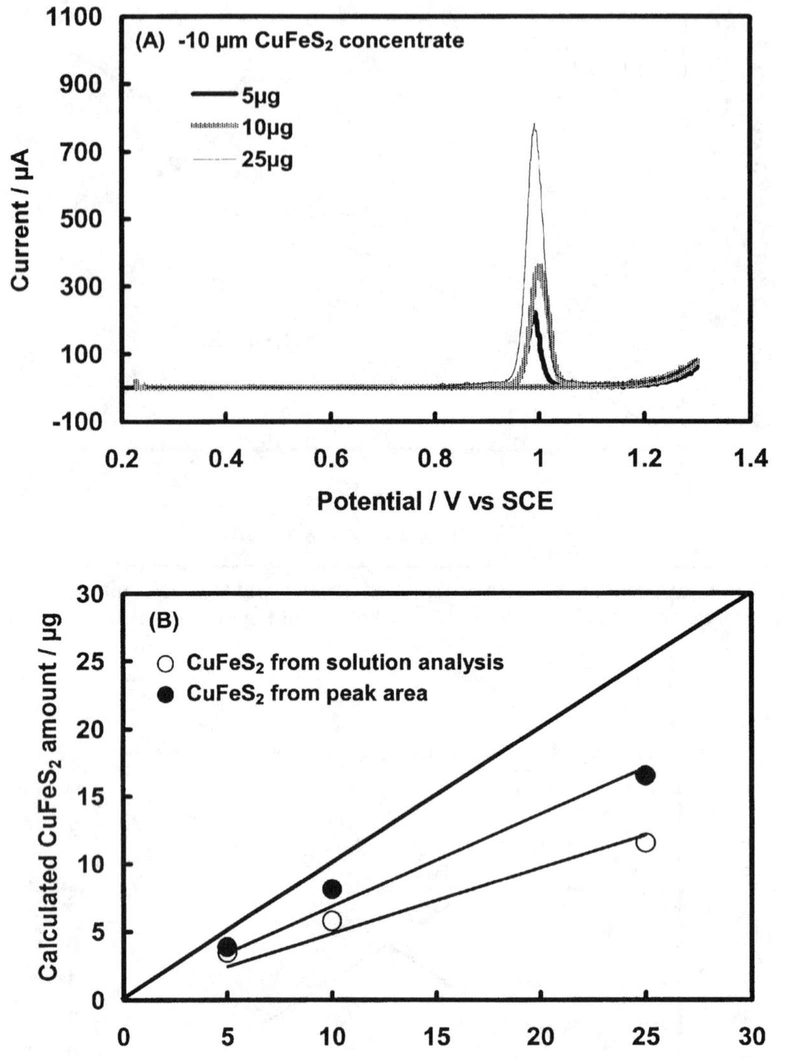

FIGURE 11 Comparison of the mass of $CuFeS_2$ (B) estimated from the charge and the solution analysis for the voltammograms for chalcopyrite concentrate (A)

In addition to the use of relatively pure mineral samples, a sample of a chalcopyrite concentrate was also used and the results obtained with various amounts of −10 µm material in 1 mol d^{-3} HCl solution is summarized in Figure 11(A). After each of these experiments, the solution was analysed for copper and the amount of chalcopyrite estimated from both the anodic charge and the mass of copper dissolved assuming that all the copper in the concentrate is in the form of chalcopyrite. These calculated amounts are plotted against the actual mass added to the electrode in Figure 11(B).

It is obvious that, although the trend with increasing mass added is as expected, the amount of mineral estimated from the charge and from the amount of copper dissolved is lower than that expected from the mass added. This is due in part to the fact that the concentrate is not pure chalcopyrite (approximately 80% $CuFeS_2$) and contained small quantities of other sulphide minerals and quartz. Further work is required to establish whether it will be possible to establish sample purity by this method.

Comparison of the results in Figure 6 with those in Figure 7, reveals an interesting difference between the behaviour of pyrite and the other minerals in that only the former displays two peaks in the presence of high concentration of chloride ions. This second peak at 1.22 V is not observed in the case of chalcopyrite and arsenopyrite. As has already been discussed, it is likely that the first anodic peak corresponds to oxidation of the mineral to ferric ions and elemental sulfur as in reaction Equation 3. It is possible that the second peak can be attributed to the oxidation of sulfur in the presence of high concentrations of chloride ions by reaction Equation 4 for which the standard reduction potential is 1.23 V versus SHE (Zhdanov 1973).

$$2S + 2Cl^- = S_2Cl_2 + 2e^- \qquad \textbf{(EQ 4)}$$

The charges involved in these reactions are at least consistent with this interpretation in that the ratio of the charge (21.1 mC) involved in the first peak to that in the second peak (16.2 mC) is close to that of 3/2 expected in terms of reactions 4 and 6. The reason for the absence of a second peak in the case of the other minerals could be related to the nature of the sulfur produced by anodic oxidation. It is possible that the initial product of oxidation is S_2 in the case of pyrite but a less reactive S_8 form in the case of chalcopyrite and arsenopyrite. This aspect will be explored in more detail in a later study.

CONCLUSIONS

This study has shown that an extension of a simple technique involving the quantitative attachment of microgram quantities of fine mineral particles to a graphite electrode can yield interesting and potentially useful results based on voltammetry.

1. High sensitivity and reproducibility can be achieved in comparison with the conventional approach using composite electrodes of powdered minerals and graphite.
2. The resulting voltammograms provide the opportunity for a fingerprint of mineral composition based on the peak potentials which are characteristic of each mineral.
3. Quantitative analysis can be carried out using the charge under each peak which can generally be resolved into Gaussian peaks in the case of peak overlap.
4. Additional information on the reactivity of the anodic oxidation products has revealed important differences between minerals.

REFERENCES

Biegler, T., and D.A. Swift. 1979. Anodic Electrochemistry of Chalcopyrite. *Journal of Applied Electrochemistry*. Vol. 9. 545–554.

Lázaro, I., N. Martínez-Medina, I. Rodríguez, E. Arce, and I. González. 1995. The use of carbon paste electrodes with non-conducting binder for the study of minerals: chalcopyrite. *Hydrometallurgy*. Vol. 38. 277–287.

Lázaro, I., R. Cruz, I. González, and M. Monroy. 1997. Electrochemical oxidation of arsenopyrite in acidic media. *International Journal of Mineral Processing*. Vol. 50. 63–75.

Lu, Z.Y., M.I. Jeffery, and F. Lawson. 2000. An electrochemical study of the effect of chloride ions on the dissolution of chalcopyrite in acidic solutions. *Hydrometallurgy*. Vol. 56. 145–155.

Sholz, F., and B. Meyer. 1998. Voltammetry of solid microparticles immobilized on electrode surfaces. *Electroanalytical Chemistry*. Vol. 20. 1–86.

Zhdanov, S.I. 1973. Sulfur. In *Encyclopaedia of Electrochemistry of the Elements*. Vol. 4. Edited by A.J. Bard. New York: M. Dekker. 273–360

Mathematical Modeling of Leaching Kinetics with Diminishing Driving Forces

Patrick R. Taylor* **and Edgar E. Vidal***

Standard textbook equations for heterogeneous kinetics typically utilize a constant driving force to solve for rates and times of reactions. In order to more closely model real systems, equations and solutions are presented for particle leaching kinetics in which the driving force is decreased as the reactions proceeds. In addition, models for continuous co-current and continuous counter current aqueous-solid reactions are presented.

INTRODUCTION

The ability to describe metallurgical fluid-solid reactors mathematically requires knowledge of the various model components as well as the types of reactors used. One way to visualize mathematical model development is shown in Figure 1 (Taylor and Martins 1986). Over the past twenty years, there has been an increasing professional awareness of the value of applied mathematics tools in the design and operation of metallurgical plants. Tools such as: statistical analysis, operations research, numerical methods and fundamental modeling have become a part of the Extractive Metallurgist's toolbox. Figure 1 shows that the development of a useful mathematical model includes both a fundamental understanding of the phenomena occurring and experimental and/or operational validation. Useful models are developed through an iterative process where observations of real systems determine the incorporation of more complexity into the mathematical description of the system.

* Kroll Institute for Extractive Metallurgy, Department of Metallurgical and Materials Engineering, Colorado School of Mines, Golden, Colorado

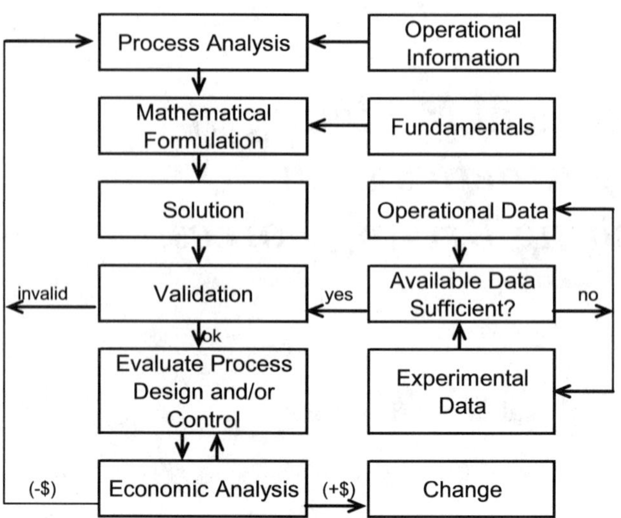

FIGURE 1 Schematic flow sheet for process model development

The first stage in a mathematical model development is the observation and attempt to quantify all of the variables, parameters, phenomena and geometric considerations that are thought to be important in a system. Many assumptions are typically made initially, in order to reduce the complexity of the system, as most extractive metallurgy systems are somewhat complex. Thus the simple models are formulated and used often; but do not have the needed complexity to accurately describe real systems. These assumptions are subsequently validated and/or eliminated (through additional mathematical complexity) based upon attempts to determine that the model represents the real system.

Once the models are formulated, then they must be solved. Programs such as Mathematica and MathCad have greatly simplified the solutions of the more complex systems and numerical methods have been shown to be very useful in solving complex models rapidly. Once the model is solved, it must be compared to real data (either experimental and/or operational). It is through this attempt at validation that the utility of the model is identified. Some of the ways that a model may be invalid include: making an invalid assumptions, failure to include and important variable or variable interaction, or failure to describe the process phenomena accurately.

The obvious value of modeling is concerned with economics. If it is possible to accurately describe a process mathematically, then it is also possible to optimize, control or improve a process by manipulation of the model. The other value of modeling is that it requires the engineer to attempt to understand all of the important phenomena occurring during the operation and thus improve the overall understanding of the process.

REACTOR DESIGN MODELS

Bartlett (1971, 1972) developed mathematical model based design curves for extraction in batch, plug flow and perfect mixed fluid-solid reactions, with and without a distribution of solid particle sizes and for single or multiple reactors. He utilized the shrinking core model (Levenspiel 1972) to describe the rain reaction kinetics and assumed a constant fluid reactant driving force. His equations follow.

The general method to incorporate a distribution of solid particle sizes, for uniform residence times, into an extraction curve is given in the following equation:

$$1 - X = \int_0^{r^*} (1 - X_1) \cdot f(R_o) dr_o \qquad \textbf{(EQ 1)}$$

where

X is the overall fraction converted
r^* is the maximum particle radius (cm)
X_I is the fraction converted for a particle of size r_o
r_o is the particle radius (cm)
$f(r_o)$ is the particle size probability density function (cm^{-1})

X_I is determined by assuming the type of shrinking core, or other, grain reaction equation and expressing it as a function of particle size.

For a G.G.S. distribution, the particle size distribution can be expressed as:

$$f(r_o) = (b(r_o)^{b-1})/(r^*)^b \qquad \textbf{(EQ 2)}$$

where b is a constant.

The equation that incorporates a residence time distribution, for uniform size particles, into the extraction curves is as follows:

$$1 - X = \int_0^{t^*} (1 - X_J) g(t) dt \qquad \textbf{(EQ 3)}$$

where
t^* is the time to react the largest particle (sec)
X_J is the fraction converted in time t
$g(t)$ is the particle residence time probability density function (sec^{-1}).

X_J is determined by assuming the type of shrinking core, or other, grain reaction and expressing it as a function of time.

For a perfect mixed reactor, the residence time probability density function is:

$$g(t) = \frac{\exp\left(-\frac{t}{\tau}\right)}{\tau} \qquad \textbf{(EQ 4)}$$

where τ is the mean residence time in the reactor (sec).

For multiple perfect mixed reactors:

$$g(t) = n^n \left(\frac{t}{\tau}\right)^{n-1} \exp\left(\frac{-n(t/\tau)}{\tau(n-1)!}\right) \qquad \textbf{(EQ 5)}$$

where:

n is the number of tanks
t is the total time (sec)
τ is the total mean residence time (sec)

Loveday (1975) developed extraction equations for batch and continuous fluid-solid reactions where the solids react completely. He included a distribution of particle sizes, chemical control and mass distributions. Sepulveda and Herbst (1978) developed a population balance model approach. Their equations can be shown to reduce to those given below.

ADDITIONAL REACTOR DESIGN EQUATIONS

In addition to equations (1) through (5) the following equations are presented.

An equation can be written to incorporate both particle size distributions and residence time distributions:

$$1 - X_N = \int_0^{t^*}\left[\int_0^{r^*}(1 - X_N)(r_o, t)f(r_o)dr_o\right]g(t)dt \qquad \textbf{(EQ 6)}$$

where

X_N is the fraction converted in tank N
$X_N(r_o,t)$ is the fraction converted as a function of particle size and time (obtained from batch tests and assuming grain reaction behavior)
$f(r_o)$ defined previously and obtained by screening or other particle size distribution tests
$g(t)$ defined previously and obtained by tracer tests

It is also possible to replace the residence time distribution with a grain reaction interface position probability distribution, if we assume we know the form of reaction control:

$$1 - X_N = \int_{r_o}^{r_I} I[1 - X_N(r_I)]p(r_I)dr \qquad \textbf{(EQ 7)}$$

For chemical control, $p(r_I)$ is defined as:

$$p(r_I) = -\frac{\exp(-n_s\rho_s(r_o - r_1)/kC\tau)}{(\tau kC)/(n_s\rho_s)} \qquad \textbf{(EQ 8)}$$

If we consider N reactors in series, with individual residence time distributions: $G_1(t)$, $G_2(t)$..., which can all be different, the overall system can be defined as follows:

$$G_N(t) = \int_{s=0}^{s=t} g_N(t - s)G_{N-1}(s)ds \qquad \textbf{(EQ 9)}$$

where N = 2, 3, 4, ...

$$G_1(t) = g_1(t) \qquad \textbf{(EQ 10)}$$

If we would like to incorporate a diminishing driving force into the individual grain reaction equations for chemical controlled shrinking core behavior, then the following equations could be used:

$$\frac{dr_I}{dt} = -\frac{n_s k}{\rho_B}\left[C^o_{Aq} - \frac{\dot{n}}{\dot{V}n_s}\left(1-\left(\frac{r_I}{r_o}\right)^3\right)\right] \qquad \text{(EQ 11)}$$

$$\frac{dr_I}{r_o dt} = -\frac{n_s k}{r_o \rho_B}\frac{\dot{n}}{\dot{V}b}\left[C^o_{Aq}\frac{\dot{V}n_s}{\dot{n}}\left(-1+\left(\frac{r_I}{r_o}\right)^3\right)\right]$$

$$x = \frac{r_I}{r_o} \qquad \tau_C = \frac{r_o \rho_B}{n_s k C^o_{Aq}} \qquad \lambda_C = C^o_{Aq}\frac{\dot{V}n_s}{\dot{n}}$$

$$\int_1^0 \frac{dx}{(\lambda_C - 1 + x^3)} = \frac{t}{\tau_C \lambda_C} \qquad \text{(EQ 12)}$$

where

$\dot{V}$ is the volume or volume flow rate
$\dot{n}$ is the solid molar volume or flow rate
C^o_{Aq} is the initial fluid reactant concentration (mol/cm^3)
r_I is the grain reaction interface (cm)
r_o is the grain radius (cm)
t is time (sec)
k is the chemical reaction rate constant (cm/sec)
$\bar{t}$ is the mean residence time in the reactor (sec)
ρ_B is the solid reactant molar density (mol/cm^3)
n_s is the stoichiometric ratio (moles solid reactant/moles fluid reactant)

For batch or steady state continuous reactors, equation (12) can be easily integrated, or equation (11) could be used in equation (1, 3 or 6).

Similarly, the effect of a diminishing driving force in fluid solid reactions shrinking core model under Fluid Film Control can be derived as follows:

$$r_I^2\frac{dr_I}{dt} = -\frac{n_s h C_{Aq}}{\rho_B} r_o^2$$

$$C_{Aq} = C^o_{Aq} - \frac{\dot{n}}{\dot{V}n_s}X_B = C^o_{Aq} - \frac{\dot{n}}{\dot{V}n_s}\left[1-\left(\frac{r_I}{r_o}\right)^3\right]$$

$$\frac{r_I^2}{r_o^2}\frac{dr_I}{dt}\frac{1}{r_o} = \left(-\frac{n_s h}{\rho_B r_o}\right)\left[C^o_{Aq} - \frac{\dot{n}}{\dot{V}n_s} + \frac{\dot{n}}{\dot{V}n_s}\left(\frac{r_I}{r_o}\right)^3\right]$$

$$x^2\frac{dx}{dt} = -\frac{n_s h}{\rho_B r_o \dot{V} n_s}\frac{\dot{n}}{}\left(\frac{C^o_{Aq}\dot{V}n_s}{\dot{n}} - 1 + x^3\right) \quad \textbf{(EQ 13)}$$

$$\int_o^x \frac{x^2}{\lambda - 1 + x^3} = \frac{n_s h C^o_{Aq} t}{\lambda_F \rho_B r_o 3} = -\frac{t}{3\lambda\tau_F} \quad \textbf{(EQ 14)}$$

where

$$\lambda_F = \frac{C^o_{Aq}\dot{V}n_s}{\dot{n}}$$

h is the fluid film mass transfer coefficient (cm/sec)

The solution is as follows and can be used for batch or steady state continuous reactors:

$$\frac{1}{3}\log\left|\frac{\lambda}{\lambda - 1 + x^3}\right| = \frac{t}{3\lambda\tau_F}$$

$$\frac{1}{3}\log\left|\frac{\lambda}{\lambda - 1}\right| = \frac{\tau_N}{3\lambda\tau_F}$$

$$\tau_N = \lambda\tau_F \log\left|\frac{\lambda}{\lambda - 1}\right|$$

where τ_N is the time for completion of the reaction.

GENERAL APPROACH TO REACTOR DESIGN FOR FLUID-SOLID REACTORS

A general approach that may be used to model various types of reactors is outlined below. The first step is to draw a reactor schematic. The number of stages and material flow is noted on the diagram. The system of equations to be developed is:

A Molar Rate Balance on Each Stage

This equation is based upon conservation of mass and stoichiometry and may be written as follows (stoichiometric ratio of moles of fluid reactant and solid reactant reacted):

$$\pm n_s\left(\frac{\text{moles}_{\text{in}}}{\text{time}} - \frac{\text{moles}_{\text{out}}}{\text{time}}\right)_{\text{Aqueous}} = \left(\frac{\text{moles}_{\text{in}}}{\text{time}} - \frac{\text{moles}_{\text{out}}}{\text{time}}\right)_{\text{Solid}} \quad \textbf{(EQ 15)}$$

A Grain Reaction Equation for Each Stage

This equation relates the amount of solid reaction as a function of time and the reaction mechanism. For chemical control, the equation is:

$$\frac{dr_I}{dt} = -n_s\frac{kC_I}{\rho_s} \qquad r_1(t = 0) = r_o \quad \textbf{(EQ 16)}$$

An Overall Reactor Design Equation for Each Stage

This equation incorporates the type of fluid flow behavior, distribution of particle sizes, etc and utilizes equations (1), (3) or (6). For perfect mixed fluid flow and uniform sized particles may be written as:

$$1-\bar{X}_I = \int_0^{t^*} \frac{(1-X_I)\exp(-t/\bar{t})}{\bar{t}}dt \qquad \textbf{(EQ 17)}$$

where X_I is a function of time, the concentration driving force and the type of controlling mechanism. The concentration driving force can be obtained from the molar rate balance, Equation (15).

It is useful to redefine the time to be between 0 and 1. This is done by defining the time for complete reaction and normalizing the time with respect to that quantity. For example, for one stage under fluid film control:

$$r_I^3 - r_0^3 = -\frac{3n_shC_Itr_o^2}{\rho_s} \text{ or } X_I = \frac{3n_shC_It}{\rho_sr_o} \qquad \textbf{(EQ 18)}$$

$$t_{X_I=1} = \frac{\rho_sr_o}{3n_sC_Ih} \text{ so } t^* = \frac{t3n_sC_Ih}{\rho_sr_o} \qquad \textbf{(EQ 19)}$$

The overall reactor design equation is then:

$$1-\bar{X}_I = \int_0^1 (1-t^*)\exp\left(-\frac{t^*\rho_sr_o}{3n_sC_Ih\bar{t}}\right)\left(\frac{\rho_sr_o}{3n_sC_Ih\bar{t}}\right)dt^* \qquad \textbf{(EQ 20)}$$

This is solved in conjunction with the molar balance equation:

$$n_s(C_o - C_I) = \rho_s(1-\bar{X}_I) \qquad \textbf{(EQ 21)}$$

If we assume the reactor is at steady state and is perfect mixed, the value of the concentration driving force (C_I) is constant with respect to time and is uniform throughout the reactor and at the exit, thus equations (20) and (21) can be solved analytically or numerically.

CONTINUOUS CO-CURRENT FLUID SOLID REACTORS

Consider the circuit shown schematically in Figure 2. Assuming the system is at steady state, fluid film control grain kinetics and perfect mixed behavior in each stage, the following system of equations can be obtained. The same set of equations could be used for leaching kinetics with a simple change in nomenclature.

Molar Balance on Each Stage

$$QC_{I-1} - QC_I = Q_C\rho_I - Q_C\rho_{I+1} \qquad \textbf{(EQ 22)}$$

where

- Q is the flow rate of the solution (cm^3/sec)
- C_I is the concentration of gold in Stage I (mol/cm^3)
- Q_C is the flow rate of the carbon (cm^3/sec)
- ρ_I is the concentration of gold on carbon in Stage I (mol/cm^3)

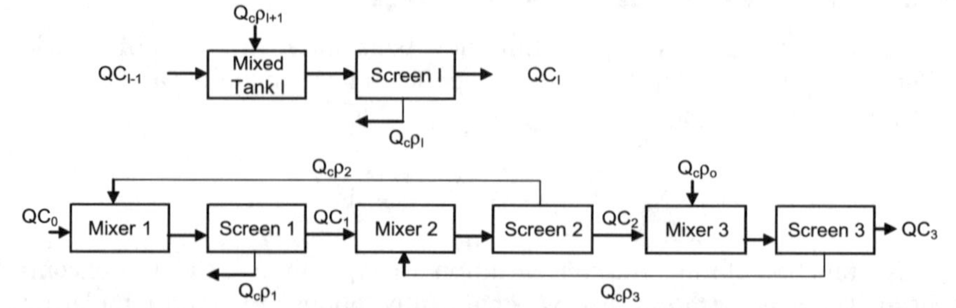

FIGURE 2 Continuous co-current reactor schematic

Transforming the equations to dimensionless form using:

$$C_I^* = \frac{C_I}{C_o} \quad E_I = \frac{\rho_I}{\rho_e} = 1 - \left(\frac{r_I}{r_o}\right)^3 \quad r_I^* = \frac{r_I}{r_o} \quad \tau = \frac{t \cdot h}{r_o} \quad \gamma = \frac{C_o}{\rho_e} \quad \lambda_I = \frac{Q_C}{Q} \quad \lambda_2 = \frac{r_o}{h\bar{t}}$$

where:

C_o is the inlet concentration (mol/cm³)
r_I is the grain reaction interface (cm)
r_o is the grain radius (cm)
t is time (sec)
h is the mass transfer coefficient (cm/sec)
is the mean residence time in the reactor (sec)
ρ_e is the equilibrium loading of gold on carbon (mol/cm³)
E_I is the extent of reaction in Stage I

Equation (22) then becomes:

$$C_{I-1}^* - C_I^* = \lambda_1 \frac{E_I - E_{I+1}}{\gamma} \tag{EQ 23}$$

Grain Reaction Interface Equation for Each Stage

$$r_{I+1}^3 - r_I^3 = \frac{3r_o^2 h C_I t}{\rho_e} \tag{EQ 24}$$

in dimensionless form:

$$(1 - E_{I+1}) - (1 - E_I) = 3C_I^* \gamma \tau \tag{EQ 25}$$

Reactor Design Equation for Each Stage

$$1 - \bar{E}_I = \int_0^t (1 - E_I) \exp \frac{(-t/\bar{t})}{\bar{t}} dt \tag{EQ 26}$$

where

t^* is the time for complete reaction (sec)

$\overline{E_I}$ is the overall extent of reaction in Stage I

defining the time for complete reaction from equation (25):

$$\tau_{E_{I=1}} = \frac{1 - E_{I+1}}{3\gamma C_I^*} \quad \text{(EQ 27)}$$

and

$$\tau_I^* = \frac{3C_I^*\gamma\tau}{1 - E_{I+1}} \quad \text{(EQ 28)}$$

Equation (26) becomes:

$$1 - \overline{E_I} = \int_0^1 [(1 - E_{I+1}) - (1 - E_{I+1})\tau_I^*]\lambda_s\lambda_x \exp(-\lambda_2\lambda_x\tau_I^*)d\tau_I^* \quad \text{(EQ 29)}$$

where:

$$\lambda_x = \frac{1 - E_{I+1}}{3\gamma C_I^*}$$

Then, for a three-stage system the equations become:

Stage 1

$$C_I^* = 1 - \frac{\lambda_I(E_I - E_2)}{\gamma} \quad \text{(EQ 30)}$$

$$1 - \overline{E_I} = \int_0^1 \{[(1 - E_2) - (1 - E_2)\tau_I^*]\lambda_2\lambda_3 \exp(-\lambda_2\lambda_3\tau_I^*)\}d\tau_I^* \quad \text{(EQ 31)}$$

$$\lambda_3 = \frac{1 - E_2}{3C_I^*\gamma} \quad \text{(EQ 32)}$$

Stage 2

$$C_2^* = C_I^* - \frac{\lambda_I(E_2 - E_3)}{\gamma} \quad \text{(EQ 33)}$$

$$1 - \overline{E_2} = \int_0^1 \{[(1 - E_3) - (1 - E_3)\tau_2^*]\lambda_2\lambda_4 \exp(-\lambda_2\lambda_4\tau_2^*)\}d\tau_2^* \quad \text{(EQ 34)}$$

$$\lambda_4 = \frac{1 - E_3}{3\gamma C_2^*} \qquad \text{(EQ 35)}$$

Stage 3

$$C_3^* = C_2^* - \frac{\lambda_I E_3}{\gamma} \qquad \text{(EQ 36)}$$

$$1 - \overline{E_3} = \int_0^1 \{(1 - \tau_3^*)\lambda_2\lambda_5 \exp(-\lambda_2\lambda_5\tau_3^*)\} d\tau_3^* \qquad \text{(EQ 37)}$$

$$\lambda_5 = \frac{1}{3\gamma C_2^*} \qquad \text{(EQ 38)}$$

The analytical solution is:

$$1 - \overline{E_3} = 1 - \frac{1}{\lambda_2\lambda_3} + \exp(-\lambda_2\lambda_3)\frac{1}{\lambda_2\lambda_3} \qquad \lambda_3 = \frac{1}{3\gamma C_3^*}$$

$$1 - \overline{E_2} = (1 - \overline{E_3}) - \frac{1 - \overline{E_3}}{\lambda_2\lambda_4} + (1 - \overline{E_3})\exp(-\lambda_2\lambda_4)\frac{1}{\lambda_2\lambda_4} \qquad \lambda_4 = \frac{1 - \overline{E_3}}{3\gamma C_2^*}$$

$$1 - \overline{E_1} = (1 - \overline{E_2}) - \frac{1 - \overline{E_2}}{\lambda_2\lambda_5} + (1 - \overline{E_2})\exp(-\lambda_2\lambda_5)\frac{1}{\lambda_2\lambda_5} \qquad \lambda_5 = \frac{1 - \overline{E_2}}{3\gamma C_I^*}$$

Example solutions are given in Figures 3 and 4.

Example—Chemical Controlled Co-current Reactors:

For the case of co-current fluid solid reactors, the schematic is shown in Figure 5.
The molar balance equation for each stage can be written as:

$$QC_{I-1} - QC_I = Q_C(\rho_I - \rho_{I-1}) \qquad \text{(EQ 39)}$$

The grain reaction interface equation, for chemical controlled shrinking core behavior, for each stage can be written as:

$$(1 - E_I)^{1/3} - (1 - E_{I-1})^{1/3} = \frac{kC_I t}{r_o \rho_e} \qquad \text{(EQ 40)}$$

The overall reactor design equations for each stage, for uniform sized particles and perfect mixed fluid behavior can be written as:

$$1 - \bar{E}_I = \int_0^{t^*} \frac{(1 - E_I)\exp(-t/\bar{t})}{\bar{t}} dt \qquad \text{(EQ 41)}$$

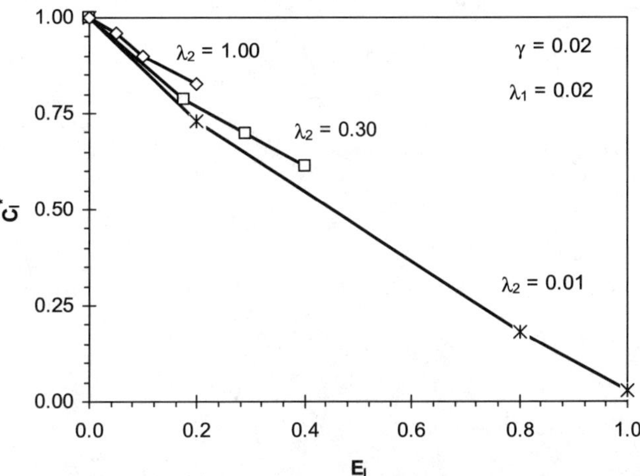

FIGURE 3 Fluid concentration leaving each stage versus metal loading at each stage for various values of λ_2

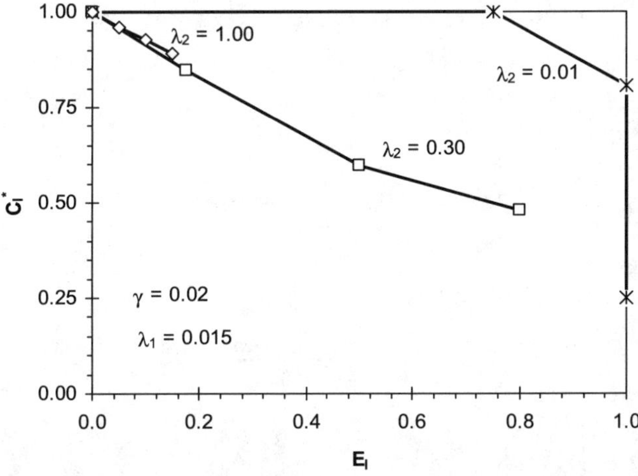

FIGURE 4 Fluid concentration leaving each stage versus metal loading at each stage for various values of λ_2, effect of changing λ_1

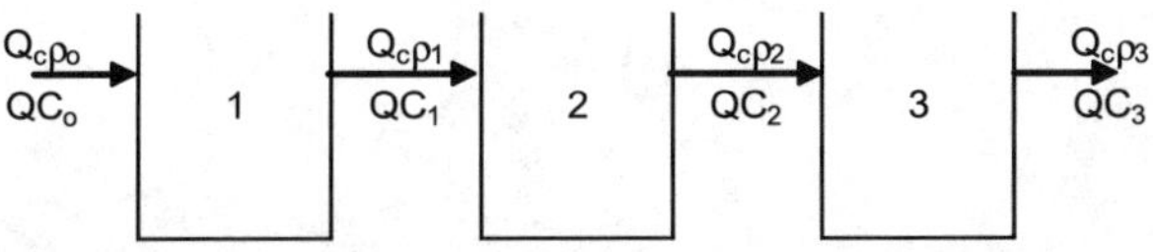

FIGURE 5 Co-current fluid-solid reactors

Time is redefined to normalize with respect to the time for complete reaction:

$$t_{E_I = 1} = \frac{(1 - E_I)^{1/3} r_o \rho_e}{k C_I} \quad \text{(EQ 42)}$$

The grain reaction equation then becomes:

$$(1 - E_I) = (1 - E_{I-1})(1 - \tau_I^*)^3 \quad \text{(EQ 43)}$$

where

$$\tau_I^* = \frac{t}{t_{E_I = 1}}$$

Transforming to dimensionless form:

Stage 1

$$C_I^* = 1 - \frac{\lambda_I E_I}{\gamma} \quad \text{(EQ 44)}$$

$$1 - \bar{E}_I = \int_0^1 [(1 - \tau_I^*)\lambda_2 \lambda_7 \exp(-\lambda_2 \lambda_7 \tau_I^*)] d\tau_I^* \quad \text{(EQ 45)}$$

where

$$\lambda_2 = \frac{r_o}{k\bar{t}} \text{ and } \lambda_7 = \frac{1}{\gamma C_I^*}$$

Stage 2

$$C_2^* - C_I^* = 1 - \frac{\lambda_I (E_2 - E_3)}{\gamma} \quad \text{(EQ 46)}$$

$$1 - \bar{E}_2 = \int_0^1 [(1 - E_I)^{1/3} - (1 - E_I)^{1/3} \tau_2^*]^3 \lambda_2 \lambda_8 \exp(-\lambda_2 \lambda_8 \tau_2^*) d\tau_2^* \quad \text{(EQ 47)}$$

where

$$\lambda_8 = \frac{(1 - E_I)^{1/3}}{\gamma C_2^*}$$

Stage 3

$$C_3^* - C_2^* = \frac{\lambda_I (E_3 - E_2)}{\gamma} \quad \text{(EQ 48)}$$

$$1-\bar{E}_3 = \int_0^1 [(1-E_2)^{1/3} - (1-E_2)^{1/3}\tau_3^*]^3 \lambda_2\lambda_9 \exp(-\lambda_2\lambda_9\tau_3^*)d\tau_3^* \quad \text{(EQ 49)}$$

where

$$\lambda_8 = \frac{(1-E_2)^{1/3}}{\gamma C_3^*}$$

Solving iteratively:

$$1-\bar{E}_1 = 1 - \frac{3}{\lambda_2\lambda_7} + \frac{6}{(\lambda_2\lambda_7)^2} - \frac{6(1-\exp(-\lambda_2\lambda_7))}{(\lambda_2\lambda_7)^3}$$

$$1-\bar{E}_2 = (1-\bar{E}_1)\left[1 - \frac{3}{\lambda_2\lambda_8} + \frac{6}{(\lambda_2\lambda_8)^2} - \frac{6(1-\exp(-\lambda_2\lambda_8))}{(\lambda_2\lambda_8)^3}\right]$$

$$1-\bar{E}_3 = (1-\bar{E}_2)\left[1 - \frac{3}{\lambda_2\lambda_9} + \frac{6}{(\lambda_2\lambda_9)^2} - \frac{6(1-\exp(-\lambda_2\lambda_9))}{(\lambda_2\lambda_9)^3}\right]$$

Figures 6 and 7 present some example results.

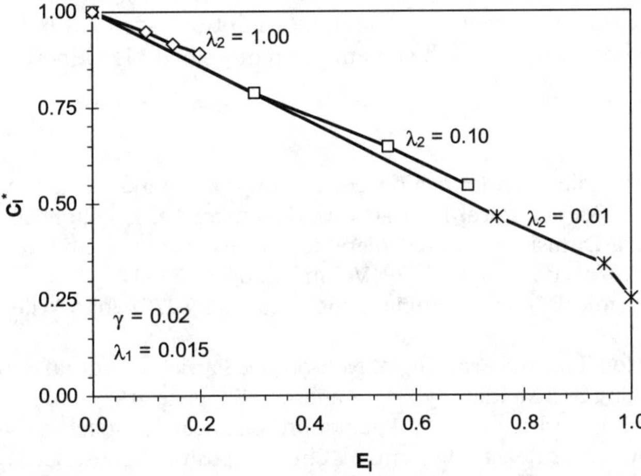

FIGURE 6 Fluid concentration leaving each stage versus extraction for various values of λ_2

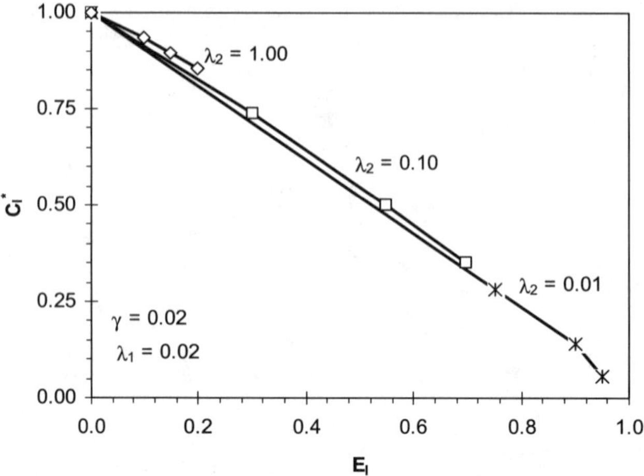

FIGURE 7 Fluid concentration leaving each stage versus extraction for various values of λ_2, effect of changing the value of λ_1.

CONCLUSIONS

A general view of process modeling is given. Equations are presented that describe how to incorporate distributions of particle sizes, distributions of residence time and diminishing driving forces into models for fluid-solid reactions. Also shown are methods to incorporate diminishing driving forces into batch, continuous co-current or continuous counter-current multi-stage reactors. The model solutions are for single step controlling shrinking core behavior but could be extended to more complex kinetics.

REFERENCES

Bartlett, R.W., Conversion and Extraction Efficiencies for Ground Particles in Heterogeneous Process Reactors, Metallurgical Transactions, November 1971, Volume 2, pp. 2999–3006

Bartlett, R.W., Pore Diffusi9on-Limited Metallurgical Extraction from Ground Ore Particle, Metallurgical Transactions, April 1972, Volume 3, pp. 913–917

Levenspiel, O., Chemical Reaction Engineering, 2nd Ed., 1972, John Wiley and Sons, New York, pp. 357–399

Loveday, B.K., A Model for the Leaching of Non-porous Particles, Journal of the South African Institute of Mining and Metallurgy, September 1975, pp. 16–19

Sepulveda, J.E. and Herbst, J.A., A Population Balance Approach to the Modeling of Multistage Continuous Reactor Systems, AIChE Symposium Series, Bo. 173, 1978, Vol. 74, pp. 41–65

Taylor, P.R. and Martins, G.P., Reactor Design for Aqueous-Solid Reactions, Hydrometallurgical Reactor Design and Kinetics, Bautista, R.G., et al. Editors, TMS, Warrendale PA 1986, pp.105–119

Innovations in Gold Leaching Research and Development

M.I. Jeffrey,* I. Chandra,* I.M. Ritchie†, G.A. Hope,‡ K. Watling,‡ and R. Woods‡

This paper presents two aspects of gold leaching research and development. The first section of the paper covers the development of innovative gold leaching processes. The research in this area has concentrated on the development of new thiosulfate leaching systems which use Fe(III) stabilized by ligands as the oxidant in place of copper(II)+ammonia. The second section will cover the development of innovative techniques such as surface enhanced Raman scattering (SERS) to study the leaching of gold in cyanide solutions. It will be shown that the use of in-situ SERS yields information which cannot be obtained using other methodologies.

INTRODUCTION

In keeping with the symposium theme, the work presented is concerned with two aspects of innovations in gold hydrometallurgy, namely: innovative new gold leaching processes; and innovative methods used for the research of current leaching processes.

Innovative New Gold Leaching Processes

The development of thiosulfate based leaching processes has been at the forefront of gold leaching hydrometallurgy research for the past two decades. The majority of this research has focussed on the copper-ammonia-thiosulfate leaching process, with the

* Dept. Chemical Engineering, Monash University, Vic 3800, Australia

† Dept. Chemistry, Murdoch University, Murdoch WA 6150, Australia

‡ School of Science and CRC microTechnology, Griffith University, Nathan, Qld 4111, Australia

goal of minimising thiosulfate oxidation, and optimising leaching for a variety of ore materials. However the copper-ammonia-thiosulfate process has a number of significant problems which have impeded its implementation industrially. These include:

1. High rates of thiosulfate oxidation in the presence of oxygen (Byerley, Fouda and Rempel 1975; Breuer and Jeffrey 2003).
2. Generation of polythionates which hinder the resin adsorption process (Nicol and O'Malley 2002).
3. The requirement for relatively high concentrations of ammonia, and hence the environmental impact of ammonia/ammonium discharges needs to be considered.

The problems of the copper-ammonia-thiosulfate process have prompted research into alternative thiosulfate leaching systems. In the present paper, the application of the ferric-oxalate-thiosulfate process will be discussed as an innovative new leaching process.

Innovative Methods of Studying Current Leaching Processes

Despite being in industrial use for over a century, the fundamental aspects of gold leaching in cyanide solutions are still not fully understood. It has recently been shown that pure gold does not leach in very pure cyanide solutions (Jeffrey and Ritchie 2001), and that traces of metals such as silver in either the solid or solution phase activate the leaching process (Jeffrey and Ritchie 2000b). A number of authors have suggested that in cyanide solutions the gold is passivated by some type of cyanide film (Kirk, Foulkes and Graydon 1980; Nicol 1980; Jeffrey and Ritchie 2001). It has also been suggested that lead, a common cyanidation additive, is able to enhance gold leaching by depositing onto the gold and hence modify the surface (Sheveleva and Kakovskii 1979; Nicol 1980; Weichselbaum, Tumilty and Schmidt 1989; Jeffrey and Ritchie 2000a). This results in much higher gold leach rates when lead is added to a cyanide solution. Despite these studies, there is yet to be evidence presented as to the nature of the film which causes low dissolution rates in the absence of impurities such as a lead and silver.

Surface Enhanced Raman Scattering (SERS) spectroscopy has been demonstrated as an excellent surface sensitive technique for studying the gold surface (Baltruschat and Heitbaum 1983; Gao and Weaver 1986). Using SERS, it is possible to obtain vibrational spectra of species in the first few layers at the solid-electrolyte interface. Simultaneous in situ electrochemical and SERS techniques may therefore assist in the elucidation of the dissolution mechanism(s), particularly through the identification of surface species and the determination of the nature and strength of ligand–gold atom interactions. The second part of the present paper will focus on the application of SERS to establish the mechanism for passivation of gold in cyanide solutions.

EXPERIMENTAL METHODS

Thiosulfate Leaching

All the thiosulfate leaching experiments were carried out using solutions prepared from analytical reagents and Millipore water. Unless specified, the solutions contained 0.1 M ammonium thiosulfate (ATS) and 5 mM thiourea (Tu). The ferric oxalate complex was prepared by mixing ferric nitrate nonahydrate and sodium oxalate. All experiments were carried out at room temperature (20–21°C), and conducted at a rotation rate of 300 rpm. All experiments were carried out at the natural pH of the solution, which was typically pH 4.5–6.. Mass changes were measured using the rotating electrochemical quartz crystal microbalance which is described elsewhere (Jeffrey, Zheng and Ritchie 2000). Prior to each gold leaching experiment, the gold was electroplated onto the quartz electrode at 25 A m^{-2} from a solution containing 0.02 M potassium dicyanoaurate, 0.23 M potassium cyanide, and 0.086 M potassium carbonate. All potential measurements were recorded relative to a saturated calomel electrode (0.242 V vs SHE), but are reported on the SHE scale.

Powder leaching experiments were performed at room temperature in a conical flask. Agitation was supplied with a magnetic stirrer rotating at 500 rpm. The gold powder was prepared by the reductive precipitation of gold from gold chloride solutions using sodium disulfite. The powder leaching experiments were performed using 10 mg of gold with 0.1 L of solution. Thus, when the leaching was complete, the solution contained 100 ppm of gold. The concentration of gold in solution was measured as a function of time using a Varian SpectrAA-400.

Gold Cyanide Spectroelectrochemistry

Raman spectroelectrochemical experiments were performed using a custom-designed five-necked borosilicate electrochemical cell with an optically flat transparent window. A conventional three electrode system was used, consisting of a 0.5 mm diameter platinum wire auxiliary electrode; a silver / silver chloride reference electrode (Cypress systems EE008, 3.0 mol dm^{-3} KCl / sat. AgCl electrolyte); and a fine gold (Johnson Matthey 99.5 %) working electrode with an exposed area of 1 cm^2. The gold working electrode was abraded to expose a fresh surface using 15 µm grit size P1200 SiC paper, and rinsed with ultrapure water. Following this procedure, the electrode was sonicated in ultrapure water and placed in a furnace at 450°C for 30 min to remove any adsorbed species from the surface. The auxiliary electrode was located 6 cm to the rear of the cell, with the reference electrode close to the gold. Potential control of the working electrode was established using an ADInstruments potentiostat controlled by a Maclab/4e analog-digital converter interfaced to a PC running ADInstruments EChem software V.1.5.2. Potentials are reported relative to the SHE. Prior to experiments, the cell was purged by bubbling high purity nitrogen (Linde, 4.0) for 20 min to remove oxygen from the system. A positive nitrogen pressure was maintained over the cell during the measurement phase by withdrawing the nitrogen bubbler from solution and flowing the gas above the solution surface.

SERS Activation

The SERS electrode was activated by applying oxidation / reduction cycling (ORC) in a 1.0 mol dm^{-3} KCl solution, acidified with HCl to pH 1, in equilibrium with air. The potential regime used consisted of cathodic cleaning for 30 s at –0.7 V, followed by the application of 0.5 s pulses for 5 min between −0.5 V and 1.0 V, ending at the lower value.

Raman Spectroscopy

SERS spectra were acquired on a Renishaw system 100 Raman fibre optic spectrometer, using 632.8 nm excitation from a HeNe laser. The laser light was delivered through an optical fibre and focused through the collection lens to a measured spot size of 50 μm, with 6 mW power recorded at the sample. Raman scattered radiation was collected through an ultra-long working distance x20 Olympus LMP Plan Fl lens with a numeric aperture of 0.4 and focussed through two super notch filters onto the spectrometer receiving fibre. The Raman shifted signals from the single pass grating spectrometer were detected using a Peltier-cooled CCD detector with spectral resolution of 2 cm^{-1}. The spectrometer was referenced to the 520 cm^{-1} lattice vibration of silicon. SERS spectra were accumulated in static mode, centred at 670 or 750 cm^{-1}.

RESULTS AND DISCUSSION

Innovative Gold Leaching Processes: Ferric Oxalate Thiosulfate Leaching

It is well known that in the absence of complexing ligands, the reaction between iron(III) and thiosulfate occurs very rapidly (Page 1960). This is indicated by a strongly intermediate violet colour formation which fades away as the reaction proceeds. The violet colour is attributed to the formation of an intermediate complex between iron(III) and thiosulfate, $FeS_2O_3^+$ (Page 1960). Therefore it would be anticipated that gold would not leach in ferric thiosulfate solutions, as the oxidant, Fe(III), is reduced rapidly by thiosulfate. This was verified by using the rotating electrochemical quartz crystal microbalance (REQCM) to study the leaching of gold in ferric thiosulfate solution. The basis of the REQCM is that small mass changes can be measured in-situ by monitoring the frequency of oscillation of the quartz crystal on which the gold has been coated. As the gold is leached, the mass decreases, and the rate of decreases in mass can be related to the dissolution rate according to Equation 1.

$$r = \frac{1}{AM}\frac{dm}{dt} \quad \textbf{(EQ 1)}$$

where r is the leaching rate (mol m^{-2} s^{-1}), M is the atomic mass of gold (197 g mol^{-1}), A is the area of the gold electrode (m^2), and t is the time elapsed (s). Using this method, a gold leach rate of 0.45×10^{-6} mol m^{-2} s^{-1} was obtained after allowing the gold to leach for 10 mins for a solution containing 5 mM Fe(III) and 100 mM thiosulfate. This rate is at least an order of magnitude slower than the rates obtained for gold-silver alloys in cyanide solutions (ca. $3–6 \times 10^{-5}$ mol m^{-2} s^{-1}) (Jeffrey and Ritchie 2000b) or gold in copper-ammonia-thiosulfate solutions (ca. $1–3 \times 10^{-5}$ mol m^{-2} s^{-1}) (Breuer and Jeffrey 2000; Jeffrey 2001; Jeffrey, Breuer and Choo 2001).

It is well known that ligands can be used to stabilize ferric ions in solution, and hence minimize its reaction with reductants such as thiosulfate. In this study oxalate was chosen as the ligand for the following reasons:

a. Oxalate forms a strong complex with iron(III): log β ($Fe(C_2O_4)^+$) = 7.53; log β ($Fe(C_2O_4)_2^-$) = 13.64; and log β ($Fe(C_2O_4)_3^{3-}$) = 18.49 (Smith and Martell 1976).
b. Oxalate is produced as a waste product from Alumina refineries, and hence is cheap.
c. Ferric oxalate undergoes photocatalytic decomposition to form Ferric/Ferrous oxide, CO_2 and CO_3^{2-} (Dudeney and Tarasova 1998). Thus it could be discharged to a tailings dam without causing a serious threat to the environment.

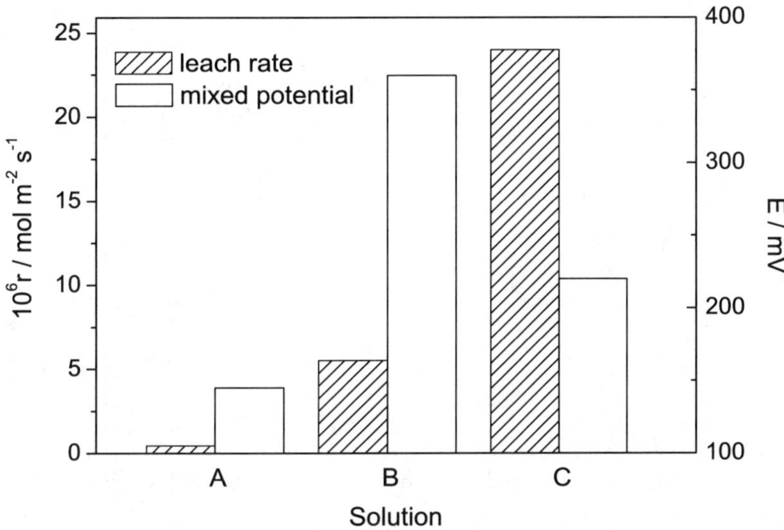

FIGURE 1 Gold leach rate and mixed potential for the Fe(III)-thiosulfate system. Solution A: 5 mM Fe(III) + 100 mM ammonium thiosulfate; Solution B: 5 mM Fe(III) + 100 mM ammonium thiosulfate + 12.5 mM oxalate; Solution C: 5 mM Fe(III) + 100 mM ammonium thiosulfate + 12.5 mM oxalate + 5 mM thiourea

The leaching of gold in ferric oxalate solutions was studied using the REQCM, and Figure 1 shows the gold leach rate and mixed potential obtained (solution B). The respective values in the absence of oxalate are also shown in Figure 1. It should be clear that the addition of oxalate to the solution increases both the mixed potential and the gold leach rate. Such a result is consistent with the oxalate forming a complex with the ferric ions. However the gold leach rate obtained, 5.5×10^{-6} mol m^{-2} s^{-1}, is still lower than that obtained in cyanide or copper-ammonia-thiosulfate solutions.

It has been previously shown that the addition of thiourea to thiosulfate solutions substantially improves the oxidation of gold to the gold thiosulfate complex (Chandra and Jeffrey 2004). Thus leaching experiments were also performed where 5 mM thiourea was added to solution. As can be seen from Figure 1, the rate at which gold leaches in ferric-oxalate-thiosulfate solutions is enhanced in the presence of thiourea. In addition, the mixed potential is more negative for the solution containing thiourea, which is consistent with an improvement to the gold oxidation half reaction. At steady state in the presence of thiourea, the gold leach rate was calculated to be 24×10^{-6} mol m^{-2} s^{-1}, which is more than about 4 times higher than the rate calculated for gold in the absence of thiourea.

In the absence of oxalate, the reaction between iron and thiosulfate is very rapid, with the iron being consumed within seconds of mixing with thiosulfate. In the presence of oxalate, the reduction of iron by thiosulfate is much slower. The solution potential, E_H, is thus dependent upon the ratio of iron(III) to iron(II), and the E_H decreases as the reaction between iron(III) and thiosulfate takes place. Therefore the change in solution E_H was used as a guide to the reactivity of ferric oxalate with thiosulfate.

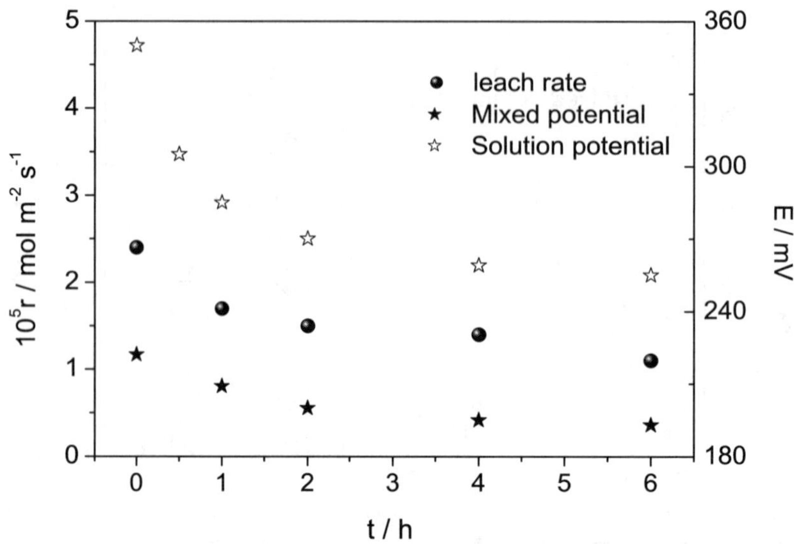

FIGURE 2 Gold leach rate, mixed potential and solution potential measured as a function of time using the REQCM for a solution containing 100 mM ammonium thiosulfate, 5 mM Fe(III), 12.5 mM oxalate and 5 mM thiourea

Figure 2 shows the solution E_H as a function of time for the 100 mM thiosulfate solution containing 12.5 mM oxalate, 5 mM Fe(III) and 5 mM thiourea. It can be seen that the E_H is initially very high, but decreases rapidly in the first 2 hours of the experiment. After 2 hours, the E_H approaches a steady value. The gold leaching rate and mixed potential were also measured during the experiment, and Figure 2 shows that both of these decrease with time as well. Initially the mixed potential was 220 mV, but this decreased as iron(III) reacts with thiosulfate to form iron(II). After 6 hours the mixed potential had decreased to 195 mV, and after 12 additional hours it decreased to 190 mV (not shown in the figure 1) and the gold leach rate under these conditions is approximately half the rate obtained initially.

The results shown in Figure 2 suggest that even in the presence of oxalate, at least some of the ferric ions are being reduced by thiosulfate. It should be remembered that for these experiments, the ratio of Fe(III) to oxalate was 1:2.5, and hence some of the ferric ions will be in the most stable $Fe(C_2O_4)_3^{3-}$ form, whilst some will be in the less stable $Fe(C_2O_4)_2^-$ form. Experiments were also conducted using a ratio of Fe(III) to oxalate of 1:3, and in this instance, the solution E_H was more negative initially, but was stable with time. These results are consistent with $Fe(C_2O_4)_3^{3-}$ being unreactive towards thiosulfate, and the decrease in solution potential for the Fe(III) to oxalate ratio of 1:2.5 being a result of the reaction between $Fe(C_2O_4)_2^-$ and thiosulfate.

Effect of Oxalate Concentration. Oxalate is important to the leaching of gold in thiosulfate using iron(III) as an oxidant as it complexes with the iron in the ferric oxidation state. To establish the role of oxalate in the thiosulfate leaching process, experiments were performed using different concentrations of oxalate ranging from 10 mM to 15 mM. These concentrations were selected in order to have the ratio of iron to oxalate between 1:2 to 1:3, and thus to illustrate the effect of the two dominant oxalate

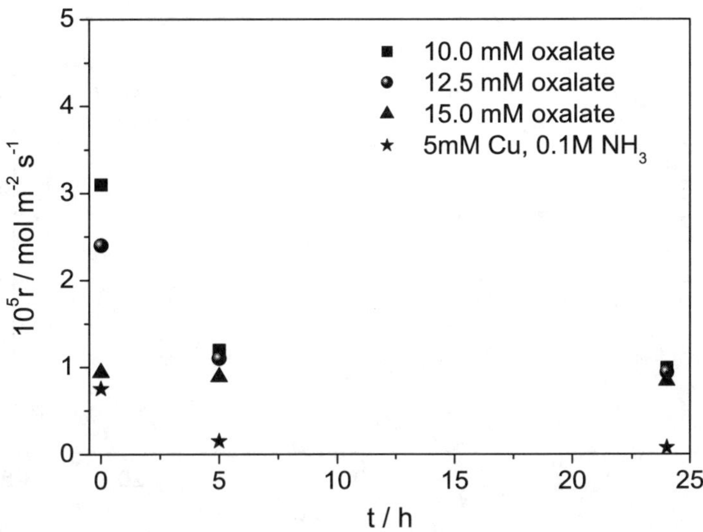

FIGURE 3 Effect of oxalate concentration on the gold leach rates in thiosulfate solution containing 5 mM Fe(III), 5 mM thiourea and oxalate. Also shown is the gold leach rate in 100 mM ammonium thiosulfate containing 5 mM Cu(II) and 100 mM ammonia.

complexes, $Fe(C_2O_4)_3^{3-}$ and $Fe(C_2O_4)_2^-$ on the gold leaching kinetics. Figure 3 shows the effect of oxalate concentration on the gold leach rate. It can be seen that when the iron to oxalate ratio is 1 to 2, initially the gold leaching is very fast. However, the leach rate rapidly decreases with time. It can also be seen from Figure 3 that when the iron to oxalate ratio is 1:2.5, a significant drop in gold leach rate occurred within the first 5 hours. However, for both the 1 Fe(III):2.5 oxalate, and 1 Fe(III):2 oxalate, the gold leach rate reached a steady state after 5 hours. When the iron to oxalate ratio of 3 was used, it can be seen from Figure 3 that the even though the gold leach rate was initially lower than that obtained for the lower oxalate concentrations, it remained constant with time. After 24 hours, the gold leach rates for all three oxalate concentrations were essentially the same. These results are consistent with $Fe(C_2O_4)_3^{3-}$ being the dominant ferric species present after 24 hours of leaching. When the Fe(III) to oxalate ratio is initially less than 1:3, the $Fe(C_2O_4)_2^-$ reacts with thiosulfate until this ratio becomes 1:3 and hence all the ferric becomes complexed as $Fe(C_2O_4)_3^{3-}$.

Leaching experiments were also carried out for solutions containing 5 mM copper(II), 100 mM ammonium thiosulfate and 100 mM ammonia, and the gold leach rate as a function of time is shown in Figure 3. It should be clear that at room temperature, the gold leach rate obtained in the ferric oxalate leaching solution is considerably higher than that obtained in this copper ammonia thiosulfate solution. Such a result indicates that the in terms of leaching kinetics, the ferric-oxalate processes is a viable alternative to the copper-ammonia process.

Effect of Ferric Concentration. Figure 4 shows the gold leach rates with different total iron(III) concentrations as a function of time. In each experiment, the Fe(III) to oxalate ratio was kept constant at 1:2.5. It should be clear that increasing the iron concentration increases the initial gold leach rates substantially. However as the solution is

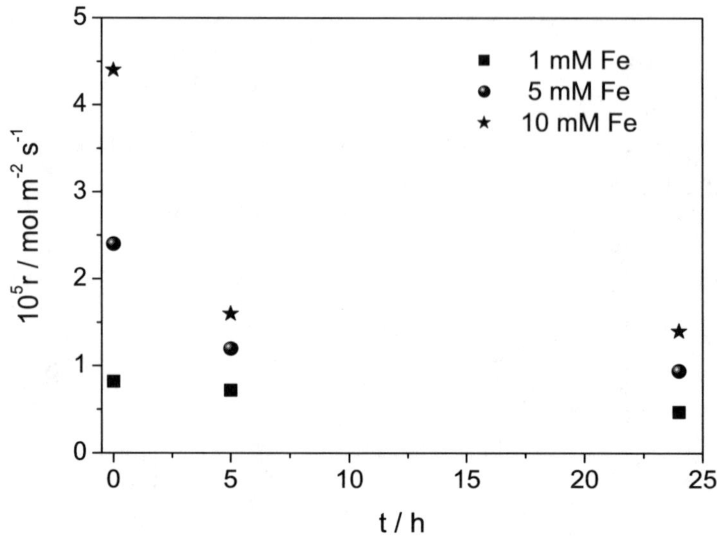

FIGURE 4 Effect of iron concentration on the gold leach rates in thiosulfate solution 100 mM ammonium thiosulfate and 5 mM thiourea. The ratio of Fe:Ox was kept at 1:2.5.

allowed to age, the leach rate drops and approaches a steady value after 5 hours for each of the Fe(III) concentrations. At steady state, there appears to be much less dependence of the leach rate on the Fe(III) concentration. These results indicate that for gold leaching process, there is little benefit in operating at the higher oxalate and iron concentrations, and hence the reagent consumption for Fe(III) and oxalate may be quite low.

Powder Leaching Studies. It has been demonstrated above using the REQCM that leaching of gold in thiosulfate utilising iron (III) as an oxidant occurs at a reasonable rate. However, most hydrometallurgists are interested in knowing the rate of gold extraction from a particular ore sample. In order to expand on the fundamental study carried out above, gold powder leaching experiments were performed in a batch type environment. Gold powder leaching is useful as a means of illustrating the effects observed above without the interactions of other minerals often present in the ores. In powder leaching experiment, the gold is exposed to the leaching solution for the entire period of experiment. This is different to the experiments performed using REQCM where the gold sample was always freshly electroplated for each of the experiments.

The powder leaching experiment was carried out in sodium thiosulfate and ammonium thiosulfate solutions. The results of the powder leaching in 100 mM thiosulfate, 10 mM Fe(III), 30 mM oxalate, and 10 mM thiourea are shown in Figure 5. It can be seen that the gold leaches rapidly with about 80% of the gold extracted within the first 5 hours and almost all the gold being extracted in 8 hours. Figure 5 also shows that the gold leaching kinetics are not affected by the cation in the thiosulfate salt used in the experiments, as the ammonium and sodium thiosulfate experiments gave almost identical extraction profiles. Such a result is important, as it means that sodium thiosulfate can be used instead of ammonium thiosulfate, and hence the environmental issues of ammonium discharge are avoided. At present, the cost of sodium thiosulfate is significantly

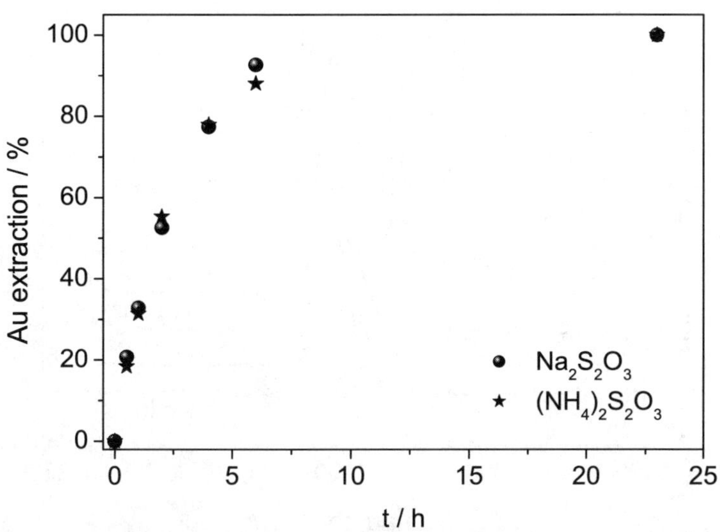

FIGURE 5 Extraction of gold from powder as a function of time in 100 mM ammonium and sodium thiosulfate solutions containing 10 mM Fe(III), 30 mM oxalate, and 10 mM thiourea

more than ammonium thiosulfate, but if a sodium-based process were adopted industrially, the cost of sodium thiosulfate may decrease.

Innovative Research: SERS of Gold Cyanide

Before presenting results on the SERS of gold in cyanide, it important to outline the relevant Raman bands which have been previously encountered in Raman and SERS studies of gold and cyanide. The CN stretching frequency, ν(CN), of cyano compounds gives a characteristic sharp vibrational band at 2200 – 2000 cm^{-1} (Nakamoto 1997). The cyanide ion in sodium and potassium and related simple cyanides absorbs at 2080–2070 cm^{-1}, with the ν(CN) of free cyanide in solution being 2080 cm^{-1} (Nakamoto 1997). The ν(CN) of the cyanide ion shifts to higher frequencies when coordinated to a metal, and bridged cyano compounds cause the ν(CN) to shift to a higher frequency (Nakamoto 1997). Raman vibrations of concentrated dicyanoaurate in solution have been studied extensively (Jones 1963). The linear dicyanides have seven normal vibrations, three of which are Raman active. Fundamental modes at 2164, 445–450 and 304 cm^{-1} were attributed to ν(CN), ν(AuC) and δ(AuCN) modes (Jones 1963).

Solid state Gold(I) cyanide, AuCN, is a linear compound in which the gold is bound to the carbon of one ligand and the nitrogen of the next, i.e., [-C≡N-Au-C≡N-Au-]. The crystal structure is hexagonal, composed of infinite linear chains arranged in parallel lines within the crystal (Sharpe 1976). Infrared bands observed by Bowmaker, Kennedy and Reid (1998) for AuCN at 2236, 598, 358 & 224 cm^{-1} were assigned as (CN), ν(AuC,AuN), δ(AuCN), and δ(NAuC). Raman assignments of 2232, 471 & 229 cm^{-1} for ν(CN), ν(AuC) & δ(AuCN) in AuCN have been reported by Murray and Bodoff (1986).

The adsorption of cyanide on gold has been studied by a number of authors (Baltruschat and Heitbaum 1983; Dorain and Von Raben 1985; Gao and Weaver 1986;

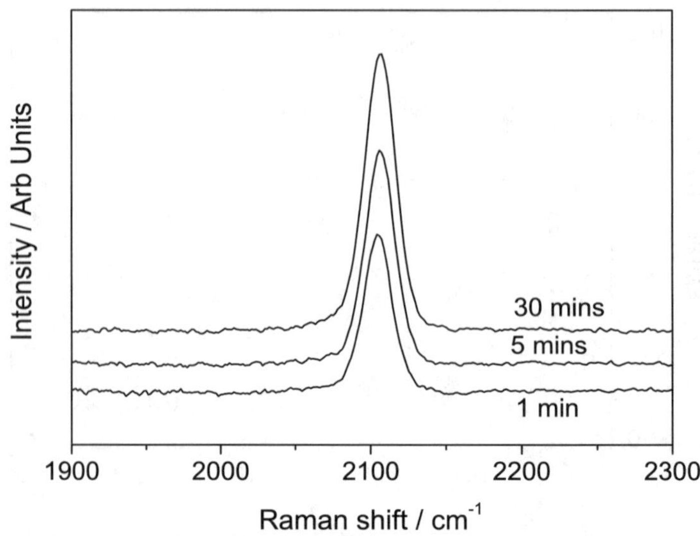

FIGURE 6 SERS spectra centred at 2100 cm^{-1} for a gold electrode in 1 mM NaCN held for the marked times at –800 mV

Yamada, Sekine and Sawaguchi 2000). Baltruschat and Heitbaum (1983) observed the potential dependence of ν(CN) of adsorbed cyanide on gold from solutions of 0.2 mol dm^{-3} KCl and 0.01 mol dm^{-3} NaCN, finding values of 2113, 375 & 290 cm^{-1} for ν(CN), ν(AuC) & δ(AuCN) at −0.8 V vs SCE. Linear Stark-tuning at a rate of 40 cm^{-1} was observed between −1.3 and −0.7 V vs SCE. Gao and Weaver (1986) report a Stark-tuning rate of 12 cm^{-1} V^{-1} for ν(CN) of cyanide on gold between −0.6 and +0.4 V vs SCE The respective ν(CN) values for the upper and lower potentials were 2122 & 2135 cm^{-1} respectively.

Spectroelectrochemical studies of gold in cyanide solutions were carried for two model systems:

1. Gold in AR Grade cyanide. For this system, gold leaching is very slow, and it has been suggested in the past that the gold surface is passivated by some form of surface film, possibly AuCN (Jeffrey and Ritchie 2001).
2. Gold in AR Grade Cyanide + Lead. For this system, gold leaching is rapid (diffusion controlled). It is known that lead is deposited on the surface, probably as an underpotentially deposited monolayer, and it is this surface modification which prevents passivation of gold in cyanide solutions at more negative potentials. However at higher overpotentials, the lead can be stripped off the gold surface, leading to re-passivation (Jeffrey and Ritchie 2000a).

The SERS spectra obtained for gold in AR grade 1 mM cyanide solutions at −800 mV is shown in Figure 6. This potential is outside the stability region of $Au(CN)_2^-$, and hence it would not be expected that gold will be oxidised. As shown in Figure 6, the characteristic vibrational band for ν(CN) can be seen at 2107 cm^{-1}. However at this potential, it was found that the ν(AuC) and δ(AuCN) associated with bonding of the cyanide to gold are not evident. Such a finding suggests that at this potential, the cyanide is not chemisorbed

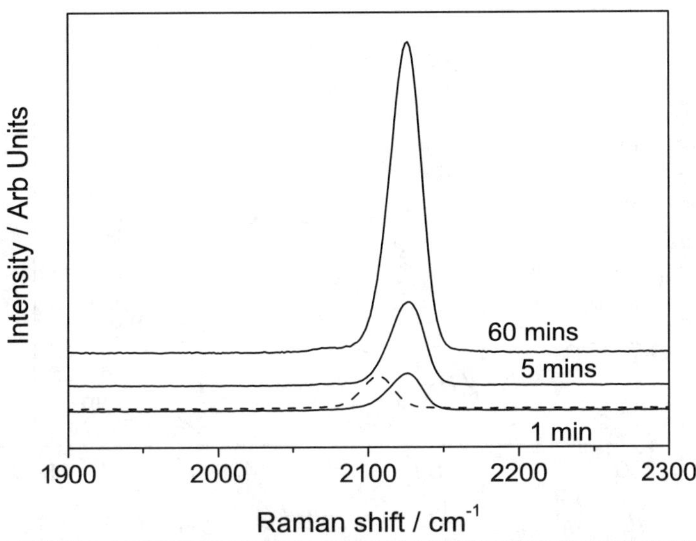

FIGURE 7 SERS spectra centred at 2100 cm^{-1} for a gold electrode in 1 mM NaCN held for the marked times at –400 mV. Also shown as a dashed line is the spectra obtained after 30 mins at –800 mV.

on the gold surface, but is more likely specifically adsorbed. Spectra obtained after holding the potential at –800 mV for 5 and 30 minutes are also shown in Figure 6, and it can be seen that there is a slow increase in intensity from the initial spectra, which most likely corresponds to an increase in the surface coverage of the adsorbed cyanide.

When the potential was stepped to –400 mV, a positive current was observed, indicating the oxidation of gold to gold(I). This potential is in the region in which gold leaches in cyanide solutions, and hence the SERS spectra obtained under these conditions are directly applicable to the cyanidation process. Figure 7 shows that initially, the spectra obtained are not that different to that shown in Figure 6, with the exception that the peak position is red-shifted by 20 wavenumbers (the 30 min spectra from Figure 6 is shown in Figure 7 as a dashed line). From the spectra alone, it is not possible to determine whether this effect is due to bonding of the cyanide to gold, or whether it is simply a Stark shift due to the increase in potential.

A spectrum obtained after holding the potential at –400 mV for 5 and 60 minutes is also shown in Figure 7, and in contrast to Figure 6, it can be seen that there is a significant increase in intensity of the ν(CN) band with time. Such a result suggests that cyanide is continually accumulating on the gold surface as the gold is being oxidised. It is worth noting that the frequency of this band, 2127 cm^{-1}, is too low to be assigned to that of the polymeric AuCN, for which ν(CN) is 2232. However the SERS spectra centred at 670 cm^{-1}, which is shown in Figure 8, clearly indicates that the cyanide is bonded to the gold. The band at 380 cm^{-1}, which can be attributed to ν(AuC), and the band at 293, which can be attributed to δ(AuCN) are clearly visible. In addition, the intensity of these bands, in particular the δ(AuCN), increases with time as well. These results are consistent with the formation of some type of gold cyanide film on the gold surface which continues to grow with time as the gold is oxidised.

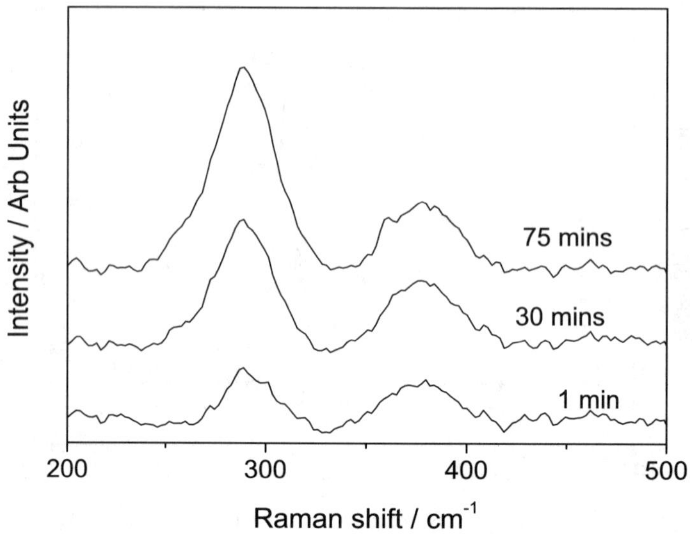

FIGURE 8 SERS spectra centred at 670 cm^{-1} for a gold electrode in 1 mM NaCN held for the marked times at –400 mV

SERS Studies of the Effect of Lead. It is well established that the addition of trace amounts of lead to solution result in activation of the gold surface in cyanide solutions. In fact, it has been suggested that the only reason gold leaches at all in AR grade cyanide is due to the trace amounts of lead (Jeffrey and Ritchie 2001). Initially the SERS spectra at −600 mV were obtained for a cyanide solution containing 10 ppm lead, and the results are shown in Figure 9. The initial spectrum obtained is very similar to that shown in the absence of lead (Figure 6), indicating the adsorption of cyanide onto the gold surface. However, after 12 mins, the band n(CN) had almost completely disappeared. Therefore it appears that when the gold surface is active, it is barely covered with cyanide. Such a result is not surprising when it is considered that the oxidation of gold in cyanide solutions containing lead approaches diffusion control, and hence the rate limiting step in the oxidation reaction is the transfer of cyanide through the diffusion layer. This leads to a surface concentration of cyanide which approaches zero. The SERS spectra centred at 670 cm^{-1} also showed the absence of the ν(AuC) and δ(AuCN).

The potential of the gold electrode in the cyanide + lead solution was then stepped to 0 mV, which is in the region where gold re-passivates in solutions containing lead. As can be seen from Figure 10, the ν(CN) returns, and starts to grow with time. It can also be observed that ν(CN) red-shifts with time, which is consistent with the formation of a stronger gold-cyanide bond. In addition, it was observed that the SERS spectra centred at 670 cm^{-1} showed the presence of the ν(CN) and δ(AuCN) at 0 mV. These results are important, as they indicate that the gold surface is still SERS active, and hence the lack of the SERS spectra obtained at −400 mV was not due to the deactivation of the gold surface. Thus it appears that when the gold is passive, strong ν(CN), ν(AuC) and δ(AuCN) bands are observed which grow in intensity, but when the gold is active, these bands are very weak.

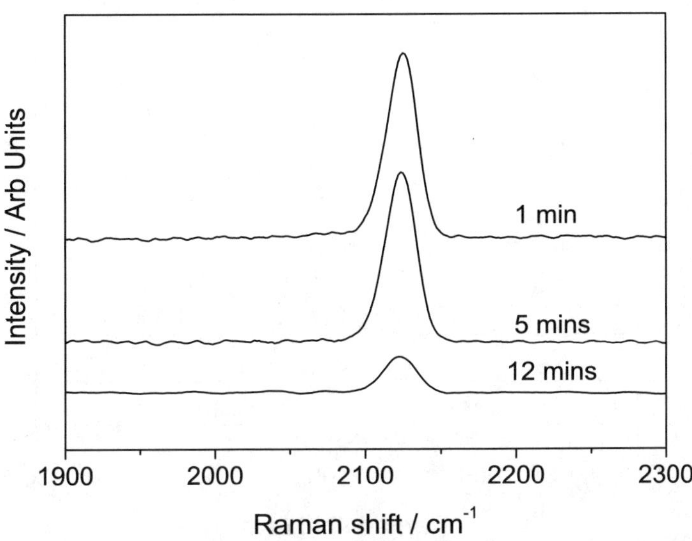

FIGURE 9 SERS spectra centred at 2100 cm^{-1} for a gold electrode in solution containing 1 mM NaCN and 10 ppm Pb^{2+} held for the marked times at –400 mV

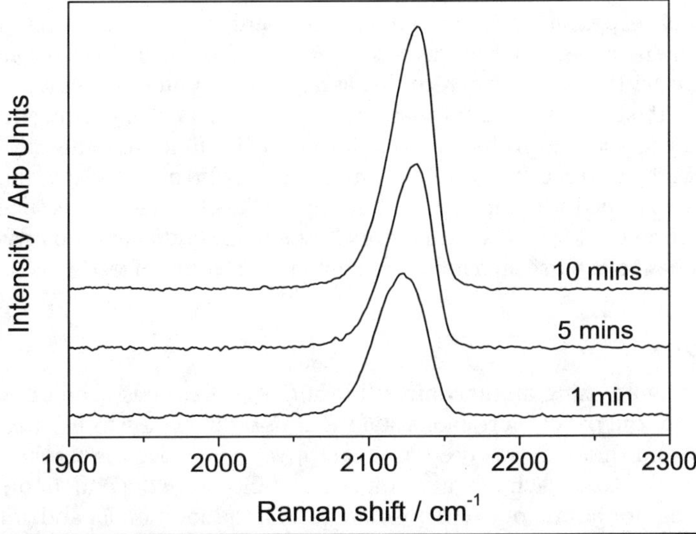

FIGURE 10 SERS spectra centred at 2100 cm^{-1} for a gold electrode in solution containing 1 mM NaCN and 10 ppm Pb^{2+} held for the marked times at 0 mV

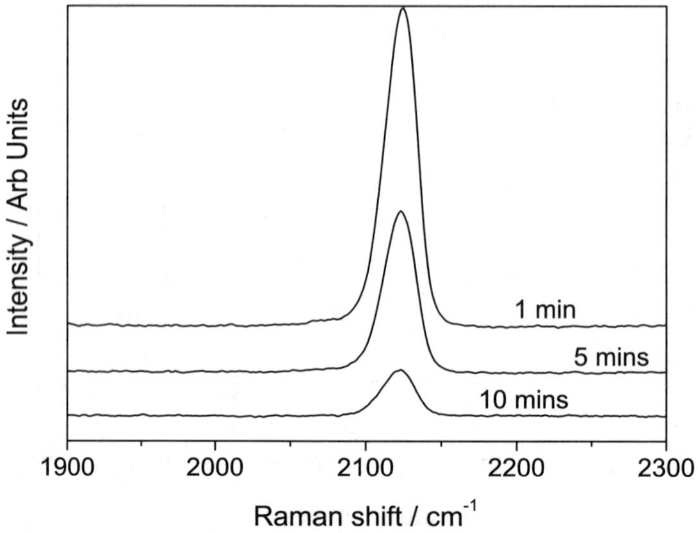

FIGURE 11 SERS spectra centred at 2100 cm^{-1} for a gold electrode held at –400 mV for 60 mins in solution containing 1 mM NaCN. The marked times indicate the time elapsed after 10 ppm lead was added to the solution.

An additional experiment was also performed where the gold was oxidised at −400 mV for 1 hour. At this point, lead nitrate was added to the solution to give a lead concentration of 10 ppm, and the SERS spectra were recorded as a function of time. The results are shown in Figure 11. It can be seen that initially the strong ν(CN) is observed, indicating the presence of some gold cyanide species on the surface. However with time, the intensity of this peak diminishes once the lead activates the gold surface. It was also observed that the ν(AuC) and δ(AuCN) also diminished with time after the addition of lead. These results suggest that lead can not only prevent gold passivation in cyanide solutions, but it can also activate the gold surface once it has been passivated in the solution in the absence of lead.

CONCLUSIONS

The leaching of gold using an innovative thiosulfate process has been demonstrated in the 1st section of the paper. Ferric ions stabilised using oxalate was used as the oxidant in this system, and thiourea was used as an additive to improve the leaching kinetics. It was shown that the gold leach rate and solution stability were dependent on the ferric to oxalate ratio, and for a ratio of Fe(III):oxalate of 1:3, the solution E_H and gold leach rate were stable with time. At lower Fe(III):oxalate, some of the Fe(III) is reactive towards thiosulfate. In the second section of the paper, SERS was used to identify surface species formed during the leaching of gold in cyanide solutions. It was shown that for pure gold in AR grade cyanide, a film of some gold cyanide species is formed slowly with time. However, when lead is added to solution, this species was found to diminish with time, resulting in a gold surface almost free of cyanide.

ACKNOWLEDGEMENTS

The authors thank the Australian Research Council and Griffith University for funding for this project.

REFERENCES

Baltruschat, H. and J. Heitbaum. 1983. On the potential dependence of the CN stretch frequency on Au electrodes studied by SERS. *Journal of Electroanalytical Chemistry* 157(2): 319–326.

Bowmaker, G.A., B.J. Kennedy and J.C. Reid. 1998. Crystal structures of AuCN and AgCN and vibrational spectroscopic studies of AuCN, AgCN and CuCN. *Inorganic Chemistry* 37: 3968–3974.

Breuer, P.L. and M.I. Jeffrey. 2000. Thiosulfate leaching kinetics of gold in the presence of copper and ammonia. *Minerals Engineering* 13(10–11): 1071–1081.

Breuer, P.L. and M.I. Jeffrey. 2003. Copper catalysed oxidation of thiosulfate by oxygen in gold leach solutions. *Minerals Engineering* 16(1): 21–30.

Byerley, J.J., S.A. Fouda and G.L. Rempel. 1975. Activation of copper(II) amine complexes by molecular oxygen for the oxidation of thiosulfate ions. *Journal of the Chemical Society, Dalton Transactions*: 1329–1338.

Chandra, I. and M.I. Jeffrey. 2004. An electrochemical study of the effect of additives and electrolyte on the dissolution of gold in thiosulfate solutions. *Hydrometallurgy* 73: 305–312.

Dorain,P.B. and K.U. Von Raben. 1985. SERS measurement of oxidation-reduction reactions of adsorbates on gold microstructures. *Surface Science* 160(1): 164–70.

Dudeney, A.W.L. and I.I. Tarasova. 1998. Photochemical decomposition of trisoxala-toiron(III): A hydrometallurgical application of daylight. *Hydrometallurgy* 47: 243–257.

Gao, P. and M.J. Weaver. 1986. Metal-Adsorbate Vibrational Frequencies as a Probe of Surface Bonding: Halides and Pseudohalides at Gold Electrodes. *Journal of Physical Chemistry* 90: 4057 - 4063.

Jeffrey, M.I. 2001. Kinetic aspects of gold and silver leaching in ammonia-thiosulfate solutions. *Hydrometallurgy* 60: 7–16.

Jeffrey, M.I., P.L. Breuer and W.L. Choo. 2001. A kinetic study that compares the leaching of gold in the cyanide, thiosulfate and chloride systems. *Metallurgical Metallurgical & Materials Transactions B.* 32B: 979–986.

Jeffrey, M.I. and I.M. Ritchie. 2000a. The leaching of gold in cyanide solutions in the presence of impurities I. The effect of lead. *Journal of the Electrochemical Society* 147(9): 3257–3262.

Jeffrey, M.I. and I.M. Ritchie. 2000b. The leaching of gold in cyanide solutions in the presence of impurities II. The effect of silver. *Journal of the Electrochemical Society* 147(9): 3272–3276.

Jeffrey, M.I. and I.M. Ritchie. 2001. The leaching and electrochemistry of gold in high purity cyanide solutions. *Journal of the Electrochemical Society* 148(4): D29-D36.

Jeffrey, M.I., J. Zheng and I.M. Ritchie. 2000. Development of a rotating electrochemical quartz crystal microbalance for the study of metal leaching and deposition. *Measurement Science and Technology* 11(5): 560–567.

Jones, L.H. 1963. Raman spectra of the linear dicyanides. *Spectrochimica Acta* 19: 1675–1681.

Kirk, D.W., F.R. Foulkes and W.F. Graydon. 1980. Gold passivation in aqueous alkaline cyanide. *Journal of the Electrochemical Society* 127(9): 1962–1969.

Murray, C.A. and S. Bodoff. 1986. Cyanide adsorption on silver and gold overlayers on island films as determined by surface enhanced Raman scattering. *Journal of Chemical Physics* 85(1): 573–584.

Nakamoto, K. 1997. *Infrared and Raman Spectra of Inorganic and Coordination Compounds, Part B: Applications in Coordination, Organnometallic, and Bioinorganic Chemistry*. New York, Wiley-Interscience.

Nicol, M.J. 1980. The anodic behaviour of gold. *Gold Bulletin* 13: 105–111.

Nicol, M.J. and G. O'Malley. 2002. Recovering gold from thiosulfate leach pulps via ion exchange. *JOM* 54(10): 44–46.

Page, F.M. 1960. Ferric thiosulfate reaction. III. Mechanism of the reaction. *Transactions of the Faraday Society* 56: 398–406.

Sharpe, A.G. 1976. *The Chemistry of Cyano Complexes of the Transition Metals*. London, Academic Press.

Sheveleva, L.D. and I.A. Kakovskii. 1979. Lead compounds and dissolution of gold in cyanide solutions. *Tsvetnye Metally* 7: 100–102.

Smith, R.M. and A.E. Martell. 1976. *Critical Stability Constants. Volume 3: Other Organic Ligands*. New York, Plenum Press.

Weichselbaum, J., J.A. Tumilty and C.G. Schmidt. 1989. The effects of sulfide and lead on the rate of gold cyanidation. *The Aus.I.M.M. Annual Conference Perth/Kalgoorlie*. Melbourne, Aus.I.M.M.: 221–224.

Yamada, T., R. Sekine and T. Sawaguchi. 2000. Ultrahigh-vacuum multitechnique study of AuCN monolayers on Au(111) formed by electrochemical deposition. *Journal of Chemical Physics* 113(3): 1217–1227.

Thiohydrometallurgical Processes for Gold Recovery

R.Y. Wan,* J.D. Miller,† and J. Li†

Thiohydrometallurgical systems including thiosulfate, thiocyanate and thiourea have been studied as alternative lixiviants for the leaching of gold. Research and development of thiohydrometallurgical technology has been prompted by environmental concerns with the use of cyanide and problems associated with the processing of difficult-to-treat ores. These thiosystems have some chemical similarities, enabling gold to be leached in a practical oxidizing environment. However, the solution chemistry of the thiosystems is complicated due to instability of the thioreagents and many other reactions which occur during the gold leaching reaction. In this paper, interference reactions of selected metal ions and sulfide minerals are examined in order to understand important aspects of solution chemistry which control the stability of the thioleaching reactions. Applications using thioreagents as lixiviants for oxide and refractory gold ores are presented.

INTRODUCTION

Until now, most of the gold production from primary resources is still achieved by cyanidation. The introduction of carbon-in-leach technology has eliminated the need for solid/liquid separations associated with Merrill-Crowe processing and reduced operating costs. The development of heap leach technology for low- grade ores has extended the world resources base considerably. The cyanidation process is remarkably robust and the gold-cyanide complex ion is very stable and gives very few operation problems. In addition to valuable results from research and development, industrial operation experiences have made the cyanidation process a great success.

However, the toxicity of cyanide is a concern. With increasing regulations on the use of cyanide and the lowering of acceptable cyanide discharge levels, the gold industry

* Metallurgy Consultant, 9634 Kalamere Court, Highlands Ranch, Colorado

† Metallurgical Engineering Department, University of Utah, Salt Lake City, Utah

faces more challenges than ever before. Some parts of the world, such as Turkey and Greece, are preventing the development of gold deposits until an alternative to cyanide is developed. Several states in the US have prepared, or are preparing, legislation against the development of any new gold mines that rely on the cyanide process. In addition, cyanide is unable to effectively extract gold from certain difficult-to-treat ores leading to poor gold recovery from carbonaceous refractory ores and lack of selectivity in the treatment of copper bearing gold ores. In these cases, cyanide consumption is generally high and tailings discharge more difficult.

Many research activities have been carried out on alternative processes using noncyanide lixiviants (Senanayake 2004). Sparrow and Woodcock (1995) listed over 25 potential alternative reagents including thioreagents and polysulfides, but only a few have attracted significant attention. Alternative reagents that have been seriously considered include the halide system (chlorine, bromine and iodine), the thiosystem (thiosulfate, thiocyanate and thiourea), the polysulfide system (S_x^{2-}), and the ammonia system (ammonia and ammonium copper-cyanide). All these reagents form complexes with gold, enabling gold to be leached in a practical oxidizing environment. The recovery of gold from noncyanide solutions has been reviewed (Wan et al.1993). Ritchie et al. (2001) reviewed non-cyanide gold processes to compare and contrast potential lixiviants on the basis of environmental issues, economic factors, toxicity of reagents and physical-chemical aspects of gold extraction and gold recovery. It was concluded that only thiosulfate has a reasonable chance of replacing cyanide as a relatively non-toxic lixiviant at the present time.

Selectivity of alternative reagents for gold recovery depends on ore mineral characteristics and many other technical and economical factors. For example, the halogens are reactive with gold and operations at Carlin and Jerritt Canyon (both in Nevada) commercially recovered gold from carbonaceous ores for several years using chlorine (Marsden and House 1992). However, the halogens proved to be too reactive with other ore minerals especially sulfide minerals. Depending on ore type and content of sulfide minerals, operating costs can be became too expensive. Ideally, the leaching reagent should not be expensive, or alternatively it should be recyclable. It should be selective for gold over other metals, less reactive with sulfide minerals, non-toxic, and compatible with downstream recovery processes. In practice, meeting all these criteria is difficult.

Thioreagents include thiosulfate, thiocyanate and thiourea. These thiosystems have some chemical similarities. All the thioreagents react with gold forming soluble complexes. Their leaching reactions require an oxidant. The solution chemistry of the thiosystem is complicated due to instability of the thioreagents and many other reactions that occur during leaching. Unlike cyanide, the thioligands are susceptible to degradation either in an oxidizing or reducing environment which leads to high reagent consumption, decreasing gold dissolution rate, formation of intermediate oxidation products, and/or further decomposition to undesirable insoluble products such as elemental sulfur, sulfide and/or other insoluble compounds. The degradation products not only passivate gold surfaces and retard gold dissolution but also interfere in subsequent unit operations for gold recovery. In this paper, interference reactions of selected metal ions and sulfide minerals are examined in order to understand important aspects of solution chemistry, which control the stability of the thioleaching reactions. Applications using thioreagents as lixiviants for oxide and refractory gold ores are presented.

TABLE 1 Stability constants of gold thiocomplexes and cyanide complex at 25°C

Ligand	Complex Ion	Log $ß_n$	Standard Potential (V)
Thiosulfate	$Au(S_2O_3)^{3-}$	28.7	0.15
Thiocyanate	$Au(SCN)_2^-$	17.2	0.66
	$Au(SCN)_4^-$	43.9*	0.66
Thiourea	$Au(SC(NH_2)_2)_2^+$	23.3	0.38
Cyanide	$Au(CN)_2^-$	38.3	–0.57

Stability constant data from Smith and Martell 1989.
*Calculated from thermodynamic data

GOLD THIOLEACHING AND INTERFERENCE REACTIONS

Thioreagents as alternative lixiviants for gold recovery from ores and concentrates have been studied over the world. Several efforts to implement thiolixiviants on a small scale in industry have been reported during the past decades. The solution chemistry of thio-systems for gold recovery has been discussed in the literature (Groenewald 1976, Hiskey and Atluri 1988, Barbosa-Filho and Monhemius 1994, Aylmore and Muir 2001).

Ritchie et al. (2001) used a schematic Eh/pL diagram to illustrate the stability of the formation of gold complexes. The larger the value of the stability constant, the greater the driving force for gold leaching. The stability constants for gold thiocomplexes are tabulated and compared with the stability constant for the gold cyanide complex in Table 1.

The thioreagents form complexes with gold. The stability order of the Au (I) complexes with these ligands is: $CN^- > S_2O_3^{2-} > SC(NH_2)_2 > SCN^-$. Also, as shown in Table 1, the thioreagents enable gold to be leached in a practical Eh range. However, under ambient condition, the gold dissolution in thiosystems in the presence of oxygen has been found to be very slow, and other oxidants are required for the thioleaching reaction. The thioligands are not as stable as cyanide. Depending on the operating conditions, decomposition of thioreagents and various side reactions occur in the leach system. In addition, thioreagents form complex ions with a variety of metals; and sulfide minerals in the ores affect the system stability. Therefore, understanding the decomposition of these thioligands is of critical importance in the commercial development of alternative non-cyanide leaching systems for gold recovery from primary ores.

Copper-Ammonia-Thiosulfate System

Gold leaching in thiosulfate solution is carried out under alkaline conditions. The advantage of leaching in alkaline solution is the minimization of the solubility of impurities, particularly the iron compounds. However, in order to treat some refractory gold ores, such as sulfide ores, it is often necessary to have a pre-oxidation process, which yields highly acidic solutions, and the system must then be neutralized prior to gold leaching. Ammonium thiosulfate is an inexpensive, nontoxic reagent, used mostly for fertilizer. Acceptable gold leaching rates using thiosulfate are achieved in the presence of ammonia with cupric ion acting as an oxidant. Ammonia concentration is critical in the gold leaching process since it stabilizes the oxidant, cupric ion (Li et.al.1996, Breuer and Jeffrey 2000). Gold leaching in copper catalyzed ammonia-thiosulfate solution involves an electrochemical reaction, with the half cell reactions being the oxidation of gold to gold thiosulfate and the reduction of Cu (II) ammonia complex to the Cu (I)-thiosulfate complex.

$$Au + 2S_2O_3^{2-}\ Au(S_2O_3)_2^{3-} + e^- \quad \textbf{(EQ 1)}$$

$$Cu(NH_3)_4^{2+} + 3S_2O_3^{2-} + e^-\ Cu(S_2O_3)_3^{5-} + 4NH_3 \quad \textbf{(EQ 2)}$$

The cupric complex is regenerated in the presence of oxygen,

$$2Cu(S_2O_3)_3^{5-} + 8NH_3 + 0.5O_2 + H_2O\ 2Cu(NH_3)_4^{2+} + 6S_2O_3^{2-} + 2OH^- \quad \textbf{(EQ 3)}$$

Generally speaking, thiosulfate stabilizes gold in solution, while ammonia and copper accelerate the leaching rate. The concentrations of ammonia and Cu(II) ion in the thiosulfate leaching system are important factors which govern thiosulfate stability and reagent management. Consumption of thiosulfate due to degradation of tetrathionate occurs according to the following oxidation reaction.

$$2Cu(NH_3)_4^{2+} + 8S_2O_3^{2-} = 2Cu(S_2O_3)_3^{5-} + S_4O_6^{2-} + 8NH_3 \quad \textbf{(EQ 4)}$$

The oxygen content in the thiosulfate system affects the leaching kinetics. Jeffrey et al. (2003) reported the importance of controlling oxygen addition during the thiosulfate leaching of gold ores; as too much oxygen results in slow leaching kinetics due to the formation of an intermediate oxy-sulfur species associated with thiosulfate oxidation. The undesirable oxidation of thiosulfate not only remarkably increases reagent consumption but also affects subsequent process for gold recovery. At low oxidizing potentials where oxidants are deficient, such as in heap leaching systems, the decomposition of thiosulfate leads to the precipitation of black copper sulfides. In this study, gold leaching kinetics has been examined in copper-ammonia-thiosulfate solutions using a gold foil without air sparging. After the foil has been leached for a number of leaching cycles, the leaching rate is found to decrease. The surface of the gold foil became tarnished and passivated. Sulfur and copper sulfides were identified by XPS and microprobe analysis.

Stability of Sulfide Minerals in Ammonium Thiosulfate Solution. In the copper-ammonia-thiosulfate solution, many reactions may occur among the system constituents, not to mention that the system itself is thermodynamically unstable. Most of the common sulfide minerals which exist in refractory gold ores, such as copper and iron sulfides, react with thiosulfate and affect gold leaching kinetics and thiosulfate stability. Thermodynamically, both copper and iron sulfide minerals are not stable in ammonium thiosulfate solutions as long as the solution potential is high enough, as shown from Eh-pH diagrams. Figure 1 indicates that common sulfide minerals such as pyrite, arsenopyrite, chalcopyrite, chalcocite, and covellite can be dissolved under proper conditions. The oxidizing potentials required for gold leaching in thiosulfate solutions are able to oxidize copper and iron sulfide minerals in thiosulfate solutions. Therefore, thermodynamically, the iron and copper sulfide minerals are unstable in the thiosulfate solutions under gold leaching conditions.

Electrochemical Measurements. The dissolution of iron and copper sulfides in ammonium thiosulfate solutions has been examined by electrochemical studies. Steady-state polarization curves were measured on sulfide mineral electrodes with the potentiodynamic technique at a scan rate of 1 mV/s under an argon atmosphere. The scan started from the open circuit potential (rest potential) of the working electrode and usually ended at 600 mV versus SCE. The polarization curves of pyrite, arsenopyrite, chalcopyrite, chalcocite, and covellite are shown in Figure 2. For comparison, the polarization curves of metallic gold are also included in Figure 2.

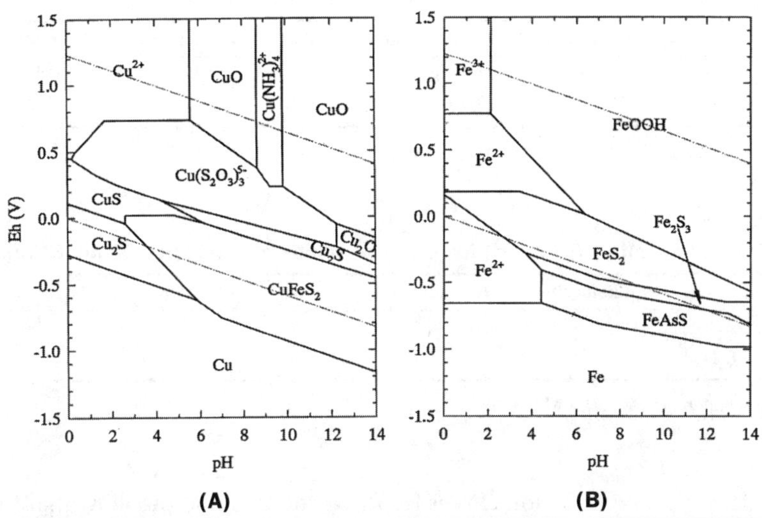

FIGURE 1 Eh-pH diagram of (A) copper-iron-ammonia-thiosulfate system, and (B) iron-arsenic-ammonia-thiosulfate system at 25°C, 0.1 M ammonia, 0.1 M ammonium thiosulfate, 5×10^{-4} M copper, 1×10^{-6} M iron and 1×10^{-5} M arsenic

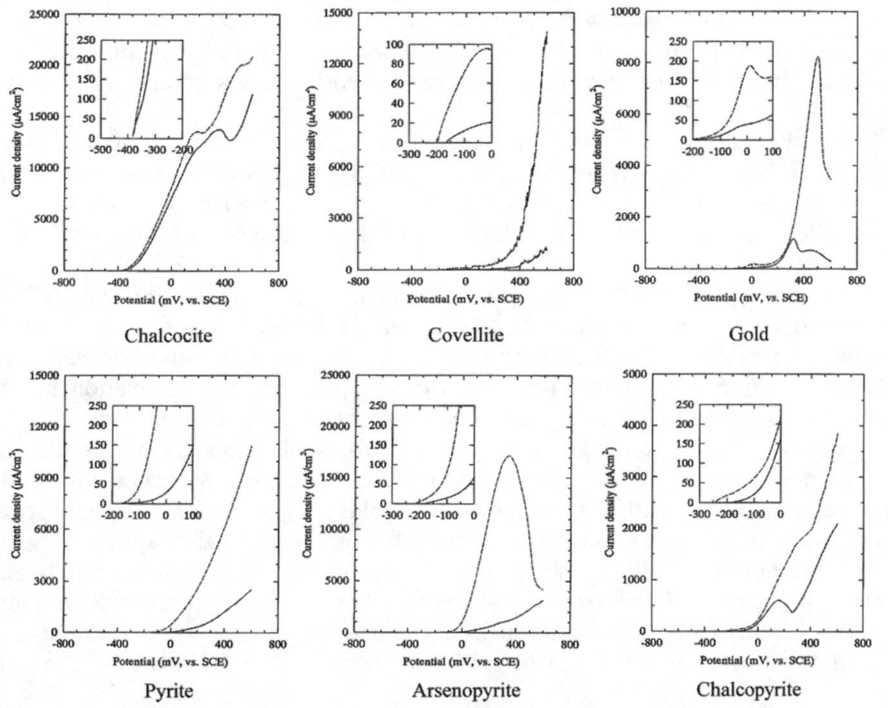

FIGURE 2 Polarization of sulfide minerals in thiosulfate solution with 0.2M NH_4OH. Solid line 0.5M $Na_2S_2O_3$; dash line 0.5M $(NH_4)_2S_2O_3$.

TABLE 2 Rest potentials (mV vs. SCE) of gold and sulfide minerals

	Chalcocite	Gold	Chalcopyrite	Arsenopyrite	Covellite	Pyrite
0.5 M $Na_2S_2O_3$	–381	–280	–183	–178	–165	–164
0.5 M $(NH_4)_2S_2O_3$	–377	–285	–240	–214	–196	–171

* Thiosulfate solution contains 0.2 M NH_4OH

TABLE 3 Current densities (μA/cm²) for gold and sulfide minerals at 100 mV (SCE)

	Chalcocite	Arsenopyrite	Pyrite	Chalcopyrite	Covellite	Gold
0.5 M $Na_2S_2O_3$	9938	278	143	484	22	64
0.5 M $(NH_4)_2S_2O_3$	11750	5292	1335	709	225	163

* Thiosulfate solution contains 0.2 M NH_4OH

Electrochemical polarization curves for these sulfide minerals show that the current densities for anodic oxidation of sulfide minerals are of a substantial magnitude. The measured rest potentials, shown in Table 2, indicate the thermodynamic stability is in the following order: chalcocite< gold< chalcopyrite < arsenopyrite < covellite < pyrite.

Thermodynamically, the most difficult mineral to be oxidized in thiosulfate solution is pyrite, while gold has a relatively strong tendency to be dissolved. Though the physical properties of these pure sulfide specimens, used as working electrodes, could be different from the minerals associated with gold in refractory gold ores, their rest potentials should give an indication of relative thermodynamic stability under the indicated experimental conditions. Materials with a lower rest potential have a greater tendency to be oxidized.

The current densities, which can be used to estimate the reaction rate, are much higher for sulfide mineral oxidation than for gold dissolution. From Figure 2, based on the current densities for gold and sulfide minerals measured at 100 mV (SCE), the kinetic order at 100 mV (SCE) is as follows (Table 3): chalcocite > arsenopyrite > pyrite > chalcopyrite > covellite > gold.

It should be noted that the measured anodic oxidation current includes two reactions: the decomposition of the sulfide mineral (or the dissolution of gold) and the decomposition of thiosulfate. Further research is needed to differentiate between sulfide oxidation and thiosulfate decomposition. However, no obvious passivation has been observed for minerals at low overpotentials where the leaching of gold normally occurs. On the contrary, gold becomes passivated easily. Nevertheless, results indicate that these sulfide minerals will be oxidized in thiosulfate leaching solutions. Results also indicate that the ammonium thiosulfate solution attacks gold and sulfide minerals more strongly than the sodium thiosulfate solution, especially for sulfide minerals.

The effect of these sulfide minerals on gold leaching has also been studied. In experiments, a piece of gold foil was suspended in a copper-ammonia-thiosulfate solution with the addition of 50 grams of sulfide minerals in one-liter of solution. Compared with the gold leaching in the absence of sulfide minerals, results indicate all the sulfide minerals depress gold leaching. Arsenopyrite, pyrite and chalcocite have a particularly significant effect on decreasing the rate of the gold leaching reaction.

Many applications for gold leaching of sulfide ores, complex ores and copper-gold ores using the thiosulfate lixiviant have been reported (Li et al. 1995, Aylmore 2001, Molleman and Dreisinger 2002, Navarro et al. 2002). When sufficient sulfide mineral dissolution has occurred, finely disseminated gold in the sulfide mineral matrix may be exposed and/or liberated for subsequent leaching. However, besides exposing/liberating gold, the dissolution of sulfide minerals may have other influences on thiosulfate leaching, such as the consumption of thiosulfate in solution and the interaction of products with dissolved gold. The formation of colloidal iron hydroxides can lead to the adsorption of a gold complex and block mass transport, while a high concentration of dissolved copper may change the system equilibrium. The content of sulfides in a refractory ore generally is much higher than that of gold, the side reactions and the side effects due to these reactions can be very important. In leaching practice, the sulfide minerals also react during the gold leaching reaction, affecting the thiosulfate stability, and resulting in high reagent consumption. The rate of thiosulfate decomposition is an important issue in the development of thiosulfate leach technology.

Effect of Copper, Iron and Activated Carbon on Thiosulfate Decomposition. Though leaching of gold in thiosulfate solution is considered to be affected less by foreign cations when compared to cyanidation, their presence involves some interaction with thiosulfate and may retard gold dissolution. Feng and Deventer (2002) studied the effect of heavy metals ions on gold dissolution behavior in the ammonium thiosulfate system. The retardation effect was found strongly dependent on ion types and concentrations. They found that zinc had a slightly positive effect on gold dissolution at very low concentrations and would retard gold dissolution at high concentrations. Other metals, Cd, Co, Cr and Ni, retarded gold dissolution to different extents at all concentration levels. The free thiosulfate concentration decreased in the presence of heavy metals, especially at high concentrations.

The formation of metallic colloids and a decrease of free thiosulfate concentration contribute to the lower gold dissolution rates in the presence of heavy metal ions. The influences of copper, iron, and activated carbon on thiosulfate decomposition were studied in an one-liter stirred reactor, where the thiosulfate concentration was measured directly as a function of time. Tests were performed under two oxidation conditions; one with a very limited air supply at the solution surface and the other with sparged air at a flow of 10 ml/min into the solution. The rate of thiosulfate degradation is given as thiosulfate decomposition in g/day•liter for comparison. The effect of copper ion concentration on thiosulfate decomposition is shown in Table 4. When the initial copper concentration (copper was added as cupric sulfate) was varied between 0.0005 M and 0.01 M with limited air flow at the surface, the rate of thiosulfate decomposition remained about the same; only at higher cupric concentration was a slightly higher thiosulfate loss observed even at the beginning of the reaction. A higher rate of oxygen supply accelerated the rate of thiosulfate decomposition.

Compared with the effect of copper ion concentration, the iron concentration had a small but clear impact on thiosulfate decomposition. With ammonium thiosulfate solution of 0.1 M $(NH_4)_2S_2O_3$, 0.1 M NH_4OH and 0.0005 M Cu, and limited air flow at the solution surface, the rate of thiosulfate decomposition increased from 0.112 (without iron addition) to 0.168 g/day•liter with 2g/L (0.011 M) iron addition as ferrous sulfate.

TABLE 4 Effect of copper ion concentration on thiosulfate decomposition

Initial Cu^{2+}, M		0.0005	0.001	0.002	0.004	0.005	0.01
Rate	Limited air at surface	0.112	0.110	0.128	0.126	—	—
g/day·L	Sparged air, 10 ml/min	0.90	0.982	—	—	0.984	1.169

Initial solution: 0.1 M ammonium thiosulfate, 0.1 M ammonia, pH 9.2

TABLE 5 Effect of activated carbon on thiosulfate decomposition

Carbon Amount, g/L		0	2	20
Rate	Limited air at surface	0.112	0.572	—
g/day·L	Sparged air, 10 ml/min	0.90	3.708	5.403

Initial solution: 0.1 M ammonium thiosulfate, 0.1 M ammonia, 0.0005 M Cu, pH 9.2

When iron hydroxides precipitated from solution, thiosulfate degradation changed from a homogeneous condition to a heterogeneous condition as inert solids may have stimulated oxygen discharge by providing catalytic sites.

It has been realized that oxygen discharges on activated carbon. The presence of activated carbon would be expected to increase the rate of thiosulfate decomposition. Results of thiosulfate decomposition in the presence of activated carbon (d_{80} = 38 µm) are given in Table 5. An increase in the amount of activated carbon increased the surface area and thus the decomposition rate. Though it has been found that activated carbon has a very low affinity for gold-thiosulfate complex ion, the thiosulfate consumption in the treatment of carbonaceous oxide ores was found to be much higher than for oxide ores under similar operating conditions.

In summary, common copper and iron sulfide minerals can dissolve in ammonium thiosulfate solutions. Their presence in the leaching system lowers the oxidation potential and decreases the rate of gold dissolution. Thiosulfate ions are metastable and tend to undergo chemical decomposition in aqueous solutions. Numerous factors affect the system stability. Depending on the solution potential and pH value, thiosulfate can decompose either oxidatively or reductively. The above results of oxidation thiosulfate decomposition from reactor testing confirm that any factor that accelerates the transport of oxygen will increase the rate of thiosulfate decomposition.

Acidic Ferric-Thiocyanate System

Thiocyanate has been extensively studied for gold leaching in acidic solutions (Fleming 1986, Broadhurst and du Preez 1993, Barbosa-Filho and Monhemius 1989, 1994, Munoz and Miller 2000, Kholmogorov et al. 2002). One of the important features in using thiocyanate is that the leaching can be performed in acidic media, thus avoiding problems related to neutralization and material handling for ores containing sulfide minerals. Gold dissolves in thiocyanate solution forming aurothiocyanate complex, $Au(SCN)_2^-$ and the aurithiocyanate complex ion, $Au(SCN)_4^-$.

$$Au + 2SCN^- = Au(SCN)_2^- + e^- \qquad \textbf{(EQ 5)}$$

$$Au + 4SCN^- = Au(SCN)_4^- + 3\,e^- \qquad \textbf{(EQ 6)}$$

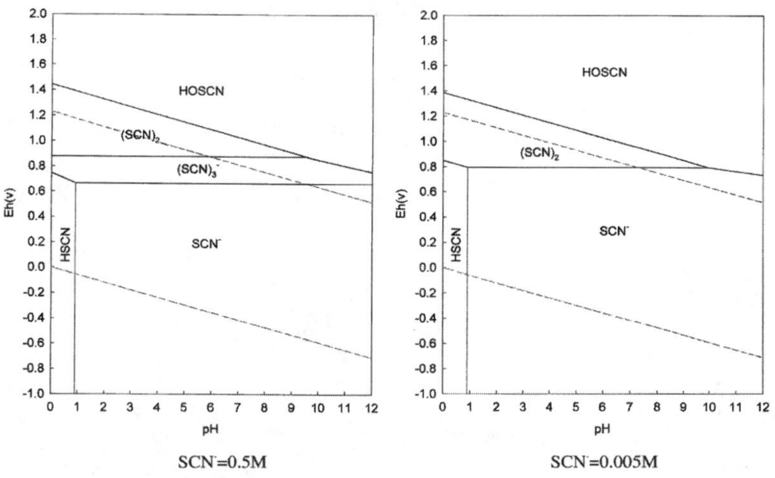

FIGURE 3 Metastable Eh-pH diagram for the SCN-H_2O system

Gold dissolves in thiocyanate solutions at potentials at which the oxidation of thiocyanate also occurs. Thiocyanate can be oxidized according to the following reactions; and ultimately to the oxidation products given by reaction (9).

$$SCN^- + 4\,H_2O = SO_4^{2-} + CN^- + 8\,H^+ + 6\,e^- \qquad \textbf{(EQ 7)}$$

$$SCN^- + 5\,H_2O = SO_4^{2-} + CNO^- + 10\,H^+ + 8\,e^- \qquad \textbf{(EQ 8)}$$

$$SCN^- + 7\,H_2O = NH3 + CO_3^{2-} + SO_4^{2-} + 11\,H^+ + 8\,e^- \qquad \textbf{(EQ 9)}$$

The oxidation of SCN^- proceeds through the formation of several intermediate species, in particular trithiocyanate $(SCN)_3^-$ and thiocyanogen $(SCN)_2$, which both act as oxidants and complexants for gold. As shown in Figure 3, at high concentrations of thiocyanate, $[SCN^-] \geq 0.1M$, the thiocyanate can be oxidized to trithiocyanate $(SCN)_3^-$ and then to thiocyanogen $(SCN)_2^-$ with an increase in the oxidation potential. Decreasing the thiocyanate concentration to 0.0005 M, the region for trithiocyanate disappears and the predominant intermediate product of thiocyanate oxidation is thiocyanogen.

In the thiocyanate system, gold can be dissolved in the acidic region, which allows the use of several oxidizing agents that are suitable for acidic solutions. Almost all the reports on thiocyanate leaching of gold refer to ferric ion as the most suitable oxidizing agent. Practically, using ferric ion as an oxidant is most appropriate for gold leaching following preoxidation treatment. High amounts of ferric ion in an acidic solution would exist after a preoxidation treatment such as biooxidation or pressure oxidation. However, thiocyanate forms complexes with both ferric and ferrous ions. The effect of these iron complexes should be considered in any actual leaching system.

The reaction mechanism of gold dissolution by thiocyanate with ferric sulfate has been reported in the literature (Broadhurst and du Preez 1993; Barbosa-Filho and Monhemius 1994). It was concluded that an autoreduction process takes place for the Fe(III)-thiocyanate solution in which Fe(III) is reduced to Fe(II) with the simultaneous

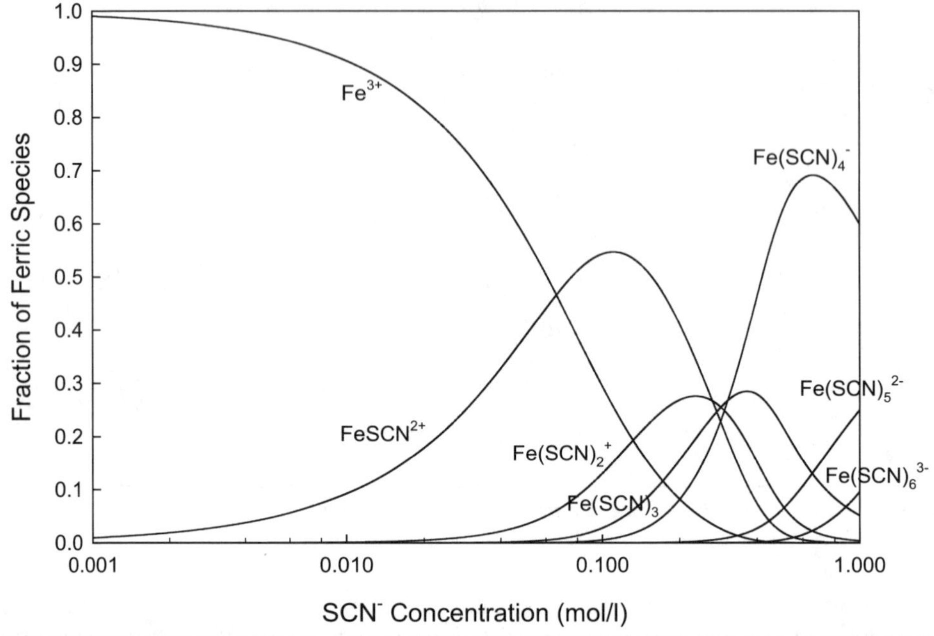

FIGURE 4 Species distribution diagram for the Fe(III)-SCN-H_2O system at [Fe(III)] 0.1M, [SCN^-] 1×10^{-3} – 1.0 M, pH 2.0 and 25°C

oxidation of thiocyanate. The oxidation of SCN^- by Fe(III) proceeds through the formation of $(SCN)_3^-$ and $(SCN)_2$, followed by their hydrolysis to more stable oxidation products. The intermediate oxidation species, particularly $(SCN)_3^-$ and $(SCN)_2$ play important roles in the dissolution of gold. It was postulated that these intermediate species act as the primary oxidizing agents in the dissolution of gold (Barobosa and Monhemius 1994). Therefore, proper control of the autoreduction process is the key to the optimization in the leaching of gold by ferric-thiocyanate solution.

The addition of ferric ion also results in a decrease in the free thiocyanate ion concentration, since thiocyanate forms a series of complexes with both ferric and ferrous ions. From the stability constants for the thiocyanate complex ions, it is found that the iron-thiocyanate complexes are less stable than gold-thiocyanate complexes. Thermodynamically, the thiocyanate ions prefer to form gold–thiocyanate complexes even in the presence of iron. With regard to the kinetics for gold leaching in an acid thiocyanate solution, there is a competition between thiocyanate oxidation and gold leaching for the free thiocyanate. The rates are expected to be significantly influenced by leaching conditions and the composition of gold bearing minerals.

Iron, Silver and Copper Thiocyanate Complexes. In a ferric-thiocyanate solution, the formation of ferric-thiocyante complexes depends on the SCN^- concentration and [SCN^-] to [Fe(III)] mole ratio, as shown in Figure 4. At low [SCN]/[Fe(III)] mole ratio, low concentrations of free thiocyanate lead to the formation of cationic iron-complexes, whereas high thiocyanate concentrations yield anionic complexes. The relative amount of Fe(III)-SCN complexes with high ligand number increases when total amount of SCN^-

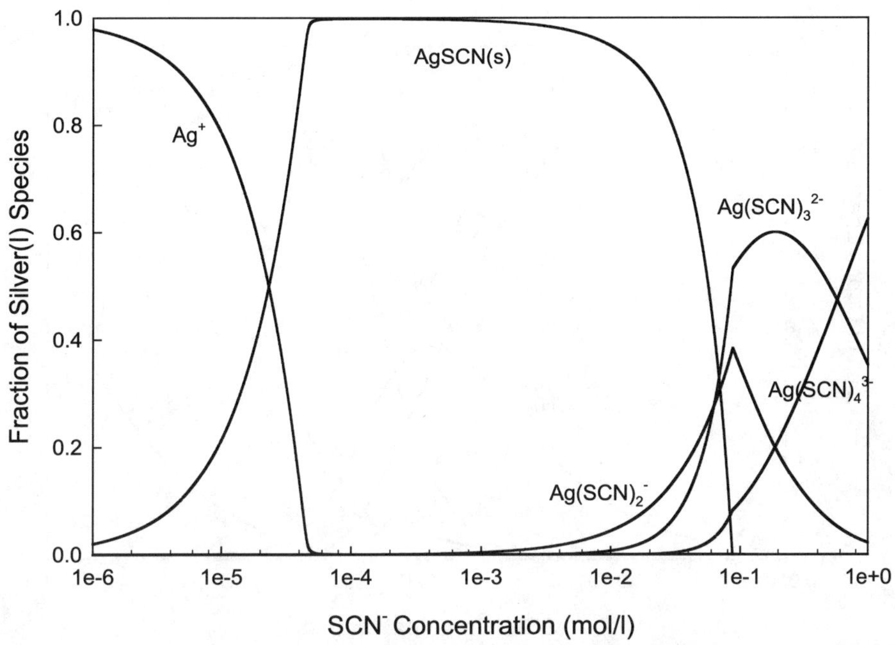

FIGURE 5 Species distribution diagram for the Ag-SCN-H_2O system at [Ag(I)] 5.0 ppm, [SCN^-] 1×10^{-6} – 1.0M, pH 2.0, 25°C and in potential range of Ag(I) species stability

concentration increases or [SCN] to [Fe(III)] mole ratio increases. On the other hand, with low [SCN]/[Fe(III)] mole ratio, the Fe(III) becomes the predominant species in the system. For the successful thiocyanate leaching of gold, it is important to identify appropriate solution compositions whereby the gold oxidation rate would be maximized with minimum degradation of thiocyanate.

In acidic oxidizing environments, some sulfide minerals become unstable and oxidize. The oxidation of sulfides consumes the oxidants such as ferric ion as well as thiocyanate. Thiocyanate is a pseudohalide, whose related pseudohalogen is thiocyanogen, $(SCN)_2$, and whose related hydropseudohalic acid is thiocyanic acid, HSCN (Barbosa-Filho and Monhemius 1989). The pseudohalide behavior of thiocyanate results in the formation of insoluble salts when the thiocyanate reacts with silver, mercury, lead and copper ions under certain conditions. As shown in Figure 5, for the distribution diagram of Ag-SCN-H_2O system at pH 2, the insoluble $AgSCN_{(S)}$ is predominant, even at a [SCN^-]/[Ag(I)] mole ratio of 200.

The cupric ion stability in thiocyanate solution is illustrated in Figure 6. At [SCN^-] = 0.02M and [SCN^-]/[Cu^{2+}] = 6.35, the predominant cupric species is $Cu(SCN)_3^-$. For cuprous ion, at the same copper and thiocyanate concentrations, the predominant species is insoluble $CuSCN_{(S)}$, as shown in Figure 7. When the thiocyanate concentration increased from 0.2 M to 0.5 M with [SCN^-]/ [Cu^+] =160, the soluble $Cu(SCN)_2^-$ becomes predominant. The insoluble compounds such as AgSCN and CuSCN may precipitate at the gold surface and retard gold dissolution.

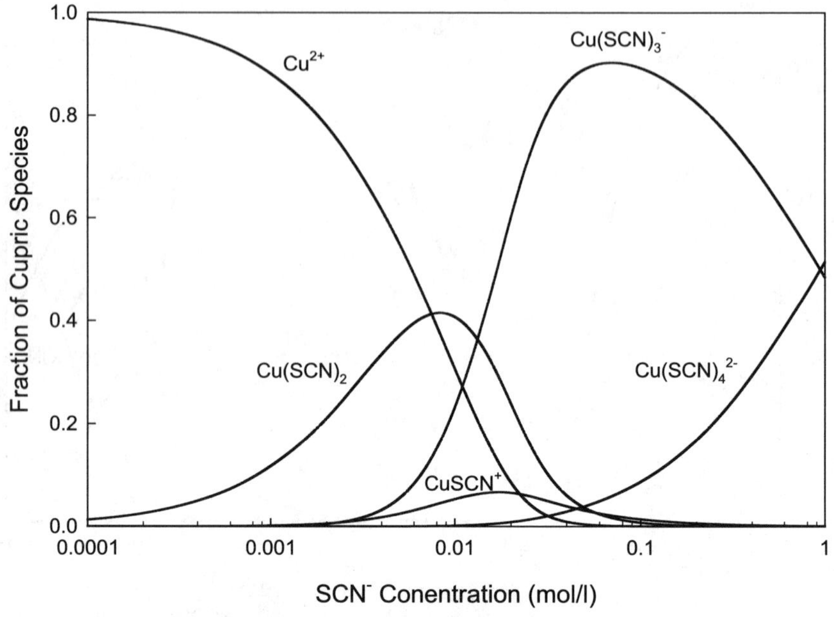

FIGURE 6 Species distribution diagram for the Cu(II)-SCN-H_2O system at [Cu(II)] 0.2g/L, [SCN^-] 1×10^{-4}–1.0M, pH 2.0, 25°C and in potential range of cupric species stability

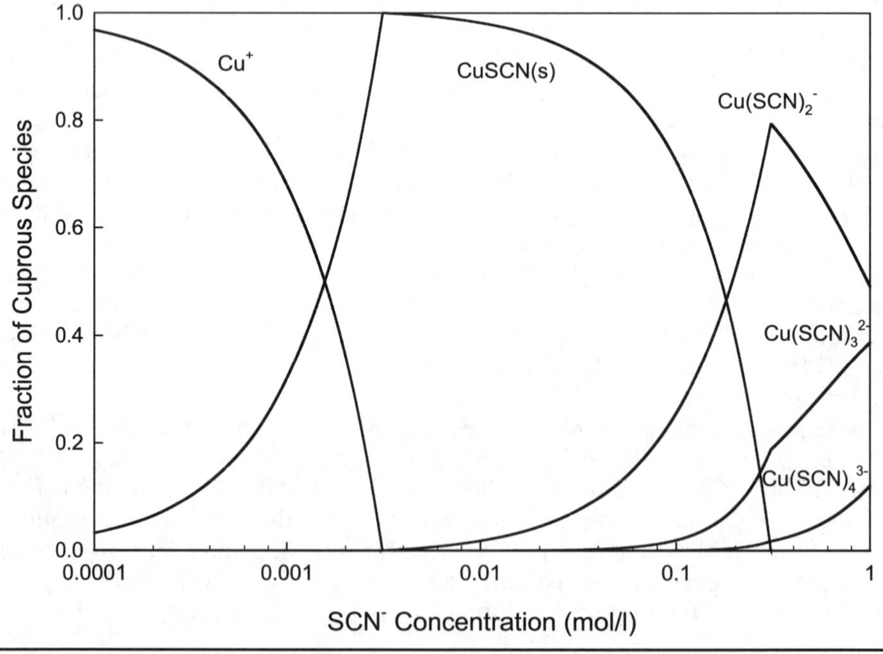

FIGURE 7 Species distribution diagram for the Cu(I)-SCN-H_2O system at [Cu(I)] 0.2 g/L, [SCN^-] 1×10^{-4}-1.0M, pH 2, 25°C and in potential range of cuprous stability

TABLE 6 Current densities (μA/cm²) of sulfide minerals at 358 mV (SCE) and 458 mV (SCE) in pH 2 solutions

Potential, mV	Chalcocite	Bornite	Covellite	Enargite	Chalcopyrite	Pyrite
358	11,000	3,800	85	36	2.2	2.5
458	16,000	7,500	80	60	16.6	12

Stability of sulfide minerals in acidic thiocyanate solutions. It is expected that typical sulfide minerals that exist in gold bearing ores would not be stable in acidic solutions in the presence of ferric ion. The stability of selected sulfide minerals, including pyrite (FeS_2), chalcopyrite ($CuFeS_2$), covellite (CuS), chalcocite (Cu_2S), bornite (Cu_5FeS_4) and enargite (Cu_3AsS_4) was compared in electrochemical studies. Chronoamperometry measurements on these sulfide mineral electrodes at pH 2 in sulfuric acid with and without thiocyanate at two potential levels (358 mV and 458 mV vs. SCE) were made. The potentials were selected based on thermodynamic analysis, which indicated that gold can be leached in ferric-thiocyanate system at potentials of 358 mV to 458 mV (SCE) at pH 1 to 3. The anodic current density for sulfide oxidation is defined with a mean value for a certain experimental time during which the current density reaches a relatively stable level. The current densities, which can be used to compare the oxidation rate of these sulfide minerals, are given in Table 6.

From Table 6, the oxidation rate of these sulfide minerals in sulfuric acid solution is in the following order: chalcocite>bornite>covellite>enargite>chacopyrite>pyrite.

Results indicate that in the potential region for gold leaching, these sulfide minerals will oxidize. Among these sulfides, chalcocite and bornite appear to be most active under test conditions; while chalcopyrite and pyrite are quite stable.

In the presence of thiocyanate, it is expected that the complexation of thiocyanate with the base metal ions may accelerate the rate of oxidation of sulfide minerals. On the other hand, thiocyanate is thermodynamically unstable and the sulfides may catalyze thiocyanate oxidation under the leaching conditions. In this regard, the oxidation rate of selected sulfide minerals, including pyrite, chalcopyrite, covellite and enargite in the presence of thiocyanate was measured by linear sweep voltammetry in an acid solution at pH 2 and a thiocyanate concentration of 0.05 M. Results indicate that current densities from the anodic polarization curves for the oxidation of these sulfide minerals with thiocyanate are significantly higher than those without thiocyanate, when the oxidation potential is above 500 mV (SCE). However, within the range of potentials for gold leaching (358–458 mV SCE), the difference is not distinct. It appears that the rate of thiocyanate decomposition in the presence of these minerals is quite slow at the oxidation potential required for gold leaching. Therefore, for an actual gold leaching process, control of the redox potential for the ferric-thiocyanate system is very important in order to minimize the degradation of thiocyanate.

Gold Dissolution and Metal Ion Interference. Gold leaching kinetics using a gold rotating disk in acidic ferric-thiocyanate solution was studied to determine the effect of system variables such as thiocyanate concentration, ferric ion concentration, and the interference of silver, cupric and cuprous ions. The disk rotational speed for these experiments was controlled at 800 rpm, under which condition gold dissolution was not found to be limited by mass transfer of the reactants. Results indicate that at ambient temperature and at an initial ferric concentration of 0.2 g/L (0.004 M) and pH 2, gold leaching kinetics increase very slightly with an increase in thiocyanate concentration from 0.05 M to 0.2 M. To examine the effect of ferric ion concentration, the rate of gold dissolution was measured with variation of ferric ion concentrations from 0.1 to 1.0 g/L (0.002 M to 0.018 M) at an acidic thiocyanate concentration of 0.1 M. Results demonstrate that the rate increases slightly with an increase in the ferric ion concentration under the conditions studied. Further tests using high mole ratios of Fe(III)/SCN^- (from 1 to 5) with thiocyanate concentration of 0.02 to 0.1 M show that excessive ferric ion does not influence the gold dissolution rate. In all these experiments, it was found that the kinetics of gold dissolution is nonlinear, and the rate decreases with an increase in the leaching time. This phenomenon is similar to that found in actual leaching systems reported by previous researchers (Fleming 1986, Munoz and Miller 2000, Kholmogorov et.al. 2002).

Selected metal ions, such as cupric, cuprous and silver, were added in the acidic ferric-thiocyanate solution to examine their effect on anodic gold dissolution. Results indicate that the presence of cupric ion does not influence the rate of gold leaching. Both cuprous and silver ion form an insoluble salt with thiocyanate depending on thiocyanate concentrations as described previously which results in a negative effect on gold dissolution. Further research using electrochemical measurements to study the interference of metal ions on gold leaching kinetics and thiocyanate oxidation rate is in progress.

The gold rotating disk surface, before and after the leaching experiments with and without the addition of Cu(II), Cu(I) or Ag(I) ion, was examined with an optical microscope and by FTIR spectroscopy. The images from the microscope show that the surfaces are not significantly changed after leaching when compare with the surface state before reaction. The FTIR spectra demonstrate that thiocyanate and its complexes with the metal ions such as Fe(III), Cu(II), Cu(I) and Ag(I) have a very weak adsorption on the gold surface. No definitive information was obtained from the FTIR spectra. However, Munoz and Miller (2000) found the presence of elemental sulfur on the gold surface after ferric-thiocyanate leaching using scanning electron microscopy and energy-dispersed spectroscopy (SEM-EDS). The sulfur reaction product appears to cause gold passivation.

Acidic Ferric-Thiourea System

The strong interest in thiourea is attributed to the fast leaching kinetics of gold in acidic thiourea solutions. Comparison of gold dissolution using acidic thiourea and cyanide solution indicates that the rate for the thiourea system is approximately an order of magnitude greater than cyanide (Hiskey and Atluri 1988). Gold leaches in acidic thiourea solution according to the following anodic reaction,

$$Au + 2\,CS(NH)_2\;Au(CS(NH_2)_2)_2^+ + e^- \qquad \textbf{(EQ 10)}$$

The oxidizing potential for this reaction (E^o=0.38 V) suggests that only mildly oxidizing conditions are required to dissolve gold in thiourea. It has generally been acknowledged that the use of ferric ion in sulfuric acid solution is the most effective oxidant. Also, similar to the other thiosystems, thiourea is unstable and can be oxidized in a number of reaction steps as show by the following reactions. In the thiourea system, an intermediate oxidation product of formamidine disulfide $(CSNH_2NH)_2$ is formed; the oxidation reaction may be similar to the thiocyanate system forming the intermediate product of thiocyanogen, $(SCN)_2$.

$$2CS(NH_2)_2 = (CSNH_2NH)_2 + 2\,H^+ + 2\,e^- \qquad \textbf{(EQ 11)}$$

$$(CSNH_2NH)_2\ CS(NH_2)_2 + \text{Sulfinic Compound} \quad \textbf{(EQ 12)}$$

$$\text{Sulfinic Compound}\ CN{\cdot}NH_2 + S^{o} \quad \textbf{(EQ 13)}$$

Reaction (11) is a reversible reaction between thiourea and formamidine disulfide. Formamidine disulfide decomposes to yield thiourea and a sulfinic compound, which further decomposes to cyanamide and elemental sulfur. Controlling the redox potential is very important in order to prevent the chemical degradation of thiourea, which not only increases reagent consumption, but also impacts gold recovery from the acidic thiourea leach solution. The oxidation reaction products of thiourea with ferric ion have been identified by HPLC and UV/Visible spectroscopy (Bukka, et al. 1992). Results indicated that the oxidation reaction of thiourea by ferric ion does not go to completion and appears to achieve equilibrium when 50% of the thiourea is converted to formamidine disulfide. The oxidation of formamidine disulfide further forms elemental sulfur along with other oxidation products at relatively longer reaction times.

Thiourea leaching has been studied by several researchers. It has been demonstrated that it is technically feasible to use thiourea as a lixiviant to leach gold from its ores (Groenewald 1976, Brierley and Wan 1990, Deschenes and Ghali 1988, Li and Miller 2002). The fast leaching rate compared to cyanide leaching is one advantage of thiourea leaching. However, a high reagent consumption and gold leaching passivation make the system impractical. Experimental results indicate that degradation of thiourea varies with the gold bearing minerals and conditions of leaching operation. It is generally believed that the high thiourea consumption mainly results from oxidation decomposition, thermal degradation, complexation with base metal ions and adsorption by mineral particles. The decomposition of thiourea also causes gold passivation. The decomposition product of elemental sulfur causes significant problems in a continuous operation.

Thiourea Oxidation by Ferric Ion. Gold dissolution in thiourea solution requires an oxidant. Ferric sulfate has been widely used in previous studies and proven to be an effective oxidant. The reaction between thiourea and ferric ion appears to be one of the major causes of thiourea degradation. In the presence of ferric ion, thiourea decomposes via the following reaction:

$$Fe^{3+} + SC(NH_2)_2 = Fe^{2+} + \tfrac{1}{2}\,NH(NH_2)C{-}S{-}S{-}C(NH_2)NH + H^{+} \quad \textbf{(EQ 14)}$$

The formamidine disulfide further decomposes or is oxidized irreversibly, which will cause a net consumption of thiourea.

Thiourea Oxidation by Ferric Ion in the Presence of Minerals. Li (2004) compared the effect of different minerals on thiourea oxidation in the presence of ferric ion. The selected minerals for testing included quartz (SiO_2), galena (PbS), sphalerite (ZnS or ZnS–nFeS), hematite (Fe_2O_3), chalcopyrite ($CuFeS_2$), and pyrite (FeS_2). Comparison of these mineral effects on the rate of thiourea oxidation with ferric ion is presented by the plots of residual thiourea concentration (g/L) as a function of time as shown in Figure 8. The results demonstrate that quartz does not have any effect on thiourea oxidation. The effect from galena and sphalerite is also limited. Hematite, chalcopyrite and pyrite significantly catalyze the oxidation reaction. In the presence of chalcopyrite and pyrite, thiourea was completely oxidized within 10 minutes. It is evident that the rate of reaction between thiourea and ferric ion is related to the mineral type. The sulfide minerals can provide a catalytic effect depending on the type of sulfide. It was also found that the rate of the degradation reaction is sensitive to particle size, the smaller the particle size the greater the catalytic effect.

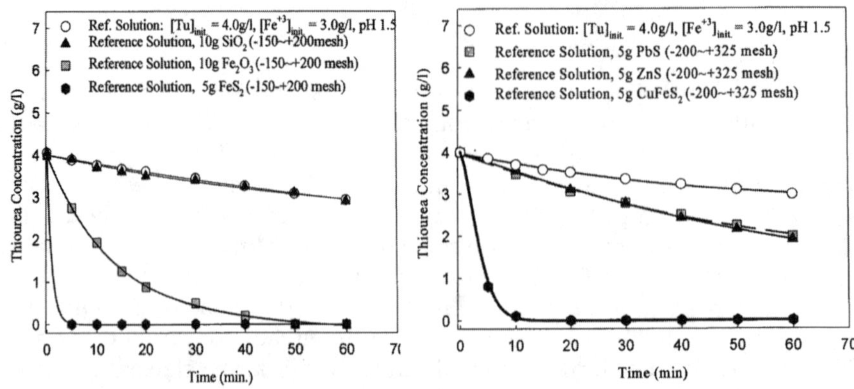

FIGURE 8 Effect of different minerals on the rate of thiourea oxidation by ferric ion at [Fe(III)] 3.0 g/L, thiourea 4.0 g/L, pH 1.5 and 18°C

Effect of Base Metal Ions on Thiourea Degradation and Gold Dissolution. Most gold bearing ores contain base metal minerals, which would be dissolved in the acidic solution especially with the presence of ferric ion. Though it has been reported that the coordination of base metal ions with thiourea is much weaker than with cyanide (Fang and Muhammed 1992), it is important to examine the interference of base metal ions in the acidic ferric-thiourea system. Experiments were performed with the addition of 2 g/L base metal ion in acidic thiourea solution with a concentration of 0.054 M thiourea and 0.054 M of ferric ion at pH 1.5. Thiourea degradation is presented by plots of residual thiourea concentration versus time. Results indicate that Zn(II), Pb(II), Ni(II) or Co(II) do not cause thiourea decomposition nor do they affect the rate of the oxidation reaction between thiourea and ferric ion. On the other hand Cu(II) has a strong effect, and the oxidation rate of thiourea by Cu(II) was found to be very fast as shown in Figure 9. Reducing the cupric ion addition from 2 g/L to 0.5 g/L yielded a similar effect. The results demonstrate that the rate of thiourea decomposition is fast during the first ten minutes and then becomes relatively slow, equivalent to the rate of the ferric-thiourea solution without the addition of cupric ion. Though the coordination between thiourea and cupric is quite strong (log $_4$ = 15.4), the complex was reported to be unstable and that the cupric ion oxidizes thiourea (Krzewska et al., 1980). The oxidizing reaction is described as follows:

$$Cu^{2+} + (n+1)\ SC(NH_2)_2 = Cu(SC(NH_2)_2)_n^{\ +} + \int (NH_2(NH_2)CSSC(NH_2)NH_2)^{2+} \quad \textbf{(EQ 15)}$$

In order to examine the effect of these base metals on the gold leaching kinetics in an acidic ferric-thiourea system, experiments on gold leaching were initiated at an initial thiourea concentration of 0.054 M with 0.01 M [Fe(III)] at pH 1.5 and 25°C. Base metal ions including zinc sulfate, [Zn (II)] 1.0 g/L, nickel sulfate, [Ni(II)] 1.0 g/L, and cupric sulfate, [Cu(II)] 1.0 g/L, were added separately at the beginning of individual leaching experiments. Results indicate that zinc and nickel ions do not show any influence on gold leaching kinetics. However, cupric ion quickly and completely limits the gold leaching process in less than 10 minutes. During the experiments a yellow glutinous product from thiourea decomposition was found in the presence of cupric ion and the product adhered

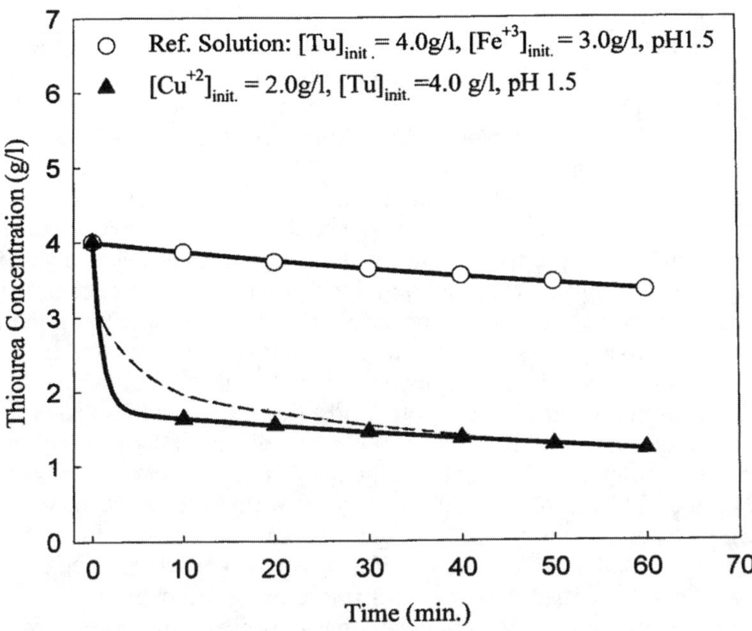

FIGURE 9 Effect of cupric ion on thiourea decomposition at [Cu(II)] 2.0 g/L, thiourea 4.0 g/L, pH 1.5 and 18°C

to the gold surface and to the reactor wall. The yellow glutinous product was found to be elemental sulfur. It appears that the cupric ion does not only oxidize thiourea at a fast rate, but also catalyzes the oxidation of thiourea by ferric ion. In addition, the presence of cupric ion accelerates the decomposition of formamidine disulfide to elemental sulfur which passivates the gold surface. These results may help to explain the fact that the rate of gold leaching in actual systems decreases with an increase in leaching time.

Obviously, the ferric-thiourea system would not be stable in actual gold leaching environments. High thiourea consumption is expected if an excess of ferric sulfate is used. In a simple ferric-thiourea solution with low concentrations of ferric ion, thiourea oxidation is rather slow. However, the redox reaction is catalyzed by typical sulfide minerals found in gold bearing ores. In addition, cupric ion has a deleterious effect on thiourea decomposition.

APPLICATION OF THIOHYDROMETALLURGICAL PROCESSES

Among these three thiosystems, the thiosulfate and thiocyanate systems are considered to be less hazardous than cyanide. The reagent unit price of ammonium thiosulfate is far cheaper than sodium cyanide ($0.2/kg vs. $1.2/kg). Therefore even with a high reagent consumption, the application of ammonium thiosulfate could be economical and compete with cyanidation. Under acidic conditions, thiocyanate would be a potential lixiviant for leaching of gold, though no practical industrial application has been reported.

Under proper control, the thiocyanate consumption may reach an economically acceptable level; though thiocyanate consumption is generally higher than that of cyanide. Synthesis of thiocyanate is reported to be relatively simple and it would be expected to have the similar price as cyanide, (quoted by manufacture in the States, Norcross 2002). In general, thiourea consumption is relatively high; with the current market price of ~$1.98/kg and the potential carcinogenic properties, it has generally been eliminated from further consideration by industry.

Aylmore and Muir (2001) gave a summary from selected literature of the various thiosulfate leaching conditions that have been applied for leaching ores and concentrates. A wide range of conditions appeared to have been used, including relatively severe conditions, compared with those used in the cyanidation process. Lower reagent consumption was achieved from the tests using lower thiosulfate concentrations. Newmont has developed a heap leach process using ammonium thiosulfate for gold recovery from low-grade carbonaceous sulfidic gold ores. A combined biooxidation-thiosulfate heap process was developed for high sulfide carbonaceous ores and a direct thiosulfate leach process for low sulfide carbonaceous ores. From 1996 to 1999, a total of 1.24 million tonnes of low-grade carbonaceous refractory ores were heap leached at Carlin, Nevada, with ammonium thiosulfate (Wan and LeVier, 2003). Gold recovery from thiosulfate pregnant solution was achieved with copper cementation. After gold recovery the barren solution was recycled back to the heap. A total of 55,790 ounces of gold was produced from the thiosulfate system. Bhakta (2003) described details of this thiosulfate heap operation. In 1999, direct thiosulfate heap leaching on low sulfide carbonaceous ores at minus 2.5-cm size, recovered 12,400 ounces of gold over the 176-day leach cycle at an average cash cost of $175 per ounce gold. The major reagent consumption based on kg per tonne-ore included ammonium thiosulfate 9.7 kg/t, ammonium hydroxide 0.75 kg/t, copper sulfate septahydrate 0.03 kg/t and copper dust 0.015 kg/t. Based on the demonstration plant results, implementation of the thiosulfate leach technology to a commercial operation at Carlin was initiated. However, the construction of a commercial plant has been postponed due to the limited low-grade carbonaceous ore reserves in Nevada.

The efficiency of the gold–thiosulfate leaching system depends on ore type and mineral characteristics. Langhans et al. (1992) investigated the feasibility of thiosulfate leaching on low-grade oxide ores. Under optimized conditions, 83% gold extraction was obtained while 0.4 kg $S_2O_3^{2-}$ was consumed per ton of ore. In comparison, 86% gold extraction was achieved using standard cyanidation methods with the cyanide consumption being 0.21 kg CN /t ore. Recently, selected oxide ores were studied (Acar 2002) using ammonium thiosulfate solution in laboratory bottle roll tests under the following test conditions: $(NH_4)_2S_2O_3$ 0.1M, NH_3 0.1 M, Cu(II) 50 ppm, pH ~9, at 20°C and 40% solids slurry for 24 hours. The sample particle size was minus 78 µm. Gold recoveries range from 86% to 93% that are comparable to cyanide. Thiosulfate consumption depends on mineral composition. Samples with high silver minerals consume more thiosulfate under the same test conditions as shown in Table 7.

Gold leaching using thiocyanate is performed in an acidic environment. For refractory ores, a pretreatment process is used to oxidize the sulfide minerals prior to leaching. In general, after sulfide oxidation, the ore residues are very acidic for most of the oxidation processes. An acidic gold leaching process following acidic pretreatment eliminates the need for neutralization and prevents problems related to material handling and hydroxide/sulfate precipitation. In addition to refractory gold ores, the economic processing of low-grade copper-gold ores remains one of the most challenging. Early commercial

TABLE 7 Ammonium thiosulfate leach results on selected oxide ores

	Feed Ore			Au Extraction	Consumption
No.	Au g/t	Ag g/t	Cu ppm	%	$S_2O_3^{2-}$ kg/t
1	25.83	1.05	22	92.9	0.77
2	17.25	19.35	253	86.2	5.63
3	8.62	1.00	21	93.3	1.08
4	2.82	1.20	16	92.4	0.64
5	1.79	1.05	19	89.7	0.37
6	1.26	14.18	62	91.9	3.31

applications of bioleach technology processed submarginal grade copper-bearing rock in dumps. Recent applications of this technology used engineered bioleach heaps. For the bioheap leach application, gold associated with the copper ores cannot be extracted economically with cyanide due to the massive amount of lime required for neutralization and material handling problems. Thus, using thiocyanate for gold leaching in an acidic environment would be beneficial.

Munoz and Miller (2000) reported gold leaching using thiocyanate for an auriferous pyrite ore from Ecuador. A maximum of 84.6% of the gold present in the ore can be leached in 24 hours after chemical pretreatment compared to 49.5% gold extraction without pretreatment. Newmont has evaluated gold recovery from low grade refractory gold ores using ferric thiocyanate following biooxidation for an acidic heap leaching approach (Wan et al. 2003). Two thiocyanate column leaching tests were performed on biooxidized ore at a particle size of minus 38mm. For comparison, a cyanide leach column was also tested. Gold extractions from thiocyanate column using SCN^- concentration of 0.01 and 0.02M were 52% after 16 days leaching, 10% higher than that obtained from the cyanide leach. Both thiocyanate and cyanide extractions continued to increase at the termination of the leach tests. Consumption of potassium thiocyanate ranged from 0.6 to 0.8 kg/tonne-ore, which was higher than sodium cyanide consumption (0.33 kg/t).

Recently, thiocyanate leaching of an oxide ore was evaluated using stirred reactors in laboratory experiments. The oxide ore contained 9.79 g Au/t and 0.09% sulfur. The sample was directly leached in acidic ferric-thiocyanate solution at pH ~2, ~25% solids slurry, and 23°C for 6 hours. The gold leaching efficiency with variation in the thiocyanate and ferric ion concentrations was investigated and the results are given in Table 8. Gold extractions ranged from 95% to 97% after six hours based on feed and residue analyses which compared to 94% gold extraction from cyanidation for 24 hours. Thiocyanate (SCN^-) consumption ranged from 0.14 to 0.43 kg/t ore.

From Table 8, it is evident that mole ratios of SCN^-/Fe(III) do not influence gold extraction for this oxide ore. In this regard, gold leaching using acidic thiocyanate solutions may accommodate a wide variation in thiocyanate and ferric concentration. Results indicate that thiocyanate is much more stable in the leaching of oxide ores compared to its stability in the presence of sulfide minerals.

SUMMARY

Thiohydrometallurgical systems include thiosulfate, thiocyanate and thiourea as alternative lixiviants for gold recovery from primary ores. All the thioreagents react with gold forming soluble gold complexes in the presence of an appropriate oxidant. The solution

TABLE 8 Acidic ferric-thiocyanate leaching on an oxide ore

	1	2	3	4	5	6	7
SCN^-, M	0.02	0.02	0.02	0.02	0.05	0.05	0.05
Fe(III), M	0.01	0.02	0.05	0.10	0.01	0.025	0.05
SCN^-/Fe(III)	2	1	0.4	0.2	5	2	1
Au Extraction %	94.8	95.9	95.4	95.7	96.9	97.5	97.1

chemistry of the thiosystems is complicated due to instability of the thioreagents and many other reactions which occur during the leaching process. Control of the redox potential for thioleaching systems is very important in order to minimize the degradation of thioreagents.

In the alkaline copper-ammonia-thiosulfate system, most common sulfide minerals which exist in refractory gold ores, such as copper and iron sulfides, become unstable in thiosulfate solution under gold leaching conditions. The oxidation of iron and copper sulfides in ammonium thiosulfate solution has been demonstrated from electrochemical studies. The presence of certain sulfide minerals accelerates thiosulfate degradation, consumes thiosulfate, and retards gold dissolution. Leaching experimental results confirm that thiosulfate is much more stable in the leaching of oxide ores compared to its stability in the leaching of sulfidic and/or carbonaceous ores.

Thiocyanate is an effective lixiviant for gold leaching in an acidic environment. Compared with thiourea, thiocyanate offers the advantage of much greater stability against oxidation decomposition. In acidic solution, especially in the presence of ferric ion, the iron and copper sulfides become unstable and oxidize. However, the rate of thiocyanate decomposition in the presence of sulfide minerals was found to be quite slow at the oxidation potential required for gold leaching. Thiocyanate is a pseudohalide. The pseudohalide behavior of thiocyanate results in the formation of insoluble salts when the thiocyanate reacts with silver, mercury, lead and copper ion under certain conditions depending on the thiocyanate concentration. At low thiocyanate concentrations, the formation of insoluble compounds such as AgSCN and CuSCN may precipitate on the gold surface and retard gold dissolution.

For the ferric-thiourea system, a high reagent consumption and gold passivation make the system impractical for commercial operation. The degradation of thiourea varies with the sulfide minerals and gold leaching conditions. It was also found that cupric ion has a strong effect on thiourea oxidation. The presence of cupric ion further accelerates the decomposition of formamidine disulfide to elemental sulfur which passivates the gold surface.

ACKNOWLEDGEMENT

The authors would like to extend their appreciation to their colleagues who have worked on these projects. Newmont Mining Corporation's support is greatly appreciated. Special appreciation is extended to the Newmont Mining Inverness Technical Facility staff for their technical assistance.

REFERENCES

Acar S. 2002. Newmont Metallurgical Services Research Report, June 26, 2002

Aylmore, M.G. 2001. Treatment of a Refractory Gold-Copper Sulfide Concentrate by Copper Ammoniacal Thiosulfate Leaching. *Minerals Engineering*. Vol. 14, No. 6, 615–637.

Aylmore, M.G. and D.M. Muir. 2001. Thiosulfate leaching of gold—a review. *Minerals Engineering*. Vol. 14, No. 2, 135–174.

Barbosa-Filho, O. and A.J. Monhemius. 1989. Thermochemistry of thiocyanate systems for leaching gold and silver ores. *Precious Metals '89*. Edited by M.C. Jha and S.D. Hill, TMS, 307–339.

Barbosa-Filho, O. and A.J. Monhemius. 1994. Leaching of gold in thiocyanate solutions—Part 1: Chemistry and thermodynamics; Part 2: Redox processes in iron (III)-thiocyanate solutions; Part 3: Rates and mechanism of gold dissolution. *Transactions of the Institute of Mining and Metallurgy (Section C)*. Vol. 103, C 105–125.

Bhakta, p. 2003. Ammonium Thiosulfate Heap Leaching. *Hydrometallurgy 2003. Proceedings of the 5th International Symposium Honoring Prof. I.M. Ritchie*. edited by C. Young et al., TMS, Vancouver, Canada, 259–268.

Breuer, P.L. and M.I. Jeffrey. 2000. Thiosulfate leaching kinetics of gold in the presence of copper and ammonia. *Minerals Engineering*. 13 (10–11), 1071–1081.

Brierley, J.A. and R.Y. Wan. 1990. Enhanced recovery of gold from a refractory sulfidic-carbonaceous ore using bacterial pretreatment and thiourea extraction. *Gold '90. Proceeding of the Gold '90 Symposium*. Edited by D.M. Hausen, SME, 463–466.

Broadhurst, J.L. and J.G.H. du Preez. 1993. A thermodynamic study of the dissolution of gold in an acidic aqueous thiocyanate medium using Fe(III) sulfate as an oxidant. *Hydrometallurgy*. 32, 317–344.

Bukka, K., J.E. Hooper, M.A. McGuire, R.Y. Wan and J.D. Miller. 1992. A study of thiourea oxidation reactions by HPLC and UV spectroscopy. *Unpublished Research Report, University of Utah.*

Deschenes, D. and E. Ghali. 1988. Leaching of gold from a chalcopyrite concentrate by thiourea. *Hydrometallurgy*. 20, 179–202

Fang, Z. and M. Muhammed. 1992. Leaching of precious metals from complex sulfide ores: on the chemistry of gold lixiviation by thiourea. *Mineral Processing and Extractive Metallurgy Review*. Vol. 11, 39–60.

Feng, D. and J.S.J. van Deventer. 2002. The role of heavy metal ions in gold dissolution in the ammoniacal thiosulfate system. *Hydrometallurgy*. 64, 231–246.

Fleming C.A. 1986. A process for the simultaneous recovery of gold and uranium from South Africa ores. *Gold 100*, Vol. 2, South African Institute of Mining and Metallurgy, Johannesburg, 301–319.

Groenewald, T. 1976. The dissolution of gold in acidic solutions of thiourea. *Hydrometallurgy*, 1, 277–290.

Hiskey, J.B. and V.P. Atluri. 1988. Dissolution chemistry of gold and silver in different lixiviants. *Mineral Processing Extractive Metallurgy Review*. 4, 95–134

Jeffery M.I., P.L. Breuer and C.K. Chu. 2003. The importance of controlling oxygen addition during the thiosulfate leaching of gold ores. *International Journal of Mineral Processing*. Vol. 72, No.1–4, 323–330.

Kholmogorov, A.G., O.N. Kononova, G.L. Pashkov and Y.S. Kononov. 2002. Thiocyanate solutions in gold technology. *Hydrometallurgy*. Vol. 64, 43–48.

Langhans, J.W., K.V.P. Lei, and T.G. Carnahan. 1992. Copper-catalyzed thiosulfate leaching of low-grade gold ores. *Hydrometallurgy*. Vol. 29, 191–203

Li, J.S. 2004. Reaction Kinetics of Gold Dissolution in Acid Thiourea Solutions. *Ph.D. Dissertation*. Department of Metallurgical Engineering, University of Utah.

Li, J. S. and J.D. Miller. 2002. Reaction kinetics for gold dissolution in acid thiourea solution using formamidine disulfide as oxidant. *Hydrometallurgy*. 63, 215–223.

Li, J., J.D. Miller, and R.Y. Wan. 1996. Important solution chemistry factors that influence the copper-catalyzed ammonium thiosulfate leaching of gold. *Presented at the 125th SME Annual Meeting,* Phoenix, Arizona, SME.

Li, J. J.D. Miller, R.Y. Wan and K.M. LeVier. 1995. The ammoniacal thiosulfate system for precious metal recovery. *Proceedings of XIX International Mineral Processing Congress*. 7, 37–42.

Marsden, J. and L. House. 1992. *The Chemistry of Gold Extraction*. England: Ellis Horwood Limited.

Molleman, E. and D. Dreisinger. 2002. The treatment of copper-gold ores by ammonium thiosulfate leaching. *Hydrometallurgy*. 66, 1–21

Muir, D.M. and M.G. Aylmore. 2004. Thiosulfate as an alternative to cyanide for gold processing-issues and impediments. *Mineral Processing and Extractive Metallurgy (Trans. Inst. Min. Metall. C)*. March 2004, Vol. 113, C2-C12.

Munoz, G.A. and J.D. Miller. 2000. Noncyanide leaching of an auriferous pyrite ore from Ecuador. *Minerals and Metallurgical Processing*. Vol.17, No.3, 198–204.

Navarro, P., C. Vargas, A. Villarroel, and F.J. Alguacil. 2002. On the use of ammoniacal / ammonium thiosulfate for gold extraction from a concentrate. *Hydrometallurgy*.65, 37–42

Norcross, R. 2000. *Private Communication with Author,* Degussa-Huls Corporation. August 2, 2000, Adlendale, NJ.

Ritchie, I.M., M.J. Nicol and W.P. Staunton. 2001. Are there realistic alternatives to cyanide as a lixiviant for gold at the present time? In *Cyanide: Social, Industrial and Economic Aspects*. Edited by C. Young et al., TMS, 427–440.

Senanayake, G. 2004. Gold leaching in non-cyanide lixiviant systems: critical issues on fundamentals and applications. *Mineral Engineering*. 17, 785–801.

Smith, R.M. and A.E. Martell 1989. *Critical Stability Constants*. Volume 5. Plenum Publishing UK.

Sparrow, G.J. and J.T. Woodcock. 1995. Cyanide and other lixiviant leaching systems for gold with some practical applications. *Mineral Processing Extractive Metallurgy Review*. 14, 193–247.

Wan, R.Y., J.A. Brierley, S. Acar and K.M. LeVier. 2003. Using thiocyanate as lixiviant for gold recovery in acidic environment. *Hydrometallurgy 2003. Proceedings of the 5th International Symposium Honoring Prof. I.M. Ritchie.* Edited by C. Young et al., TMS, Vancouver, Canada, 105–122.

Wan, R.Y. and K.M. LeVier. 2003. Solution chemistry factors for gold thiosulfate heap leaching. *Internatinal Journal of Mineral Processing*. Vol.72, No. 1–4, 312–322.

Wan, R.Y., K.M. LeVier and J.D. Miller. 1993. Research and development activities for the recovery of gold from noncyanide solutions. *Proceedings Wadsworth Symposium, Hydrometallurgy: Fundamentals, Technology and Innovation.* Edited by J.B. Hiskey and G.W. Warren, SME, 415–436.

. . . .

Raman Spectroelectrochemical Investigations of the Leaching of Gold in Chloride and Thiosulfate Media

Ronald Woods,* Gregory A. Hope,* Kym Watling,* and Matthew I. Jeffrey†

Surface enhanced Raman spectroscopy carried out in real time has been applied to identify surface species formed during gold leaching in chloride and thiosulfate media. Chloride ion is adsorbed prior to the gold dissolution potential, but its coverage is significantly diminished at the beginning of leaching. A gold chloride band is again observed when dissolution is rapid, together with bands characteristic of $AuCl_4^-$ ions in solution. With thiosulfate, a thin gold sulfide layer is formed slowly in the potential region where leaching occurs. At higher potentials, thiosulfate is oxidized to sulfur and sulfate, both products being identified by Raman spectroscopy.

INTRODUCTION

Leaching in chloride media is the oldest of the precious metal dissolution processes and was first applied commercially for gold extraction in the Plattner process in 1848 (Habashi 1970) using chlorine as the oxidant. This approach was superceded by cyanidation following the patenting of the cyanide process by MacArthur and the Forrest brothers in 1887. Cyanide leaching remains the main process for gold recovery today, but perceived environmental concerns have led to a great amount of research to identify effective alternative leachants. In this search, chloride is again being considered for the recovery of gold (Ferron et al 2003). A chloride medium with air as the oxidant is utilized

* School of Science and CRC microTechnology, Griffith University, Nathan, Queensland, Australia

† Department of Chemical Engineering, Monash University, Victoria, Australia

in the Intec Gold Process that is being marketed for the treatment of refractory gold ores (Wood 2004).

Thiosulfate has received considerable attention as a leachant for gold and appears to be the most promising alternative to cyanide. The gold leach rate is significantly enhanced when copper(II) ions are present in the leach solution (Breuer and Jeffrey 2003). Since leach solutions are alkaline, ammonia is added to stabilize the copper as a copper(II) amine complex. Successful large-scale pilot plant testing of thiosulfate leaching of gold has been carried out by Newmont Mining Corporation (Bhakta 2003) and by Barrick Gold Corporation (Fleming et al 2003) for the processing of carbonaceous gold ores that exhibit preg-robbing characteristics.

In gold leaching processes, intermediates formed on the metal surface in the dissolution process, and surface species formed in side reactions, play a significant role in determining the leaching rate. Surface enhanced Raman scattering (SERS) spectroscopy has been demonstrated as an excellent surface sensitive technique for studying the gold surface (Baltruschat and Heitbaum 1983; Gao and Weaver 1986). Using SERS, it is possible to obtain vibrational spectra of species in the first few layers at the solid-electrolyte interface. Simultaneous *in situ* electrochemical and SERS techniques may therefore assist in the elucidation of the dissolution mechanism(s), particularly through the identification of surface species and the determination of the nature and strength of ligand–gold atom interactions.

In the present paper, SERS investigations are reported of gold dissolution in chloride and thiosulfate media. Studies in thiosulfate are focused on the passivation reaction that results from decomposition of thiosulfate (Breuer and Jeffrey 2003). Other aspects of the leaching of gold are presented by Jeffrey et al. in this Volume.

EXPERIMENTAL

Electrochemistry

All reagents and reference compounds used were of AR grade and solutions were prepared using high purity water from a Permutit Australia Hi-Pure Water System with typical resistivity of 18 MΩ cm^{-1}. The electrochemical cell and all other glassware were cleaned by soaking in 0.8 mol dm^{-3} nitric acid (BDH Aristar). Raman spectroelectrochemical experiments were performed using a custom-designed five-necked borosilicate electrochemical cell with an optically flat transparent window. A conventional three electrode system was used, consisting of a 0.5 mm diameter platinum wire auxiliary electrode; a silver / silver chloride reference electrode (Cypress systems EE008, 3.0 mol dm^{-3} KCl / sat. AgCl electrolyte); and a fine gold (Johnson Matthey 99.5 %) working electrode with an exposed area of 1 cm^2. The gold working electrode was abraded to expose a fresh surface using 15 μm grit size P1200 SiC paper, and rinsed with ultrapure water. Following this procedure, the electrode was sonicated in ultrapure water and placed in a furnace at 450°C for 30 min to remove any adsorbed species from the surface. The auxiliary electrode was located 6 cm to the rear of the cell, with the reference electrode close to the gold. Potential control of the working electrode was established using an ADInstruments potentiostat controlled by a Maclab/4e analog-digital converter interfaced to a PC running ADInstruments EChem software V.1.5.2.

The ambient room temperature was 22°C, resulting in a silver / silver chloride reference potential of 0.210 V on the standard hydrogen electrode (SHE) scale (Galster 1991). Potentials are reported relative to the Ag/AgCl reference unless otherwise specified. Prior to experiments, the cell was purged by bubbling high purity nitrogen (Linde, 4.0) for 20 min to remove oxygen from the system. A positive nitrogen pressure was maintained over the cell during the measurement phase by withdrawing the nitrogen bubbler from solution and flowing the gas above the solution surface.

Voltammograms from a rotated gold electrode were recorded at ambient temperature (20–21°C) using a Pine MSRX rotator.

SERS Activation

The SERS electrode was activated by applying oxidation / reduction cycling (ORC) in a 1.0 mol dm^{-3} KCl solution, acidified with HCl to pH 1, in equilibrium with air. The potential regime used in chloride media consisted of cathodic cleaning for 30 s at −0.7 V, followed by three cycles of potential switching between −0.5 V and 1.0 V, holding at each potential for 30 s before returning to −0.7 V for 30 s to displace adsorbed chloride from the surface. The electrodes were withdrawn whilst still under potential control and rinsed with ultrapure water. For experiments with thiosulfate, the ORC procedure was carried out by automatically applying 0.5 s pulses for 5 min between the same potentials, ending at the lower value.

Raman Spectroscopy

SERS spectra were acquired on a Renishaw system 100 Raman fibre optic spectrometer, using 632.8 nm excitation from a HeNe laser. The laser light was delivered through an optical fibre and focused through the collection lens to a measured spot size of 50 μm, with 6 mW power recorded at the sample. Raman scattered radiation was collected through an ultra-long working distance x20 Olympus LMP Plan Fl lens with a numeric aperture of 0.4 and focused through two super notch filters onto the spectrometer receiving fibre. The Raman shifted signals from the single pass grating spectrometer were detected using a Peltier-cooled CCD detector with spectral resolution of 2 cm^{-1}. The spectrometer was referenced to the 520 cm^{-1} lattice vibration of silicon. SERS spectra were accumulated in static mode, with scans centered at 670 or 750 cm^{-1}. For experiments in chloride media, spectra were collected for 2 s during potentiodynamic scans, with a spectrometer rest time of 1 s between scans. The number of scans acquired over the potential ramp was between 60 and 126, with the initial scan commenced while the electrode was held at the lower potential limit. The potential interval between each scan was between 33 and 50 mV. This combined in-situ electrochemistry–Raman spectroscopy technique is referred to as CV-SERS in this paper. As repeated voltammetric cycling was found to alter both the electrochemical response and SERS intensity (presumably due to re-structuring of the gold surface), the first voltammetric run was considered to be best representative of the system. Fresh ORC electrode preparation was therefore required for CV-SERS investigations of high and low frequency spectral regions. Recording spectra as a function of time from a potentiostatted electrode was found to be more informative for investigations of thiosulfate leaching.

RESULTS AND DISCUSSION

Raman Spectra of Gold Chloride Compounds

Raman spectra for the gold chloride anions have been well characterized (Braunstein and Clark 1973; Puddephatt 1978; Pan and Woods 1991; Murphy et al. 2000). In the present study, a band was observed at 327 cm^{-1} from $[AuCl_2]^-$ in tetrabutyl ammonium gold(I) dichloride; this is similar to the value of 326 cm^{-1} observed by Murphy et al. (2000) on heating a $[AuCl_4]^-$ solution with gold metal to 250 °C and assigned to the ν(Au-Cl) of $[AuCl_2]^-$. Braustein and Clark (1973) assigned a band at 329 cm^{-1} to Au-Cl stretching for salts containing the $[AuCl_2]^-$ ion. There is good agreement between the different authors for the band positions from $[AuCl_4]^-$; those reported by Murphy et al. are 348 cm^{-1} and 324 cm^{-1} from the ν_1 and ν_3 Au-Cl symmetric stretching modes and 171 cm^{-1} from the bending vibration δ(ClAuCl). Our Raman spectra from the $[AuCl_4]^-$ ion in the potassium compound gave bands at 350, 327 and 178 cm^{-1}, which agrees with these literature values. Surface bonded gold-chlorides were observed by Gao and Weaver (1986) using SERS. These authors observed the adsorption of halides and pseudohalides on silver and gold electrodes and report a gold-chloride stretch band ranging from 245 to 275 cm^{-1} within a potential range of −400 and 500 mV vs. SCE. These investigations did not, however, extend into the potential region where gold actively dissolves, which is of particular interest in gold leaching.

CV-SERS of a Gold Electrode in Chloride Media

SERS spectra were recorded in real time during voltammetric scans in 1 mol dm^{-3} HCl, 1 mol dm^{-3} KCl, 1 mol dm^{-3} NaCl, 1 mol dm^{-3} KCl adjusted to pH 1 with HCl, 1 mol dm^{-3} NaCl adjusted to pH 1 with HCl, and 1 mol dm^{-3} KCl adjusted to pH 12 with KOH. The ionic strength of these solutions is such that the electric field from the electrode decays rapidly into the solution and vibrational Stark shifting (Ashley and Pons 1988) will only be observed for species at the interface. A constant feature of the CV-SERS results was the observation that when dissolution commences there was a diminution in the intensity of the band that is assigned to $\nu(AuCl^-_{ads})$ from adsorbed chloride ion on gold. The diminution commenced when there was only a small anodic current flowing.

This is illustrated in Figure 1, which presents Raman shift data overlaid on the voltammogram for (a) a gold electrode run at 5 mV s^{-1} in 1 mol dm^{-3} KCl, adjusted to pH 1 with HCl and (b) a gold electrode run at 5 mV s^{-1} in 1 mol dm^{-3} KCl. In this figure, the area of each extracted data point indicates the intensity of the Raman band acquired during a cyclic voltammogram. In Figure 1a, a Raman band is observed at 235 cm^{-1} shortly after starting the sweep. This is assigned to the $\nu(AuCl^-_{ads})$ band from a chloride ion adsorbed on a gold atom at the surface. The band is substantially lower than the ν(AuCl) values observed for $[AuCl_2]^-$ complexes, and reflects the difference between the bond energy of the dichloroaurous species and that of the adsorbed chloride anion on a gold atom located at the negatively charged surface. The behaviour in neutral solution is similar (Figure 1b), except that the band initially appears at ~280 cm^{-1} before falling to ~240 cm^{-1} at −200 mV. Adsorption of chloride on gold at similar potentials was observed by Lipkowski (1998) from 10^{-3} mol dm^{-3}, chloride solution and by Gao and Weaver (1986) for 0.1 mol cm^{-3} KCl. As the potential is increased in Figure 1, the band intensifies, presumably as a result of an increase in chloride coverage, as there are no surface-blocking products forming or re-structuring of the gold surface occurring. The band also

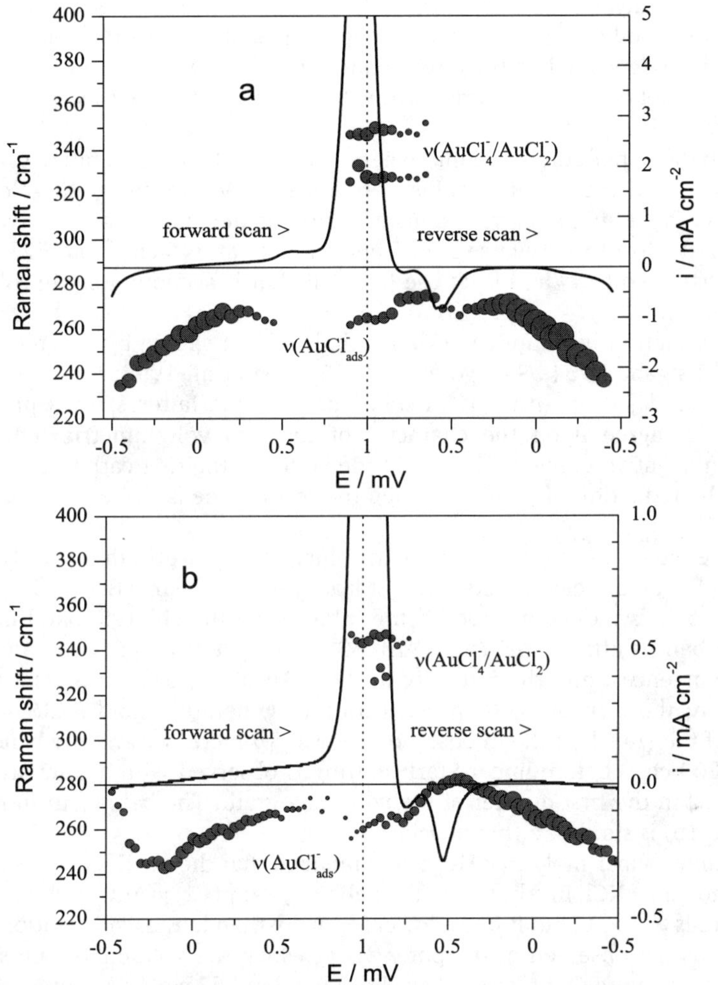

FIGURE 1 CV-SERS for gold in (a) 1 M KCl at pH1 and (b) 1 M KCl. Scan rate 10 mV s^{-1}, Raman band intensity proportional to spot area

progressively increases in Raman shift as the gold electrode potential increases. This Stark tuning for the pH 1 section of the experiment was 52 cm^{-1} V^{-1} and 53 cm^{-1} V^{-1} in neutral KCl.

The species considered to be intermediates in the gold-dissolution process are $AuCl^-_{ads}$, $AuCl_{ads}$ and $[AuCl_2]^-_{ads}$ (Gallego et al. 1975). The first one is adsorbed chloride that generates the 240 cm^{-1} Raman band at potentials prior to dissolution. The other two species would be expected to generate Raman spectra, but are not observed. Loss in intensity of Raman bands can usually be attributed to either a gold surface containing no Raman active bonds, such as non-bonded chloride, or factors that alter the SERS activity of the gold surface, typically formation of thick product layers, and there is no dissolution of the active substrate. Non-bonded chloride would seem to be unlikely and there is no thick product layer observed on the gold surface. Dissolution of the active substrate would also seem to be ruled out by the reappearance of the adsorbed chloride spectrum before gold is re-deposited on the return scan. It would appear, therefore, that the loss of signal at the onset of gold dissolution is due to the absence of species at the gold surface that generate SERS spectra.

Close to the upper potential limit, where $[AuCl_4]^-$ is being generated, the $\nu(AuCl^-{}_{ads})$ band is again observable, but remains of relatively low intensity until gold dissolution ceases. In the high anodic current density region at the end of the positive-going scan and the initial part of the reverse scan, bands appear at 345 cm^{-1} and 325 cm^{-1}. These bands, are assigned to v_1 and v_3 of the $[AuCl_4]^-$ that is accumulating in solution at the gold surface.

Gold dissolution in dilute chloride media has been shown by a number of authors (Evans and Lingane 1964; Gallego et al. 1975; Diaz et al. 1993; Kolics et al. 1996) to form both $[AuCl_2]^-$ and $[AuCl_4]^-$ in solution, with the latter species predominating. These authors agree upon the character of the two voltammetric reduction peaks observed on negative-going scans in chloride systems; the first cathodic current peak is related to the reduction of $[AuCl_2]^-$, while the second one is associated with the reduction of $[AuCl_4]^-$.

It can be seen from the current-potential curves in Figure 1a that $[AuCl_2]^-$ which has accumulated near the electrode surface starts to reduce near 0.8 V and with the commencement of this reduction process, the intensity of the chloride band increases and the Raman band shifts to 275 cm^{-1}. At 0.68 V, gold deposition from $[AuCl_4]^-$ ions in solution commences, and the intensity of the $\nu(AuCl^-{}_{ads})$, decreases. These conditions are similar to those in the ORC process used to generate a gold SERS surface. With depletion of the $[AuCl_4]^-$, the intensity of $\nu(AuCl^-{}_{ads})$ increases and the band eventually returns to 265 cm^{-1} before linear Stark shifting is observed with a comparable slope to that observed in the first quarter of the voltammogram. The behaviour in neutral solution (Figure 1b) is similar to that in acid.

The behavior in 1 mol dm^{-3} HCl was similar to that the acidified KCl solution. With neutral 1 mol dm^{-3} KCl, in contrast to the behaviour at pH 1, $\nu(AuCl^-{}_{ads})$ was observed at scan potentials ≥ -0.4 V, but it had a lower intensity than in the acid solutions. As before, this band was not observed at the potential at which the voltammogram showed gold dissolution to begin taking place. A $(AuCl^-{}_{ads})$ band at 263 cm^{-1} developed at a potential close to the upper limit of the sweep (1.0 V). A broad band near 350 cm^{-1}, due to solution $[AuCl_4]^-$, was observed at 0.85 V shortly after the onset of gold dissolution.

When the experiment was repeated at pH 12, the spectra observed on the positive-going scan were featureless (Figure 2) until a potential of ~0.8 V was reached at which point a band at ~550 cm^{-1} became evident. This band can be assigned to ν(AuO) of a surface, oxide layer formed on the gold surface; it Stark shifts, a property that is expected for a chemisorbed layer. The low current density and rapid electrochemical passivation observed in the voltammogram are also consistent with the presence of oxide on the surface. In 1 mol dm^{-3} acid solution, it is known that oxygen begins to adsorb on gold at ~0.4 V and reaches limiting chemisorption coverage at ~0.8 V (Woods 1976); these values shift to ~0.7 V and 1.1 V, respectively, in pH 12 solutions. The development of multilayer gold oxide requires more severe oxidizing conditions than those operative on a 10 mV s^{-1} sweep to 1.1 V (Woods 1976). The Raman band arising from the oxide layer is present until the potential reaches ~0.2 V on the return scan. This

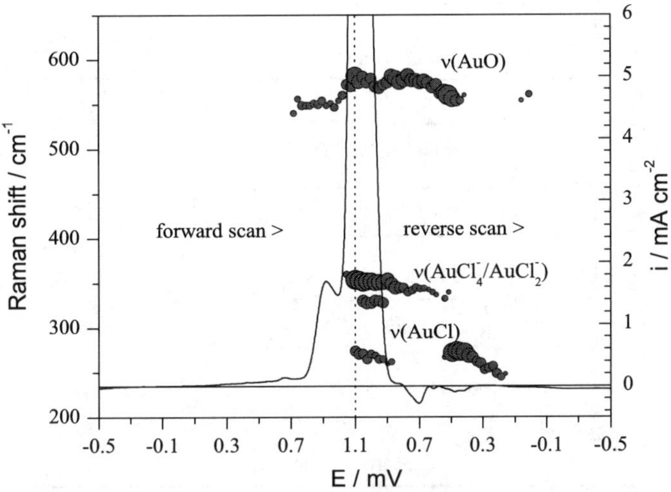

FIGURE 2 CV-SERS for gold in 1 M KCl at pH12. Scan rate 10 mV s^{-1}, Raman band intensity proportional to spot area

potential is then in the region in which chemisorbed oxygen is desorbed from a gold surface at pH 12 (Woods 1976). It is apparent that in the presence of the oxide layer on the positive-going scan blocks the adsorption of chloride ion. Gold dissolution is not prevented, however; this is shown by the development of a band near 355 cm^{-1} arising from the $[AuCl_4]^-$ in solution when the potential reached ~1.0 V. This band increases in intensity as the positive-going scan proceeds.

Raman Spectra of Compounds Relevant to Gold Leaching in Thiosulfate

Figure 3 presents Raman spectra from solid sodium thiosulfate and a 0.1 mol dm^{-3} solution of this compound. Spectra from 0.1 mol dm^{-3} solutions of $Na_3Au(S_2O_3)_2$ and of the known oxidation products of thiosulfate, sodium tetrathionate and sodium sulfate, are also included in this figure. In addition, spectra are presented from solid $Na_3Au(S_2O_3)_2$.

It can be seen from Figure 3 that the solid sodium thiosulfate spectrum displays a number of bands. Those of most interest to the present work are the ones at 431 cm^{-1} and 1019 cm^{-1}, which are assigned to ν(S-S) and ν_{sym} ($-SO_3$) (Gabelica 1980; Haigh et al. 1993), respectively. The values found here for these bands are similar to those of 434 cm^{-1} and 1017 cm^{-1} reported in the literature (Haigh et al. 1993).

The bands shift to 445 cm^{-1} and 999 cm^{-1}, respectively, when the thiosulfate ion is in solution and no longer influenced by the sodium cation. Note that the broad band at low wavenumbers is only observed in water. The ν(S-S) and ν_{sym} ($-SO_3$) bands appear at 415 cm^{-1} and 1040 cm^{-1}, respectively, from the gold thiosulfate complex in the solid sodium compound and at 419 cm^{-1} and 1019 cm^{-1}, respectively, from the gold complex ion in solution. The difference between ν(S-S) for the gold thiosulfate complex and the thiosulfate ion reflects the strengthening of the S-S bond when bonded to gold. The values of ν(S-S) and ν_{sym} ($-SO_3$) for the tetrathionate ion in solution were 388 cm^{-1} and 1040 cm^{-1}, respectively. Sulfate exhibits a sharp ν_{sym} (SO_4^{2-}) band at 983 cm^{-1}.

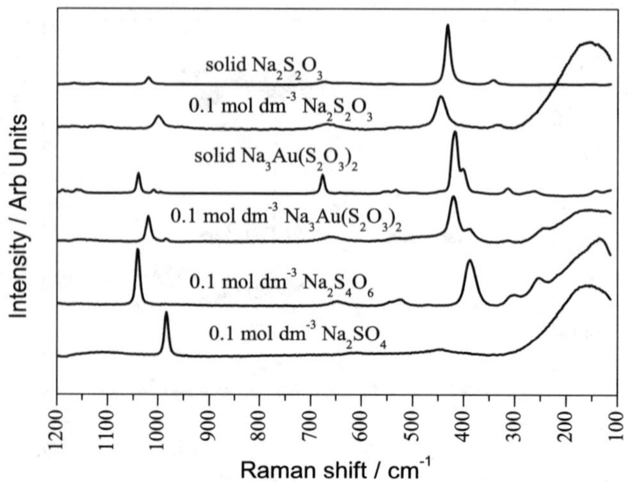

FIGURE 3 Raman spectra of various compounds of relevance to the investigation of gold leaching in thiosulfate media

Thiosulfate leaching of gold would be more acceptable as a commercial process if gold dissolution were not hindered by the growth of a passive film on the gold surface during the process (Breuer and Jeffrey 2003). It has been suggested (Chen et al. 1996; Molleman and Dreisinger 2002; Zhang and Nicol 2003) that the passive layer is sulfur. Thus, it is important to be able to identify sulfur layers on a gold surface.

The upper curve in Figure 4 is a Raman spectrum from elemental sulfur and it can be seen that there are three major bands, a S-S stretching band at 471 cm^{-1} and bands corresponding to S-S-S bending modes of the S_8 molecule at 216 cm^{-1} and 152 cm^{-1} (Eckert and Steudel 2003). The second spectrum in this figure is a Raman spectrum from gold sulfide prepared by mixing solutions of gold(I) sodium thiosulfate and sodium sulfide. It displays a broad band centred at 332 cm^{-1}, which appears to be a composite of three bands. X-ray crystallography (Ishikawa et al. 1995) has shown that Au_2S has a cuprite-type lattice in which sulfur atoms are in a body-centred cubic lattice with the space group *Pn3m*. The three bands could be due to lattice vibrations as well as vibration of the gold/sulfur bond. The broadness of the Raman band observed here would suggest a disordered structure.

Figure 4 also shows SERS spectra observed from a gold electrode held at different potentials in a 2×10^{-4} mol dm^{-3} Na_2S solution of pH 9.2. The potential of −0.6 V is in the stability region of the HS^- ion (Pourbaix 1963) and hence no surface species are detected on the gold surface. Voltammetry has shown (Buckley et al. 1987) that a prewave occurs in the oxidation of Na_2S on gold and studies using XPS has confirmed that this is due to the formation of a gold sulfide surface layer. The potential of −0.5 V is in the prewave region and hence the band evident at 305 cm^{-1} at this potential can be assigned to $\nu_{Au\text{-}S}$. As the coverage of the underpotential sulfur layer grows with increase in potential, the band shifts from 267 cm^{-1} to 310 cm^{-1}; both these wavenumbers are within the range of the broad band displayed by Au_2S.

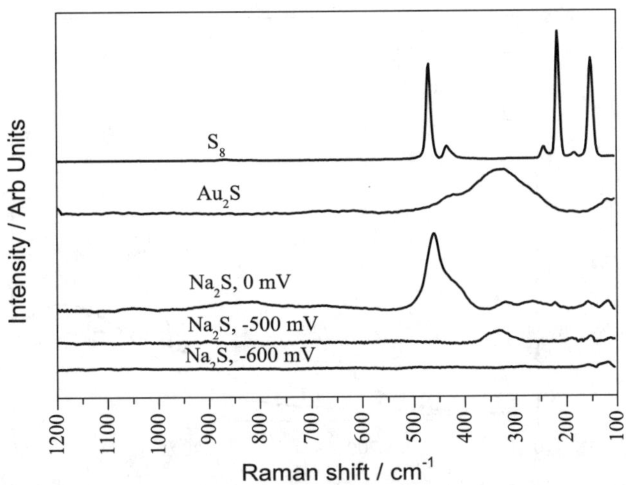

FIGURE 4 Raman spectra of S_8 and Au_2S and SERS spectra from a gold electrode in a 2 × 10^{-4} mol dm^{-3} Na_2S solution of pH 9.2 held at the marked potentials

At higher potentials, elemental sulfur becomes the thermodynamically stable sulfur species (Pourbaix 1963), and significant voltammetric currents have been observed at gold electrodes in sulfide solutions at such potentials (Buckley et al. 1987). XPS has found (Buckley et al. 1987), however, that the deposit has a lower binding energy and volatility than sulfur in its S_8 configuration. It was suggested that the sulfur layer had polysulfide characteristics involving bonding of sulfur to the underlying gold sulfide monolayer. It can be seen from Figure 4 that the Raman spectrum from the layer formed at 0 V has only a broad ν(S-S) band centred at 460 cm^{-1}. This indicates that there is a range of S-S bond lengths in the surface sulfur layer. Also, the absence of bands resulting from S-S-S bending modes suggests that the layer is disordered.

As the surface sulfur layer grows, the position of ν(S-S) also shifts; it is initially at 457 cm^{-1} when the band first appears and gradually moves to 462 cm^{-1}. As the sulfur layer thickens, the bending modes of S_8 appear and eventually the Raman spectrum from the sulfur-covered gold surface becomes the same as for S_8 in Figure 4. Similar findings were reported by Gao and Weaver (1992a,b).

Voltammograms recorded for a rotated gold electrode in 0.1 mol dm^{-3} solutions of $(NH_4)_2S_2O_3$ and $Na_2S_2O_3$, and in the latter solution containing 0.4 mol dm^{-3} ammonia, are presented in Figure 5. The voltammograms will not be discussed in detail here.

It can be seen from Figure 5 that the dissolution of gold in thiosulfate solution commences near 0 mV, and that the potential dependence of the dissolution current differs between the sodium and ammonium compound and is strongly influenced by the presence of ammonia. Finally, of significance to the present work, the characteristics of each curve in Figure 5 are indicative of inhibition by a surface species formed during the gold dissolution reaction.

Figure 6 presents SERS spectra from a gold electrode in 0.1 mol dm^{-3} $Na_2S_2O_3$ solution in equilibrium with air at open circuit (0 mV) and at the marked potentials after holding the electrode sequentially at the set potentials for 2 min. The SERS spectrum

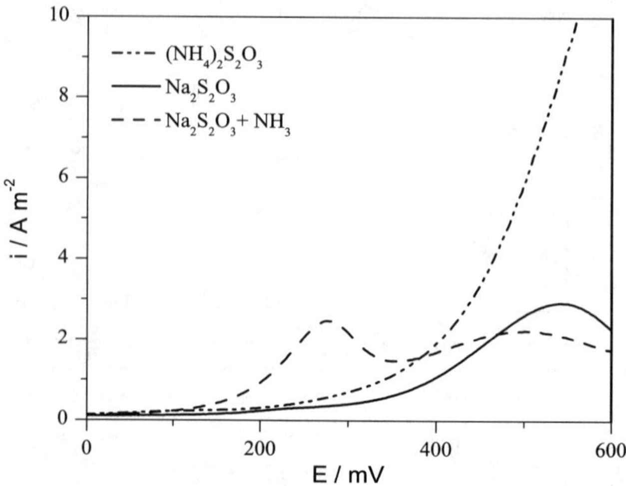

FIGURE 5 Voltammograms recorded at 1 mV s^{-1} with a gold electrode rotated at 300 rpm: the concentration of the thiosulfate is 0.1 mol dm^{-3}, and the ammonia, 0.4 mol dm^{-3}. E is potential vs Ag/AgCl.

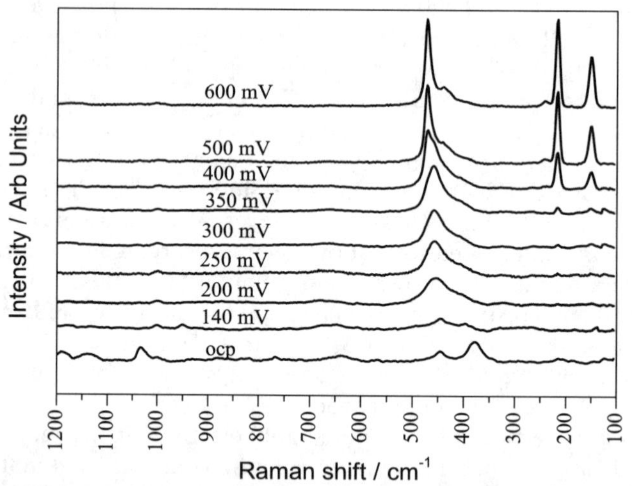

FIGURE 6 SERS spectra from gold electrode in 0.1 mol cm^{-3} $Na_2S_2O_3$ recorded at the open circuit potential and after holding at the applied potential for 2 min

recorded at open circuit displays bands at 445 cm^{-1} and 999 cm^{-1} due to ν(S-S) and ν_{sym} ($-SO_3$), respectively, of thiosulfate present at the gold / solution interface. In addition, bands occur at 378 cm^{-1} and 1033 cm^{-1} assigned in this study to ν(S-S) and ν_{sym} ($-SO_3$), respectively, of tetrathionate formed on the gold surface by oxidation of thiosulfate. These bands are blue shifted by 10 cm^{-1} and 7 cm^{-1}, respectively from the corresponding values for the tetrathionate ion in solution. This indicates that the tetrathionate is bonded to the gold surface rather than present in the double layer.

When the potential was held at 140 mV for 2 min, ν(S-S) and ν_{sym} ($-SO_3$) from thiosulfate are again observed while those arising from tetrathionate are diminished. A weak band, hardly discernible in Figure 6, was also apparent at ~300 cm^{-1} that can be assigned to the formation of a thin layer of gold sulfide (see Figure 4). At 200 mV, ν_{sym} ($-SO_3$) of thiosulfate is clearly visible, but a broad band now appears in the ν(S-S) wavenumber region. This band, which is centred at 455 cm^{-1}, is analogous to that shown in Figure 4 for sulfur deposited on gold from sodium sulfide solution at 0 mV. As the potential is increased, the intensity of the broad band increases and shifts to higher wave numbers; ν(S-S) is 460 cm^{-1} at 300 mV. Again, this is the same behaviour as that observed for gold in sulfide solution discussed above. At higher potentials, the bending modes of elemental sulfur are also evident in Figure 6, confirming that elemental sulfur is formed on the gold surface by oxidation of thiosulfate.

Extended polarization at 500 mV in 0.1 mol dm^{-3} $Na_2S_2O_3^{2-}$ at pH 9.2 was carried out to identify the other species formed when sulfur is deposited from thiosulfate. A spectrum of the solution after polarization for 70 h showed only bands from sulfur and residual thiosulfate, and one at 983 cm^{-1} that can confidently be assigned to sulfate. This indicates that the oxidation of thiosulfate at 500 mV is:

$$S_2O_3^{2-} + H_2O \rightarrow S^0 + SO_4^{2-} + 2H^+ + 2e^- \qquad \textbf{(EQ 1)}$$

The procedure used to acquire the spectra presented in Figure 6 was also used to study the anodic oxidation of 0.1 mol dm^{-3} thiosulfate in 0.05 mol dm^{-3} sodium tetraborate (pH 9.2) and in 0.031 mol dm^{-3} $NaHCO_3$ plus 0.034 mol dm^{-3} NaOH (pH 10.5). The spectra recorded were very similar to those in Figure 6. This indicates that the rate-determining step in the deposition of sulfur does not involve H^+ or OH^- ions. The first oxidation product of thiosulfate is tetrathionate:

$$2S_2O_3^{2-} \rightarrow S_4O_6^{2-} + 2e^- \qquad \textbf{(EQ 2)}$$

and this process is pH independent. Also, adsorbed tetrathionate is observed at low potentials in Figure 6. To assess the possibility that reaction (1) proceeds via reaction (2), the same procedure as that used to produce Figure 6 was applied to study the oxidation of tetrathionate on a gold surface. Bands characteristic of tetrathionate in solution, *viz.* ν(S-S) and ν_{sym} ($-SO_3$) at 388 cm^{-1} and 1040 cm^{-1}, respectively, were observed at all potentials studied. In addition, a band at 461 cm^{-1}, analogous to those observed with thiosulfate and assigned to the initial sulfur layer on gold, was observed when the potential was 350 mV. At 400 mV and at higher potentials, the bands characteristic of S_8 became the dominant feature of the spectrum. Thus, tetrathionate behaves in a similar manner to thiosulfate on a gold surface, being anodically oxidized to sulfur. The appearance of this product occurred, however, at higher potentials than that for thiosulfate oxidation. Furthermore, no shifted bands were observed characteristic of adsorbed tetrathionate. Thus, the rate limiting process step in reaction (1) would not appear to be reaction (2).

Oxidation of thiosulfate to tetrathionate is expected to proceed via:

$$S_2O_3^{2-} \rightarrow S_2O_3^{2-}{}_{ads} \qquad \textbf{(EQ 3)}$$

$$S_2O_3^{2-}{}_{ads} \rightarrow S_2O_3^{-}{}_{ads} + e^- \qquad \textbf{(EQ 4)}$$

$$2S_2O_3^{-}{}_{ads}\ S_4O_6^{2-}{}_{ads} \qquad \textbf{(EQ 5)}$$

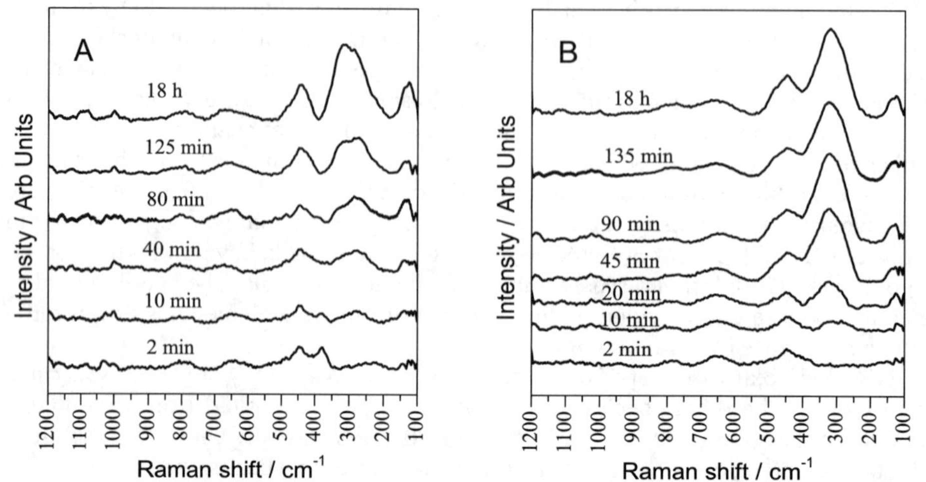

FIGURE 7 SERS spectra from gold electrode in 0.1 mol dm^{-3} $Na_2S_2O_3$ held for the marked times at A, 0 mV, and B, 100 mV

$$S_4O_6{}^{2-}{}_{ads} \rightarrow S_4O_6{}^{2-} \quad \textbf{(EQ 6)}$$

It is possible that reaction (4) is the rate determining step in both reactions (1) and (2). In the formation of sulfur from thiosulfate, reaction(1), however, reactions (5) and (6) are probably not involved, but $S_2O_3{}^{-}{}_{ads}$ reacts further by a different route.

Raman spectra were also recorded as a function of time with a gold electrode held at 0 mV and 100 mV in 0.1 mol dm^{-3} $Na_2S_2O_3$ and $(NH_4)_2S_2O_3$ solutions. These potentials are significant because they are in the potential region in which gold dissolves in thiosulfate in leaching practice. Figure 7 A and B present the SERS spectra recorded at the two respective potentials.

The spectrum in Figure 7A recorded after 2 min shows ν(S-S) for both thiosulfate and adsorbed tetrathionate. As time progressed, the tetrathionate band diminished and a band at 282 cm^{-1} that can be assigned to a gold-sulfur bond developed. At longer times, this band grows and shifts to 317 cm^{-1}, indicating the formation of a thin gold sulfide layer on the gold surface. In addition to this band, one at 443 cm^{-1} develops that overlaps with ν(S-S) for thiosulfate. The above discussion in relation to Figure 4 would suggest that this is a thin sulfur layer together with gold sulfide. The spectrum from Au_2S (Figure 4) shows, however, intensity in the region of 443 cm^{-1}. Thus, it cannot be ruled out that the intensity in the higher wavenumber region observed with thiosulfate oxidation originates from a gold-sulfur environment in a disordered Au_2S.

Similar behavior to that discussed above was observed at 100 mV (Figure 7B), but the additional bands develop faster. The gold-sulfur bands did not grow after ~ 45 min and there would appear to be a limiting thickness of the Au_2S layer which is not substantial.

SERS spectra were also recorded at 0 mV and at 100 mV with an electrolyte of 0.1 mol dm^{-3} $(NH_4)_2S_2O_3$ solution. The results at 100 mV are presented in Figure 8 and it can be seen that bands indicative of gold sulfide formation occur analogous to those observed in $Na_2S_2O_3$ solution. Spectra recorded at 0 mV were also similar to those at the same potential in $Na_2S_2O_3$ solution (see Figure 5).

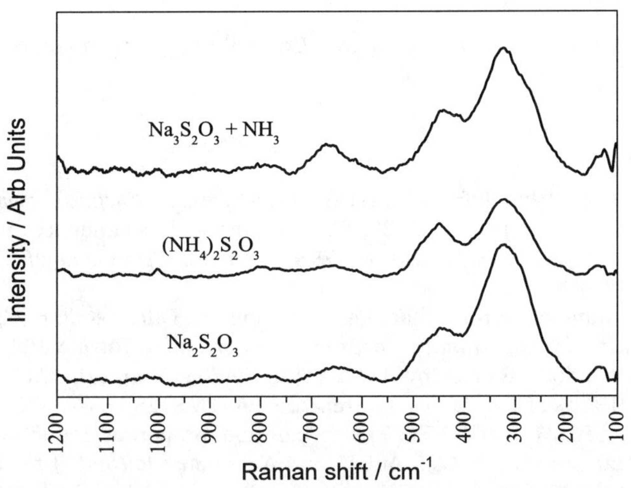

FIGURE 8 SERS spectra recorded at 100 mV for (lower) 0.1 mol dm^{-3} $Na_2S_2O_3$ after 135 min, (middle) 0.1 mol dm^{-3} $Na_2S_2O_3$ after 202 min, (upper) 0.1 mol dm^{-3} $Na_2S_2O_3$ plus 0.4 mol dm^{-3} NH_3 after 160 min

Ammonia was added to 0.1 mol dm^{-3} $Na_2S_2O_3$ solution to make it 0.4 mol dm^{-3} in NH_3. The SERS spectrum observed at 100 mV is also presented in Figure 6. Again, the spectrum shows a gold sulfide layer to be formed. Similar findings were made at 0 mV. Thus, the formation of gold sulfide does not depend on the thiosulfate cation or the presence of ammonia.

CONCLUSIONS

SERS spectra have been recorded from gold electrodes under potential control in chloride and thiosulfate solutions. They have identified surface species formed during the leaching of gold in these media. In each chloride solution studied, a Stark-shifted Raman band was observed over a wide potential range that is assigned to the chemisorption of the chloride ion onto the gold. This band diminished in intensity when gold dissolution commenced. The loss of Raman signal is explained by the absence of species at the gold surface that generate SERS spectra. Thiosulfate has been shown to oxidize in the potential region where leaching of gold occurs to deposit a thin gold sulfide layer on the electrode surface. The thickness of the sulfide layer grows slowly with time and leads to a limiting value, which is not substantial. At higher potentials than those at which gold sulfide develops, thiosulfate is oxidized to sulfur and sulfate, both products being identified by Raman spectroscopy. The rate of oxidation of thiosulfate in the oxidation to gold sulfide and to sulfur is independent of pH and this and related findings indicate that the rate-determining step in the process is the first electron transfer from adsorbed thiosulfate ions.

ACKNOWLEDGEMENTS

The authors thank the Australian Research Council and Griffith University for funding for this project.

REFERENCES

Ashley, K, and S. Pons. 1988. Infrared spectroelectrochemistry. *Chemical Reviews*. 88: 673.

Baltruschat H., and J. Heitbaum. 1983. On the potential dependence of the CN stretch frequency on Au electrodes studied by SERS. *Journal of Electroanalytical Chemistry*. 157: 319.

Bhakta, P. 2003. Ammonium thiosulfate heap leaching. In *Hydrometallurgy 2003, Proceedings of the International Symposium in Honor of Professor Ian M. Ritchie*. Vol. 1: Leaching and Solution Purification. Edited by C.A. Young, A.M. Alfantazi, C.G. Anderson, D.B. Dreisinger, B. Harris and A. James. Warrendale, PA: TMS. 259–267.

Braunstein, P., and R.J.H. Clark. 1973 The preparation, properties and vibrational spectra of complexes containing the $AuCl_2^-$, $AuBr_2^-$ and AuI_2^- ions. *Journal of the Chemical Society, Dalton Transactions*. 1845.

Breuer, P.L., and M.I. Jeffrey. 2000. A review of the chemistry, electrochemistry and kinetics of the gold thiosulfate leaching process. In *Hydrometallurgy 2003, Proceedings of the International Symposium honoring Professor Ian M. Ritchie*. Vol. 1: Leaching and Solution Purification. Edited by C.A. Young, A.M. Alfantazi, C.G. Anderson, D.B. Dreisinger, B. Harris and A. James. Warrendale, PA: TMS. 139–154.

Buckley, A.N., I.C. Hamilton, and R. Woods. 1987. An investigation of the sulfur(-II)/sulfur(0) system on gold electrodes. *Journal of Electroanalytical Chemistry*. 216: 213.

Chen, J.Y., T. Deng, G. Zhu and J. Zhao. 1996. Leaching and recovery of gold in thiosulfate based system–a research summary at ICM. *Transactions of the Indian Institute of Metals*. 49: 841.

Diaz, M.A., G.H. Kelsall, and N.J. Welham. 1993. Electrowinning coupled to gold leaching by electrogenerated chlorine. I. Au(III)-Au(I)/Au kinetics in aqueous Cl_2/Cl^- electrolytes. *Journal of Electroanalytical Chemistry*. 361: 25.

Eckert B., and R. Steudel. 2003. Molecular spectra of sulfur molecules and solid sulfur allotropes. *Topics in Current Chemistry*. 231: 31.

Evans, D.H., and J.J. Lingane. 1964. Chronopotentiometric Investigation of the Reduction of the Chloride Complexes of Gold. *Journal of Electroanalytical Chemistry*. 8: 173.

Ferron, C.J., C.A. Fleming, D. Dreisinger, and T. O'Kane. 2003. In *Hydrometallurgy 2003, Proceedings of the International Symposium honoring Professor Ian M. Ritchie*. Vol. 1: Leaching and Solution Purification. Edited by C. Young, A. Alfantazi, C. Anderson, A. James, D. Dreisinger and B. Harris. Warrendale, PA: TMS. 89–104.

Fleming, C.A., J. McMullen, K.G. Thomas, and J.A. Wells. 2003. Recent advances in the development of an alternative to the cyanidation process: thiosulfate leaching and resin in pulp. *Minerals and Metallurgical Processing*. 20: 1.

Gabelica, Z. 1980. Structural study of solid inorganic thiosulfates by infared and Raman spectroscopy. *Journal of Molecular Structure*. 60: 131.

Gallego, J.H., C.E. Castellano, A.J. Calandra, and A.J. Arvia. 1975. The electrochemistry of gold in acid aqueous solutions containing chloride ions. *Journal of Electroanalytical Chemistry*. 66: 207.

Galster, H. 1991. *pH measurement: Fundamentals, methods, applications, instrumentation* VCH, Weinheim. p 76.

Gao, P., and M.J. Weaver. 1986. Metal-adsorbate vibrational frequencies as a probe of surface bonding: halides and pseudohalides at gold electrodes. *Journal of Physical Chemistry*. 90: 4057.

Gao, X., Y. Zhang, and M.J. Weaver. 1992a. Observing surface chemical transformations by atomic-resolution scanning tunneling microscopy: sulfide electrooxidation on Au(III). *Journal of Physical Chemistry*. 96: 4156.

Gao, X., Y. Zhang, and M.J. Weaver. 1992b. Adsorption and electrooxidative pathways for sulfide on gold as probed by real-time surface-enhanced Raman spectroscopy. *Langmuir,* 8: 668.

Habashi, F. 1970. *Principles of Extractive Metallurgy–Hydrometallurgy*. New York: Gordon and Breach.

Haigh, J.A., P.J. Hendra, A.J. Rowlands, I.A. Degen, and G.A. Newman. 1993. Raman spectroscopy of thiosulphates. *Spectrochimica Acta*, 49A: 723.

Ishikawa, K., T. Isonaga, S. Wakita, and Y. Suzuki, Structure and electrical properties of Au_2S. *Solid State Ionics*, 79: 60.

Molleman, E., and D. Dreisinger. 2002. The treatment of copper-gold ores by ammonium thiosulfate leaching. *Hydrometallurgy*, 66: 1.

Kolics, A., A.E. Thomas, and A. Wieckowski. 1996. ^{36}Cl labeling and electrochemical study of chloride adsorption on a gold electrode from perchloric acid media. *Journal of the Chemical Society, Faraday Transactions*. 92: 3727.

Lipkowski, J., Z. Shi, A. Chen, B. Pettinger, and C. Bilger. 1998. Ionic adsorption at the Au(111) electrode surface. *Electrochimica Acta*, 43: 2875.15.

Murphy, P.L., G. Stevens, and M.S. LaGrange. 2000. The effects of temperature and pressure on gold-chloride speciation in hydrothermal fluids: a Raman spectroscopic Study. *Geochim. Cosmochim. Acta*. 64: 479.

Pan, P., and S. A. Wood. 1991. Gold-chloride complexes in very acidic aqueous solutions at temperatures 25–300°C: a laser Raman spectroscopic study. *Geochimica et Cosmochimica Acta*. 55: 2365.

Pourbaix, M. 1963. *Atlas D'Equilibres Electrochimiques*. Paris, France: GauthienVillars.

Puddephatt, R.J. 1978. *The Chemistry of Gold*. Elsevier, Amsterdam pp 211, 213.

Wood, P. 2004. The Intec gold process for refractory gold concentrates, *Breaking New Ground Mining Technology Conference*, 26 May 2004, Fremantle, Western Australia.

Woods, R. 1976. Chemisorption at electrodes: hydrogen and oxygen on noble metals and their alloys. In *Electroanalytical Chemistry: A Series of Advances*. Vol. 9. Edited by A.J. Bard. Dekker, New York. 1–162.

Zhang, S., and M.J. Nicol. 2003. An electrochemical study of the dissolution of gold in thiosulfate solutions Part I: Alkaline solutions. *Journal of Applied Electrochemistry*. 33:767.

Leaching Behavior of Gold in Ammoniacal Solutions in the Presence of Bromine as an Oxidant

Peter Nam-soo Kim* and Kenneth N. Han*

The extraction behavior of gold in ammoniacal solutions in the presence of bromine as an oxidant has been studied. The effect of concentration of ammonia, bromine, and pH of the solution on the rate of dissolution of gold was investigated. The dissolution behavior has been analyzed using electrochemical method, and its results were compared with direct leaching tests using gold powder.

The overall dissolution of gold in ammoniacal solution containing bromine was found be affected significantly by pH of the solution and the concentration of bromine. The rate expression found in this study has the following form.

$$\text{Rate} = k_o(\text{H}^+)^{0.375}(\text{Br}_2)^{1.570}\frac{\text{mol}}{dm^3 s} \quad \text{where } k_o = 1.19\times 10^{-4}\frac{dm^{2.835}}{mol^{0.945} s}$$

INTRODUCTION

Gold is traditionally extracted in solution using cyanide since the technology was first introduced by MacArthur and Forrest (1889). However, in recent years many alternative technologies have been proposed to replace the cyanide technology. These include leaching with thiourea (Deshenes and Ghali, 1988), thiosulfate (Zipperian et al., 1988;

* Department of Materials and Metallurgical Engineering, South Dakota School of Mines and Technology, 501 E. St. Joseph Street, Rapid City, South Dakota

Alymore and Muir, 2001; Breuer and Jeffrey, 2004), and halogens (McGrew and Murphy, 1985; Qi and Hiskey, 1991). Some of these technologies have found its usefulness in the industry but there are still barriers for these technologies to become commercialized.

Ammonia is widely used in leaching of metals including copper (Halpern et al., 1959; Guan and Han, 1995), nickel (Morioka et al., 1967; Bhuntumkomol et al., 1980), cobalt (Vu and Han, 1977), and silver (Grybos and Samontus, 1984; Guan and Han, 1994). All of these metals are soluble in ammoniacal solutions without subjecting to high pressure and temperature but in the presence of appropriate oxidants.

Skibsted and Bjerrum (1974a,b) first recognized that gold is soluble in ammonia. Han and Meng (1992,1994) demonstrated that gold can be dissolved from ores using cupric ammine as an oxidant in an autoclave. More detailed studies on leaching behavior of gold in ammoniacal solutions followed (Meng and Han, 1995; Dasgupta et al., 1997; Guan and Han, 1996). Their study revealed that gold dissolved in ammoniacal soluitons with various oxidants such as oxygen, hypochlorite, and hydrogen peroxide. Gold was also found to be amenable to dissolution in ammoniacal solutions with iodine as an oxidant (Meng and Han, 1997). However, most of these leaching systems had to utilize high temperature in an autoclave for an effective dissolution of gold.

In this study, gold has been leached in ammoniacal solutions at ambient temperature with bromine as an oxidant. The chemistry of bromine system has been examined first to help understand the effect of bromine in extracting gold in ammoniacal solutions. Direct leaching of gold particles was compared with the results obtained by electrochemical analysis of gold leaching using a potentiometer. The leaching mechanism of gold in ammoniacal solutions with bromine has been established and the rate of dissolution was determined in view of pH of the system and concentrations of ammonia and bromine.

THEORETICAL CONSIDERATIONS

It is well known that when bromine is added into water, many bromide-complexes are formed due to primarily the following chemical equations, Eqs. 1 through 5. There exist at least the following seven species including Br^-, Br_2, Br_3^-, HBrO, BrO^-, BrO_3^- and $HBrO_3$.

$$HBrO \leftrightarrow H^+ + BrO^- \quad \textbf{(EQ 1)}$$

$$HBrO_3 \leftrightarrow BrO_3^- + H^+ \quad \textbf{(EQ 2)}$$

$$Br_2 + H_2O \leftrightarrow H^+ + Br^- + HBrO \quad \textbf{(EQ 3)}$$

$$Br_2 + Br^- \leftrightarrow Br_3^- \quad \textbf{(EQ 4)}$$

$$3Br_2 + 3H_2O \leftrightarrow HBrO_3 + 5\,Br^- + 5\,H^+ \quad \textbf{(EQ 5)}$$

The distribution of these species was calculated as a function of pH using "go-seek/solver" in the Excel spread sheet. The initial bromide and bromine concentration ratio was kept constant at 4:1 over the entire pH range considered. The concentrations of these species are calculated and plotted as a function of pH in Figure 1. The thermodynamic data for these calculations were taken from Pourbaix (1974).

In all these analyses, the total bromine added was 0.0052 mol/dm^3 and the concentration of the total bromide-species was kept constant at 0.0312 mol/dm^3. It is interesting to note that the predominant species at relatively low pH, less than 5, was found to be Br^-, while at high pH, higher than 8, the dominant species was BrO_3^-.

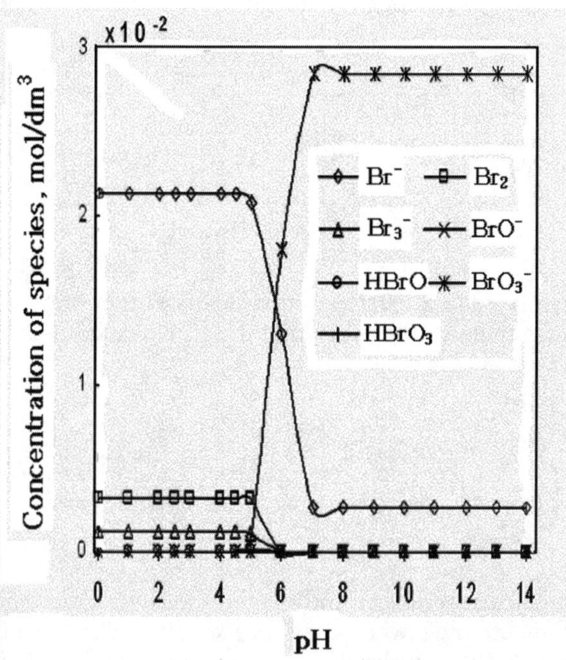

FIGURE 1 Distribution of bromide containing species as a function of pH (Bromine: 0.0052 mol/dm^3; Total bromide-species kept constant 0.0312 mol/ dm^3)

The dissolution of a conductive metal in aqueous media is a redox reaction taking place at the metal-solution interface. One of the main advantages of using an electrochemical method is the fact that the overall reaction can be split into anodic and cathodic reactions and hence these individual steps can be examined separately. The dissolution rate is easily obtained electrochemically without determining dissolved metal ion in solution, which is another advantage to electrochemical testing.

Activated overpotential can be simplified as an additional barrier caused by an added potential. The rate can be expressed by:

$$i_c = nFC_+k_ce^{\frac{-\beta VF}{RT}} \quad \textbf{(EQ 6)}$$

$$i_a = nFk_ae^{\frac{-(1-\beta')VF}{RT}} \quad \textbf{(EQ 7)}$$

At the mixed potential, V_{mix}, electrical charge conservation principles necessitate that $|i_c| \times A_c = |i_a| \times A_a$.

$$nFA_cC_+k_ce^{\left[\frac{-\beta FV_{mix}}{RT}\right]} = nFA_ak_ae^{\left[\frac{-(1-\beta')FV_{mix}}{RT}\right]} = i \times A \quad \textbf{(EQ 8)}$$

Eq. 8 can be rearranged in terms of V_{mix},

$$V_{mix} = \frac{RT}{F(\beta + 1 - \beta')}\left[Ln\left(\frac{A_c k_c}{A_a k_a}\right) + Ln(C_+)\right] \quad \text{(EQ 9)}$$

Eqs. 8 and 9 yield the following relationship.

$$i = \frac{nFA_c k_c}{A}\left(\frac{A_c k_c}{A_a k_a}\right)^{\frac{-\beta}{\beta + 1 - \beta'}}(C_+)^{\frac{1 - \beta'}{\beta + 1 - \beta'}} \quad \text{(EQ 10)}$$

Eq. 11 is a rearrangement of Eq. 10 illustrating that the order of reaction, m simply consists of two transfer numbers. Therefore, the order of reaction can be estimated from Tafel plots.

$$i = k_o(C_+)^m \; ; \quad k_o = \frac{nFA_c k_c}{A}\left(\frac{A_c k_c}{A_a k_a}\right)^{\frac{-\beta}{\beta + 1 - \beta'}} \; ; \quad m = \frac{1 - \beta'}{\beta + 1 - \beta'} \quad \text{(EQ 11)}$$

EXPERIMENTAL

Analytical grade chemicals were used throughout the study. Ammonia, ammonium sulfate, and ammonium bromide were used as lixiviants. The pH of the solution was adjusted by changing sodium hydroxide or ammonia concentration. The polarization tests were performed in an open-system at ambient temperature.

The gold disk used for all the electrochemical experiments was made from a sheet of 99.95% gold metal. The area of the gold disk tested was 1.04 cm^2. Disk holder was made of Tefron and ensured to provide extra margin to avoid the edge effect. Electrochemical tests were carried out with a static disk electrode system. The EG&G Princeton Applied Research model 352 SoftCorr™ corrosion measurement software was used to control the EG&G PARC model 273A potentiostat to carry out potentiodynamic polarization, Tafel polarization and potentiostatic measurements.

In order to maintain the reliable current flow of electrons, the gold surface had to be cleaned enough to avoid partial-concentrated current. The mirror-like surface of the gold as the working electrode was obtained by polishing it with 1µm alumina and again abraded with 0.05 µm alumina. All potentials were reported against the saturated calomel electrode (SCE) at ambient temperature.

99.99% grade gold-powder manufactured by Johnson Matthey was also used for the direct leaching experiments. The mean particle size was found to be 45µm of diameter. Dissolution experiments were carried out in a one-liter glass vessel. One-half liter of the solution was used. Samples of solution were withdrawn at regular time intervals for chemical analysis. All chemical analyses were carried out using a Perkin-Elmer atomic absorption spectrometer, Model 5500.

RESULTS AND DISCUSSION

Effect of Ammonia

The anodic dissolution of gold in ammoniacal solution has been studied by potentiodynamic polarization experiments. The variables investigated include ammonia concentration,

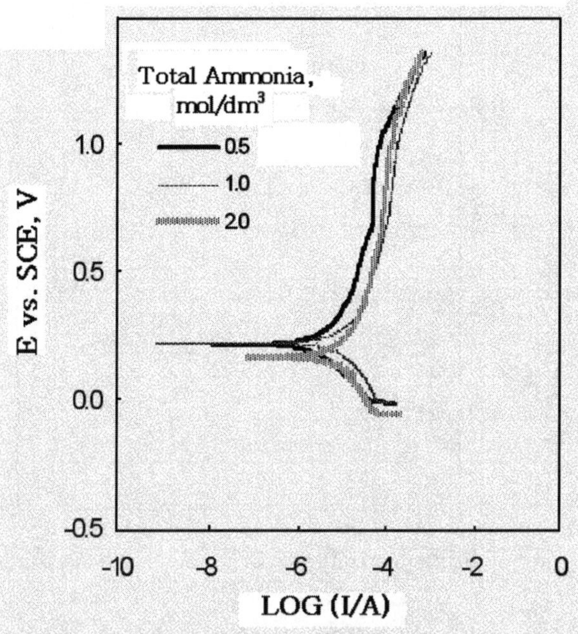

FIGURE 2 Tafel polarization curves of a static gold disk at three total ammonia concentrations. Temp.:25°C; pH:9.8; scan rate:0.2 mV/s.

bromine concentration, and pH of the solution. Figure 2 shows typical anodic potentiodynamic polarization curves of a static gold disk at three total ammonia concentrations without bromine.

Guan and Han (1996) reported an appropriate anodic half reaction for gold dissolving in the ammoniacal solution. When gold is placed in ammoniacal solutions, the dissolution of gold will take place following the following sequence.

$$Au = Au^{+} + e \quad \textbf{(EQ 12)}$$

$$Au^{+} + 2NH_3 = Au(NH_3)_2^{+} \quad \text{where} \quad K = \frac{[Au(NH_3)_2^{+}]}{[Au^{+}][NH_3]^2} \quad \textbf{(EQ 13)}$$

The overall reaction becomes:

$$Au + 2NH_3 = Au(NH_3)_2^{+} + e \quad \textbf{(EQ 14)}$$

The equilibrium potential of the anodic reaction can be written as:

$$E_a = E_a^o + \frac{RT}{F}\ln\left\{\frac{[Au(NH_3)_2^{+}]}{K[NH_3]^2}\right\} \quad \textbf{(EQ 15)}$$

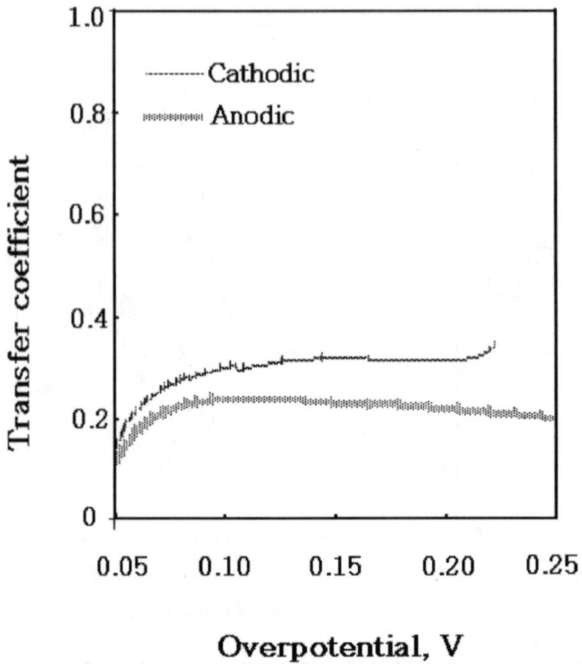

FIGURE 3 Plot of transfer coefficient vs. overpotential. Temp.: 25°C; pH: 9.8; Total Ammonia: 0.5 mol/dm³.

When the overpotential is plotted as a function of log(i), a straight-line is obtained. From the Tafel polarization curve, the transfer coefficient () was obtained. Such Tafel plots were made for the anodic and cathodic reactions of gold and the transfer coefficients obtained were plotted as a function of overpotential (Figure 3). The anodic transfer coefficient was found to be reasonably constant over the range of 0.18–0.22 when the overpotential increased from 0.05V to 1.1V. These values were less than the values reported earlier as 0.23 (Guan and Han,1996), although the temperature was 100°C and not 25°C, however.

In the absence of a specific oxidant, the dissolution of gold in ammonical solutions can be described by Eqs. 16 and 17.

$$2Au + 4NH_3 + \frac{1}{2}O_2 + H_2O = 2Au(NH_3)_2^+ + 2OH^- \quad \textbf{(EQ 16)}$$

$$Au + 2NH_3 + H_2O = Au(NH_3)_2^+ + \frac{1}{2}H_2 + OH^- \quad \textbf{(EQ 17)}$$

The cathodic reactions and their corresponding potentials are given below:

$$\frac{1}{2}O_2 + H_2O + 2e = 2OH^- \qquad E_c = E_c^o - \frac{RT}{2F}\ln\left\{\frac{[OH^-]^2}{P_{O_2}^{\frac{1}{2}}}\right\} \quad \textbf{(EQ 18)}$$

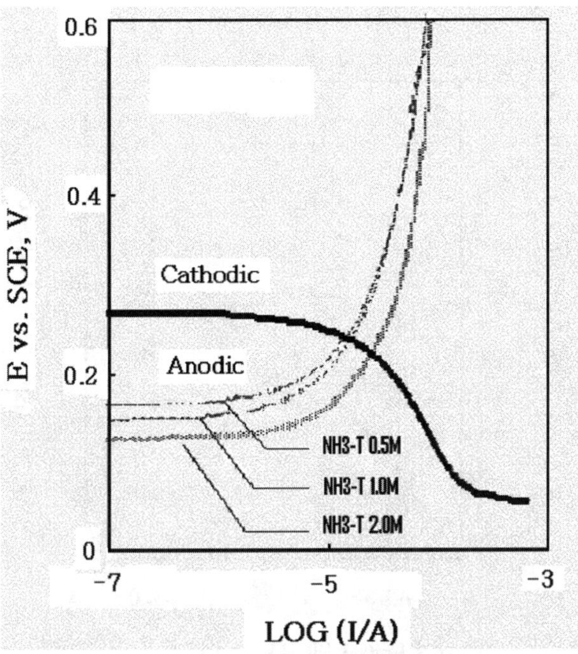

FIGURE 4 A cathodic-polarization and three anodic-polarization curves of a static gold disk at three total ammonia concentrations. Temp.: 25°C; pH: 9.8; scan rate: 0.2 mV/s.

$$H_2O + e = \frac{1}{2}H_2 + OH^- \quad E_c = E_c^o - \frac{RT}{F}\ln\left\{P_{H_2}^{\frac{1}{2}}[OH^-]\right\} \tag{EQ 19}$$

As seen from Eq. 15, the equilibrium potential of the anodic reaction decreases with the increase of the total ammonia concentration. When the system is open to air and pH is fixed, the cathodic potential, E_c does not change as indicated by Eqs. 18 and 19.

Figure 4 shows the effect of the total ammonia concentration on the gold dissolution rate and dissolution potential. The dissolution potential is located at the intersection point of the extension of the anodic Tafel line and that of the cathodic Tafel line. The dissolution potential slightly decreased with the increase of the total ammonia concentration.

However, the dissolution rate increased with the increase of the total ammonia concentration. The increase of gold dissolution rate can be explained by the decrease of anodic potential. It is well known that when complexing agents such as cyanide, ammonia, bromine and iodine are added to the solution, metal forms complexes in the solution. These complexes make a significant difference in the electromotive force (EMF) and usually lower the E° in the solution. Han (2003) reported the EMF values of gold, silver and copper in ammoniacal solutions, gold being the most noble, whereas copper is more noble than silver. Lowering the anodic potential by adding a complexing agent successfully takes advantage of the thermodynamic principles to extract metal.

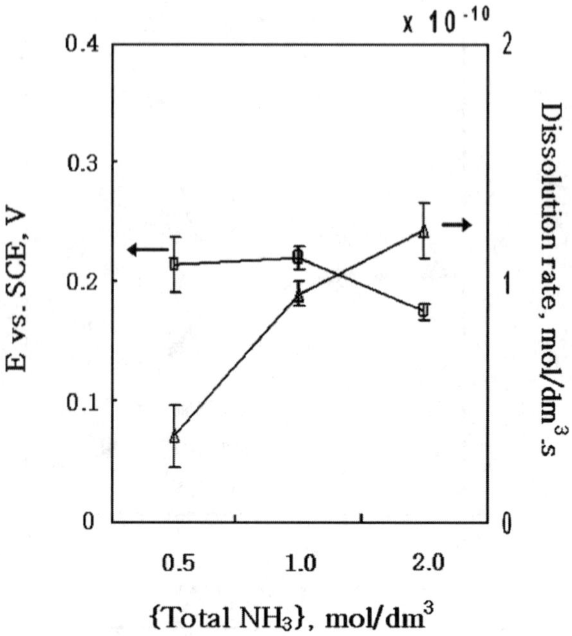

FIGURE 5 Effect of ammonia concentration on the dissolution rate of gold and dissolution potential. Temp.: 25°C; pH: 9.8.

The effect of ammonia concentration on the dissolution rate of gold and the dissolution potential by using electrochemical method is shown in Figure 5. The number of electrons (n) per mole of gold dissolved was found to be 1.04 when the applied potential is 0.5 V, vs. SCE in ammonia system. The number of electrons has been reported to be 1.07–1.22 in the presence of oxidant in ammoniacal solutions by Guan and Han (1996) and 0.94–1.47 in the cyanide system by Kirk et al. (1978). The dissolution potential was decreased from 0.22 to 0.18 V, vs. SCE and the gold dissolution rate was increased from $3.64*10^{-11}$ to $1.22*10^{-10}$ mol/dm^{3}s when total ammonia concentration increased from 0.5 to 2.0 mol/dm^3.

The rate of dissolution of gold in the presence of bromine is expected to be affected by the concentration of bromine on one hand and also by pH of the solution. As a result, the overall reaction rate can be described by:

$$\text{Rate} = k_o(H^+)^l(By_2)^m \quad \textbf{(EQ 20)}$$

It should also be noted that when bromine is added into the solution, pH is inevitably changed as indicated by Eqs. 1–5. As seen from the speciation of bromine, pH and bromine concentration are interdependent. In order to identify the functional relationship between the dissolution of gold and pH and also the concentration of bromine, experiments were conducted at a constant pH when the concentration of bromine was changed and vice versa.

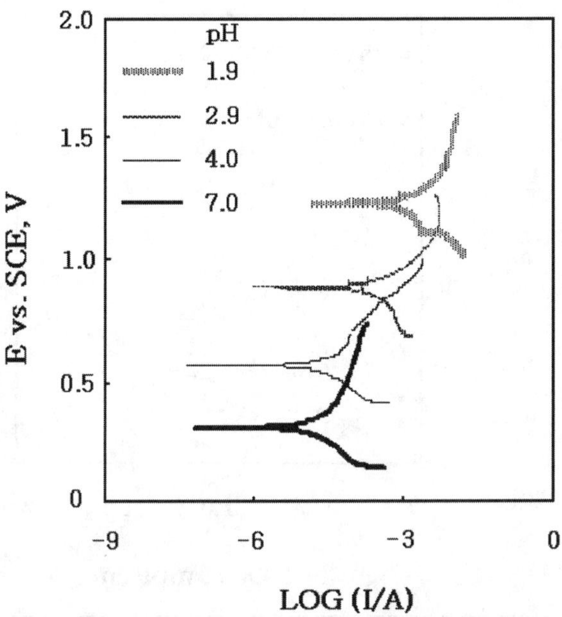

FIGURE 6 Tafel polarization curves of a static gold disk at four pH. Temp.: 25°C; bromine conc.: 0.0052 mol/dm^3; $(NH_4)_2SO_4$: 1 mol/dm^3 ;scan rate: 0.2 mV/s.

Effect of pH

The first set of experiments was conducted using the potentiodynamic polarization method while the pH of the solution was changed but the concentration of bromine was kept constant.

Tafel polarization curves of a static gold disk are plotted in Figure 6 for four different pH values. It is seen that the reaction potential and dissolution rate increased with the decrease of the pH in the pH range 1.9–7.0. Bromine acted not only as an oxidant but also as a complexing agent. When the pH is higher than 7.0, a significant amount of bromine, Br_2 was present as bromide ion, Br^-. Therefore, when the pH of the solution is higher than 7.0, due to lack of bromine, the gold dissolution curve followed the same behavior of that in the absence of bromine but with ammonia alone.

Figure 7 shows the effect of ammonia concentration (NH_4OH) on the gold dissolution current density with and without bromine. The gold dissolution current density increased with the increase of the ammonia concentration. As indicated by Eq. 21, ammonia concentration increased with the increase of the pH.

$$NH_4^+ + H_2O = NH_4OH + H^+ \qquad \textbf{(EQ 21)}$$

Therefore, at the pH range 7.0–10.3, the effect of bromine on the gold dissolution rate is not significant but the ammonia concentration significantly affects the gold dissolution rate.

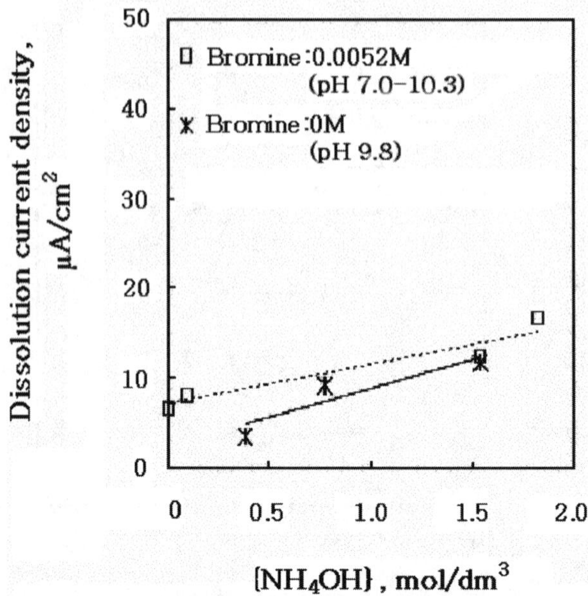

FIGURE 7 Effect of ammonia concentration (NH_4OH) on the gold dissolution current density. Temp.: 25°C; (□) Bromine conc.: 0.0052 mol/dm^3; $(NH_4)_2SO_4$: 1 mol/dm^3 ; pH : 7.0, 8.0, 9.8 and 10.3.; (*) Bromine conc.: 0 mol/dm^3; Total Ammonia : 0.5, 1.0 and 2.0 mol/dm^3; pH : 9.8.

Figure 8 shows the change of the anodic transfer coefficient with respect to overpotential. The anodic transfer coefficient was about 0.18 when pH was 10.3. This was exactly the same as the result of the anodic transfer coefficient which ranged over 0.18–0.22 when bromine was absent. The anodic transfer coefficient was initially 0.6 and 0.4 when pH was 1.9 and 2.9, respectively. The transfer coefficient decreased with the increase of overpotential. After two hours of the electrochemical analysis, the final pH of the solution was increased from 1.9 to 2.3 and 2.9 to 3.2, respectively. At low pHs of the solution, the predominant species containing bromide are bromide, Br^- and bromine, Br_2. It was anticipated that bromine would be subjected to vaporization resulting in fast depletion of bromine from the system. As bromine escaped from the solution, the pH of the system was expected to increase as shown by Eq. 3.

In order to see the effect of the bromine speciation, 0.01 mol/dm^3 of bromine were added into the solution at pH 9.8. As seen form Figure 9, the dissolution potential with the aid of bromine as an oxidant decreased with time under these conditions. It should be noted that the dissolution potential changed from about 0.4 V vs. SCE to 0.2 V in about 150 min. At this time, the potential became almost the same as the system without bromine. The absence of bromine at this time was apparent to the naked eye since the brownish color due to the presence of bromine became colorless. The pH was adjusted by adding H_2SO_4 or NaOH.

The chemical reaction order, l in Eq. 20, was obtained from the plot of the rate of dissolution of gold versus the concentration of the hydrogen ion on the log-log scale

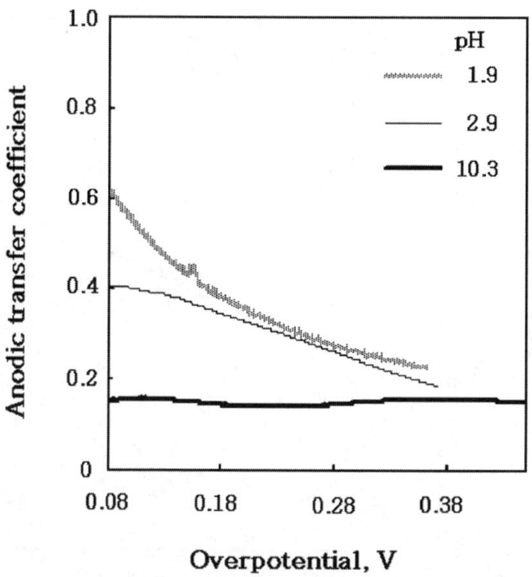

FIGURE 8 Plot of transfer coefficient vs. overpotential. Temp.: 25°C; bromine conc.: 0.0052 mol/dm^3; $(NH_4)_2SO_4$: 1 mol/dm^3.

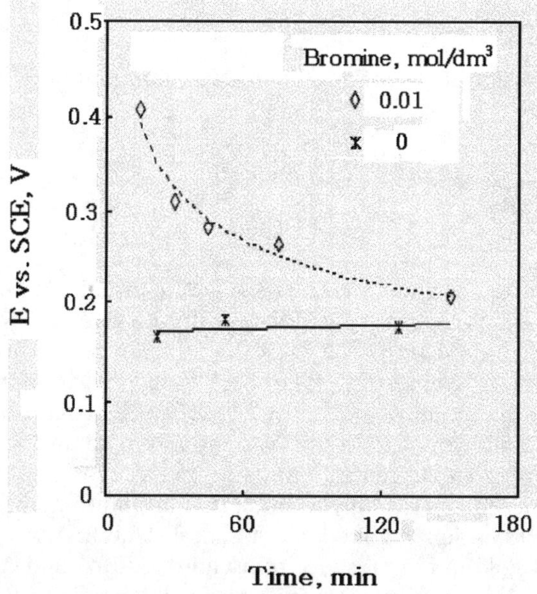

FIGURE 9 Plot of dissolution potential vs. time.Temp.: 25°C; pH: 9.8 (◇) Bromine conc.: 0.01 mol/dm^3; $(NH_4)_2SO_4$: 1 mol/dm^3 (∗) Bromine conc.: 0 mol/dm^3; $(NH_4)_2SO_4$: 1 mol/dm^3.

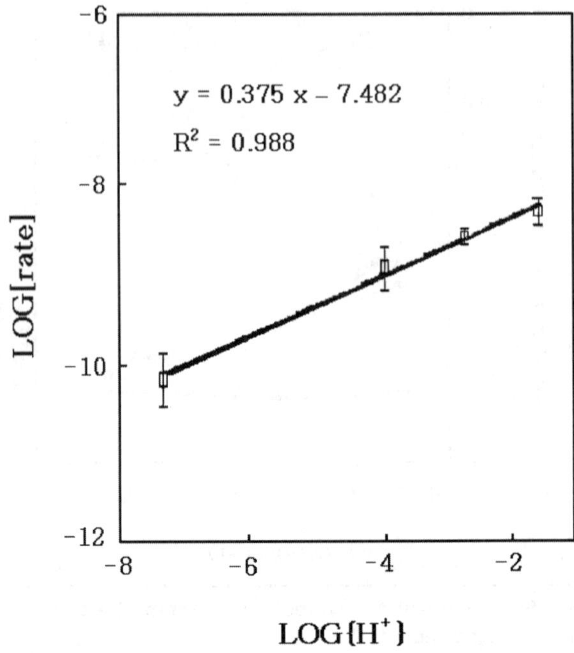

FIGURE 10 Schematic of the gold dissolution rate vs. the H^+ concentration on the log scale. 25°C; Bromine conc.: 0.0052 mol/dm^3; $(NH_4)_2SO_4$: 1 mol/dm^3; pH range 1.9–7.0.

(Figure 10). The concentration of Br_2 was kept constant at 0.0052 mol/dm^3. The order was determined to be 0.375.

$$\text{Rate} = k_o(H^+)^{0.375}(Br_2)^m \quad \textbf{(EQ 22)}$$

Effect of Bromine

Tafel polarization curves of a static gold disk in ammoniacal solutions for four different bromine concentrations were plotted and the results are given in Figure 11. The dissolution current density of gold increased with the increase of the bromine content from 0.0008 mol/ dm^3 to 0.0052 mol/ dm^3. It has been found the reaction potential increased with the increase of the bromine concentration. When the concentration was increased more than to 0.0052 mol/dm^3, the results were found to be unstable and therefore, the effect of bromine for higher concentrations were observed only by direct leaching tests.

The current experiments strongly indicate that bromine can be used successfully as an oxidant for the dissolution of gold in ammoniacal solutions. According to the thermodynamic analysis of gold in the presence of bromine, $AuBr_2^-$ and $AuBr_4^-$ were found to be stable complexes. At relatively higher concentrations of bromine and oxidational potentials, $AuBr_4^-$ is more predominant specie in solution.

Figure 12 shows the potentiodynamic anodic sweep profile in ammoniacal solution using bromine as an oxidant. When the bromine concentration was less than 0.0015 mol/ dm^3 and the pH of the solution was lower than 7, $AuBr_2^-$ seems to be

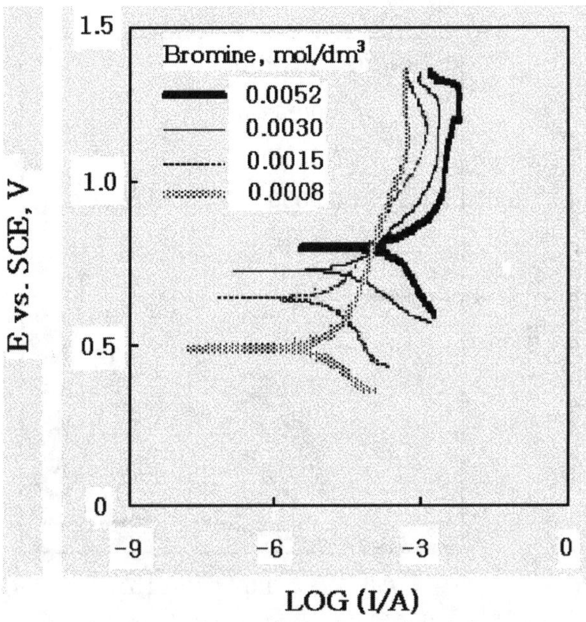

FIGURE 11 Potentiodynamic polarization curves of a static gold disk at 4 initial bromine concentrations: scan rate: 5 mV/s; temp.: 25°C; $(NH_4)_2SO_4$ conc.: 1 mol/dm^3

predominant complex. Under these conditions, the anodic equation, $Au + 2Br^- = AuBr_2^-$ $+ e$ and the cathodic equation, $2Br_2 + 2e = Br^- + Br_3^-$ can be coupled to yield:

$$2Au + 2Br_2 + 3Br^- = 2AuBr_2^- + Br_3^- \quad \textbf{(EQ 23)}$$

When the bromine concentration was higher than 0.0030mol/ dm^3 and the applied potential was less than 0.1 V vs. SCE, the "line-a" in Figure 12 represented the formation reaction of $AuBr_2^-$. Under the same condition of bromine concentration, but the applied potential was higher than 1.2 V vs. SCE, the solid "line-b" in Figure 12 represented the formation of $AuBr_4^-$. At higher bromine concentrations and oxidential potentials, $AuBr_4^-$ is more stable in the solution. Under this condition, the anodic reaction, $AuBr_2^- + 2Br^- =$ $AuBr_2^- + 2e$ and the cathodic reaction, $Br_2 + 2e = 2Br^-$ can be coupled to yield Eq. 24.

$$AuBr_2^- + Br_2 = AuBr_4^- \quad \textbf{(EQ 24)}$$

Figure 13 shows the variation of anodic transfer coefficient with the increase of overpotential. The initial anodic transfer coefficients were about 0.11, 0.22, 0.38 and 0.52 for Br_2 concentrations of 0.0008, 0.0015, 0.0030 and 0.0052 mol/dm^3, respectively. The transfer coefficient is closely related to chemical reaction order.

When the metal ion is deposited on the metal by electrochemical reaction, the anodic and cathodic transfer coefficients are usually assumed to be 0.5. Thus the chemical reaction order of metal ion, m becomes 0.5. As seen from this system, m is always less than unity in metal deposition. In this study, the cathodic reaction is related to bromine and the anodic reaction relates to gold dissolution. The concentration term, C_+

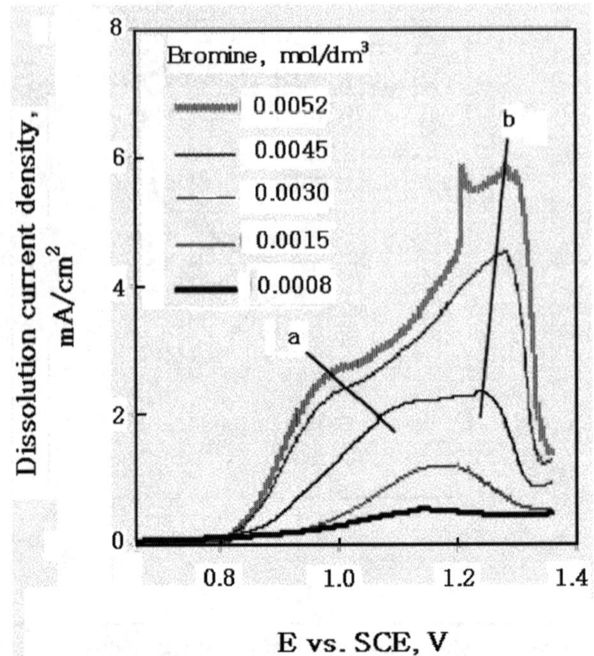

FIGURE 12 Potentiodynamic anodic sweep profile in ammoniacal solution at various bromine concentrations. Temp.: 25°C; $(NH_4)_2SO_4$ conc.:1 mol/dm^3; scan rate 1mV/s.

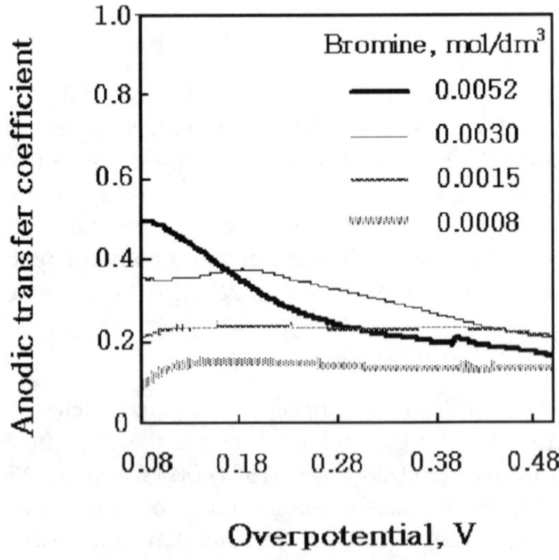

FIGURE 13 Plot of transfer coefficient vs. overpotential. Temp.: 25°C; Total Ammonia: 1.0 mol/dm^3.

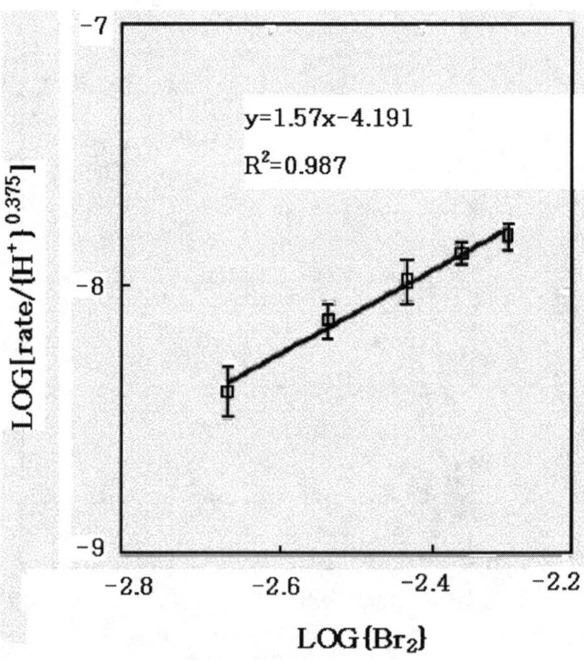

FIGURE 14 Schematic showing gold dissolution rate with respect to bromine concentration on log scale. Temp.: 25°C; Bromine conc.: 0.0022–0.0052 mol/dm^3; pH range: 2.2–3.5.

represents the bromine concentration. The anodic transfer coefficient 1 − β' also plays an important role and takes part in the determination of the chemical reaction order of bromine, m. The higher the 1 − β' value, the higher the chemical order of bromine, m. To find out empirically the chemical reaction order of {Br_2}, the rate expression was reorganized to give Eq. 25.

$$\frac{\text{Rate}}{(H^+)^{0.375}} = k_o(Br_2)^m \qquad \textbf{(EQ 25)}$$

As seen from Figure 14, the relationship between the gold dissolution rate and {Br_2} was plotted on the log scale. The slope was found to be 1.570. From this result, k_o was calculated as 1.19×10^{-4} ($dm^{2.835}/mol^{0.945}s$). The dissolution rate can be expressed as:

$$\text{Rate} = k_o(H^+)^{0.375}(Br_2)^{1.570}, \text{ unit: } \frac{mol}{dm^3 s} \quad k_o = 1.19 \times 10^{-4}\frac{dm^{2.835}}{mol^{0.945}s} \qquad \textbf{(EQ 26)}$$

In order to check the effect of bromine at higher concentrations than 0.0052 mol/dm^3, a direct leaching investigation using gold powder of 45 microns was conducted at 25°C. Four bromine concentrations, namely, 0.0375, 0.075, 0.15 and 0.3 mol/dm^3 were used while keeping the ratio of the concentrations of $(NH_4)Br$ and Br_2 at 4:1 throughout. The pH of the solution was found to be in the range 1.8–2.3. As the concentration of Br_2

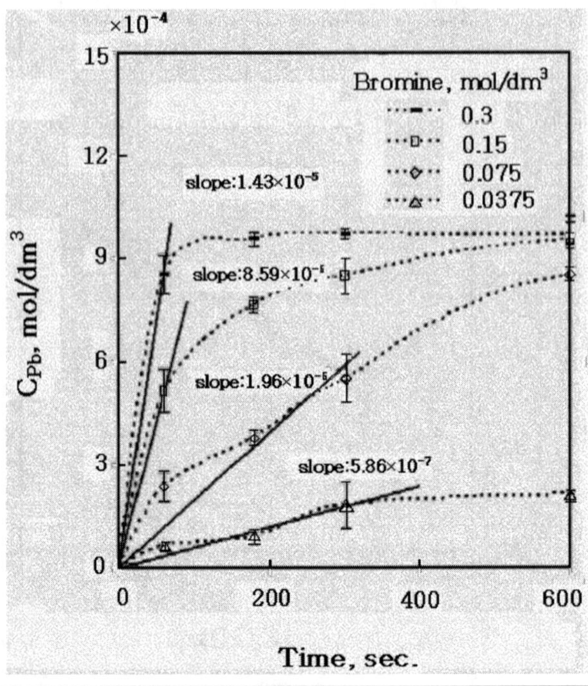

FIGURE 15 Concentration of dissolved gold versus time at various bromine concentrations; Temp.: 25°C

increased to more than 0.15 mol/dm^3, the overall dissolution of gold became significantly affected by mass transfer control and therefore, only the concentrations of Br_2 of 0.0375, 0.075 and 0.15 mol/dm^3 were used to determine the reaction order.

Figure 15 represents the rate of dissolution of gold as a function of time for the four different Br_2 concentrations. The initial rates for the three concentrations as specified earlier were used to plot log (rate) versus log (Br_2) as shown in Figure 16. The reaction order found in this study was 1.60 which is compared very favorably with the value of 1.57 for lower concentrations of Br_2 using the electrochemical method.

It was noted that at a high concentration of Br_2 such as 0.3 mol/dm^3, the rate of dissolution of gold was found to be very fast. The dissolution rate of gold was found to be about 1.1×10^{-2} mol/m^2.s for the concentration of Br_2 of 0.3 mol/dm^3 and 8.8×10^{-2} mol/m^2.s when the concentration of Br_2 was 1.0 mol/dm^3. These rates are about 120 fold and 1000 fold respectively larger than that of the cyanide system, which was 9.5×10^{-5} mol/m^2.s at 0.03 – 0.1 mol/dm^3 of NaCN and at pH 10 (Chimenos et al., 1996).

Figure 17 shows the dissolution current density vs. the bromine concentration at various pHs in ammoniacal solutions. When pH is less than 4.0, the bromine content in solution played a very important role on the gold dissolution rate, while at higher than pH 7.0, bromine concentration does not affect the gold dissolution rate. To maximize the gold dissolution rate, low pH and high bromine concentration are needed.

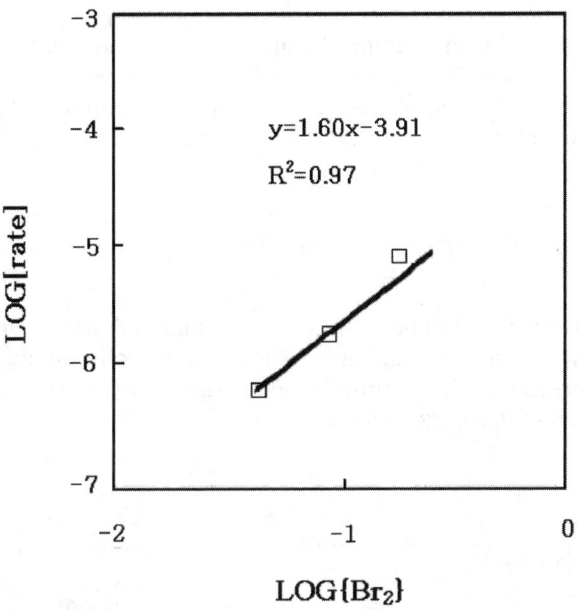

FIGURE 16 Schematic showing the gold dissolution rate determined by powder leaching test with respect to log bromine concentration. Temp.: 25°C; bromine conc.: 0.0375–0.3 mol/dm^3.

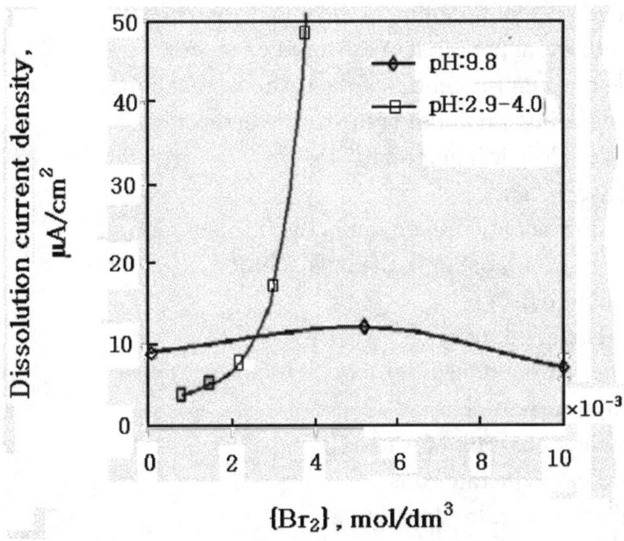

FIGURE 17 Plot of dissolution current density vs. bromine concentration at various pH. Temp.: 25°C;Total Ammonia: 1.0 mol/dm^3.

CONCLUSIONS

The leaching behavior of gold in ammoniacal solutions in the presence of bromine as an oxidant has been studied. It has been found that a combination of the anodic and cathodic transfer numbers serve as the reaction order. Such predicted reaction order was proven to be consistent with experimentally found values. The overall reaction rate was given by:

$$\text{Rate} = k_o(H^+)^{0.375}(Br_2)^{1.570}, \text{ unit: } \frac{mol}{dm^3 s} \quad k_o = 1.19\times 10^{-4}\frac{dm^{2.835}}{mol^{0.945} s}$$

Bromine has been found to be an excellent oxidant and also a complexing agent for gold leading to fast reaction rate. The lower the pH of the solution, the faster the dissolution rate. However, low pH helps bromine evaporate and results in significance loss of bromine in a relatively short time period.

NOMENCLATURE

A	area of sample
A_a	anodic area
A_c	cathodic area
$\{Br_2\}$	initial concentration of bromine in the bulk solution
C_+	concentration of the metal ion or oxidant in the bulk solution
C_{Pb}	molar concentration of products at bulk
E_a^o	the standard half-potential of the anodic reaction
E_a	the equilibrium potential of the anodic reaction
E_c^o	the standard half-potential of the cathodic reaction
E_c	the equilibrium potential of the cathodic reaction
E_r	reversible electrical potential
F:	Faraday constant
$\{H^+\}$	concentration of hydrogen ion in the bulk solution
[i]	activity of i component in the bulk solution
i	net current density
i_a	anodic current density
i_c	cathodic current density
i_o	exchange current density
K	the stability constant of $Au(NH_3)_2^+$
k_c	forward cathodic rate constant
k_a	forward anodic rate constant
k_o	overall rate constant
k_o'	overall rate constant coefficient
l,m	reaction orders with respect to individual species
n	the number of electron per mole of gold dissolved

R	gas constant
T	temperature in Kelvin
V	applied potential [v]
V_{mix}	mixed potential
β	cathodic transfer coefficient
1-β'	anodic transfer coefficient

REFERENCES

Aylmore, M.G. and Muir, D.M., 2001, Thermodynamic analysis of gold leaching by ammoniacal thiosulfate using Eh/pH and speciation diagrams," Min. and Metal. Proc. Vol. 18, pp. 221–227.

Breuer, P.L. and Jeffrey, M.I., 2004, The reduction of copper(II) and the oxidation of thiosulfate and oxysulfur anions in gold leaching solutions" Hydrometallugy, Vol. 70, pp. 163–173.

Bhuntumkomol, K., Han, K.N. and Lawson, F., 1980, Leaching behavior of metallic nickel in ammonia solutions, Trans. Inst. Min. Metall. Sect. C, pp. 7–13.

Chimenos, J.M., Segarra, M., Guzman, L., Karagueorguieva, A., and Espiell, F., 1996, Kinetics of the reacton of gold cyanidation in the presence of a thallium salt, Hydrometallurgy, Vol. 44, pp. 269–286.

Dasgupta, R., Guan, Y.C., and Han, K.N., 1997, "The electrochemical behavior of gold in ammoniacal solution at 75°C," Metall. Mater. Trans. B, Vol. 28, pp. 5–12.

Deschenes, G. and Ghali, E.,1988, "Leaching of gold from a chalcopyrite concentrate by thiourea," Hydrometallurgy, Vol. 20, pp. 179–202.

Grybos R. and Samotus A., 1984, Potential-pH diagram for the silver-water-ammonia-sulfuric acid system at elevated temarature," J. Less-Common Met. Vol. 98, No. 1, pp. 131–140.

Guan, Y. and Han, K.N., 1994, The dissolution behavior of silver in ammoniacal solutions with cupric ammine," J. of the Electrochemical Soc., Vol. 141, pp. 91–96.

Guan, Y.C. and Han, K.N., 1995, "An electrochemical study on the dissolution of copper and silver from silver copper alloys," J. Electrochem. Soc. Vol. 142, No. 6, pp. 1819–1824.

Guan, Y. and Han, K.N., 1996, "The electrochemical study on the dissolution behavior of gold in ammoniacal solutions at temperatures above 100°C" J. Electrochemical Soc., Vol. 141, No.6, pp. 1875–1880.

Halpern, J., Milants, H. and Wiles, D.R., 1959, "Kinetics of the solution of copper in oxygen-containing solutions of various chelating agents," J. Electrochem. Soc. Vol. 106, pp. 647–650.

Han, K.N., 2003, The interdisciplinary nature of hydrometallurgy," Met. Trans. B, Vol. 34, pp. 757–767.

Han, K.N. and Meng, X., 1992, "Extraction of Gold/silver from refractory ores using ammoniacal solutions," in proceedings of Randol gold Forum, Randol Intern., pp. 213–218.

Han, K.N. and Meng, X., 1992, 1994, "Ammonia extraction of gold and silver from ores and other materials," U. S. Patent No.5,114,687, U.S. Patent No.5,308,381.

Kirk, D.W., Foulkes, F.R., and Graydon, W. F., 1978, A study of anodic dissolution of gold in aqueous alkaline cyanide," J. Electrochemical Soc., Vol. 123, No. 9, pp. 1436–1443.

McArthur, J.W. and Forrest, R.W.,1889, "Process of obtaining gold and silver from ores," US Patents 403,202 and 418,137.

McGrew, K.J. and Murphy, J.W., 1985, "Iodine leach for the dissolution of gold," US patent 4,557,759.

Meng, X. and Han, K.N., 1994, Pub. 1995, " Adsorption of gold from iodide solution by activated carbon," Trans. Soc. Min. Metall. Explor. Vol. 296, pp. 31–36.

Meng, X. and Han, K.N., 1997, The dissolution kinetics of gold in moderate aqueous potassium iodide solutions with oxygen under pressure," Miner. Metall. Proc., Vol. 14, pp. 1–9.

Morioka, S., Sawada, Y., Shimakage, K. and Tsui, C.K., 1967, "Ammonia pressure leaching of metallic copper and nickel powder," Tohoku Daigaku Senko Seiren Kenkyusho Iho, Vol. 23, No. 2, pp. 155–166.

Pourbaix, M., 1974, *Atlas of Electochemical Equilibria in Aqueous Solution*, NACE, pp. 644.

Qi, P.H. and Hiskey, J.B., 1991, "Dissolution kinetics of gold in iodide solutions," Hydrometallurgy, Vol. 27, pp. 47–62.

Skibsted, L.H. and Bjerrum, J., 1974, "Gold complexes. I. Robustness, stability, and acid dissociation of the tetramminegold(III) ion," Acta Chem. Scand. Ser. A, Vol. 28, No. 7, pp. 740–746.

Skibsted, L.H. and Bjerrum, 1974, Gold complexes. II. Equilibrium between gold(I) and gold (III) in the ammonia system and the standard potentials of the couples involving gold, diammine gold(I) and tetramminegold(III)," Acta Chem. Scand. Ser. A, Vol. 28, No. 7, pp. 764–770.

Vu, C. and Han, K.N., 1977, Leaching behavior of cobalt in ammonia solutions," Trans. Instn. Min. Metall., Vol. 86, pp. 119–125.

Zipperian, D., Raghavan, S.,and Wilson, J.P., 1988, Gold and silver extraction by ammoniacal thiosulfate leaching from a rhyolite ore," Hydrometallurgy, Vol. 19, pp. 361–375.

Applied Separations—Energy Resource Recovery

Water Treatment Plant Modifications at Turquoise Ridge Joint Venture

John B. Ackerman* **and John G. Mansanti***

INTRODUCTION

Placer Dome's Turquoise Ridge Joint Venture operation is located some 75 km northeast of Winnemucca, a community in north central Nevada. Formerly known as the Getchell Gold Corporation and First Miss Gold, the operation mines refractory ore from the Getchell and Turquoise Ridge gold deposits located on the east flank of the Osgood Mountain Range above the Kelly Creek Basin in the Potosi Mining District. Mining has been ongoing at Getchell since shortly after its discovery in 1934, and was the nation's second leading producer of gold after Homestake in the early 1940s. Over its history the Getchell property has been mined for gold, tungsten, molybdenum and arsenic; mining methods varied from open pit mining to underground mining; processing options included conventional milling, flotation, roasting, heap leaching, and most recently pressure oxidation followed by CIL (carbon-in-leach). As an historical note, the US Bureau of Mines developed the use of carbon for gold recovery at Getchell during the 1940s (Horton 1999).

Placer Dome acquired the property in May 1999. The pressure oxidation plant was mothballed in July 1999 and the property was placed on care and maintenance in February 2002 pending further exploration results. The property is currently ramping up to full production status. A joint venture agreement (75% Placer Dome, 25% Newmont) was established with Newmont Mining in 2003 and ore is processed at their nearby Twin Creeks autoclave plant. Mine dewatering is a key element to success. Arsenic is a ubiquitous mineral in many of the mineralized areas of northern Nevada, necessitating treatment of water prior to re-infiltration.

---

* Turquoise Ridge Joint Venture, Golconda, Nevada

WATER TREATMENT SYSTEM

The Water Treatment Plant (WTP) was initially constructed in 1994 to treat 315 L/sec (5,000 gpm) of arsenic bearing mine waters produced from dewatering activities. Ferric hydroxide precipitation and adsorption was used as the primary treatment reagent. Soluble ferric iron is added to the mine water, which quickly precipitates as a ferric hydroxide in the near neutral pH environment. The ferric hydroxide has a tendency to adsorb ions and is particularly effective in removing arsenic ions, which is then removed by a liquid-solids separation step. The plant was quite straightforward and consisted of the following:

- An on-site surge and storage pond that accepted the mine water flows prior to the WTP
- Ferric sulfate reagent injection into the reaction tank feed stream
- A flocculant dosing system to promote particle growth and settling of the very fine ferric hydroxide precipitate
- A self-agitated process reaction tank, 7.3 m × 7.3 m, providing about 30 minutes of residence time. The agitation is provided by feeding the tank tangentially with the incoming flow.
- A settling pond to collect the resulting ferric hydroxide precipitate and adsorbed arsenic ions
- A pipeline, cement lined channels and natural creek drainage to transport treated solution to two sets of infiltration basins located in the Kelly Creek basin.
- Precipitated sludge from the settling pond was periodically pumped to the tailings impoundment

The surge pond ahead of the plant allowed for a constant flow to the plant; the only instrumentation needed was a flow meter recording plant throughput. The ferric sulfate and flocculant reagent pumps were adjusted manually when the flow rate was modified or the chemistry fluctuated.

The water chemistry was ideally suited for the process in that the mildly alkaline mine water was lowered to a pH of 6.5 to 7.0 by the acidic ferric sulfate reagent. This pH range was both suitable for arsenic removal and discharge compliance regulations. See Figure 1.

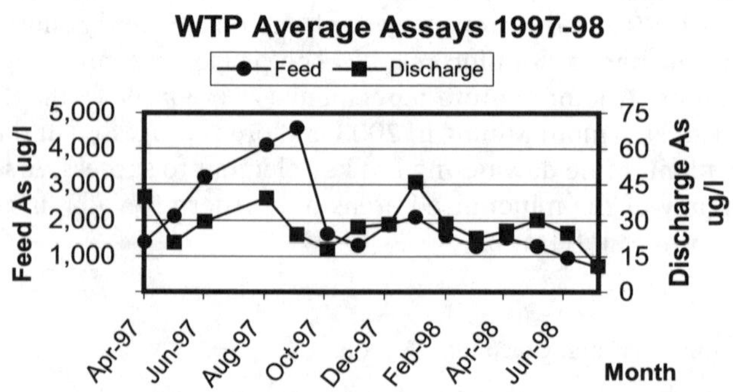

FIGURE 1 Feed & discharge to the WTP prior to modifications

Details of the plant parameters at this time were:

- The plant feed averaged 1,800 ug/l (micrograms per liter, ppb) arsenic.
- The arsenic was in the pentavalent state, which is ideal for the ferric sulfate treatment process.
- The surge pond, upstream of the WTP, provided residence time, and allowed for natural oxidation of any trivalent arsenic to the pentavalent state and also clarified the plant feed.
- The acidic ferric sulfate reagent typically lowered the process pH to 6.5, the minimum regulation for discharge.
- The 50 ug/l arsenic concentration requirement for discharge was achieved.

In mid 1998, a thickener was added ahead of the process to collect the mine sediments so the surge pond could eventually be retired in 2000 and reclaimed; the mine waters were then fed directly to the treatment plant. With the addition of the thickener, the process scheme remained much the same, but without the surge control the mechanical and chemical parameters became more difficult to control.

- The feed rates now modulated randomly with the cycling of the mine pumps. Flows varied from 0 to 190 L/sec (3,000 gpm).
- Without the benefit of blending and mixing in the surge pond the arsenic concentration state could vary from one thousand to several thousand ug/l on a daily basis.
- Without the opportunity to oxidize naturally in the surge pond, the trivalent arsenic concentration fluctuated.
- The Turquoise Ridge Mine was producing water of greater variability as development progressed.

The major changes with the addition of the thickener were:

- The 20 m diameter (65 ft) thickener directly accepted the varying mine water flows and removed the mine water sediments from the WTP feed stream. The overflow was then pumped to the process reaction tank using two variable speed clear water pumps that modulated along with the incoming flows. The collected mine sediments reported to the tailings impoundment.
- An upgrade to the flocculant system, to provide for automatic floc mixing.
- Addition of a simple Programmable Logic Controller (PLC) instrument. Both ferric sulfate and flocculant dosing varied directly with flow to the plant by varying the motor speed of the pumps. The ratio control (liters of reagent/1000 liters of water) remained under manual control and was adjusted by either changing the concentration of the flocculant mixture or the stroke length of the ferric sulfate feed pumps.

The metallurgists and operators were about to learn more about the ferric sulfate process of arsenic removal. This paper will share those insights, which include results from bench and pilot studies, the present operational results and the likely direction the process will follow in the near future.

During this time the reagent flow rates varied directly with the feed to the plant, but the dosage rate (L/1000L) was manually controlled. Fortuitously, the one flocculant, a medium molecular weight, low anionic charged polymer, served the dual purpose of settling the solids in both the thickener and the settling pond. With the removal of sediments at the thickener and the controlled addition of ferric sulfate treatment, the arsenic removal was largely successful. The feed concentrations were generally reduced from a range of 1,000 to 2,500 ug/l to a discharge concentration of less than 50 ug/l. This seemed to be the case until the early spring of 2000, shortly after the surge pond was abandoned and mine flows were pumped directly to the thickener.

UPSET CONDITIONS

In the spring of 2000 arsenic concentrations in excess of permitted levels began to appear in weekly and monthly compliance samples from an outside laboratory. Normal time lags in shipping samples to a certified outside laboratory and the concomitant delay in receiving assay results complicated the investigation process. This was further exacerbated due to internal assays indicating the plant performance was well within permit levels and quite steady.

While the cause of the upset was unknown, it became apparent that our analytical procedures were in question. Due to the lack of a readily available AA graphite furnace, water samples were allowed to accumulate at room temperature, for a week or more at a time. Given this period of time, arsenic in the water naturally reduced in concentration, giving a false impression that the plant was performing satisfactorily. Because the samples were analyzed for soluble arsenic and filtered prior to assay, any precipitation of arsenic that might have occurred in the samples was eliminated. This situation was resolved by purchasing a second AA and dedicating the original one to water treatment plant assays. This eliminated the time consuming set-up time required in switching a single machine from flame to graphite furnace operation. In addition, when samples could not be assayed the same day they were placed in refrigerated storage, which proved to be a straightforward method of limiting analytical discrepancies by slowing down the precipitation of the arsenic species.

This was a confusing time for the operation, in that these analytical issues had not been evident in the past, and had shown a reasonable variation of results averaging between zero and 50 ug/l arsenic. Operational issues at the lab in January and February delayed timely sample assaying, and led to the belief that the plant was generally doing well. The addition of the Turquoise Ridge Mine flows did not seem to have an immediate negative impact on the plant. However, coupled with the analytical issues it was difficult to determine precisely when the plant began to exhibit elevated arsenic discharges and whether the Turquoise Ridge Mine waters, caused immediate upsets or caused problems somewhat later as the chemistry of these waters may have changed. In either event, it was clear by March and April that the existing process was not performing as required (Figure 2).

After an initial investigation the upset conditions were suspected to have not only analytical, but operational and chemical origins. The plant performance was inconsistent, which implied the water treatment was successful much of the time. In retrospect the upsets were related to a series of issues that masked each other's effect.

Operational concerns influencing the plant performance included:

- Surging flows varying from zero to 190 L/sec
- Inconsistent feed concentrations varying an order of magnitude from 500 to 5,000 ug/l arsenic

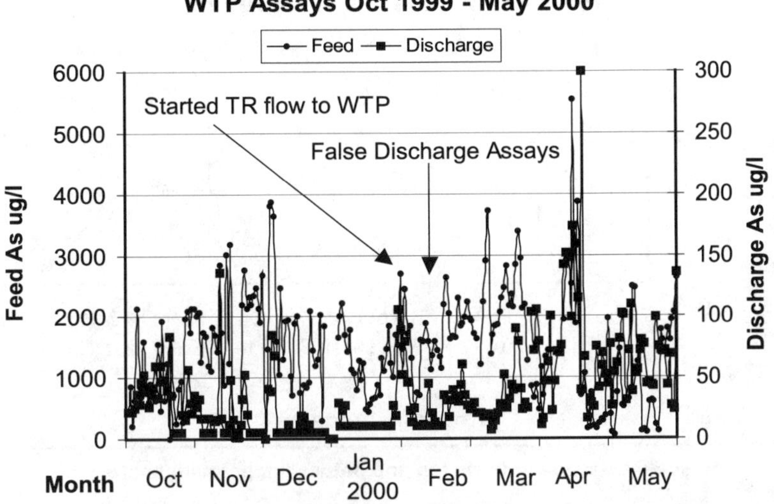

FIGURE 2 Upset conditions began to appear in April 2000 after the Turquoise Ridge flows were introduced to the water treatment plant

- Variable arsenic speciation which, qualitatively, ranged from zero to 30% plus three arsenic
- Sediment in the feed water and thickener overflow
- WTP retention time
- Intermittent thickener upset issues causing mine waters to overflow directly to the settling pond and by-passing the treatment process
- Dosage rate limitations and problems with the ferric feed pumps
- Quality of the ferric sulfate reagent
- Increased contribution of Turquoise Ridge waters from 40% to 70% of the feed
- New water sources from the Turquoise Ridge underground that were both three times as high as the average feed in arsenic and apparently 100% present as trivalent arsenic
- pH fluctuations in the feed from 7.5 to almost 11.0
- Suitable conditions for arsenic adsorption by the ferric hydroxide
- A high inventory of precipitated sludge in the sedimentation pond

FINDINGS AND RESOLUTION

While it was difficult to single out the factors responsible for the fluctuating results at the WTP, it appeared that the most likely culprit, in addition to mechanical shortcomings, was the presence of trivalent arsenic. It is well known that the tri-valent arsenic ion does not respond nearly as well to treatment as the penta-valent ion (Wang, Nishimura and Umetsu 2000). With a feed level of several thousand ug/l arsenic, even a relatively small percentage of trivalent arsenic could impact the plant performance requirement of

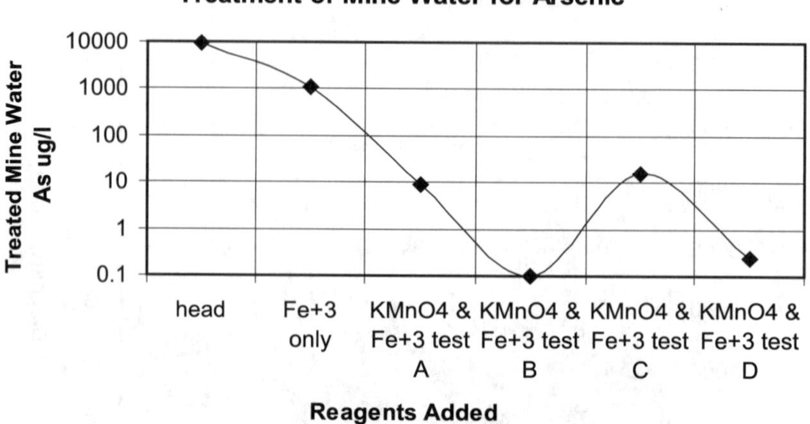

FIGURE 3 Bench scale tests using ferric ion and potassium permanganate

50 ug/l. Bench testing of potassium permanganate as an oxidant began as soon as speciation was considered as a problem. Potassium permanganate had been used successfully by one of the authors at Santa Fe's nearby Lone Tree operation and it seemed the most likely candidate causing the erratic discharge results. Also, information obtained from the Minor Elements Symposium at the 2000 SME Annual meeting was very timely for Getchell's needs.

The reaction for the oxidation of tri-valent arsenic with potassium permanganate is given as (Carus 2004):

$$3As^{+3} + 2MnO_4^{-1} + 4H_2O = 3As^{+5} + 2MnO_2 + 8OH^{-1}$$

Applicable bench scale testing was directly related to the presence of the trivalent arsenic ion, in that, water free of this species was relatively easy to treat both in the plant and on the bench. Because the processing problems were intermittent, it was challenging to procure plant feed water samples that were difficult to treat. At the same time, plant feed samples containing the trivalent arsenic ion, stored for one or two days also appeared to oxidize naturally and became easily treated, so the test work had to be done when suitable samples were obtained. Initial results indeed indicated the issue was trivalent arsenic that randomly upset the plant performance.

Figure 3 summarizes this testing and illustrates several points:

- By testing a sample with and without permanganate, one can qualitatively determine if both species of arsenic are present. If only the pentavalent species is present the treatment will not improve with permanganate addition.
- The case shown here is an example of exceptionally high arsenic in the mine water. It is relatively high in trivalent arsenic, since 10% of the arsenic remains after ferric treatment.
- The permanganate is apparently successful in oxidizing the trivalent arsenic, since the treatment is very successful following permanganate addition.

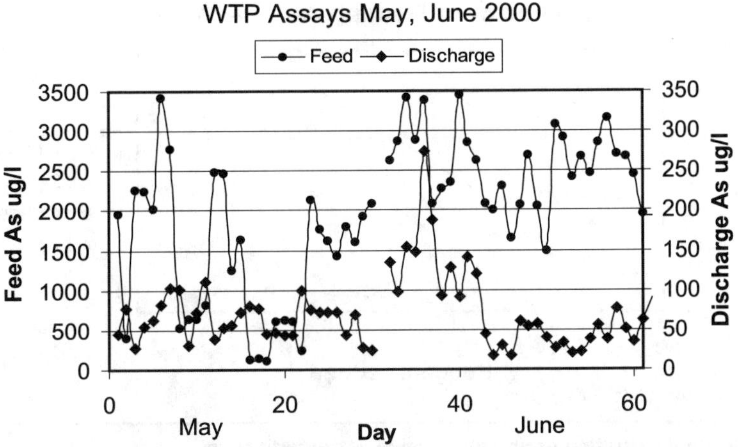

FIGURE 4 Graph of May & June 2000, showing limited but encouraging results, after the potassium permanganate pilot system was installed in early June 2000

- The experimental test results can be challenging to reproduce in the zero to 10 ug/l range. However, the results are encouraging in that both experimental and plant results have shown that arsenic can be lowered to less than 10 ug/l.

Figure 4 shows the natural oxidation of arsenic that can occur in the mine waters with time. A fresh sample requires the addition of potassium permanganate prior to ferric treatment to lower the arsenic level to 10 ug/l and less. A sample aged for several days, shows little difference with and without oxidant use.

In collaboration with the Nevada Department of Environmental Protection (NDEP) a pilot project, adding potassium permanganate to the plant feed, was commissioned in June 2000, to establish proof of concept. Potassium permanganate was batched at 3% solution by weight and injected into the Turquoise Ridge water line as it emerged from the shaft. This had the advantage of providing a distance of one kilometer to promote mixing and provide residence time for the reaction. While the mine waters at this point were sediment laden, the bench scale test work had not shown this to be a particular issue. The Getchell underground mine water was not treated as it had not exhibited the treatment problems indicative of trivalent arsenic.

The results were encouraging, with the water treatment plant showing improvements, producing more consistent results below the 50 ug/l arsenic level, but the results continued to be erratic (Figure 4). Indeed, only the trained eye of a diligent observer could notice the improvement from May to June. Equipment problems, coupled with other factors listed above easily masked the improvement. When all was running as expected, one could see extended times when the arsenic in the discharge was held below 50 ug/l, even when the feed concentrations were relatively high in tri-valent arsenic.

Figure 5 also illustrates a qualitative test used on site to determine the presence of the trivalent arsenic ion. When side-by-side tests using only ferric are compared with a test using both potassium permanganate and ferric, one can determine if appreciable trivalent arsenic is present. If the assay results are similar then the oxidant is not necessary and the speciation is primarily pentavalent arsenic. A large difference provides a

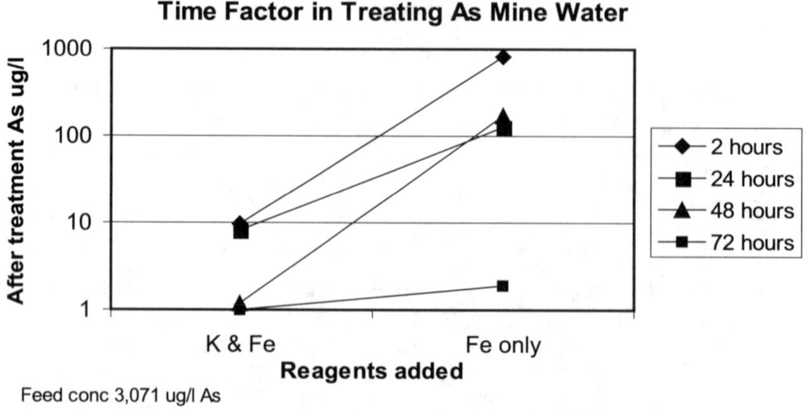

FIGURE 5 Natural oxidation of arsenic in mine water with time

qualitative reference for the presence of reduced arsenic species and the impact it may have on the process.

One of the advantages of piloting this system with actual feed streams, was the results could be monitored over a period of time to assess the performance on actual mine water feed. The expected improvement however was not long lived as shown in Figure 6. Sudden and unexplainable spikes continued to plague the plant along with periods of noticeable improvement. Some of these upsets could be explained by mechanical problems with the permanganate reagent pump and possibly the sediment loading was a factor, not apparent in the bench scale work. One perceptive operator noticed a seldom used dewatering well appeared to have an impact on the plant performance. He also noticed it had a distinct hydrogen sulfide smell. Indeed this became the obvious culprit contributing reducing conditions to the flow, and when turned off, the plant performance improved (Bennett 2000). This was a process variable that had not been fully appreciated up to this point. Reducing conditions could be a fatal flaw to plant performance, and could lead to high oxidant use in that sulfide ions, in this case, would have to be oxidized as well as the trivalent arsenic ion, in order to achieve the pentavalent arsenic oxidation state.

The settling pond had not been dredged since inception. Prior to the knowledge of oxidation the only tool available was ferric sulfate, the iron dosage had been increased significantly during the year, and the pond, already short of available settling volume, quickly reached its limit. A contract dredge was brought in during August and September of 2000, to remove the settling pond sediments. This disturbance contributed to erratic discharge results and it was not until early October that the addition of permanganate could realistically be evaluated. As portrayed in Figure 7, the plant performance evened out during the last quarter of 2000, with most assays below 50 ug/l, and many below 10 ug/l. This occurred during a time when the feed concentrations were on the rise.

The pilot set-up provided minimal control at the mine location, which was remote from the WTp. With the impact of minor amounts of free sulfide ions now evident, discussions with others and some results from early test work prior to 1994, it became obvious that free sulfide minerals in the mine water sediment could be a factor. Additional test work showed the residence time required for the permanganate oxidation was very

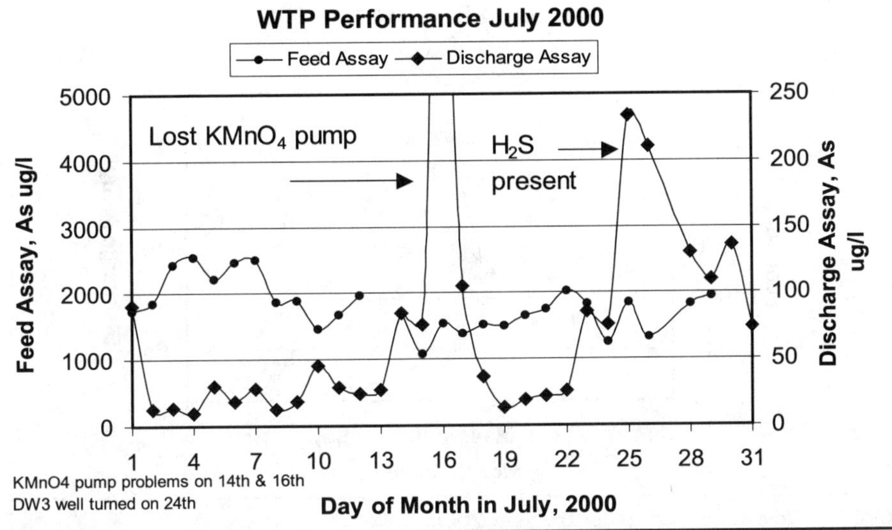

FIGURE 6 Upset conditions in July 2000 following addition of the permanganate system

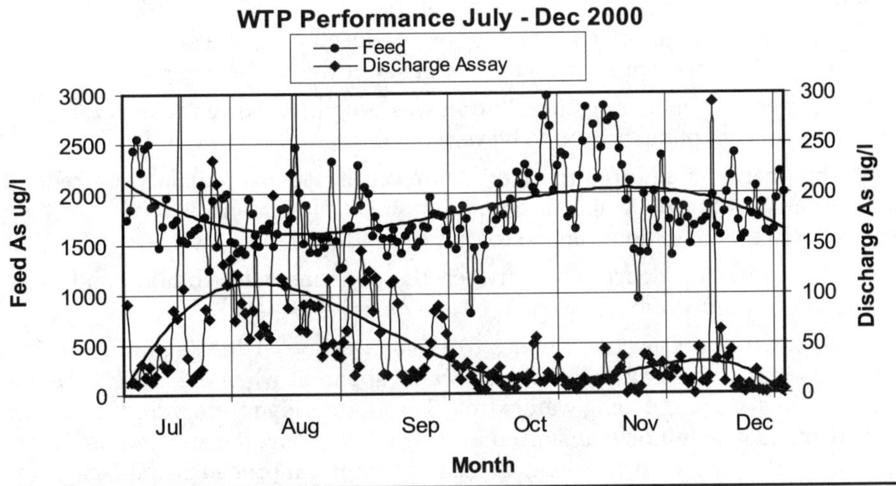

FIGURE 7 WTP performance for latter half of 2000

short–on the order of seconds. With no apparent need for additional tankage and residence time at the WTP, the potassium permanganate pilot unit was relocated to the WTP in early October, where the reagent was added to the thickener overflow on the suction side of the transfer pump.

The original permanganate reagent pump, scavenged from a boiler feed water installation, was replaced with a more appropriate reagent pump and was ratio controlled to the plant feed rate. A redox probe was installed to provide for a measure of

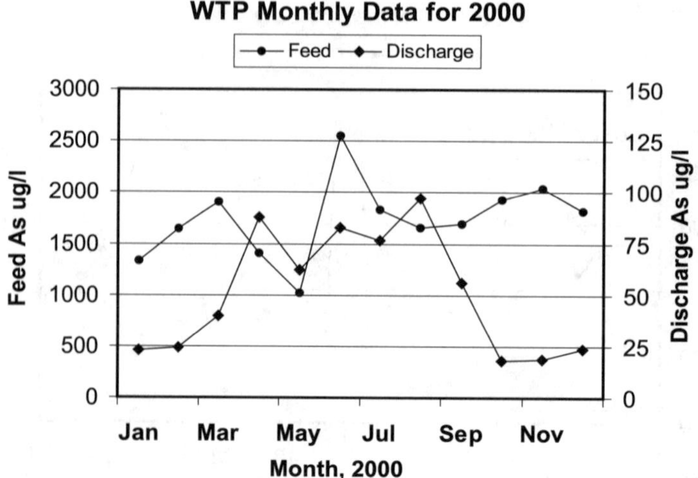

FIGURE 8 Summary of results for the year 2000

manual operator control. Again, there was an improvement in results. The reasons were probably four fold:

- The sediments in the mine water were reduced from the one percent level to a relatively consistent 30 mg/l by treating after the thickening stage
- The permanganate reagent addition was proportioned to the feed rate, and the new reagent pump proved to be very reliable.
- The operators adjusted the stroke rate on the pump to maintain a redox level around +400 mv. While this was a manual adjustment, it proved to be quite effective as conditions tended to change gradually.
- The shorter residence time between the permanganate addition and the ferric sulfate addition did not seem to have a negative impact.

Looking at 2000 in perspective (Figure 8) the low assays at the beginning of the year were due to inaccurate assays. Problems became apparent when the assay discrepancies were resolved and the reducing waters from Turquoise Ridge were introduced in February. Permanganate addition was started in June to overcome the apparent impact of the tri-valent arsenic. Between June and October, mechanical problems and lack of control limited the success of the oxidant, while settling pond dredging in August and September upset the plant performance. Finally, in October, the addition of permanganate showed its success for the remainder of the year.

The average arsenic discharge for the last three months of 2000, was 20 mg/l. Figure 7 shows the exceptionally low discharge values that occurred as well, 29 out of 92 days were under 10 ug/l arsenic.

Controlling pH became the next key factor. As development mining at the Turquoise Ridge mine continued, difficult ground conditions led to increased use of shotcrete for ground stability and cement grouting to control water seepage. For the first time, the pH of the mine waters feeding the WTP rose above 9.0. A pattern emerged, that process discharge pH values much above 7.0 resulted in higher arsenic concentrations.

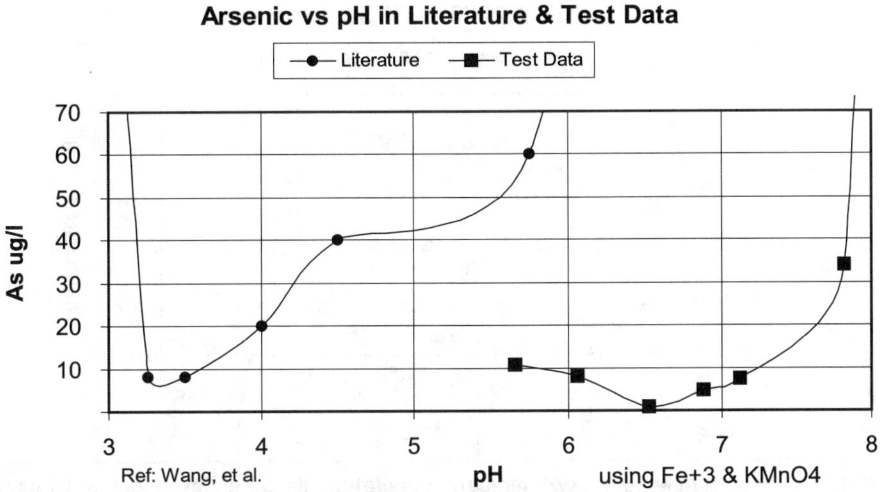

FIGURE 9 Laboratory results compared to results in the literature for the removal of arsenic in water at varying pH values

The operational results appeared to be contrary to results shown in the literature that indicated, at a pH of 6 to 7, the arsenic in solution would be well above 100 ppb and that lower pH values would be necessary to bring the arsenic concentration down to 10 mg/l (Wang, Nishimura and Umetsu 2000; Bowell 2003). A series of bench tests was performed to quantify the results of the plant. These results illustrated in Figure 9, showed a somewhat differing relationship than others had found. Both the plant data and the laboratory work showed that operating at a neutral pH up to about 7 provides suitable conditions for arsenic removal to levels below 10 mg/l. Others have also showed this as well (Nishimura and Umetusu 2000; Ball and Brix 2000; Bowell 2004). This discrepancy is one that should be further defined in optimizing the most effective and cost efficient conditions for arsenic removal.

To accommodate the higher pH feed, the acidic ferric sulfate reagent was increased to maintain control of the final discharge pH. This was not considered a viable longer-term solution and the use of carbon dioxide for pH adjustment was tested. Carbon dioxide has been widely used for pH control and was piloted at Newmont's Twin Creeks water treatment plant operation across the valley. Bench scale work was initiated and showed that the reaction for lowering the pH was effective and proceeded rapidly. This led to a pilot installation of CO_2 injection at the WTp. Since this time, elevated mine water pH feeds have been rare, and the system has had little opportunity to prove itself.

The primary reasons to consider carbon dioxide were:

- Does not add sulfates to the water, the concentrations of which were already relatively close to the permit limitations of 500 mg/l
- Requires less capital to install than a sulfuric acid system, and does not have the risks associated with handling a concentrated acid
- The pH control is more forgiving, in that excessively low pH's cannot be reached

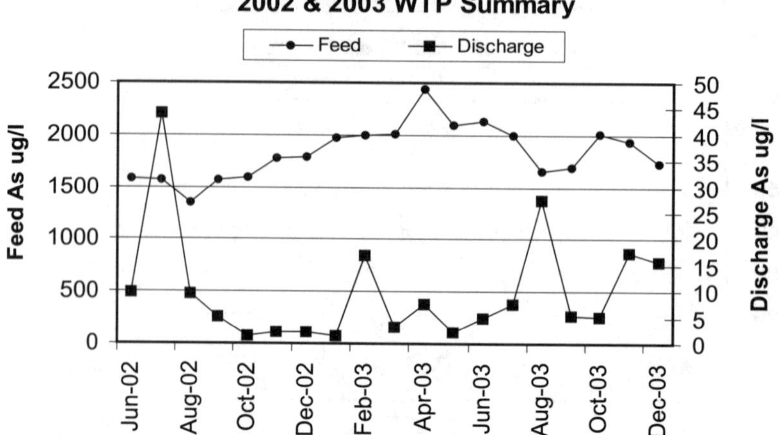

FIGURE 10 WTP performance in 2003 indicating operating discharge levels below 10 ug/l arsenic is possible

The results discussed above led us to believe that controlling the plant discharge at less than 10 mg/l arsenic might be possible. This was confirmed in both 2002 and 2003 as illustrated in Figure 10.

One factor that contributed to the enhanced performance in the years 2002 and 2003 is the reduced activity in the mines. The property ceased ore production during this time and entered into a period of exploration and development. While the water chemistry was more consistent, and the waters were generally oxidized, it showed that a reduction by more than two orders of magnitude and achieving less than 10 ug/l arsenic was certainly feasible.

The most recent data from 2004 shows the water treatment plant continues to be resilient even as the water chemistry becomes variable and challenging again (see Figure 11). The Getchell Main Underground mine is at its maximum mining rate and the Turquoise Ridge Mine is ramping up to full production. As a result, the mine water chemistry has become more variable. The average discharge level for 2004 has been 15 ug/l (Jan–Oct), and the plant has been below 10 ug/l arsenic 60% of the time. This performance has been achieved without any additional controls and at the same time upgrade construction projects at the water treatment plant this year, have occasionally caused upset conditions.

Looking back, the cause of the discharge excursions became more evident: these upsets could generally be traced back to mechanical or control problems or discharge pH values above 7.5. The important distinction to be emphasized is that the process seems to be successful if the parameters are controlled. This can be seen in the summary chart shown in Figure 12, where discharge arsenic concentrations have been held in the 10 ug/l range in a rising feed concentration environment.

The Water Treatment Plant is currently undergoing a series of improvements. For 2004 and 2005, these include the following:

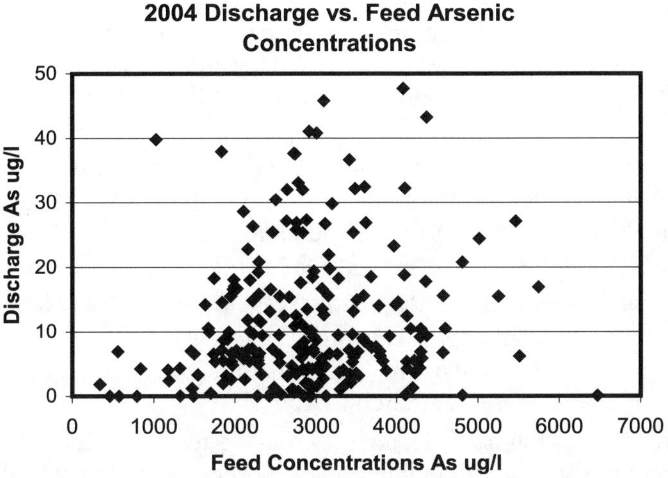

FIGURE 11 Scatter plot showing the mine water feed variability and the resulting discharge arsenic concentration

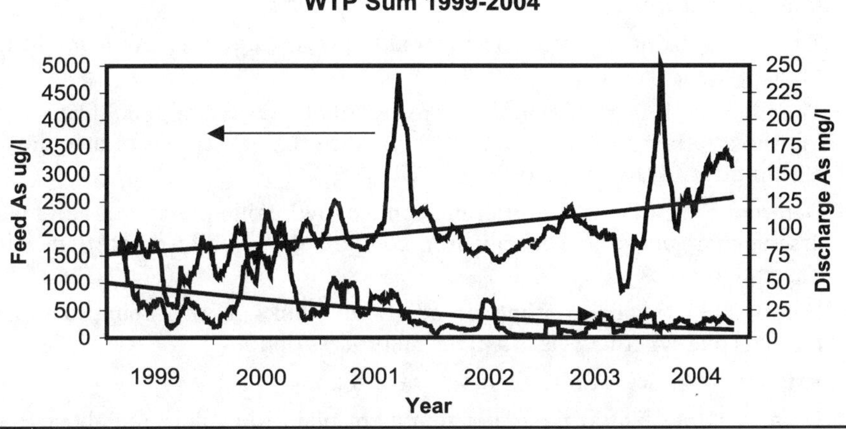

FIGURE 12 Summary of water treatment plant performance, 1999–2004

- Installation of a second settling pond to allow one to be taken off line for maintenance and cleaning. This will alleviate the upset conditions seen during past pond clean-out operations. The ponds are also arranged to allow series flow, to provide additional settling time for the ferric hydroxide precipitate.
- Addition of a thickener feed box to provide for flocculant blending with the feed and allow recycle of the underflow sediments to promote clarity of the overflow.
- Addition of a carbon dioxide injection system to lower the pH of the mine waters when excessive alkaline conditions are encountered.

- Addition of sodium hydroxide to the final discharge. This will bring the pH of the discharge within the compliance limits of 6.5 to 9.0, when and if the discharge drops below this level due to higher arsenic concentrations requiring additional ferric sulfate reagent and/or the addition of carbon dioxide to optimize the process.
- Installation of a new programmable logic controller (PLC) to monitor all aspects of the plant and provide more sophisticated control of the process parameters.
- Additional instrumentation for pH and Eh measurements. This will provide data to the PLC for improved process control.
- Additional instrumentation to quantify reagent flows. This will allow the operators to more precisely control reagent additions and costs.
- Alarm annunciation to alert the operators of a plant upset since the plant runs without full time operator attention.
- Modifications to the tank elevation to provide for gravity process flow through the system. This would eliminate plant upsets associated with pump maintenance and power interruptions.

Additional research might include the following:

- The need for additional mix tanks, primarily to control short-circuiting of flows.
- The use of ferric hydroxide sludge recirculation to enhance the arsenic adsorption. Limited work has shown this not to be effective.
- The in-situ reduction of ferric hydroxide precipitate and how it might impact final discharge results.
- The use of ferrous sulfate in lieu of ferric sulfate. Oxidizing the ferrous iron during the treatment to ferric iron may enhance the adsorption of arsenic (Wang, Nishimura and Umetsu 2000).
- Determine the role of the different iron oxy-hydroxide precipitates and whether arsenic-manganese and arsenic-iron compounds can be formed that will optimize the process.
- Determine if residence time or additional tanks in series improves arsenic removal and whether this is a practical process option.
- Explore this process for the removal of antimony.
- Install a pilot plant at the water treatment plant to explore the above items on the actual and varying plant feed.
- Develop a specific ion probe for arsenic.

Once the mechanical and control aspects of the plant become robust, it is the expectation of the authors that the process may be able to meet the more stringent discharge regulations that are likely to be implemented in the future. For mining operations with sludge disposal facilities, this type of water treatment may prove to be ideal for control of arsenic in water discharge streams. The primary flow diagram of the process is illustrated in Figure 13.

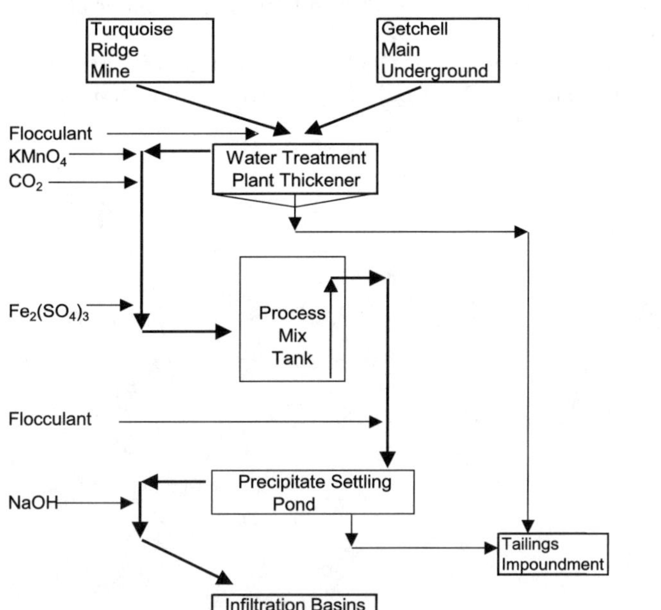

FIGURE 13 Current water treatment plant process

SUMMARY

A variety of analytical and operational issues contributed to elevated arsenic in the Getchell WTP discharge when a change to the Turquoise Ridge water chemistry was encountered. In cooperation with the Nevada Department of Environmental Protection, Getchell Gold was able to test, pilot and implement process modifications that successfully addressed the operational and chemical issues resulting in elevated arsenic levels. The utilization of potassium permanganate to oxide trivalent ions was key to this success. The control of pH is important to the success of the ferric iron, arsenic adsorption system.

REFERENCES

Ball, Brandon R. and Brix, Kevin V. Passive Treatment of Metalloids Associated with Acid Rock Drainage, Parametrix. *Minor Elements 2000,* Edited by Courtney Young. SME. 96, 97

Bennett, Vance. Getchell Gold Corp, July, 2000. Internal communication.

Bowell, R. The Influence of Speciation in the Removal of Arsenic from Mine Waters. *Land Contamination and Reclamation, 2003,* EPP Publications Ltd.

Bowell, R. 2004. Private communication.

Carus Chemical Co. Technical Brief. revised 2004.

Horton, Robert C. History of the Getchell Gold Mine, *Mining Engineering Journal.* SME. July 1999. 50–56.

Nishimura, T. and Umetusu, Y. 2000. Chemistry of Elimination of Arsenic, Antimony and Selenium from Aqueous Solution with Iron (III) Species. *Minor Elements 2000.* Edited by Courtney Young. SME. page 105–112.

Wang, Q., Nishimura, T., Umetusu, Y. Oxidative Precipitation for Arsenic Removal in Effluent Treatment. *Minor Elements 2000*, Edited by Courtney Young. SME. pp 39–52.

Removal of Selenium Oxyanions from Mine Waters Utilizing Elemental Iron and Galvanically Coupled Metals

Larry Twidwell,* Jay McCloskey,† Helen Joyce,† Eric Dahlgren,* and Andy Hadden‡

The reduction of dissolved selenium oxyanions from mine wastewater utilizing an elemental iron ce mentation technology has been studied on a laboratory scale at Montana Tech of The University of Montana and on a pilot scale at MSE-Technology Applications as a water treatment process potentially capable of removing dissolved selenium to <50 μg/L. Laboratory and pilot scale studies using elemental iron have demonstrated effective removal of dissolved selenium. Enhanced selenium reduction rates have been demonstrated when galvanically coupled iron/copper and iron/nickel are utilized. The laboratory and pilot scale results are presented and discussed.

INTRODUCTION

An extensive review of the literature has been previously performed by Montana Tech for the EPA Mine Waste Technology Program (MWTP) to identify potential technologies for the removal of selenium from a variety of wastewaters (MWTP 1999). The results of the literature review showed a need for the development of new technologies that are capable of achieving selenium removal to less than 50μg/L. The studies described in this presentation were conducted in response to the noted need.

* Montana Tech of The University of Montana, Butte, Montana

† MSE-Technology Applications, Butte, Montana

‡ Newmont Mining, Carlin, Nevada

TABLE 1 Illustrative examples of selenium sources

Mine and Process Waters	µg/L
Coal mine groundwater	3–330
Gold mines, CA, MT, UT, WY	200–33,000
Uranium mine discharge, NM	1,600
Oil Refinery	
Wastewater	170–4,900
Stripped sour water	1,000–5,000
Smelters	
Lead smelter scrubber water	3,000
Lead smelter water treatment overflow	1,600
Copper smelter groundwater	2,000
Surface water	
CA	10–4,200
MT	1–560

Selenium U.S. EPA MCL = 50 µg/L

Brief illustrative examples of potential selenium water problems are presented in Table 1. Extensive source data and references are presented elsewhere (Twidwell et al. 2002; Dahlgren 2000; MWTP 1999).

Selenium occurs in the environment in large quantities, but in low concentrations, in areas of sulfide mineralization and is a problem for non-ferrous metal producers of gold, copper, lead, uranium, and zinc. Selenium contamination problems are abundant in scrubber blowdown solutions, acid plant wastewater, and gas cleaning plant water. Additionally, selenium contamination is a problem in agricultural reservoirs (Chamberlin 1996). Unlike other anions and cations present in wastewaters, conventional lime treatment does not effectively remove selenium and further water treatment is necessary (Mizra and Ramachandran 1996; MWTP 1999; Twidwell et al. 2002). Also, the U.S. EPA's designated Best Demonstrated Available Technology (BDAT; Rosengrant and Fargo 1990) is ineffective for achieving final selenium concentrations <50 µg/L (MWTP 2001).

Selenium is found in the environment as *selenide* (H_2Se, HSe^-), *selenium* (Se^0), *selenite* ($H_2SeO_3^o$, $HSeO_3^{1-}$, SeO_3^{2-}), or *selenate* ($HSeO_4^{1-}$, SeO_4^{2-}) depending on the pH and oxidation/reduction conditions. STABCAL (Huang 2000), a thermodynamic modeling program developed at Montana Tech, was used to model the stability regions for the various selenium species shown in Figure 1. The *selenate* and *selenite* species are responsible for most selenium compound solubilities and are responsible for the contamination of most wastewater. *Selenite* is found in mildly oxidizing conditions while *selenate* is found in stronger oxidizing conditions. Both species of selenium are found over a broad pH region, but *selenate* predominates in most natural and mine waters (Chamberlin 1996).

THERMODYNAMIC CONSIDERATIONS FOR SELENIUM REMOVAL

The present studies were based on the removal of selenium, present as *selenate,* from wastewater via reduction to elemental selenium using elemental iron as the reductant. Elemental iron was chosen because of its availability, relative low cost (compared to

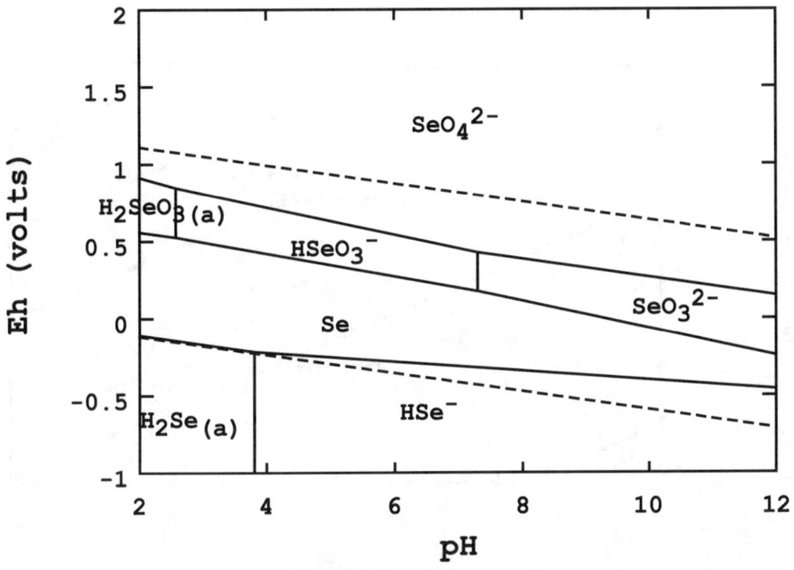

FIGURE 1 E_H/pH diagram showing the stability regions for various selenium species: 0.050 mg/L Se (Thermodynamic data from Allison et al. 1991; diagram constructed using STABCAL, Huang 2000)

other reducing agents), ease of handling, and its ability to produce a low reduction potential in an aqueous solution. Metallic iron can reduce either *selenate* or *selenite* to various solid products (FeSe, $FeSe_2$, or Se^o) depending on the pH and the oxidation/reduction potential of the solution. All experimental work described in this paper considered the reduction of *selenate* because it is the more difficult aqueous specie of selenium to reduce (Koyama et al. 2000). An E_H/pH diagram for the iron/selenium/water system is presented in Figure 2. Note that elemental iron controls the potential in the system to levels conducive for the reduction of selenium oxyanions to a selenium bearing solid.

LABORATORY STUDIES

The results of three studies will be summarized in this presentation: the use of elemental iron as the reductant for removing selenium from solutions (Dahlgren 2000; Cockhill 2002) and the use of galvanic couples for removing selenium from solutions (Hadden 2002).

Elemental Iron Reductant

Selenium occurs in solution as oxyanions. These oxyanions can be reduced by electrochemical reactions. Metals can be used as the reductant. Any metal that has a reduction potential that is less noble than the impurity reduction potential will thermodynamically reduce the impurity specie, e.g., those metals that will reduce selenium oxyanions include iron, zinc, aluminum, and magnesium. As discussed previously thermodynamically one can predict what metals are potentially acceptable for decreasing solution aqueous species

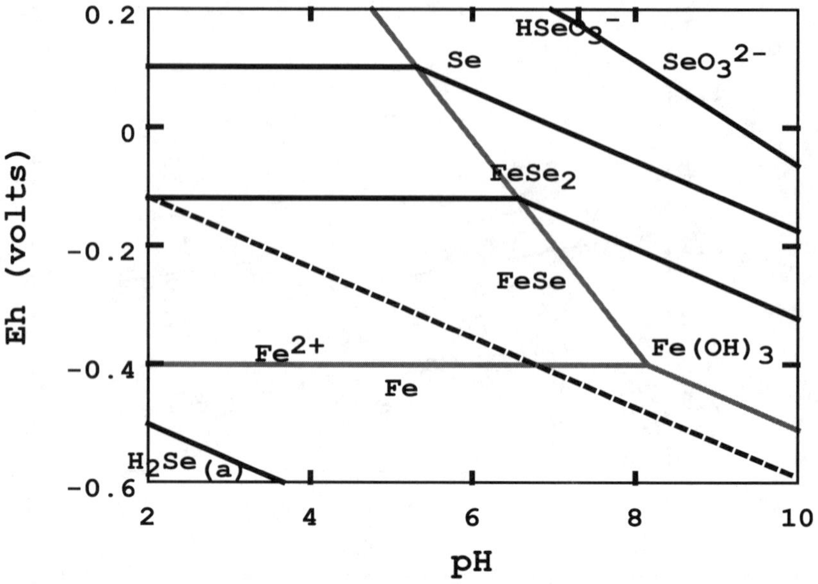

FIGURE 2 E_H/pH diagram for the iron/selenium/water system; 100 g/L Fe, 0.05 mg/L Se (Data from Allison et al. 1991; diagram constructed using STABCAL, Huang 2000)

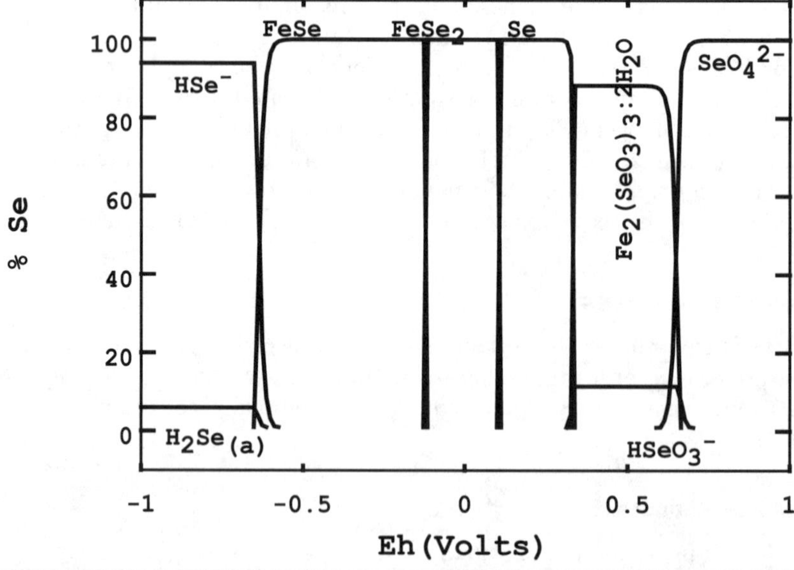

FIGURE 3 Distribution/E_H diagram for the iron/selenium/water system; 100 g/L Fe, pH 5, 0.05 mg/L Se (Data from Allison et al. 1991; diagram constructed using STABCAL, Huang 2000)

concentration. The presence of elemental iron in a slurry reactor or column reactor should provide a potential sufficiently reducing to remove selenium oxyanions as either elemental selenium or iron selenide. This effect is predicted from the diagram presented in Figure 3. Note that selenium oxyanions should be completely converted to FeSe at a pH of 5 and a potential of approximately −0.5 volts.

Dahlgren investigated the use of elemental iron to remove selenium from synthetic wastewater. Laboratory work was conducted using a two-liter reaction kettle containing one liter of solution. The agitation rate, reagent addition, pH (3–9), reduction potential, gas shrouding (N_2), and temperature (25–65°C) were controlled or monitored as a function of time. Samples were extracted as a function of time (5–60 minutes), filtered through 0.45 micron high density polyethylene filter disks, preserved with nitric acid and analyzed by Induction Coupled Plasma Atomic Emission Spectrometry (ICP-AES for iron) and Graphite Furnace Atomic Adsorption (GFAA for selenium). Experimental variables investigated included: initial selenium concentration (2–10 mg/L), temperature (25–65°C), amount of iron (1–100 g/L), size of iron (−20 mesh, −8/+20 mesh), presence of sulfate (0–10 g/L), effect of shrouding with N_2 or not, and time of exposure. The results are summarized in Table 2. Example diagrams illustrating the effect of some of the variables studied are presented in Figures 4–8 (unless otherwise stated the conditions used to generate the data presented in the figures are stated in the note associated with Table 2).

The conclusions from Dahlgren's study were: selenium was successfully removed from synthetic selenium waste solutions to <5 µg/L in 30 minutes or less. The variables that had a significant effect on the removal of selenium were: the amount of metal reductant, the particle size of the reductant, time and the treatment pH. The test results using −20 mesh iron particulate at 100 g/L over the pH range 5–7 for initial selenium concentrations of 2–10 mg/L are described by the following equation showing the importance of second order interaction effects:

$$\text{Log [Se]}, \mu\text{g/L} = 0.591 + 0.311{*}\text{pH} + 0.049{*}\text{time (minutes)} - 0.212{*}[\text{Se}_{\text{initial}}]\ \text{mg/L} - 0.011{*}\text{pH}{*}\text{time (minutes)} + 0.059{*}\text{pH}{*}[\text{Se}_{\text{initial}}]\ \text{mg/L} - 0.0033{*}\text{time}{*}[\text{Se}_{\text{initial}}]\ \text{mg/L}$$

The experimental study also demonstrated that sulfate additions and the initial selenium concentration had only a small effect on the removal of selenium from synthetic wastewater. Scanning Electron Microscopy/Energy Dispersive X-ray (SEM/EDX) studies demonstrated that the selenium reduced product occurred primarily at the iron particle edges and in crevices on the iron surface.

Cockhill (2002) conducted a preliminary investigation of the iron/selenium system using a column reactor. The column reactor, 22 mm × 35 cm, was constructed of plexiglass. The column was loosely packed with −20 mesh iron particulate. The solution phase was pH adjusted prior to entry into the column and was pumped in an upflow direction through the iron mass. The solution potential was measured continuously and sampled periodically, preserved and analyzed for iron by ICP-AES and for selenium by GFAA. Two residence time studies were conducted, one at 30 minutes and the second at 15 minutes. The conditions and results for the test work are presented in Table 3. The test results demonstrate effective removal of selenium to <1 µg/L for the entire test period.

TABLE 2 Summary of experimental test work conducted by Dahlgren (2000): characterization of the iron/selenium system

	Summary of Data and Comments	
Variable	Iron	Selenium
Agitation	Mass transfer rate maximized: 1100–1400 rpm	Mass transfer rate maximized: 400–1400 rpm
Agitation rate chosen for Dahlgren study: 1200 rpm		
pH	pH 3, 3364 mg Fe/L; pH 5, 269 mg Fe/L; pH 7, 47 mg Fe/L; pH 9, 1 mg Fe/L	pH 3, 6 μg Se/L; pH 5, 4 μg Se/L; pH 7, 6 gSe/L; pH 9, 1100 μg Se/L
pH chosen for study: 5 or 7		
Time	pH 5–60 min, 269 mg Fe/L; pH 7–30 min, 46.9 mg Fe/L	pH 5–30 min, 4 μg Se/L; pH 7–30 min, 6 μg Se/L
Time chosen for present study: samples collected at 5, 15, 30 (or 45), 60 minutes		
Initial Se		Same result for 2 mg Se/L and 10 mg Se/L, i.e., 4 μg Se/L achieved at 30 minutes
Initial Selenium chosen for present study: 2 mg/L		
Fe Size	–8/+20 mesh: 20.4mg Fe/L; –20 mesh: 269 mg Fe/L	–8/+20 mesh: 29 μg Se/L; –20 mesh: 4 μg Se/L
Fe size chosen for present study: -20 mesh		
Fe Amount	10 g Fe/L, 10.9 mg Fe/L; 100 g Fe/L, 269 mg Fe/L	10 g Fe/L, 30 μg Fe/L; 100 g Fe/L, 4 μg Fe/L
Fe amount chosen for present study: 100 g/L		
Temp	pH 7: 50°C, 95.2 mg Fe/L; 25°C, 46.9 mg Fe/L	pH 7: 50°C, 4 μg Se/L; 25°C, 6 μg Se/L Activation energy: 89 kj/mole
Temperature chosen for present study: 25 °C or ambient		
N_2	Without N_2, 199 mg Fe/L; with N_2, 269 mg Fe/L	Without N_2, 10 μg Se/L; with N_2, 4 μg Se/L
Nitrogen sparging chosen for present study		

Variables that were held constant (except for the listed variable) included the following: initial selenium 2 mg/L, agitation 1200 rpm, iron platelets 100 g/L, size –20 mesh (–0.83 mm), temperature 25°C, pH 5, time 30 minutes, 1.8 L N_2/minute. Reproducibility under constant standard constant conditions (triplicate samples): Fe 269 ± 5 mg/L; Se 4 ± 2 μg/L.

At the end of the above described twenty-three day test study the pH of the influent solution was adjusted to pH 7.0–7.3. The pH of the effluent solution was 8.4–9.0. The result of this change was to passivate the iron in the column and selenium removal stopped.

The conclusions from Cockhill's column test work were: effective selenium reduction occurs (<1 μg/L) in relatively short residence times but there are potential problems to be aware of: low pH solutions dissolve iron at the bottom of column, and since the pH in column increases with column height ferrous hydroxide may precipitate in the upper region of the column which would require pH control in the column or periodic back-flushing to remove the precipitated ferrous hydroxide.

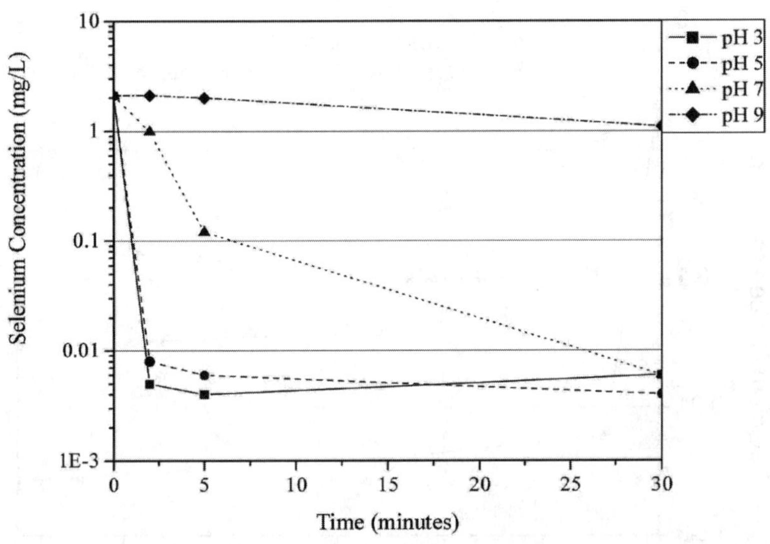

FIGURE 4 Influence of pH on final selenium concentration (Dahlgren 2000)

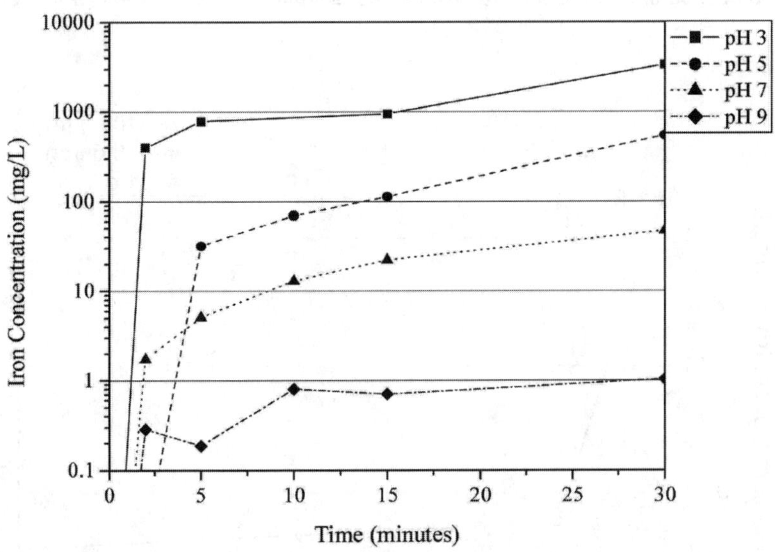

FIGURE 5 Influence of pH on final iron concentration (Dahlgren 2000)

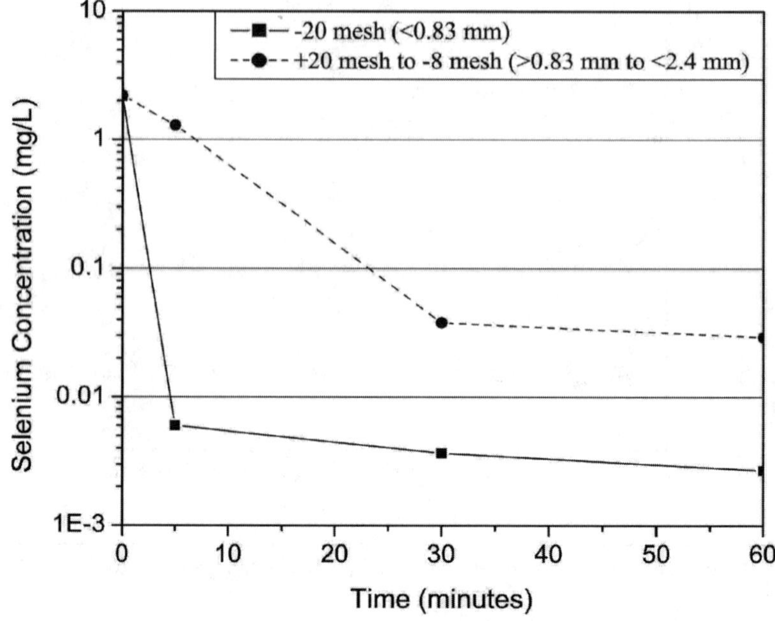

FIGURE 6 Influence of iron particulate size on final selenium concentration (Dahlgren 2000)

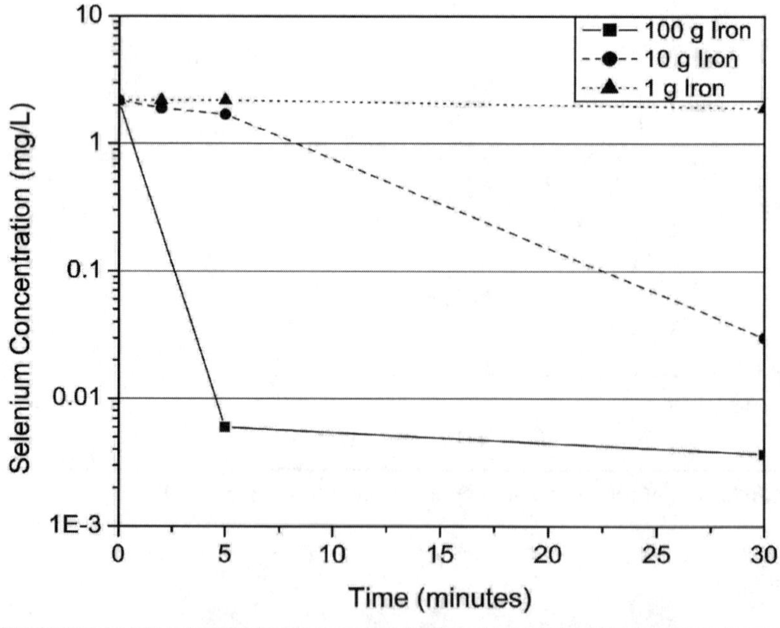

FIGURE 7 Influence of amount of iron on final selenium concentration (Dahlgren 2000)

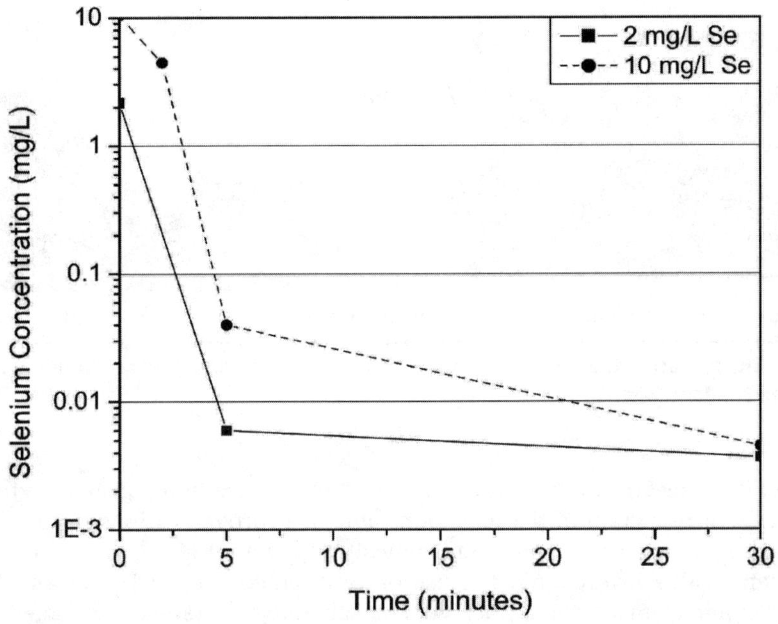

FIGURE 8 Influence of initial selenium on final selenium concentration (Dahlgren 2000)

TABLE 3 Summary of experimental column test work conducted by Cockhill (2002)

	Condition	Result	Comments
pH of influent solution	2.5–2.7		pH controlled
pH of effluent solution		5.2–5.6	
E_H, mV	–380		
Bed Volumes		1450 (426)[1]	No breakthrough
Flow, mL/min	4 (8)[1]		Residence time 30 min (15 min)[1]
Se(Initial), µg/L	4,000		
Se(Final), µg/L		<1 (<1)[1]	
Se(Total), gm		0.49	
Time, hrs	490 (72)[1]		
Fe mass in column, gm	300		

[1]Conditions and results for the 15 minute residence time study
Influent selenium concentration was 4000 µg/L.

Coupled Metals

The driving force that controls and explains the usefulness of a particular metal as the reductant for a specific aqueous specie is electron current flow. Metal dissolution creates electron flow. The metal dissolution electron generation must be balanced by electron consumming reduction reactions, i.e., the production of electrons by dissolution reactions must be balanced by the consumption of electrons by reduction reactions.

FIGURE 9 Illustration of the anodic dissolution of iron and the cathodic reduction of selenate to selenium (Hadden 2002)

Therefore, if the electron flow can be increased by the use of a galvanic metal couple then the rate of reduction of the oxyanions and/or hydrogen cations in solution must increase. The concept is illustrated schematically in Figure 9.)

Hadden (2002) investigated the use of galvanic couples of Fe/Cu and Fe/Ni to remove selenium from synthetic wastewaters. Laboratory work was conducted using an EG&G Instruments Princeton Applied Research Potentiostat/Galvanostat (ARP/G) Model 263A with an EG&G corrosion cell Model K47. The instrument variables controlled during the investigation included the initial potential, the potential step size, the range of potential to be scanned, and the rate at which the potential range was scanned. The response measured was the current density in microamps/cm^2 generated at each selected potential step.

Test solutions of various selenium concentrations were prepared so that they contained two or ten mg/L selenium as selenate and a similar ionic matrix. The reactor for the corrosion cell was filled with 920 ml of solution. The pH of the solution was adjusted to either 5 or 7, depending upon the test parameters, with 0.05 M NaOH. Carbon counter electrodes, salt bridge, and a nitrogen sparge tube were all placed in the cell. EG&G Soft Corr III software was used to control the system variables and to collect current density data for each specific experiment. After 30-minutes of nitrogen sparging a solution sample, pH reading were taken. The active electrode containing the metal sample was then placed in the solution. The Soft Corr III program was then initiated. An initial 3600-second delay specified in the software set up, allowed for an hour of conditioning to take place. The salt bridge tip was placed so that it was approximately two millimeters from the surface of the active anodic electrode.

Metal samples were cut to " diameter and then polished through 600-grit aluminum oxide. After the initial polishing, the sample was re-polished after each experiment prior to reuse. The preparation procedure consisted of polishing for three minutes on 600-grit paper. The direction of the polishing motion was changed back and forth using small clockwise and counterclockwise circular motions such that no distinct pattern of grooves formed. For preparation of the galvanic couples, a hole was first drilled in the cathodic metal to a desired diameter and then the outer dimension was cut to a 3⁄4-inch circle. The anodic metal was cut to a 3⁄4-inch circle and was placed flat against the cathodic metal.

TABLE 4 Rate of deposition of selenium on iron, iron/copper and iron/nickel substrates

System	$i_{o,\ selenate}$, amp/cm^2	mg Se deposited/minute	Time to remove 2 mg/L Se, minutes
Fe/Se	0.25E–06	0.15	13
Fe/Cu/Se	0.51E–06	0.31	6
Fe/Ni/Se	6.24E–06	3.73	0.5

i_o is the selenate reduction current density. The assumptions used to estimate the time to remove 2 mg/L Se include: 100 g/L iron, surface area of iron 0.0735 m^2/g iron, pH 7, 2 mg/L Se as selenate, vigorous agitation, 25–28°C.

This arrangement allowed for one square centimeter of anodic metal (in the center of the exposed area of the cathodic metal) to always be exposed to the solution environment.

Summary of Electrochemical Characterization Results for Selenate

Electrochemical characterization studies were performed on iron, iron/copper and iron/nickel substrates with and without the presence of two or ten mg/L selenium. The experimental results are summarized in detail elsewhere (Hadden, 2002; MWTP, 2002). The overall conclusion of the test work was that galvanic couples of Fe/Cu and Fe/Ni show significant enhancement in the rate of removal of selenate from synthetic solutions (see Table 4). The Fe/Cu galvanic couple was chosen for subsequent pilot scale demonstration test work.

DEMONSTRATION PROJECT

MSE-Technology Applications performed a pilot scale field test for selenium-bearing water in connection with the EPA MWTP in connection with this study. A demonstration field test was conducted on selenium bearing waters. The Selenium Treatment/Removal Alternatives Demonstration Project consisted of demonstrating one standard process (U.S. EPA's BDAT, for removing selenium, i.e., ferrihydrite adsorption) and two innovative processes (elemental iron reduction and biological reduction) at the Garfield Wetlands-Kessler Springs site at the Kennecott Utah Copper Corporations site in Magna, Utah (MWTP 2001). The nominal selenium composition of the test site water was 1950 μg/L (1870 selenate, 49 selenite); the pH was 7.0. The primary project objective of this demonstration was to lower the selenium concentration to <50 μg/L. The ferrihydrite adsorption technology did not meet the project goal. The elemental iron reduction and the biological reduction technologies both achieved the goal level for selenium concentration after optimizing each process. The conditions for the pilot demonstrations are presented in Table 5.

Ferrihydrite

Three iron (as ferric chloride) doping treatment conditions were investigated. The test results are presented in Table 6. Ferrihydrite adsorption removal of selenate even at extremely high additions of ferric iron was not very effective.

TABLE 5 Conditions for the EPA MWTP demonstration project (MWTP 2001)

Technology	Flowrate, gal/min	Volume, gal	Residence Time, min	Flowsheet
Ferrihydrite	5	19,100	Reactors–15 Thickener–200	Ferric chloride/lime reagents, 2–80 gal reactors, 1000 gal thickener, filter press, final pH adjust
Iron Reduction	1	10,000	Reactors–30 Thickener–15 hr	–20 mesh iron, 80 gal solution mixer, 80 gal fluidized bed recycling iron slurry reactor, 1000 gal thickener, 80 gal iron oxidation reactor, filter press
Iron Reduction Follow up	70 cc/min	240	Reactors–50	Two 1.8 gal iron filled reactors in sequence, iron recycle, iron oxidation and settler
Biological Reduction (BSeR)	1	100,000	Variable	Three test series were conducted: Series 1: 5–500 gallon reactors in series. Series 2 and 3: 3–500 gallon reactors in series

The final report (MWTP–191) for this demonstration is available on the website:
http://www.epa.gov/ORD/NRMRL/std/mtb/mwtpcontacts.htm

TABLE 6 Ferrihydrite adsorption of selenate demonstration test (MWTP 2001)

Iron loading (pH)	Fe/Se Mole Ratio	Mean Se Effluent, μg/L (n=number of samples)	Minimum Se, μg/L	TCLP Test, <1 mg/L required to pass
Low Iron, 1,400 mg/L (3.9)	1000	304±69 (n=27)	115	
Medium Iron, 3,000 mg/L (4.1)	2120	201±103 (n=13)	42	
High Iron, 4,800 mg/L (3.8)	3390	90±28 (n=5)	35	2 tests on filtercake product failed test: 1.6 and 1.1 mg/L

Iron Reduction

The iron reduction technology was demonstrated on the pilot scale at one gallon/minute for 16 days. The test results are presented in Table 7. The removal of selenium during the initial test period did not meet the project goals. However, when the pH was lowered during the later stages of the pilot study the goal level was achieved. Because of this an additional series of tests were conducted on the Kesseler Springs water at the MSE-TA site under the conditions specified in Table 5. Based on the results of the final optimized laboratory test further pilot scale test work using this technology has been proposed to the EPA MWTP.

Biological Reduction

The biological selenium reduction (BSeR) technology was demonstrated on the pilot scale at one gallon/minute for four months. The BSeR is a patented process that uses anaerobic bed reactors. Selenate is reduced to elemental selenium by specially developed bioifilms containing proprietary microorganisms. A series of tests were conducted at different residence times. The conditions and test results are summarized in Table 8.

TABLE 7 Iron reduction of selenate demonstration test (MWTP 2001)

Study (pH)	Test Period, days	Mean Se Effluent, µg/L (n=number of sample)	Minimum Se, µg/L	TCLP Test, <1mg/L required to pass
Pilot Study (~5)	16	834±204 (n=42)	193	2 tests on filtercake product passed test: 0.3 and <0.1 mg/L (TCLP required <1 to pass)
Pilot Study (~3.5)	30 hrs	35 (n=2)	26	
Laboratory Confirmation (~3.0)	17	3.0±4.4 (n=5)	<1	

TABLE 8 Biological reduction of selenate demonstration test (MWTP 2001)

Condition (pH)	Test Period, days	Mean Se Effluent, µg/L (n=number of samples)	Minimum Se, µg/L
Series 1: Residence Time/reactor 12 hr (6.3–7.5)	30	8.8±10.2 (n=17)	<2
Series 2: Residence Time/reactor 11 hr (7.0–7.7)	30	4.9±4.9 (n=16)	<2
Series 3: Residence Time/reactor 8 hr (7.0–7.7)	30	<2±2.6 (n=12)	<2
Series 2: Residence Time/reactor 5.5 hr (7.0–7.7)	30	<2±2.1 (n=26)	<2

Series 2 and 3 were identical system run concurrently except they were run at different residence times

The BSeR technology was very successful in lowering the selenium concentration in all studies to less than 10 µg/L.

CONCLUSIONS

Laboratory and pilot studies have demonstrated that:

- Elemental iron is an effective reductant for removing selenate to near detection limits from synthetic solutions; project goal levels of 50 µg/L can be obtained in a residence time of a few minutes for pH levels <7.
- Preliminary exploratory studies showed that elemental Fe/Cu and Fe/Ni galvanic couples enhance the rate of removal of selenate from synthetic solutions but further studies are necessary to evaluate their response in real industrial solutions.
- Elemental iron is effective in removing selenate to project goal levels of <50 µg/L for the Kennecott Kesseler Springs water. The effectiveness of the reduction is increased by lowering the pH to approximately 3–3.5. This may present a safety problem because of the possibility of forming hydrogen selenide gas.
- Biological inoculates in anaerobic reactors are effective in removing selenate to levels much below the EPA Minimum Concentration Level (MCL) of 50 µg/L for the Kennecott Kesseler Springs water at the normal pH level of the water (~7).

- Ferrihydrite precipitation is not effective for removing selenate even at very high ferric chloride doping concentrations for the Kennecott Kesseler Springs water at the normal pH level of the water.

ACKNOWLEDGEMENTS

Financial support from the following organizations is gratefully acknowledged. Dahlgren's M.Sc. studies were supported by MONTEC, Inc, Butte, Montana. Hadden's M.Sc. studies and the MSE-TA pilot work were supported by the EPA-DOE MWTP program, contract number DE-AC09–96EW96405. Cockhill's column reactor studies were supported by the Metallurgical and Materials Engineering Department at Montana Tech of the University of Montana.

REFERENCES

Allison, J., D. Brown, K. Novo-Gradac, 1991, MINTEQA2/PRODEFA2, A geochemical assessment model for environmental systems: version 3.0 user's manual, EPA/600/3–91/021, 105 p.

Chamberlin, P.D., 1996, Selenium removal from wastewaters–an update. In *Randol Gold Forum '96*, Denver, CO,119-27.

Cockhill, L. 2002, Reduction of selenate by elemental iron in column reactors, Metallurgical and Materials Engineering Department, Montana Tech of the University of Montana, Butte, Montana.

Dahlgren, E., 2000, Parameters affecting the cementation removal of selenium from wastewaters. *Master of Science Thesis*, Montana Tech of the University of Montana, Butte, Montana, 119 p.

Hadden, G.A., 2002, Rate of removal of oxyanions of arsenic and selenium from mine wastewater using galvanically enhanced cementation, *Master of Science Thesis*, Montana Tech of the University of Montana, Butte, Montana, 81 p.

Huang, H.H., 2000, STABCAL. Montana Tech of the University of Montana, Metallurgical and Materials Engineering Department, Butte, Montana.

Koyama, K., M. Kobayashi, A. Tsunashima, 2000. Removal of selenium in effluents of metal refineries by chemical reduction using solid iron, In: Proceedings *Minor Elements 2000, Processing and Environmental Aspects of As, Sb, Se, Te, and Bi*, Ed. C.A. Young, San Antonio, TX, SME , Littleton, CO., pp 363–70.

Mizra, A. and Ramachandran, V., 1996, Removal of arsenic and selenium from wastewaters–a review, In: *Second International Symposium on Extraction and Processing for the Treatment and Minimization of Wastes*, Ed. Ramachandran V. and C. Nesbitt, TMS, Warrendale, PA, pp. 563–82.

McGrew, K., J. Murphy, and D. Williams. 1996. Se reduction via conventional water treatment. In *Randol Gold Forum 96*, Denver, CO, pp. 53–66.

MWTP, 1999. Issues identification and technology prioritization report: Se. EPA Mine Waste Technology Program, Activity I, Vol. VII, MWTP-106, MSE Technology Applications, Butte MT, 1999, 118 p.

MWTP, 2001, Final Report-Selenium treatment/removal alternatives demonstration project, EPA Mine Waste Technology Program, Activity III, Project 20, MWTP-191, MSE Technology Applications, Butte MT, 131 p.

MWTP, 2002, , Final Report-Investigation to develop a galvanically enhanced cementation technology for removing selenium from mine waste waters, EPA Mine Waste Technology Program, Activity IV, Project 19, MWTP-213, MSE Technology Applications, Butte MT, 51 p.

Rosengrant, L, L. Fargo, 1990, Final best demonstrated available technology (BDAT) background document for K031, K084, K101, K102, characteristic as wastes (D004), characteristic Se wastes (D010), and P and U wastes containing As and Se listing Constituents. EPA/530/SW-90/059A, U.S. EPA, 124 p.

Twidwell, L.G., J. McCloskey, P. Miranda, M.G. Lee, 2000, Potential technologies for removing selenium from process and mine wastewater. In: *Proceedings Minor Elements 2000*, Ed. C.A. Young, SME, Littleton, CO., pp 1847–65.

Twidwell, L.G., J. McCloskey, M. Gale-Lee, 2002, Removal of Hazardous Constituents from Wastewater, In: *Proceedings Mineral Processing Plant Design, Control and Practice*, Vol. 2, Eds. A.L. Mular, D.N. Halbe, D.J. Baratt, Vancouver, Canada, SME, Littleton, CO., pp 1847–65.

Air-Sparged Hydrocyclone (ASH) Technology for Separation and Environmental Clean-up

J. Hupka* and J.D. Miller†

Environmental applications of the air-sparged hydrocyclone (ASH) technology are presented. The original ASH design used porous sintered steel or porous polyethylene tubes as spargers. To diminish/prevent plugging and to reduce pressure head loss in the gas path a wire screen sparger and new hydrocyclone design have been developed, which allows for wide application of the technology to solve environmental problems and to be used as a chemical reactor. Field and laboratory trials have shown that the ASH technology might be preferred for the stripping of volatile organic compounds (VOC) from contaminated water. The ASH has also been developed for absorption of SO_2, toluene vapors, and for the recovery of cyanide.

AIR-SPARGED HYDROCYCLONE TECHNOLOGY

The air-sparged hydrocyclone (ASH) has been recognized as one of the basic designs for flotation equipment in the seventh edition of Perry's Chemical Engineers' Handbook (Perry, Green, and Maloney 1997). Although the ASH is included in a paragraph on *Flotation Columns,* it functions differently from the traditional flotation column. The ASH technology was originally developed at the University of Utah for fast flotation of fine particles from a mineral suspension (Miller and Van Camp 1982). The original ASH unit consists of two concentric right-vertical tubes and a conventional cyclone header at the top. The porous inner tube is constructed of plastic, ceramic, or stainless steel and allows

* Gdansk University of Technology, Gdansk, Poland

† University of Utah, Salt Lake City, Utah

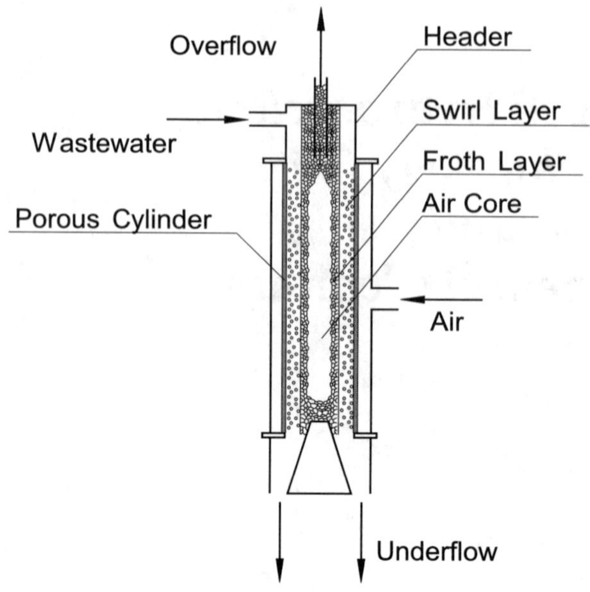

FIGURE 1 Original schematic illustrating the original concept for particulate flotation in a centrifugal force field

for the sparging of air, or other fluid. The outer nonporous tube provides an air jacket to secure even distribution of the gas phase through the porous tube.

The aqueous phase is fed tangentially at the top through the cyclone header to develop a swirl flow adjacent to the internal surface of the porous tube, leaving an air core centered on the axis of the ASH unit, see Figure 1. The high-velocity swirl flow shears the sparged gas to produce a high concentration of small bubbles. Hydrophobic particles in the slurry, after attachment to the bubbles, are significantly reduced in their tangential velocity and transported radially into a froth phase, which forms on the cylindrical axis. Hydrophilic particles generally remain in the slurry phase and are discharged as an underflow product through the annular opening created by the froth pedestal. The number density of bubbles far exceeds that of other flotation systems, and air-to-water ratios (Q^*) of 2/1 to 100/1 are routinely achieved.

In addition to its use for particle flotation, the ASH can be used as a reactor including use for stripping and/or absorption in environmental applications. Given the small bubble size, large bubble flux and the kinetic paths of the bubbles through the swirl layer, gas transfer rates are very high. Consequently the ASH can be used to remove volatile organic species from contaminated water and to absorb impurities from effluent gases. Our recent studies indicate that a high concentration of air bubbles with a directed motion within the ASH unit can be effectively applied for the removal of SO_2, CO_2, HCN and vapors of organic solvents from various industrial gas streams by absorption. Unlike the Couette-Taylor flow reactors (helicoidal flow reactors), the ASH units have no moving parts. However, significant design and processing changes of the technology were required to accommodate environmental applications.

The aim of this paper is to review major environmental separations and wastewater treatment cases for the ASH technology has been evaluated.

NEW ASH DESIGN APPROACH AND ADVANTAGES

Early investigations revealed that inlet water velocity was an important operational parameter permitting control of the flotation efficiency and the bubble-size (Miller et al. 1985). Later research indicated that the water flow rate and surfactant concentration had a greater impact on the bubble-size distribution than the pore size of the porous tube. The mean radius of the bubbles, exiting the bottom discharge, was as small as 50 to 150 μm (Lelinski et al. 1996; Bokotko 2001). The minimum bubble radius recorded was 10 μm, which represented the detection limit of the analytical procedure.

These findings led to replacement of the conventional porous tube sparger by a stainless steel wire screen. The openings could be as large as 1 or 2 mm since bubbles are sheared by the turbulent flow of the liquid phase. The trajectories of particles and the momenta of fluids being contacted in the ASH unit also have to be considered. Although this does not fundamentally change the ASH principle, the new approach expands the potential of the ASH technology.

The wire mesh tube serving as the sparger can be fabricated, from a great variety of materials, which widens the field of application of the ASH into many chemical reactor designs. Wire screens can easily be formed, welded, coated and designed for enhanced heat exchange. Depending on strength, abrasion resistance, chemical resistance, corrosion resistance and heat transfer characteristics of the wire mesh, the ASH can be used as a contactor or a chemical reactor in the oil industry, in the food and pharmaceutical industries, and of for environmental applications.

The use of a wire screen results in a dramatic drop of resistance to the flow of the gas phase (from 1×10^{-1} MPa/15 psi to 5×10^{-4} MPa/0.07 psi) thus a significant savings in energy. A tangential air inlet further reduces the pressure drop and provides uniform flow of gas through the mesh. Additionally, multi-stage and parallel arrangements of the ASH units can be used for process scale-up. Operation under a vacuum in a closed system with recycle of the gas phase has been successfully completed. Such a mode of operation is vital with poisonous gas stripping/absorption, such as HCN. Low resistance in the path of the gas phase allows for large volumetric ratios of gas to liquid ($Q^* > 100$), particularly important for solutes having low Henry's law constants.

APPLICATIONS OF ASH TECHNOLOGY

Deoiling

The first environmental application of the ASH system was for oil separation from water. A schematic drawing of the hydrocyclone in which the oil-water separation was accomplished is shown in Figure 2 (Miller and Hupka 1983). The inner diameter of the porous cylinder was 2" (5.0 cm), and the height was 25 cm (diameter-to height ratio 1/5). The air was introduced through a stainless-steel porous cylinder with an average pore size of 1 micrometer. Due to the resulting shear forces acting on the surface of the porous wall, the air-bubble size was reduced to between 0.2 mm to 0.5 mm in the absence of surfactants. The hydrocyclone was equipped with an adjustable gap to control the distribution of water to the underflow and overflow. The design shown in Figure 2 allows for the separation of oil and water without addition of frother. The original hydrocyclone fed from the top requires the generation of a froth phase to carry oil to the overflow, countercurrent to the underflow stream. Flooding of the air-sparged hydrocyclone was avoided,

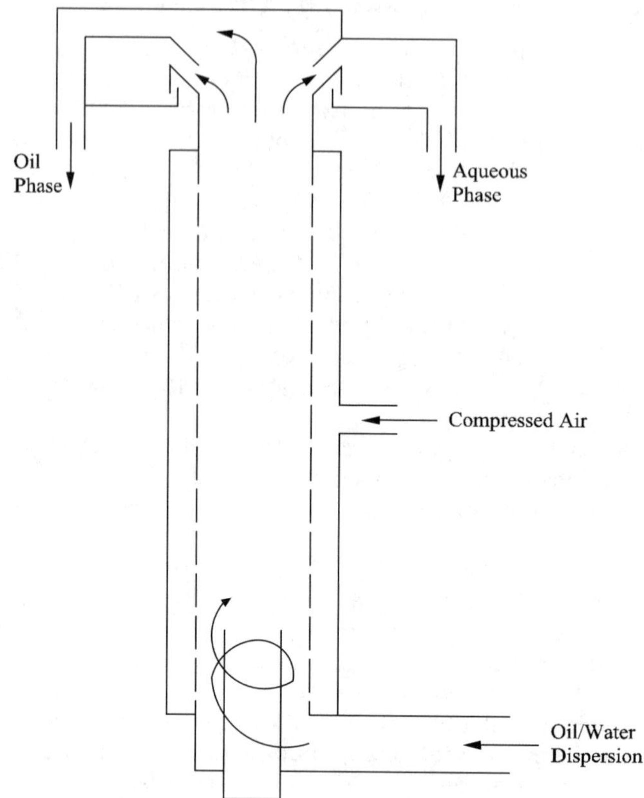

FIGURE 2 Early design of the modified air sparged hydrocyclone as applied for oil flotation from water

and the formation of a water layer at the porous cylinder wall was achieved when the water flow rate was maintained above 1.5 m^3/h. Later research proved that destabilization of the oil-water emulsion using surfactants and polyelectrolyte prior to feeding the ASH results in much better performance (Nicol and Beeby 1993; Bokotko 2001).

Deinking

The basic ASH design used for flotation in various environmental applications is shown in Figure 3. Major research has been accomplished in printing ink removal from cellulose pulp during wastepaper processing (Miller et al. 1993). The performance of ASH here was successful and led to the first commercialization of the technology by Kamyr, USA.

In addition printing ink and other hydrophobic impurities were removed from actual process water circulating in a closed system during processing of wastepaper in large pulp and paper works in northern Poland (Bokotko et al. 1996; Bokotko 2001). A battery of three ASH units was employed for flotation at the nominal process water flow of 240 dm^3/min. See Figure 4. The amount of air was kept constant throughout the testing at 180 dm^3/min. The separation process was aimed at “whitening” the aqueous

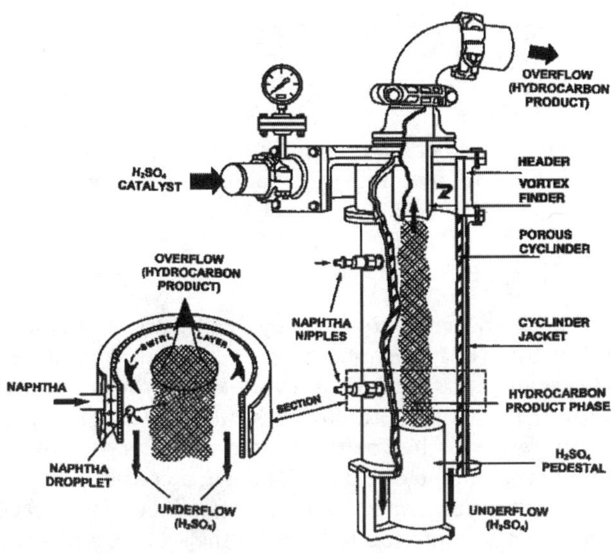

FIGURE 3 **Classic design of ASH tested for many environmental applications as a flotation unit**

FIGURE 4 **Battery of three 2" ASH units operated for four months in pulp and paper works in northern Poland**

stream prior to recycle with simultaneous minimization of water volume directed to the overflow. A single passage of wastewater through the battery resulted in 30% reduction of ink content. The average concentration ratio of ink in the overflow was 10. Microscopic analysis revealed that 90% of the ink particles were less than 10 micrometers in size; 23% below 3μm. Particles above 10 μm usually constituted aggregates of ink fines of various colors. Cellulose fibers (content $1 g/dm^3$) and kaolin particles (content 0.2 wt.%) reported mainly to the underflow and were recycled to the wastepaper digester.

Wastewater Treatment (BAF)

Further changes to the ASH design were made by ZPM/CWT, a California company. The bubble accelerated flotation (BAF) system no longer includes an underflow restriction to force the froth and contaminants through the vortex finder into the overflow (Owen et al. 1999). In the ASH, the removal of the froth through the overflow, resulted in a contaminated stream with relatively low concentration of solids. Removing the underflow restriction in the BAF improved the operational ease and consistency.

The froth together with the cleaned water are introduced into a receiving tank. The froth and bubble-particle aggregates accumulate at the surface and water is drained out of the froth, resulting in a high solids concentration (12–25%) in the froth. The hydraulic flow rate through this receiving tank was found to be more than 10 fold that through any dissolved air flotation (DAF) system. It is assumed that coagulation, flocculation and attachment of bubbles to microscopic particles had already occurred inside the ASH bubble chamber and all bubble-particle aggregates were already formed before introduction into the separation tank.

The BAF system has been successful in many applications that represent challenging environments such as high conductivity, pH or stable emulsions and complex surfactants, where traditional flotation techniques were ineffective, e.g., treatment of food processing wastewater and removal of petrochemical oils as well as vegetable and animal products which are naturally emulsified or solublized (Jovine et al. 1999).

Separation in Oil-Water-Solids System

Investigation on the recovery of bitumen from Utah oil sands, see schematic of flow diagram in Figure 5, showed the importance of complete oil displacement from the mineral surfaces in order to obtain a clean sand product in the underflow of the ASH and an oil-rich froth in the overflow (Lelinski, Miller, and Hupka 1993). Undigested oil-sand aggregates, present in the slurry, were found to report to the underflow, since the buoyancy force of the gas bubble is not able to overcome the pronounced centrifugal force developed in the ASH. But bitumen-sand agglomerates (which contain more oil phase at the surface than undigested particles) were evenly distributed to the underflow and the overflow. Free bitumen droplets reported to the overflow.

Specification of operating variables and evaluation of ASH technology for the treatment of oily soil, were considered with respect to both: oil displacement from particulates and dispersed oil separation from the aqueous phase (Niewiadomski et ai. 2000). The natural hydrophobicity of the oil phase and the density difference between the oil and the sand should facilitate the ASH flotation separation. However, the separation results become unpredictable if oil firmly adheres to sand particles, such as in the insufficiently digested oil sand slurry. Processing conditions were established to separate oil

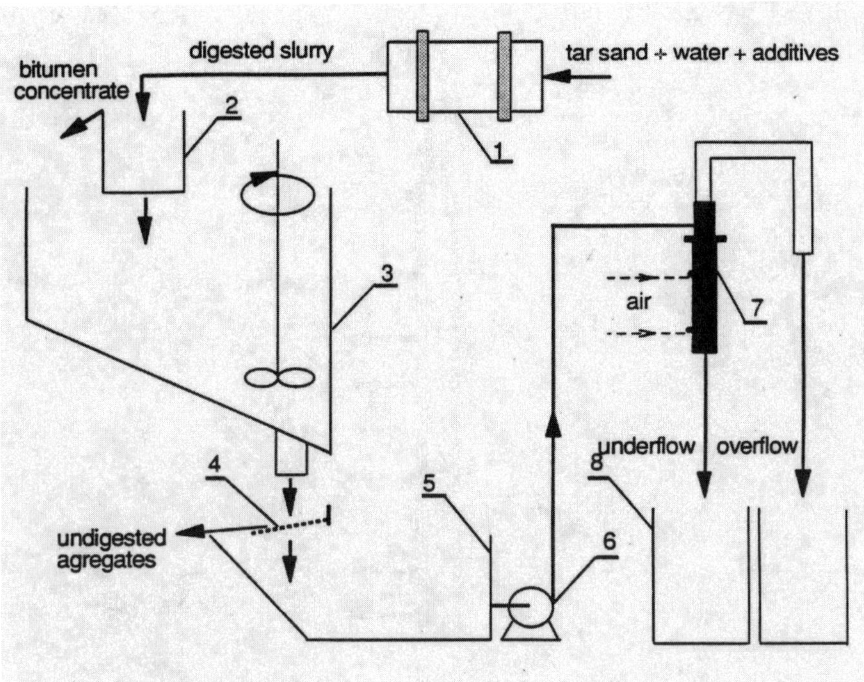

FIGURE 5 Flow diagram from bitumen separation from tar sands. 1. rotating drum for tar sand digestion; 2. settler; 3. conditinoing tank; 4. screen; 5. preparation tank; 6. feeding pump; 7. ASH unit; 8. receiving tank.

from finer part of the sand fraction (0.038 to 0.6 mm) using one step flocculation of dispersed oil and fine sand. It was found that in the centrifugal force field of the ASH the flocculated phases split. Thus flocculated oil reported to the froth phase and the flocculated sand to the underflow providing high oil recovery to the overflow. Finely dispersed oil (1 to 60 μm), stabilized by surfactants used in the washing process, did not float effectively without prior flocculation (Niewiadomski 2004).

The data clearly show the dependence of oil displacement efficiency on the contact time of the oil phase with the sand during preparation of the model system. Soil cleaning was much more effective for sand exposed to oil for less than 24 hours, and very difficult for sand which remained in contact with oil for more than 3 months. Batch tests indicated that oil recovery deteriorated from 54% to 17% when the contact time with oil was extended from 1 day to 240 days. The efficiency of soil cleaning could be controlled by combined action of temperature and surfactant concentration. Oil displacement during the conditioning of sand slurry was two times faster when the temperature of the process was raised 10°C. Of course, heating of the entire mass of the sand is not possible from an economical point of view. This situation can be compensated, to some degree, by an increase in the surfactant concentration.

Another environmental application in this area was aimed at the flotation separation of fine, unburned coke grains from coal-fueled power plant fly ash (Niewiadomski

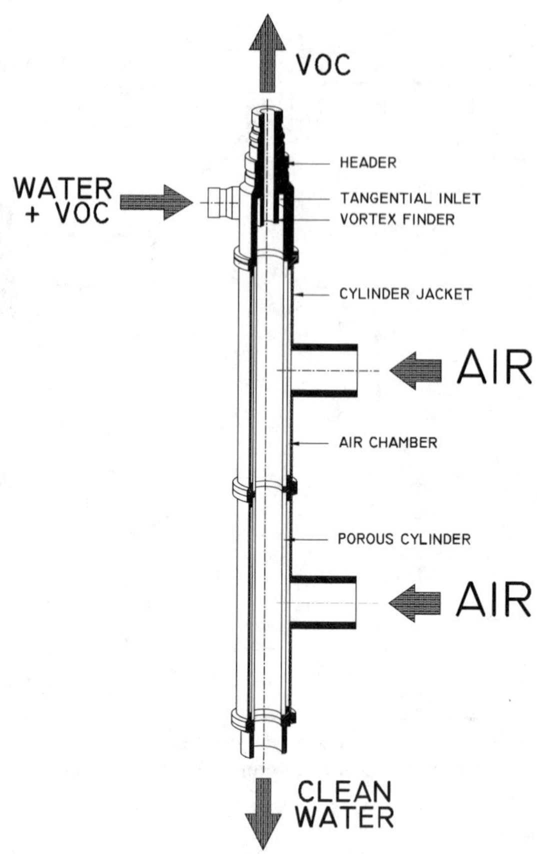

FIGURE 6 Schematic of air-sparged hydrocyclone modified for the stripping of volatile organic compounds

1996; Niewiadomski et al. 1999). Due to mechanical carryover of minute ash fines to the froth phase, the separation efficiency was limited.

VOC Removal

Relatively early the ASH technology was used for stripping of volatile organic compounds from water. The effectiveness was demonstrated through a number of experiments using an ASH unit with a polyethylene porous tube as sparger from which the pedestal was removed, (Lelinski and Miller 1996), see Figure 6. Results show that single stage contaminant removal exceeding 90 % at 24°C was achieved for trichloroethylene (TCE) contaminated drinking water. The air to water flow rate ratio (Q*) and the Henry's law constant controlled the efficiency of stripping. ASH stripping of VOC was found to be a first order process and the extent of removal independent of the feed concentration. Extensive field tests on TCE removal from a contaminated aquifer were carried out using a mobile system incorporating two 800 dm^3 and one 1200 dm^3 tanks with a set of several ASH units (LaPlante, Miller, and Lelinski 1998).

Three regions of different flow characteristics, depending on Q* were found (Lelinski 1998). A water continuous region prevails for Q* less than 40. At higher Q* values air starts to disrupt the uniformity of the swirl layer. This Q* value marks the start of the second region, the transition region, in which the fraction of air as a continuous phase increases as Q* increases. The third region starts at Q* = 100. Above Q* = 100 the aqueous phase exists in the form of droplets. It was also found that the nominal air residence time strongly depends on the air flow rate and its value can decrease from 0.264 s at 100 dm^3/min air and 70 dm^3/min water flow rates to 0.003 s at 6000 dm^3/min air and 35 dm^3/min water flow rates. Thus the multiphase flow inside the porous polyethylene tube changed from the water swirl layer at low air flow rates to a water dispersion in air at high air flow rates.

The ASH technology has also been demonstrated for the removal of MTBE and methanol from contaminated water.

Gas and Vapor Absorption

The large flow of flue gas resulting from the combustion of fossil fuels significantly influences the cost-effectiveness of the absorption method. Recently the high specific capacity ASH has been shown to be effective in such applications (Bokotko, Hupka, and Miller 2004). Tap water and alkaline solutions were used for absorption and the influence of gas flow rate, water flow rate and length of the ASH unit were investigated. The water flow rate was maintained at 75 dm^3/min. The concentration of sulfur dioxide was varied in the feed within the range of 0.8 to 19.1 g/m^3. With an increase of sulfur dioxide concentration, for constant air flow rates, the degree of absorption increased since the absorption rate is in proportion to the concentration of the reacting substance.

The research results indicate that the ASH can efficiently treat off gas emissions for the removal of SO_2. The entire SO_2 was removed in weakly alkaline solution (0.01 mol $NaOH/dm^3$). By comparison the extent of SO_2 removal with water in conventional column applications is between 18–37 percent. Only alkaline solutions allow conventional column absorption to exceed 90%. The absorption tests were completed in 10 minutes which accounts for 6 passes through the ASH.

Only a small increase in the efficiency of sulfur dioxide removal from air took place for poros tubes longer than 16 cm. A similar relationship was obtained for oxygenating water (Bokotko and Hupka 1996). The increase in length from 16 to 32 cm resulted in an increase in the extent of absorption from 96.5% to 98%. Further lengthening of the section to 47 cm had even less pronounced effect. As in the case of oxygenating water, the influence of the length of the porous section is connected with a change in the conditions of multiphase flow and generation of gas bubbles. It seems that most of the absorption occurs near the inlet where the shear force and turbulence are greatest. Further research is needed to clarify this point.

Another application is treatment of voluminous gas streams from ventilation systems of industrial facilities, discharging organic contaminants. Absorption of vapors (e.g., of hydrocarbon solvents) in hydrophobic liquids helps to avoid atmospheric pollution. An automatic mobile system was designed and used for toluene removal from ventilation air in pharmaceutical works in northern Poland. The system was equipped with two 800 dm^3 and one 1200 dm^3 cylindrical tanks and several new type ASH scrubbers operating in series or in parallel. Toluene absorption in one pass through the ASH scrubber using a light oil absorbent was within 65 to 98 percent depending on the process conditions.

Cyanide Recovery

A most advanced and complete investigation for ASH application in the area of environmental separations has been carried out for cyanide recovery from a barren solution originating from gold hydrometallurgical processes. ASH units installed in a mobile system were used as contactors for HCN stripping and re-absorption (Dabrowski 2004). The ASH system was operated under a negative pressure for safety reasons. Three rounds of field tests were performed at Newmont Mining Corporation gold cyanidation plant near Midas, Nevada, USA. From the safety and the processing efficiency standpoint most significant was the performance of ASH as a scrubber (Hupka et al. 2004). The gas phase leaving the scrubber was monitored for HCN content using Dräger monitor, adapted for on-line conditions. HCN removal efficiency exceeded 95% in one pass through the ASH equipped with stainless steel mesh.

HCN content in the discharged air was dependent on the pH of the alkaline solution in the absorption tank, and the air flow rate, and was maintained below 35 ppm for pH=13 and 2.7 m^3/min (90 scfm) flow rate, but could be suppressed further in a more alkaline environment. Using the polyethylene porous tube and comparable absorption parameters, HCN content of the off gas was less than 10 ppm, indicating a somewhat higher absorbing efficiency with the porous tube. Unlike the porous PE tube, the stainless steel mesh was not susceptible to plugging. After 65 hours of operation, the stainless steel mesh was scale free. Thus anti-scalant reagent addition was not necessary. The pressure drop in the scrubber was several centimeters of H_2O. The required pumping pressure for the alkaline solution was 0.05 MPa (7 psi), which corresponds to (or is less than) the pumping head necessary for absorption in packed towers.

The total pressure drop in the entire system already allows for operation in a closed circuit for the gas phase which alleviates concerns about atmospheric air pollution. The system requires a bleed air system to maintain a vacuum in the circuit.

Photochemical Degradation

In the last decade advanced oxidation processes have been gaining importance for destruction of waste as a cleaner method enabling conversion of organic contaminants to harmless species. Photodegradation processes in the presence of hydrogen peroxide (H_2O_2), ozone (O_3) and titanium dioxide (TiO_2) are most frequently offered for wastewater treatment. The novel gas-sparged reactor equipped with an UV lamp, see Figure 7, proved to be an efficient system for photochemical degradation (Artuna and Hupka 2000; Kowalska et al. 2004). In a large scale stationary ASH setup 80 ppm aqueous solution of phenol was treated for 40 min. in a recycled mode of operation. 13 % of the phenol was eliminated in the UV/air system and 73 % in the UV/H_2O_2/air system at 80 dm^3/min. air flow rate. The sparged air did not strip phenol to any noticeable extent (Kowalska 2004).

Scale Precipitation

Extensive water evaporation, possible rapid reactions in a multiphase system, and suspension of fine solids can contribute to plugging of the ASH sparger. Indeed, slow buildup of sediments inside the porous tubes has been experienced in field tests. In Figures 8 and 9 three porous polyethylene tubes removed from a battery of ASH units are shown in Figure 4 after two weeks of continuous operation for cleaning wastepaper processing water. While the tops of the tubes were still permeable to the gas phase, the bottom parts had experienced gradual plugging.

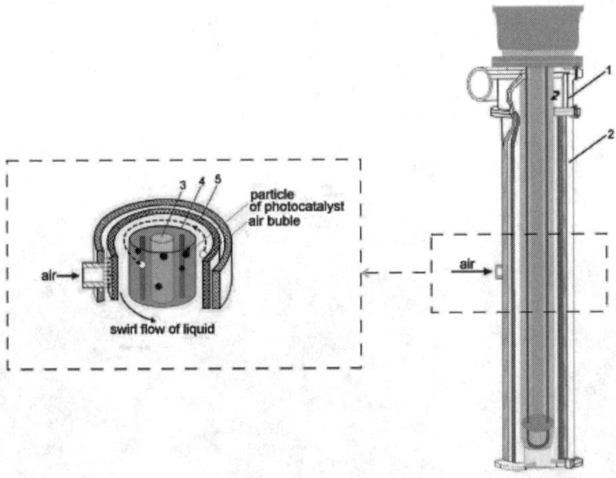

FIGURE 7 ASH unit equipped with 2500 W UV lamp used for degradation of organic compounds

FIGURE 8 Upper part of ASH porous tubes used for treatment of process water from waste paper recycle plant

In many applications the sparging gas cleans and protects the porous tube from scaling and fouling. Fouling of the porous tube was one of the most troublesome problems faced during experiments with the mobile unit in cyanide recovery field tests (Hupka et al. 2004). Although plugging of the porous PE tubes was not significant for ASH stripping with the synthetic cyanide solution, made with tap water, plugging became severe during field tests. To alleviate the problem a new design of the ASH units was required with the following features: screen mesh tube, tangential air inlet, larger space between tubes, and spraying nozzles installed in the cyclone casing.

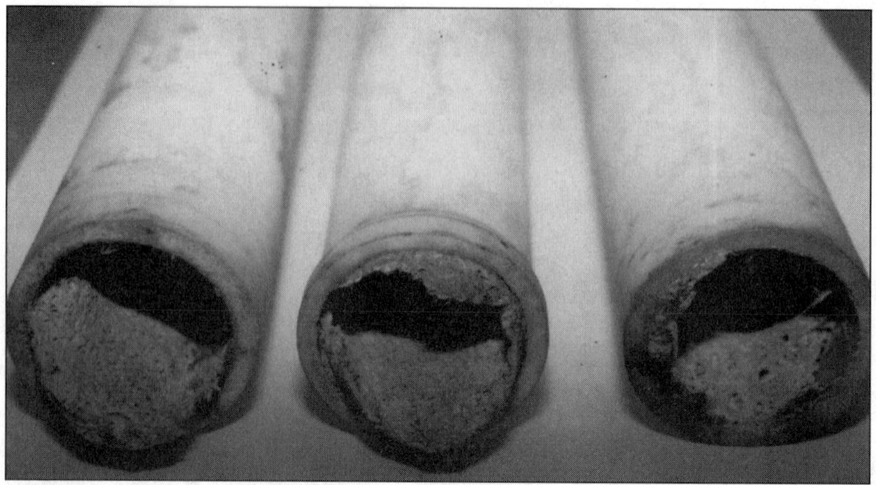

FIGURE 9 Bottom part of porous tubes used as spargers for ink flotation after several weeks of operation

FINAL COMMENTS

The ASH is now recognized as a unique flotation cell, and has been intensely investigated for environmental separations and chemical reactions. The paper pertains mainly to those findings and developments which resulted from collaboration between the University of Utah and the Gdansk University of Technology. Dissertations and theses used as references contain almost complete, up to date, list of papers pertaining to the subject which include other academic institutions doing research on fundamental and technological aspects of ASH performance.

A new, low pressure drop (5–7 cm H_2O; 0.1 psi), ASH unit has been developed, constructed and tested for the stripping and absorption of HCN. The 2" ASH performed much better than the 6" ASH with respect to residual HCN content and the amount of alkaline solution required. Both the stripping and absorbing ASH units were equipped with stainless steel wire screen spargers. During testing using actual process water plugging was found to be minimal–thus justifying design of fully automated system, which probably can be used for treatment of slurries.

The new ASH design allows for larger gas flow rate (easy adjustment of Q*), allows for handy replacement of the porous tube, reduced energy demand, and closed system operation thus avoiding atmospheric pollution. The ASH absorbing/stripping system is relatively small and readily transportable, particularly in comparison with the conventional technology that requires the use of large packed towers and the resultant large concrete foundations and footprint. An ASH system could be fabricated and assembled completely off-site, then transported and made operational in a short time period with a small, temporary footprint, which is preferable to several large concrete and steel structures. A transportable ASH mobile system can be moved from one site at the end of operations to another site at little cost. These results and the results from other research programs demonstrate the utility of the ASH for environmental applications.

ACKNOWLEDGEMENTS

The investigations were supported by NSF, Polish Committee for Scientific Research (KBN), US EPA, SCERP, Foundation for Polish Science (FNP), University of Utah, Gdansk University of Technology, ZPM/CWT Inc., Newmont Mining Corporation, International Paper Poland and other companies which agreed to test the technology. Lukasz Hupka from University of Utah is acknowledged for assistance in editing the paper and figures.

REFERENCES

Artuna, E., and J. Hupka. 2000. H_2O_2/UV/air Oxidation of Organic Contaminants in the Gas-Sparged Cyclone Reactor, Central European Journal of Public Health, *8*(44), 88–89.

Bokotko R. 2001. Investigation of Separation and Mass Transfer Efficiency in ASH Flotator Focusing on Bubble Size Distribution, Ph.D. dissertation, Gdansk University of Technology.

Bokotko, R., and J. Hupka. 1996. Efficiency of Water Aeration in Air-Spparged Hydrocyclone, Inzynieria i Aparatura Chemiczna, No 2, 15–17.

Bokotko, R., J. Hupka, and J.D. Miller. 2004. Flue Gas Treatment for SO_2 Removal with Air-Sparged Hydrocyclone Technology, Environmental Science and Technology (accepted).

Bokotko, R., J. Hupka, D. Lelinski, J.D. Miller, J.D. 1996. Separation of Oil-Containing Particles from Water in Cyclone Flotation Machine, Chemistry for the Protection of the Environment 2, ed. Pawlowski L., et al., New York, NY, Plenum Press, 155–164.

Dabrowski, B. 2004, Air Stripping for the Recovery of Cyanide Using Air-Sparged Hydrocyclone (ASH) Technology, M.Sc. thesis, University of Utah.

Hupka, J., B. Dabrowski, J.D. Miller, and D. Halbe. 2004. Air-Sparged Hydrocyclone (ASH) Technology for Cyanide Recovery, Minerals and Metallurgical Processing, (accepted).

Jovine, R., M. Colic, D.E. Morse, J.D. Miller. 1999. Free and Emulsified Oil Removal by Bubble Accelerated Flotation (BAF), Proceedings of the *Environmental Technology for Oil Pollution AUZO 1999* UEF Conference, Aug.29-Sept.3, Jurata, Poland, vol. I, 106–114.

Kowalska, E. 2004. Investigation of Photochemical Degradation of Organic Compounds, Ph.D. dissertation, Gdansk University of Technology, 2004.

Kowalska, E., M. Janczarek, M. Grynkiewicz, J. Hupka. 2004. H_2O_2/UV Enhanced Degradation of Pesticides in Wastewater. *Water Science & Technology*, *49*(4), 261–266.

LaPlante, L., J. D. Miller, and D. Lelinski. 1998. Removal of VOCs from Groundwater Using Air–Sparged Hydrocyclone Stripping Technology, *Physical, Chemical, and Thermal Technologies, Remediation of Chlorinated and Recalcitrant Compounds,* Proceedings of the *First International Conference on Remediation of Chlorinated and Recalcitrant Compounds,* Monterey, California, 18–21 May, G.B. Wickramanayake and R.E. Hinchee, eds., Battelle Press, Columbus, Ohio, 223–229 (1998).

Lelinski, D. 1998. Air-Sparged Hydrocyclone Treatment of Water Contaminated with Volatile Organic Compounds, Ph.D. dissertation, University of Utah.

Lelinski, D., and J.D. Miller. 1996. Removal of Volatile Organic Compounds from Contaminated Water Using Air–Sparged Hydrocyclone Stripping Technology, Proceedings of the *International Conference on Analysis and Utilization of Oily Wastes, AUZO '96*, Supplement, Gdansk University of Technology, Poland, 8–12 Sept. 53–62.

Lelinski, D., J.D. Miller, and J. Hupka. 1993. The Potential for Air–Sparged Hydrocyclone Flotation in the Removal of Oil from Oil–in–Water Emulsions, Proceedings of the *First World Congress on Emulsions,* Oct. 19–22, Paris, France, 1–22–016/01 to 1–22–016/96.

Lelinski, D., R. Bokotko, J. Hupka, and J.D. Miller. 1996. Bubble Generation in Swirl Flow during Air–Sparged Hydrocyclone Flotation, Minerals and Metallurgical Processing Journal, May, 87–92.

Miller J.D., and J. Hupka. 1983. Water Deoiling in an Air-Sparged Hydrocyclone, Filtration and Separation, 280, July/August.

Miller J.D., A. Das, D. Lelinski, and J.W. Chamblee. 1993. Multiphase Flow Characteristics of Gas Sparged Hydrocyclone Flotation Deinking, Proceedings, 1993 TAPPI Engineering Conference, *Fluid Mechanics of Deinking and Recycling,* Orlando, Florida, 417–428.

Miller, J.D., K.R. Upadrashta, D. J. Kinneberg, and S. Gopalakrishnan. 1985. Fluid-Flow Phenomena in the Air-Sparged Hydrocyclone, Proceedings of *XV International Mineral Processing Congress*, Cannes, France, June 2–9, 1985, vol. II, 87.

Miller, J.D., and M.C. Van Camp. 1982. Fine Coal Flotation in a Centrifugal Field with an Air Sparged Hydrocyclone, Mining Engineering, November, 1575.

Nicol, S.K., and J.P. Beeby. 1993. Concentration of Oil-in-Water Emulsion Using the Air-Sparged Hydrocyclone, Filtration and Separation, *30*(2), 141–146.

Niewiadomski, M. 1996. Recovery of Fine Coke from Fly Ash Using Foam Flotation, M.Sc. thesis, Gdansk University of Technology.

Niewiadomski, M. 2004. Air Sparged Hydrocyclone Flotation for Oily Wastewater Treatment, Ph.D. dissertation, University of Utah.

Niewiadomski, M., J. Hupka, R. Bokotko, and J.D. Miller. 1999. Recovery of Coke Fines from Fly Ash by Air Sparged Hydrocyclone Flotation, *Fuel*, 161–168.

Niewiadomski, M., J. Nalaskowski, J. Hupka, and J.D. Miller. 2000. Removal of Oil Contaminants from Soil by Flotation, Proceedings of the *XXI International Mineral Processing Congress*, ed. Paolo Massacci, Rome 2000, Elsevier, *Developments in Mineral Processing*, Vol. B, 13, 31–37.

Owen, J.J., D.E. Morse, W.O. Morse, and R. Jovine. 1999. Advances in Flotation Technology, B. K. Parekh and J.D. Miller, Eds., SME, Littleton, CO, 381–389.

Perry R.H., D.W. Green, and J.O. Maloney, Eds. 1997. Perry's Chemical Engineers' Handbook, Seventh Edition, McGraw-Hill, 19–63.

Water-Fuel Emulsions for Energy Application

M. Misra,[*] S. Chen,[*] and P. Grimes[†]

The combustion of petroleum products produces emissions which contain oxides of nitrogen (NO_x), hydrocarbons (HC), and particulate matter (PM) which have adverse health and environmental effects. In recent years there has been increasing interest in the use of water-fuel emulsions as a substitute for conventional fuels. Emulsified fuels are emulsions of water in fuel, which are typically made of 10 to 20 percent of water mixed with certain additives and fuel. Studies have shown that the emulsified diesel can significantly reduce the emissions of NO_x, PM and carbon monoxide and increase the fuel efficiency. In order to produce a stable diesel emulsion, the development of an effective surfactant additive is critical. The stability of diesel emulsions is influenced by many factors, including size distribution of the droplets, the viscosity of the continuous phase, the physical properties of the interfacial film, and temperature. This paper describes a method to produce emulsified diesel and combustion results obtained from various tests. It has been found that the diesel emulsion made with this additive package was stable during the test, and the emissions were lower in NO_x, CO, and PM compared to the baseline diesel.

INTRODUCTION

Diesel engines are widely used in transportation, mining, and power plants. Compared to gasoline engines, diesel engines have the advantages of high thermal efficiency, high power-to-weight ratio and good fuel economy [Pischinger, 1998]. The current soaring fuel prices make the diesel engines even more attractive. While diesel engines have the obvious advantage in fuel economy over gasoline engines, there are concerns associated with their emissions. The principal pollutants generated from diesel engines include particulate matter (PM), nitrogen oxides (NO_x), carbon monoxide (CO) and hydrocarbons.

* Metallurgical & Materials Engineering, University of Nevada, Reno, Nevada

† Clean Fuels Technology, 210 Gentry Way, Reno, Nevada

Generally, the diesel engine produces lower levels of carbon monoxide and hydrocarbons, but higher levels of particulate matter and nitrogen oxides compared to the conventional gasoline engine [NRC, 1982]. In certain situations, such as the case in the state of California, this problem prevents the wide application of diesel engines.

Diesel exhaust contains more than 400 different components, including vapors and fine particles. Exposure to this mixture may result in cancer, exacerbation of asthma, and other health problems. For the same load and engine conditions, diesel engines generate more particles than gasoline engines. As a result, diesel engines account for an estimated 26 percent of the total hazardous particulate pollution from fuel combustion sources in our air, and 66 percent of the particulate pollution from on-road sources. Diesel engines also produce nearly 20 percent of the total nitrogen oxides in outdoor air and 26 percent of the total NO_x from on-road sources, which is a major contributor to ozone production and smog. The production of photochemical smog via NO_x and HC reactions with solar ultraviolet radiation might irritate the throat, trachea, lung and eyes [Lin and Wang, 2003].

The aim of the development of diesel fuel emulsions is to provide an environmentally friendly substitute for traditional diesel fuels, which can be used in diesel engines for reduction of emissions. Typically diesel emulsions are mixtures of diesel with 10 to 20% of water and several percent of various types of surfactant additives which help to stabilize the emulsion. Compared to conventional diesel fuel, diesel emulsions have been reported to simultaneously reduce emissions of nitrogen oxides, carbon monoxide and particulate matter, without the need for any mechanical modifications [Ryan, et al. 1981].

One of the combustion characteristics of diesel emulsions is the so-called secondary atomization caused by the vigorous evaporation of the interior liquid. For diesel-fueled engines the flame temperature is relatively high during combustion (as high as 2200 °F). Under such conditions free radicals of oxygen and nitrogen are formed and chemically combined as nitrogen oxides. Use of diesel emulsion can greatly reduce the production of NO_x. When the diesel emulsion enters the combustion chamber, it is atomized into 75 to 100 μm size droplets. The emulsified water droplets within each drop of oil flash into steam, exploding it into numerous smaller droplets. This phenomenon can increase the combustible surface area per volume of fuel oil by as much as 100 times, which leads to more complete combustion. Furthermore, due to the existence of water in diesel emulsion, the combustion temperature is lowered. As a result, nitrous oxides generation is reduced. Another benefit of diesel emulsion is the lower particulate emissions due to the evaporation of added water and the simultaneous reduction of the temperature. The water addition may increase OH radicals which are very effective in the oxidation of soot precursors. The soot formation in the gas phase would be reduced as a result of the oxidation of soot by increased OH radicals [Kadota and Yamasaki, 2002].

The number of vehicles using diesel emulsions is in steady increase as an immediate solution for the reduction of polluting air emulsions, especially in Europe. It is reported that around 450 city buses in the UK, 1,000 in France, and 8,500 in Italy use emulsified diesel fuels. Diesel emulsions can also be used in marine engines, locomotives, power generation equipment and construction equipment, as well as in industrial boilers and mining equipment. A test was conducted in London where a proportion of 13% water was used in the fuel. It was found to achieve a NO_x reduction of 13% and PM10 reduction of 25% where there was no particulate trap fitted [Sadler, 2003].

By definition, emulsion is a heterogeneous system consisting of one or more liquids dispersed in an immiscible liquid in the form of droplets. Usually the diameter of the droplets is larger than 0.1 μm. Traditionally emulsions are categorized into two groups: water-in-oil (W/O) and oil-in-water (O/W). The W/O emulsion is prepared using either a mechanical, chemical, or electric homogenizing machine to stir water into micrometer droplets scattered evenly in the oil layer. A number of devices can be used to produce emulsions. These devices include rotor-stators, colloid mills, high-pressure homogenizers, and ultrasonic systems. However, without any additives, emulsions may quickly separate into two phases due to coalesce. To keep emulsions stable, it is necessary to add surfactants (emulsifiers) to emulsion [Becher, 2001].

One frequently used method for selection of emulsifiers is known as the hydrophile-lipofile balance (HLB) method. It is based on the theory that emulsifying agents contains oil-soluble and water-soluble moieties [Lisant, 1974]. This dual nature of the surfactant molecule tends to limit its solubility in both phases and causes it to concentrate at the interface. The balance between these groups will determine whether a specific surfactant would be soluble in either water or oil and the type of emulsion it would stabilize. In general, low HLB (3–6) emulsifiers favor the formation of W/O emulsions, while higher HLB (8–18) emulsifiers favor the formation of O/W emulsions.

Thermodynamically the emulsions are unstable [Becher, 1983]. This instability arises from the large interfacial area, and therefore large surface energy, associated with finely dispersed systems. Consequently, in order to reduce the total interfacial area, colloidal particles tend to either flocculate or to accumulate at an interface. The mechanism of colloidal instability involves three types of phenomena: creaming, Brownian flocculation, and sedimentation flocculation.

To produce stable diesel emulsions, the selection of appropriate additives (emulsifiers) is critical. The purpose of applying emulsifiers is to reduce the surface tension forces and to promote the affinity force between water and oil phases. It should also provide long enough shelf life for transportation and appropriate storage time.

Although a lot of research has been carried out for the development of diesel emulsions, there are only a few products currently in small-scale use. The main problems with diesel emulsion production include high cost, instability, and complex systems. Therefore new methods which can provide economic, stable, and low-polluting emulsified diesel should be developed.

In this paper, a new emulsifier package was used for the preparation of diesel emulsions. The physical properties of the emulsion were evaluated. Its stability was also studied. For evaluation of the performance of this emulsion in engines an emission test was also conducted.

EXPERIMENTAL PROCEDURE

There are four constituents in the additive used in the study: a carboxylic acid with long saturated carbon chains, an alkali, a stabilizer, and deionized water. The purpose of the addition of the stabilizer is to sustain an appropriate stability for the W/O emulsion. The additive is an amber colored fluid. Appropriate stirring is required before using it for emulsified diesel production.

In the bench scale experiment, diesel emulsion was prepared with a homogenizer. First the additive was added to diesel and the content was thoroughly mixed with the homongenizer at midrange rpm. After that deionized water was added and the mixture

TABLE 1 Engine specifications

Engine parameter	Specification
Engine type	Diesel, 4-cycle
Model	6067MK6E
Configuration	L–6
Displacement	12.7 L (778 CID)
Aspiration	Turbocharged
Rated power	430 HP at 2100 rpm
Peak Torque	1550 lb-ft at 1200 rpm
Idle speed	600 rpm
Governed idle speed	2225 rpm
Combustion system	DI, TC, ECM, CAC, EGR*
Engine family	VH603WAF

* DI: Direct injection
TC: Turbocharger
ECM: Electronic control module
CAC: Cooled air charge
EGR: Exhaust gas recirculation

was sheared by the homogenizer at midrange rpm. The volume percentage of the additive, water, and diesel were approximately 2%, 16% and 82 %, respectively.

Characterization of the product was performed by using a DT−1200 electroacoustic spectrometer (Dispersion Technology Inc.) for measurement of droplet size distribution. Incubation of the prepared diesel emulsions was carried out in order to measure the degree of phase separation of the water phase from the oil phase. The freshly prepared emulsion was placed in a 100 ml cylinder and set at room temperature. The separation heights were recorded daily.

Emissions generation, collection, and analysis were performed according to the applicable standards in the Code of Federal Regulations. A 2003 Detroit diesel engine was selected for the testing of the emulsified diesel. The specifications of the engine are described in Table 1.

RESULTS AND DISCUSSION

Droplet Size Analysis

The size distribution, or particle size distribution (PSD) of water droplets in the medium (diesel) provides critical information which can be used in the prediction of emulsion stability, and in the explanation of combustion performance. It can provide a measurement for quality control purposes as well.

Under the conditions of flocculation and creaming, the droplet sizes tend to become larger. It was reported that the stability to flocculation increases exponentially with increasing droplet size. The smaller droplets flocculate quickly to form relatively stable larger floc [Becher, 1988]. If the emulsion is stable, changes in PSD would occur slowly. From this point of view, it is possible to evaluate the emulsion stability by measurement of the change of PSD with time.

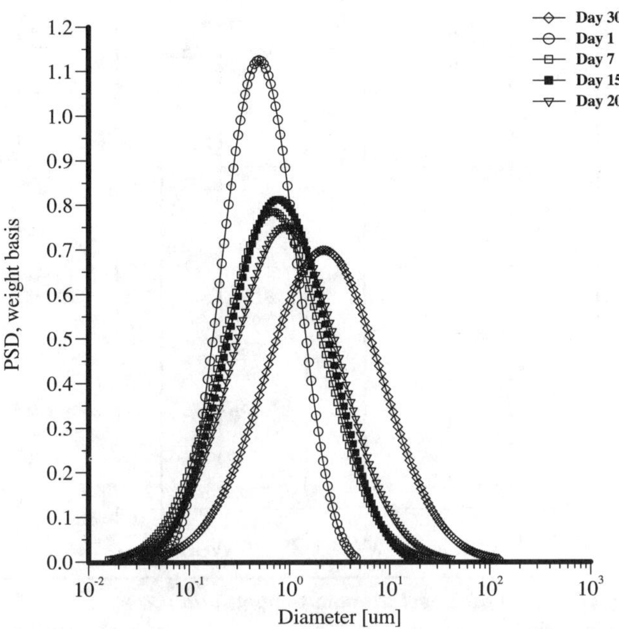

FIGURE 1 Measurement of PSD of diesel emulsion

TABLE 2 Mean droplet size of emulsified diesel

	Mean droplet size (μm)				
Sample ID	Day 1	Day 6	Day 15	Day 20	Day 30
Emulsified Diesel	1.05	1.47	1.69	1.83	2.04

Figure 1 shows the PSD measurement results of the diesel emulsion at various time intervals. In the beginning, a large part of the droplets fell into submicron range, with a narrow distribution in particle sizes. After that the distribution covered a wider range. During the first three weeks the median sizes changed very slowly. The mean droplet sizes measured by electroacoustic spectrometer during a period of one month are shown in Table 2. The initial mean droplet size was 1.05 μm. The mean size gradually increased during the period. After 30 days the mean size was 2.04 μm. Although the mean size increased, the pace was slow, meaning that the diesel emulsion made with this additive was relatively stable. These results indicated that this additive package is effective in stabilizing the W/O diesel emulsion.

Stability Test Results

The stability of diesel emulsion at room temperature is shown in Figure 2. It can be seen that the percentage of oil phase separation was 8% after a month's incubation. In the beginning the oil phase separation was more repressed, then the emulsion appeared to undergo a flocculation period. However, after three weeks, the emulsion seemed to be

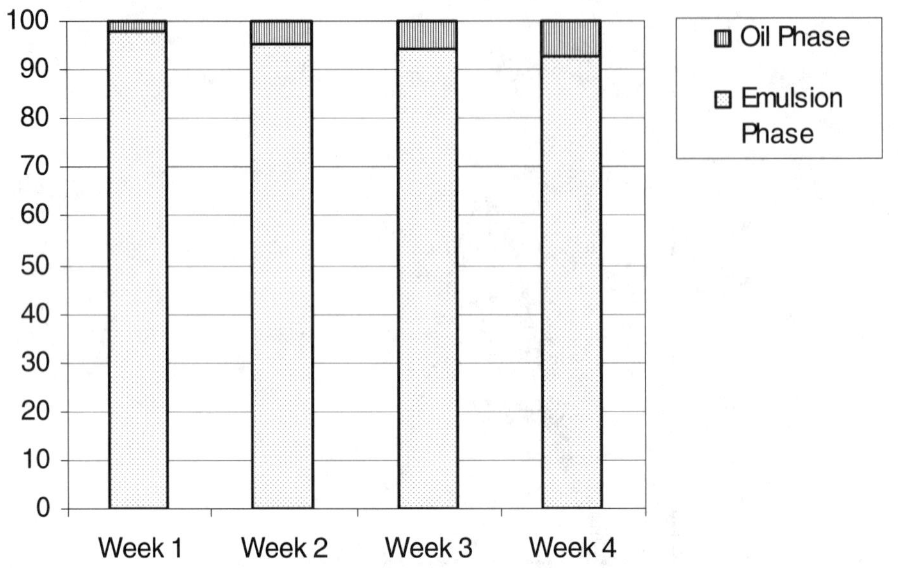

FIGURE 2 Stability of emulsified diesel at room temperature

TABLE 3 Analysis of fuel properties

Property	Base Fuel	Diesel Emulsion
API Gravity	32.4	28.3
Sulfur, wt%	0.049	0.038
Cetane Number	43.2	40.3
Cetane Index	44.1	34.6
Aromatics, vol %	37.9	34.0
Olefins, vol %	1.2	2.8
Saturates, vol %	60.9	47.2

stabilized. The percentage of oil separation on the 7th day, 14th day, and 21st day was measured to be 2%, 4.5%, and 6%, respectively. These results demonstrated a stable diesel emulsion with a good shelf life.

Emission Test

For the purpose of comparison, a baseline fuel was obtained and tested, together with the emulsified diesel. Both fuels were analyzed for their properties. Table 3 presents the measured fuel properties.

The average emission results obtained from the test of PM and NO_x using the emulsified diesel fuel were compared to the average emission results using the reference diesel fuel. The comparison is presented in Table 4. The emulsified fuel reduced PM emissions by 35% and NO_x emissions by 14% compared to the reference baseline fuel. The only emission component which did not meet the requirement of the regulation is particulates. This phenomenon may have been due to the high aromatic and sulfur content of the base

TABLE 4 Summary of composite regulated emissions

Test	Emissions Results, g/bhp-hr			
	THC	CO	NO_x	Particulate
2003 Standard	1.3	15.5	4.0	0.10
Baseline Fuel				
Test 1 Composite	0.10	0.8	2.7	0.17
Test 2 Composite	0.11	0.8	2.8	0.16
Test 3 Composite	0.11	0.7	2.8	0.17
Average of Three Composites	0.11	0.8	2.8	0.17
Emulsified Diesel Fuel				
Test 1 Composite	0.11	0.6	2.5	0.11
Test 2 Composite	0.14	0.7	2.4	0.10
Test 3 Composite	0.13	0.8	2.3	0.10
Average of Three Composites	0.13	0.7	2.4	0.11

fuel. It can be expected that an emulsion based on a diesel with a lower sulfur content may yield reduced PM emissions.

Average hydrocarbon emissions increased from 0.11 for the reference fuel to 0.13 g/hp-hr for the emulsified fuel. However, hydrocarbon emissions for the emulsified fuel remained well below the limit specified.

Carbon monoxide emissions were 12% lower with the emulsified fuel than with the reference fuel. Carbon dioxide emissions were not significantly different with the emulsified fuel.

Speciation of the volatile hydrocarbon compounds with carbon numbers from C_1 to C_{12} and andehydes, ketones, and two ethers (MTBE and ETBE) was also performed. More than 200 compounds were checked for their presence in the dilute exhaust. In general, all compounds found in the exhaust with the emulsified diesel fuel were also found in the exhaust with the baseline diesel fuel. While no additional compounds were found with the emulsified diesel fuel, thirteen compounds were detected in the exhaust of baseline diesel fuel that were not detected with the emulsified diesel fuel.

CONCLUSIONS

A new emulsifier based on long chain carboxylic acids and alkali has been developed for the production of emulsified diesel oil. An emulsified diesel was prepared by using a homogenizer. This diesel emulsion is of the W/O type which contains approximately 16% of water and 2% of additive. The stability test conducted with the emulsified diesel fuel made with this emulsifier showed that a stable diesel emulsion can be achieved by using the new additive. Particle size analysis also confirmed that this fuel emulsion presented a droplet size distribution that is relatively stable.

To determine the exhaust characteristics of the emulsified diesel fuel, an emission test was conducted on a heavy-duty diesel engine. It has been found from the test results that the emulsified diesel fuel produced results lower than the 2003 emission standards except for particulate matter. The increase in particulate matter above the standard may have been due to the high aromatic and sulfur content of the base fuel. Compared to the baseline diesel fuel, the emulsified diesel produced about 12% less CO emissions, about

14 % less NO_X, and about 35% less particulate emissions. The speciation of the hydrocarbons showed that no additional compounds, which could be attributed to the use of emulsified diesel fuel, were found in the exhaust at the detection limits for the analytical procedures.

REFERENCES

Becher, P., 2001. Emulsions: Theory and Practice", Oxford University Press, New York, p. 7

Becher, P., 1988. "Encyclopedia of Emulsion Technology, Vol. 3", Deckker, New York, p. 41.

Becher, P., 1983. "Encyclopedia of Emulsion Technology Vol. 1", Mercel Dekker, New York, p. 133.

Kadota, T and H., Yamasaki, 2002. Recent advances in the combustion of water fuel emulsion. *Progress in energy and combustion science*, 28:385–404.

Lin, C. and K. Wang, 2003. The fuel properties of three-phase emulsions as an alternative fuel for diesel engines. *Fuel*, 82:1367–1375.

Lissant, K.J., 1974. "Emulsions and Emulsion Technology", Mercel Dekker, New York, p. 77.

NRC, 1982. "Diesel Technology", National Academy Press, Washington, D.C., p. 6.

Pischinger F., 1998. Compression-ignition Engines in: Handbook of pollution from internal combustion engines. Sher E, (ed.) Academic Press, p. 261.

Ryan, T.W., J.O. Storment, B.R. Wright, R. Waytulonis, 1981. The effects of fuel properties and composition on diesel engine exhaust emissions. SAE Paper No. 810953.

Sadler, L., 2003. The air quality impact of Water-Diesel Emulsion Fuel (WDE) and Selective Catalytic Reduction (SCR) Technologies, London, p. 5.

Induction Time Measurement in Bitumen Extraction Systems

Guoxing Gu,* Ryan Lindmark,* Zhenghe Xu*, and Jacob H. Masliyah*

Induction time is often referred to the time required to form a stable three-phase contact between a gas bubble and a flat or droplet surface in a liquid medium. It can be used to quantify the attachment process in a flotation system. Various techniques have been developed to measure this important parameter. Due to its importance in flotation systems and other processes, significant advances in refining the induction time measurement techniques have recently been made. This paper summarizes these recent advances and describes a novel method developed to monitor bubble attachment to a pendant bitumen drop or bitumen-coated flat surface. This visualization-based method offers a higher sensitivity and accuracy than the previously developed apparatuses in the induction time measurement. As an example, the induction time for oxygen bubble-bitumen attachment was measured at a fixed bubble size as a function of temperature in two different aqueous solutions. The results showed that higher process temperature is favorable for oxygen bubble-bitumen attachment and hence bitumen flotation. Addition of calcium to the aqueous solution slightly decreased induction time.

INTRODUCTION

Flotation has been used to separate solids from multicomponent, multiphase complex suspensions in many industrial processes, such as, mineral recovery, waste water treatment, soil decontamination, and ink removal from recycle paper pulp. A critical step in flotation is the attachment of target solids to air/gas bubbles. Bubble–solid attachment is strongly dependent on a number of characteristics of a flotation system, including bubble size, particle size, surface properties of the dispersed phases, process temperature,

* Department of Chemical & Materials Engineering, University of Alberta, Edmonton, Alberta, Canada

physicochemical properties of the continuous phase and prevailing hydrodynamic conditions.

In a flotation process, flotation efficiency[1,2] (E_f) depends mainly on bubble-particle collision efficiency (E_c), bubble particle attachment efficiency (E_a) and bubble-particle attachment stability (E_s). In a mathematical form, $E_f = E_c E_a E_s$. E_s can in most cases be considered to be unity for stable bubble-particle aggregates. As a result, Ef may be measured experimentally or calculated if Ec and Ea are known. Ec can be calculated with a reasonable degree of confidence. Dai et al.[3] and Ralston et al.[4] published two excellent reviews on bubble-particle collision models. The prediction of E_a, on the other hand, is more difficult due to its dependence on unknown characteristics of intervening film drainage which can be measured by induction time.[5,6,7] It is well documented that an air bubble contacting a particle does not necessarily result in attachment. The bubble attachment process involves thinning of the intervening film between the two interacting phases to a critical value (h_{cr}) and rupture of the intervening film, followed by subsequent formation of a three-phase contact line (tpcl, wetting perimeter) and the expansion of tpcl to a minimal required rim diameter (r_{min}) at which the attachment force exceeds detachment forces.[4] Attachment force is mainly determined by thermodynamic properties (surface and interfacial tensions) and detachment force is created by hydrodynamic conditions (gravitational or inertial force) in the process.

Many publications[5,6,8] indicate that induction time is a readily measurable, but not well defined parameter. This statement may be true for air bubble-particle attachment, as the induction time can be easily measured using a captive bubble to pick up particles from a particle bed. But it is not valid for air bubble-bitumen attachment as will be discussed later in this paper. The definition of induction time varies in open literature. Currently, there are three main definitions of induction time: (1) time from the film drainage up to its critical thickness of rupture (film rupture not finished),[9,10] (2) the film drainage up to the nucleation of tpcl (film rupture finished),[11] and (3) the film drainage up to its required minimal rim diameter (r_{min}).[12,13] However, from an experimental point of view, measurement of the time required for the film thinning and rupture is fairly difficult. The measured induction time usually is a sum of the times of film drainage, rupture and tpcl expansion. For any shorter contact time than the induction time measured as such, a stable attachment cannot be observed or detected. The film rupture is a quick process, and the time of film rupture can be neglected for most of systems that have induction time longer than 10 ms. Nguyen et al.[14,15,16] developed setups for the measurement of gas-solid-liquid tpcl expansion time. It was found that the tpcl expansion kinetics in flotation is strongly controlled by line tension and particle hydrophobicity.

Generally, there are three methods available to measure induction time. In the first method, a captive bubble is moved down to a flat particle bed[17,18,19] or a flat bitumen surface[13] and then away from it. In the second method, a particle is allowed to contact and slide on a stationary[20,21,22] A variation of this method is a rising bubble contacting and sliding on a pendant bitumen drop[23] or an inclined, bitumen-coated flat surface. This method was developed in our laboratory. In the third method, induction time is obtained indirectly from flotation tests, especially the single bubble flotation tests.[2,7,24] To date, the mostly known apparatus for measuring induction time is fabricated by Virginia Polytechnic Institute and State University. A schematic of an induction timer based on captive bubble-particle attachment method is shown in Figure 1. This induction timer was designed to measure the contact time needed to establish attachment between two

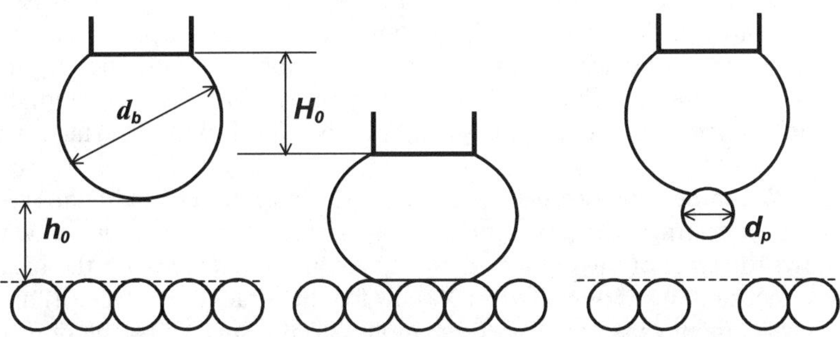

FIGURE 1 A schematic of a captive bubble-particle attachment apparatus

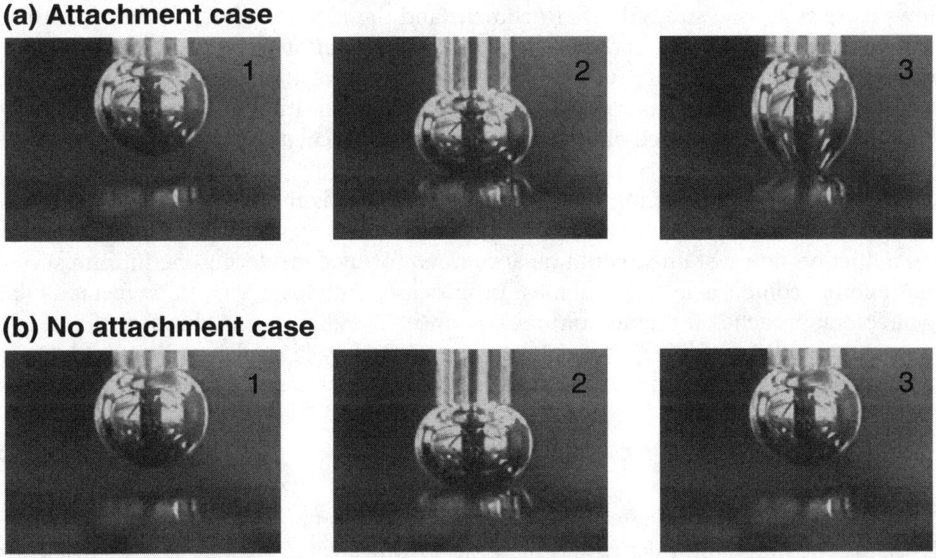

(a) Attachment case

(b) No attachment case

FIGURE 2 Photographs of air bubble-bitumen (a) attachement and (b) no attachment processes observed with a visualization-assisted induction time apparatus

phases separated by a liquid film. This is a widely adopted measurement technique in mineral flotation for air bubble-mineral particle attachments.

In a recent study, Gu et al.[13] refined this kind of induction timer[18,19,25] by using a computerized visualization system. With fine control of physical parameter in this method, the induction time, measured by moving an air bubble toward and then away from a silica particle bed, was found to be affected by the initial gap between the air bubble and the particle bed, the displacement and size of the air bubble, and the velocities of the air bubble approach to and retraction from the particle bed. Air bubble-bitumen attachment in different solutions (de-ionized water, clear process water and process water containing 0.5% fine solids) with or without calcium ion addition was studied at different temperatures. The results indicate that induction time decreased with increasing temperature. The induction time was lowest in de-ionized water and highest in process water containing 0.5% fine solids with 50 ppm calcium ions. A snap shot of the approach and retraction process of an air bubble to a flat bitumen surface is shown in Figure 2.

In Figure 2, Frame 1 represents the initial position of an air bubble above the bitumen surface in a testing solution temperature-controlled with a thermal water bath. Frame 2 shows the state of the deformed air bubble at full contact with the bitumen surface. The contact time was controlled at 300 ms for the attachment case (a) and 250 ms for the non-attachment case (b). The force applied to the surface by the deformed bubble can be calculated precisely by measuring the displacement of the capillary tube and original bubble size. After the capillary tube retracts back to its original position, attachment was observed in frame 3. It is important to note that the amount of capillary retraction can be controlled as desired to minimize the detachment force. This approach allows us to study the strength of attachment and the kinetics of detachment. With the current set up, the physical parameters used in the induction time measurement can be controlled and their effects studied systematically. More importantly, it allows us to identify the range of these parameters over which the variation of the measured induction time could be minimized and as such that the chemical and physicochemical effects be studied.

One concern raised during previous measurements is that the measured induction time was much longer than the time that bubbles could stay in a flotation cell. Such a long induction time obtained could be accounted for by considering the infinite surface of bitumen in comparison with bubbles. In this case, dimples may well be formed when air bubble approaches the plane bitumen surface. The dimple ring hinders thinning of the intervening liquid films.[26,27,28] A theory developed by Li et al.[29] can be used to estimate the induction time semi-quatitatively for bubble-particle attachment.

$$t \propto (d_b)^{2.38}\left(1 + \frac{d_b}{d_p}\right)^{-1.59} \quad \textbf{(EQ 1)}$$

Equation (1) indicates a dependence of induction time on both bubble (d_b) and particle (d_p) sizes. The induction time would increase 17 times when the bubble to particle size ratio (d_b/d_p) changes from 5 (for the case of bitumen flotation) to 0 (for the case of bubble-flat bitumen surface attachment). The unrealistic induction time value obtained with this bubble-solid configuration indicated the need to explore alternative methods for induction time measurements.

In a commercial bitumen production plant, air bubble sizes are about 0.3~0.4 mm. it is well know that electro-flotation[30,31] and dissolved gas flotation[32,33,34] are more efficient than conventional air flotation in water treatment practices. Furthermore, combined flotation by gas nuclei formed from air super saturation systems and bubbles mechanically generated produced a higher flotation recovery than by individual aeration mechanism,[35,36,37,38] since the attachment between gas nuclei (pico or micro bubbles) and flotation bubbles is considered more favorable than bubble-solid attachment. The fact that small bubbles improve flotation efficiency is widely known. The challenge in small bubble-solid attachment research is to generate small individual bubbles within a narrow size range. Bubbles that are smaller than 0.6 mm in diameter have been proven to be particularly difficult to generate and study in laboratory-scale flotation experiments.

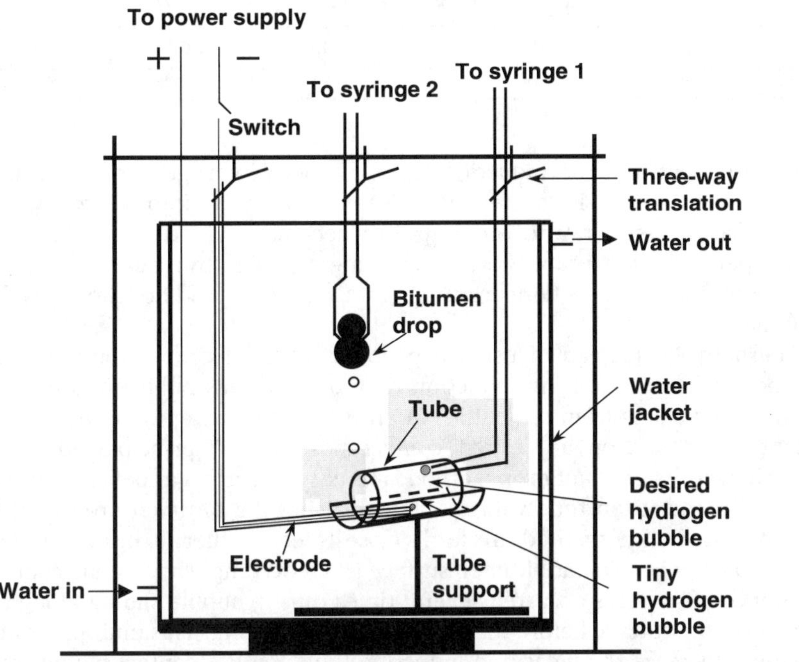

FIGURE 3 A schematic of the single bubble-bitumen drop attachment apparatus

Recently, a new apparatus (Figure 3) was built in our laboratory with a unique feature to generate individual hydrogen or oxygen bubbles at any size larger than 0.01 mm.[23] Once a gas bubble of a desired diameter is produced, it is released and rises to contact a suspended bitumen droplet. The entire process is recorded with a high-speed digital imaging system for subsequent analysis in playback mode. The induction time of hydrogen bubble-bitumen droplet attachment was measured as a function of temperature, bubble size, bubble rise velocity and aqueous phase conditions. The results of this study indicated that higher process temperatures and smaller bubble sizes were favourable for bitumen flotation. Additionally, the maximum (critical) bubble size required for effective bitumen flotation was dependent upon aqueous phase conditions, e.g. dissolved air content and ionic composition. For experiments where the aqueous phase chemistry resembled that typically found in the oil sands industry, only very small bubbles were found to attach to the bitumen droplet. Bubble-bitumen attachment in deaerated Edmonton municipal water exhibited a longer induction time than in as-received Edmonton municipal water. In as-received municipal water, the naturally dissolved air fostered the formation of air nuclei, activating the bitumen surface and bridging larger bubbles to the bitumen surface. A comparison was made between hydrogen and oxygen bubbles in gas bubble-bitumen attachment.[39] It was found that, under the same conditions, hydrogen bubbles had a higher potency to attach to the bitumen surface than oxygen bubbles. A critical gas bubble size was determined for both hydrogen and oxygen bubbles attaching to bitumen. All bubbles with sizes larger than the critical diameter did not attach to the pendant bitumen drop. Only those bubbles with sizes smaller than its critical value attached to the pendant bitumen drop. In deaerated water

TABLE 1 Major ions in deaerated Edmonton municipal water (mg/L)

Ca^{2+}	Mg^{2+}	Na^{+}	K^{+}	HCO_3^-	SO_4^{2-}	Cl^-	F^-
93	42.2	17.6	1.4	97	52.1	4.8	2.1

at room temperature, the induction times for both hydrogen and oxygen bubbles at their critical diameters is shorter than 130 ms, which fell into the real world of flotation. Increasing temperature was found to decrease induction time exponentially. A high temperature is favorable for the attachment of both hydrogen and oxygen bubbles to bitumen. But a process temperature higher than 45°C did not show any further benefit to the attachment.

Alternatively, instead of using a pendant bitumen drop, a bitumen coated slide could be used for measuring induction time of small gas bubble-bitumen attachment. This approach stems from difficulties in ensuring the contact of a released bubble to a pendant bitumen drop at desired contact positions. A principal advantage of this approach over the pendant drop method is its easy operation as delicate alignment is not needed. The induction time values obtained using the bitumen drop method and the bitumen coated slide method are fairly close to each other. It has been observed that bubble attachment to a flat bitumen surface is not accomplished in one contact, the contact-bounce off process occurred several times during bubble sliding along an inclined, bitumen-coated surface before the bubble stopped moving. This multiple contacting process would increase the measured induction time. Similar contact-bounce off processes are anticipated in bubble attaching to a pendant bitumen droplet. However, the multiple contact-bounce off process was not observed due to changing bubble sliding velocity.

EXPERIMENTAL

A sample of upgrading feedstock, known as vacuum distillation feed bitumen, was obtained from Syncrude Canada Ltd. Sub-samples from this feedstock sample were used without additional treatment to prepare bitumen coated slides for the attachment experiments. The bitumen was coated on a glass slide, about 15 mm × 15 mm in size, by a Spin-coater (P6700, Specialty Coating Systems, USA) running at 6,000 rpm for 10 minutes.

The experiments were conducted using deaerated municipal water (City of Edmonton tap water) with and without 40 ppm calcium addition. The major ions in the municipal water before deaeration are shown in Table 1. The deaeration was performed by heating the municipal water to 60°C for 4 hours, followed by cooling to room temperature prior to experimentation. A schematic illustration of the single bubble-bitumen attachment apparatus is shown in Figure 4.

Except for the bitumen slide holder, all other parts and working principle of this apparatus are the same as the apparatus described previously.[23] The vertical distance between the point where the bubble was released to the point at which the gas bubble made initial contact was controlled to be sufficiently large for the bubble to reach its free terminal velocity. The sequence of a single gas bubble rising, sliding and attaching to an inclined bitumen surface coated on a glass slide is shown in Figure 5. The attachment is judged by the location where the gas bubble stops moving along the bitumen surface. Several cycles of the bubble contacting and bouncing off the bitumen surface are observed before the gas bubble stops moving. The induction time is taken to be equal to the traveling time of bubble, which is the time required for a bubble to move

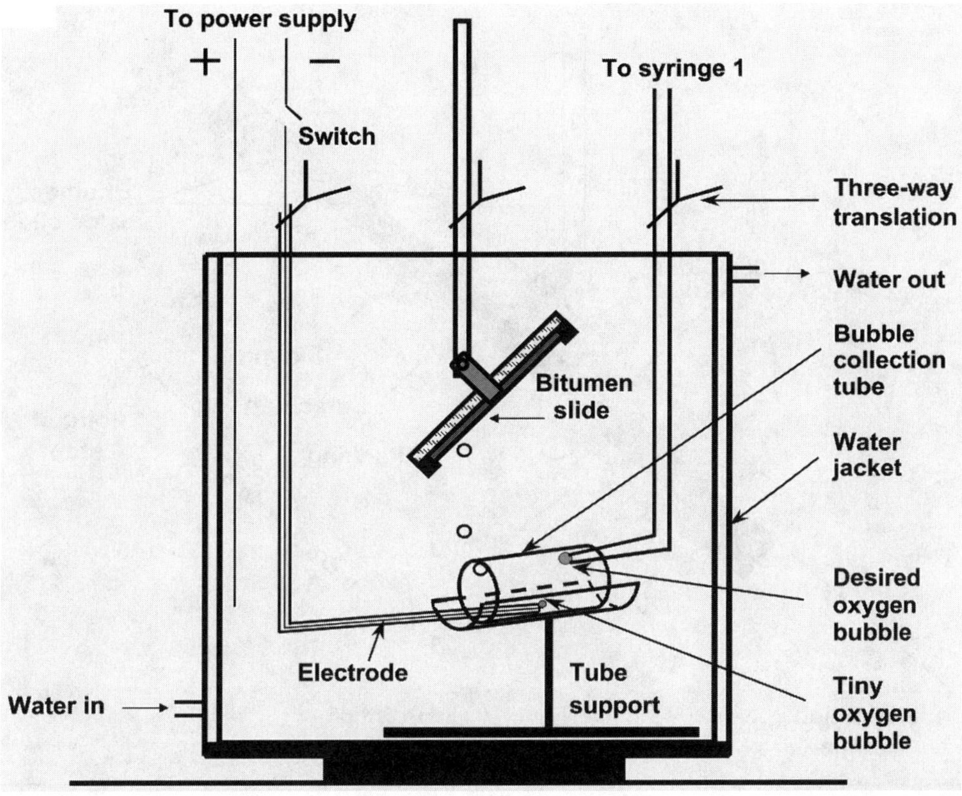

FIGURE 4 A schematic of an apparatus measuring a single bubble attachment to an inclined, bitumen-coated flat surface

from its initial contact point on the bitumen surface to the position where it stops moving. In the case where the gas bubble slides away from the bitumen surface, the contact time is said to be less than induction time as such that the three phase contact is not established. In this paper, only oxygen bubbles are used in single bubble-bitumen attachment tests and the inclination angle is fixed at 30°, unless otherwise stated. The bubble moving velocity is obtained by analyzing the recorded high-speed video. It should be noted that the bubble velocity before contacting the bitumen surface is its rising velocity, while after contact it is the bubble sliding velocity.

RESULTS AND DISCUSSION

Control of Attachment and Detachment

Ralston[4] proposed an equation to describe thermodynamic and dynamic aspects of flotation, $F_{ad} = F_a - F_d$, where F_{ad}, F_a and F_d are the net adhesive, attachment and detachment forces, respectively. Once the wetting perimeter is established, following spreading of the three-phase contact line (tpcl), the static attachment force F_a arises. A dynamic equilibrium

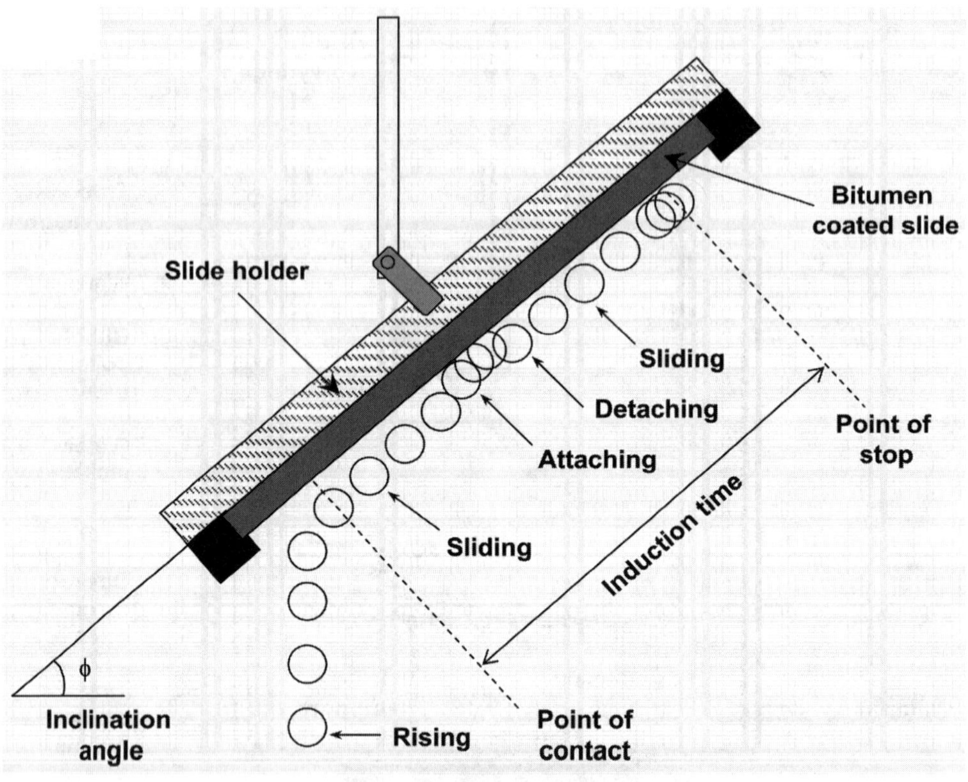

FIGURE 5 A gas bubble rising, sliding, and attaching to an inclined, bitumen-coated slide

position of the tpcl is achieved at F_{ad} = zero. The particle will not remain attached to the bubble and will return to the liquid phase if F_{ad} is negative. It is noteworthy that F_a is determined mainly by thermodynamics, while the detachment force, F_d, is largely controlled by gravitational, inertial or hydrodynamic nature of the system. F_d increases with particle volume, showing a cubic dependence on particle diameter. As a result, F_{ad} is governed by the physicochemical and hydrodynamic characteristics of the system. In our case, the bubble did not attach to the bitumen surface, instead it slid on bitumen surface if F_{ad} is negative. The observed periodic change of bubble sliding velocity along the inclined plate (Figure 5) indicates a delicate force balance during thinning intervening water films and tpcl expansion. The increase of the attachment force that is dependent on rim diameter of the tcpl was responsible for the slowing down and the detachment force to reduce the perimeter of tpcl, for the speeding up of bubble sliding velocity. This periodic change resembles stick-slip motion of two solid surfaces sliding against each other under the external driving force. The contact angle hysteresis appears to be a reason for oscillation in the rim diameter of the tpcl during bubble sliding against an inclined hydrophobic plate. Taylor and Michael showed that a hole in a film of liquid expands if the diameter of the hole is more than twice the film thickness. Otherwise, the hole would close up. The reason for the tpcl oscillation could be a time-dependent change of contact angle, a reflection of surface heterogeneity and instability.

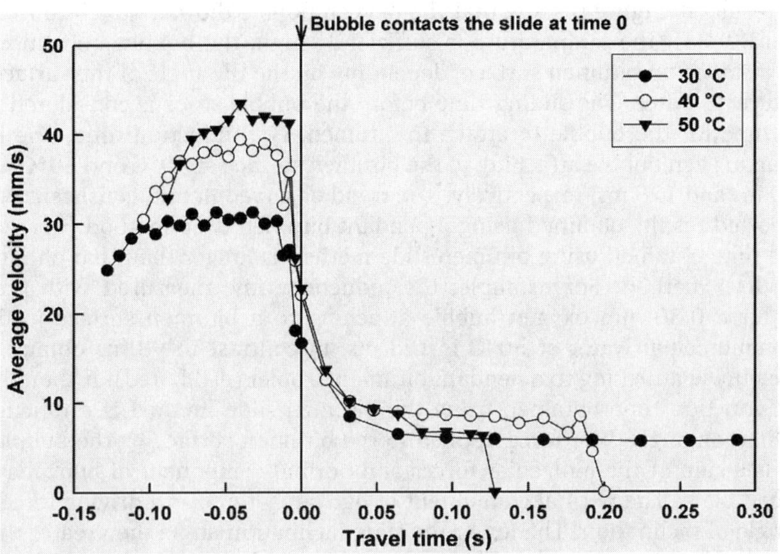

FIGURE 6 Effect of temperature on induction time of oxygen bubble (0.30 mm)—bitumen attachment

Effect of Temperature

Figure 6 shows the results of bubble-bitumen attachment in deaerated water without calcium addition at three different temperatures. The bubble (0.30 mm in diameter) average velocity profiles feature four distinct regions: (1) free rising in which the velocity increases, (2) rising at its terminal velocity in which the velocity is constant, (3) approaching and contacting the bitumen surface in which the velocity decreases sharply due to increased drag and impact under the influence of the bitumen-coated surface, and (4) sliding along the bitumen surface. The bubble terminal velocity increases with increasing temperature. The measured oxygen bubble terminal velocities of 30.5, 38.3, 42.0 mm/s at 30, 40 and 50°C, respectively, are in the same increasing trend as the values of 43.9, 48.1 and 52.0 mm/s calculated using the Turton and Levenspiel's correlation[40] modified by Dewsbury et al.[41] The presence of contaminants in the deaerated municipal water appears to be responsible for bubble rising at velocities below its calculated terminal velocity. The contaminants would accumulate at air-water interface and become part of rising bubbles. Coupled with well-known effect of added mass, the accumulated contaminants would further reduce bubble rising velocity. It is conceivable that deformation of bubble would also contribute to deviation of bubble rising velocity from its terminal velocity of equivalent spherical solid particles. However, the deformation for the contaminated bubbles in the size range encountered in our experiments and hence its effect on bubble rising velocity could be considered negligible. In these measurements, the vertical distance between the point where the bubble was released to the point at which the gas bubble made initial contact was sufficiently large to ensure that the bubble reaches its terminal velocity, indicated by a constant rising velocity. Interestingly, the steady state bubble sliding velocity before attachment does not change significantly with changing water temperatures.

The results in Figure 6 show that the oxygen bubble did not attach to the bitumen surface at 30°C. If the temperature is sufficiently high, the bubble stops moving after sliding a distance on bitumen surface, depending on the tilt angle of the surface and the temperature of water. The sliding time before the bubble stops is considered to be the required time for the bubble to attach to bitumen, i.e., induction time. The induction time for an oxygen bubble attaching to the bitumen surface at 40°C and 50°C was found to be 200 ms and 128 ms, respectively. The trend observed here is consistent with previously reported results obtained using a pendant bitumen drop method.[23] However, the induction time obtained using bitumen slide method is longer than that obtained using bitumen drop method. For example, the induction time measured with the current method, for a 0.30 mm oxygen bubble attaching to a bitumen surface in deaerated Edmonton municipal water at 50°C is 128 ms, in contrast to 98 ms obtained for the same size bubble attaching to a pendant bitumen droplet (estimated). It should be noted that the induction time obtained using the bitumen slide method is a function of the slide inclination angle. The force applied to the bitumen surface by the bubble for film thinning is a sum of the molecular forces and normal component of buoyancy force to the inclined plate. The normal component of buoyancy force as a driving force depends on the angle of inclination. The lower the slide inclination angle the greater the normal component of buoyancy force to bitumen surface and hence the shorter the induction time. In this study the inclination angle was kept at 30°. Variation of tilt angle provides a convenient way to control applied bouncy force for thinning intervening liquid films, which allows us to evaluate the role of hydrodynamic forces on intervening liquid film thinning. In the bitumen drop method, the initial angle of contact was kept at 12°, which is equivalent to a slide inclination angle of 12°. After contact, in the case of pendant bitumen drop method, the tangential angle of bubble sliding changes with time.

Effect of Calcium Addition

As an example of application, the effect of calcium addition on bubble-bitumen attachment was studied using this method. The results obtained in the deaerated Edmonton municipal water with and without the addition of 40 ppm calcium at 50°C are summarized in Figure 7.

Similar average velocity profiles were obtained in both cases, indicating a minimal influence of calcium addition on bubble rising velocity. The induction time for the case with 40 ppm calcium addition is reduced to 116 ms from the case without calcium addition (128 ms). This 9.4% reduction is considered marginal, as the experimental error with current method is approximately 5% for clear water statistically. Marginal effect of calcium addition on induction time of air bubble-bitumen attachment in deionized water was observed in our earlier work.[13]

CONCLUSIONS

A modified set up for induction time measurement is described. The induction time measured using this method is comparable with those obtained using bubble attachment to a pendant bitumen droplet. The results of the single oxygen bubble–bitumen attachment experiments show that induction time, used to quantify bubble attachment to a bitumen coated slide, is greatly reduced by increasing the solution temperature. The addition of calcium in the absence of fine clays has a marginal effect on induction time. These

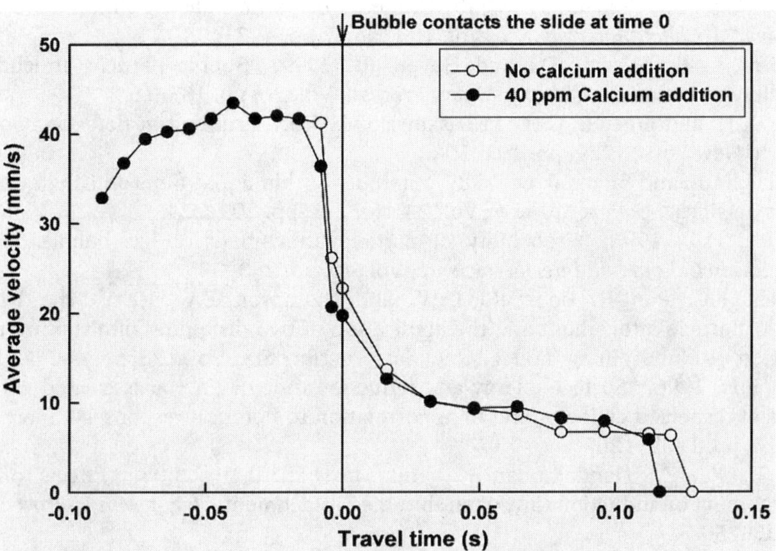

FIGURE 7 Effect of calcium addition on attachment of an oxygen bubble (0.30 mm) to a bitumen-coated slide at 50°C

findings confirm that temperature is one of the most important operating parameters in industrial bitumen flotation process. Compared with the bitumen pendant drop method, a principal advantage of this new approach is its easy operation for accurate induction time determination. In addition, the external forces for thinning an intervening liquid film can be easily controlled by varying the tilt angle of flat bitumen surfaces, which allows fundamental mechanism of film thinning to be studied.

ACKNOWLEDGEMENT

Financial support from NSERC Industrial Research Chair in Oil Sands Engineering (held by JHM) is greatly appreciated.

REFERENCES

1. Dai, Z., Dukhin, S., Fornasiero, D., and Ralston, J., 1998, "The inertial hydrodynamic interaction of particles and rising bubble with mobile surface," *J. Colloid Interface Sci.*, Vol. 197, pp. 275–292.
2. Dai, Z., Fornasiero, D., and Ralston, J., 1999, "Particle-bubble attachment in mineral flotation," *J. Colloid Interface Sci.*, Vol. 217, pp. 70–76.
3. Dai, Z., Fornasiero, D., and Ralston, J., 2000, "Particle-bubble collision models—a review," *Advances in Colloid and Interface Science*, Vol. 85, pp. 231–256.
4. Ralston, J. Dukhin, S.S., and Mishchuk, N.A., 2002, "Wetting film stability and flotation kinetics," *Advances in Colloid and Interface Science*, Vol. 95, pp.145–236.
5. Dobby, G. S., and Finch, J.A., 1987, "Particle size dependence in flotation derived from a fundamental model of the capture process," *Int. J. Miner. Process.*, Vol. 21 No. 3–4, pp. 241–260.

6. Dobby, G.S., and Finch, J.A., 1986, "A model of particle sliding time for flotation size bubbles," *J. Colloid Interface Sci.,* Vol. 109, No. 2, pp. 493–498.
7. Ralston, J., Fornasiero, D., and Hayes, R., 1999, "Bubble–particle attachment and detachment in flotation," *Int. J. Miner. Process.*, Vol. 56, pp. 133–164.
8. Yoon, R.H. and Luttrell, G.H., in Laskowski, J. (Ed.), Frothing in flotation, Gordon and Breach, New York, 1989, pp. 101–102.
9. Schulze, H.J., and Birzer, J. O., 1987, "Stability of thin liquid-films on Langmuir-Blodgett layers on silica," *Colloid Surfaces*, Vol. 24, No. 2–3, pp. 209–224.
10. Schulze, H.J., 1992, "Probability of particle attachment on gas-bubble by sliding," *Advances in Colloid and Interface Science,* Vol. 40, pp. 283–305.
11. Paulsen, F.G., Pan, R., Bousfield, D.W., and Thompson, E.V., 1996, "The dynamics of bubble/particle attachment and the application of two disjoining film rupture models to flotation.: I. Nondraining model ," *J. Colloid Interface Sci.,* Vol. 178, pp. 400–410.
12. Peng, F.F., 1996, "Surface energy and induction time of fine coals treated with various levels of dispersed collector and their correlation to flotation responses," *Energy & Fuels,* Vol. 10, pp. 1202–1207.
13. Gu, G., Xu, Z., Nandakumar, K., and Masliyah, J.H., 2003, "Effects of physical environment on induction time of air–bitumen attachment," *Int. J. Miner. Process.,*Vol. 69, pp.235–250.
14. Nguyen, A.V., Schulze, H.J., Stechemesser, H., Zobel, G., 1997, "Contact time during impact of a spherical particle against a plane gas–liquid interface: experiment," *Int. J. Miner. Process.*, Vol. 50, pp. 113–125.
15. Stechemesser, H., and Nguyen, A. V., 1999, "Time of gas–solid–liquid three-phase contact expansion in flotation," *Int. J. Miner. Process.*, Vol. 56, pp. 117–132.
16. Phan, C.M., Nguyen, A.V., and Evans, G.M., 2003, "Assessment of Hydrodynamic and Molecular-Kinetic Models Applied to the Motion of the Dewetting Contact Line between a Small Bubble and a Solid Surface," *Langmuir,* Vol. 19, pp. 6796–6801.
17. Eigeles, M.A., and Volova, M.L., 1960, "Kinetic investigation of effect of contact time, temperature and surface condition on the adhesion of bubble to mineral surfaces," Proc. 5th Int. Mineral Processing Congr. (London). Inst. Min. Metall., London, pp. 271–284.
18. Ye, Y., Khandrika, S.M., and Miller, J.D., 1989, "Induction time measurements at a particle bed," *Int. J. Miner. Process.*, Vol. 25, pp. 221–240.
19. Yoon, R.H., and Yordan, J.L., 1991, "Induction time measurements for the quartz– amine flotation system," *J. Colloid Interface Sci*. Vol. 141, No.2, pp. 374–383.
20. Wang, W., Zhou, Z., Nandakumar, K., Xu, Z., and Masliyah, J.H., 2003, "Attachment of individual particles to a stationary air bubble in model systems," *Int. J. Miner. Process.*, Vol. 68, pp. 47–69.
21. Nguyen, A.V., and Evans, G.M., 2004, "Attachment interaction between air bubbles and particles in froth flotation," *Experimental Thermal and Fluid Science*, Vol.28, pp. 381–385.
22. Nguyen, A.V., and Evans, G.M., 2004, "Movement of fine particles on an air bubble surface studied using high-speed video microscopy," *J. Colloid Interface Sci.* Vol. 273, pp. 271–277.
23. Gu, G., Sanders, R. S., Xu, Z., Nandakumar, K., and Masliyah, J.H., 2004, "A novel experimental technique to study single bubble–bitumen attachment in flotation," *Int. J. Miner. Process.*, Vol. 74, pp. 15–29.
24. Nguyen, A.V., Ralston, J., and Schulze, H.J., 1998, "On modeling of bubble–particle attachment probability in flotation," *Int. J. Miner. Process.*, Vol. 53, pp. 225–249.
25. Sven-Nilsson, I., 1934, "Effect of contact time between mineral and air bubbles on flotation," *Kolloid-Z,* Vol. 69, No. 2, pp. 230–232.

26. Joye, J.L., Miller, C.A., and Hirasaki, G.J., 1992, "Dimple formation and behavior during axisymmetrical foam film drainage," *Langmuir*, Vol. 8, pp. 3083–3092.
27. Hartland, S., Yang, B., and Jeelani, S., 1994, "Dimple formation in the thin-film beneath a drop of bubble approaching a plane surface," *Chem. Eng. Sci.*, Vol. 49, No. 9, pp. 1313–1322.
28. Tsekov, R., and Ruckenstein, E., 1994, "Dimple formation and its effect on the rate of drainage in thin liquid-films," Colloids and Surfaces A-Physicochemical and Engineering Aspects, Vol. 82, No. 3, pp. 255–261.
29. Li, D., Fitzpatrick, J.A., and Slattery, J.C., 1990, "Rate of collection of particles by flotation," *Ind. Eng. Chem. Res.*, Vol. 29, No. 6, pp. 955–967.
30. Fukui, Y., and Yuu, S., 1984, "Development of apparatus for electro-flotation," *Chem. Eng. Sci.*, Vol. 39, No. 6, pp. 939–945.
31. Raju, G., and Khangaonkar, P., 1982, "Electro-flotation of chalcopyrite fines," *Int. J. Miner. Process.*, Vol. 9, No. 2, pp. 133–143.
32. Ljunggren, M., and Jonsson, L., 2003, "Separation characteristics in dissolved air flotation–pilot and full-scale demonstration," *Water Sci. Technol.*, Vol. 48, No. 3, pp. 89–96.
33. Kiuru, H.J., 2001, "Development of dissolved air flotation technology from the first generation to the newest (third) one (DAF in turbulent flow conditions)," *Water Sci. Technol.*, Vol. 43, No. 8, pp. 1–7.
34. Kempeneers, S., Van, M.F., and Gille, L., 2001, "A decade of large scale experience in dissolved air flotation," *Water Sci. Technol.*, Vol. 43, No. 8, pp. 27–34.
35. Klassen, V.I., and Mokrousov, V.A., 1963, An Introduction to the Theory of Flotation; Butterworth: London.
36. Dziensiewicz, J., and Pryor, E.J., 1950, "An investigation into the action of air in froth flotation," *Trans. IMM., London,* Vol. 59, pp. 455–491.
37. Shimoiizaka, J., and Matsuoka, I., 1982, "Applicability of air-dissolved flotation for separation," *Proc. XIV International Mineral Processing Congres,"* Toronto, Canada, Oct. 17–23.
38. Tao, D., 2004, "Role of bubble size in flotation of coarse and fine particles–A review, " *Separation Science and Technology,* Vol. 39, No. 4, pp. 741–760.
39. Gu, G., Sanders, R.S., Xu, Z., Nandakumar, K., and Masliyah, J.H., 2004, "Hydrogen and oxygen bubble attachment to a bitumen drop," *Can. J. Chem. Eng.*, accepted.
40. Turton, R., and Levenspiel, O., 1986, "A short note on the drag correlation for spheres," *Powder Technol.*, Vol. 47, No. 1, pp. 83–86.
41. Dewsbury, K., Karamanev, D., Margaritis, A., 1999, "Hydrodynamic characteristics of free rise of light solid particles and gas bubbles in non-Newtonian liquids," *Chem. Eng. Sci.*, Vol. 54, pp. 4825–4830.

Recent Advances in Coal Flotation

Subash Chander* **and B. Arnold†**

Many developments have occurred in coal flotation with regard to increasing recovery from the flotation circuit and its impact on overall coal preparation plant recovery. This includes innovative design of flotation circuits, equipment modification and development, methods of enhancing the use of reagents, effect of slimes, and recognizing the effect of water chemistry. This paper will review recent advances and present some guidelines to improve process efficiency.

RECENT ADVANCES

Many developments have occurred in coal flotation with regard to increasing combustible matter recovery and lowering the ash content of the product. The developments include innovative design of flotation circuits, equipment modification and developments, methods of enhancing the use of reagents, and recognizing the effect of water chemistry. The process is generally employed for flotation of fine coal, that is, coal having nominal particle sizes smaller than 595 µm (28-mesh). Up until the mid–1980s, froth flotation circuits were designed to process a nominal 28-mesh top-size. In general, as particle size increases, the flotation rate decreases rather dramatically if conventional reagents and flotation machines are used. Aplan (1999) presented excellent review of the historical development of coal flotation in the United States. In addition, Laskowski (2001) has written a monograph on fine coal flotation and utilization. The focus of this article is on more recent developments.

* Penn State University, University Park, Pennsylvania

† PrepTech, Inc., Apollo, Pennsylvania

Innovative Design of Flotation Circuits

In a typical preparation plant, the size range of particles that are presented to the flotation process often depends on the performance of upstream processing units. In a typical operation, the coarse coal is deslimed with 0.5 mm wedge wire screens that tend to provide a flotation feed with a topsize of 0.7 to 1.00 mm, depending on the amount of wear on the sieve bends (Firth, 1999). Early installations of conventional froth flotation in the United States also followed this circuitry, though most 0.5 mm topsize froth flotation circuits have been abandoned in favor of circuits with 0.15 mm or 0.10 mm topsize coal. The minus 0.5 mm × 0.15 mm size fraction is often cleaned in a two-stage water-only cyclone (or hydrocyclone) circuit or on spiral concentrators.

The flotation feed passes through one or more cyclone stages and inefficiency in those separations results in a wide range of particle sizes that report to flotation. Large fluctuations in particle size distribution present additional challenges to the flotation plant operator. The presence of both the extra coarse and very fine particles present difficulties in conventional flotation and these problems may be tackled by modifying the flotation circuit.

Various froth flotation circuitry options have been discussed in the past by various researchers (Firth et al, 1979; Purcell, 1982; Arnold, 2000). Circuits such as split feed flotation, two-stage reagent addition and the ‘Grab and Run’ circuit take advantage of differences in flotation rate and recovery of impurities and coal particles of different sizes.

Today, other circuitry options are now becoming popular in commercial coal preparation plants.

Desliming Before Flotation. Separating the fine fraction by screens or hydrocyclone and recovering the combustible matter by the flotation process is the traditional method for recovering fine coal. Reportedly, British Coal found that cleaning the fine coal fraction was 3–4 times more costly than cleaning the coarse fraction (Anonymous, 1991). Fine coal is also more difficult to dewater, adding to the overall production cost. As a result many approaches have been considered for improving process efficiency and cost reduction. One of the approaches is to deslime fine coal material before it is sent to the flotation section. Some very fine coal is lost, but the process improvement can justify this operation. The carryover of ash-forming minerals with water is reduced, thereby decreasing the need to use wash water for improving froth properties and clean coal grade. Additionally, hydrophobic coal particles may improve the settling characteristics of black water thickener and reduce reagent consumption. Removal of clay fines by desliming prior to flotation improves the flotation efficiency in both conventional and column flotation cells. Desliming before flotation has the advantage of removing clay fines, which are deleterious to flotation of coarser particles, thereby reducing residence time for flotation and increasing plant throughput and capacity. With the removal of the ultrafine fraction, the moisture content of the dewatered product can be reduced to relatively low values. In addition to these advantages, the capital cost of a new fine coal circuit can almost be halved by reducing the volume of slurry feeding the froth flotation circuit (Bethell, 2004), when small diameter (say 165 mm) desliming cyclones are used prior to flotation.

Two stages of desliming cyclones prior to flotation cells have been incorporated at the Mount Thorley plant in the Hunter Valley (Firth, 1999). Desliming cyclones are now used prior to column froth flotation cells in several coal preparation plants in West Virginia and Virginia, installed in circuitry similar to that given in Figure 1. Desliming

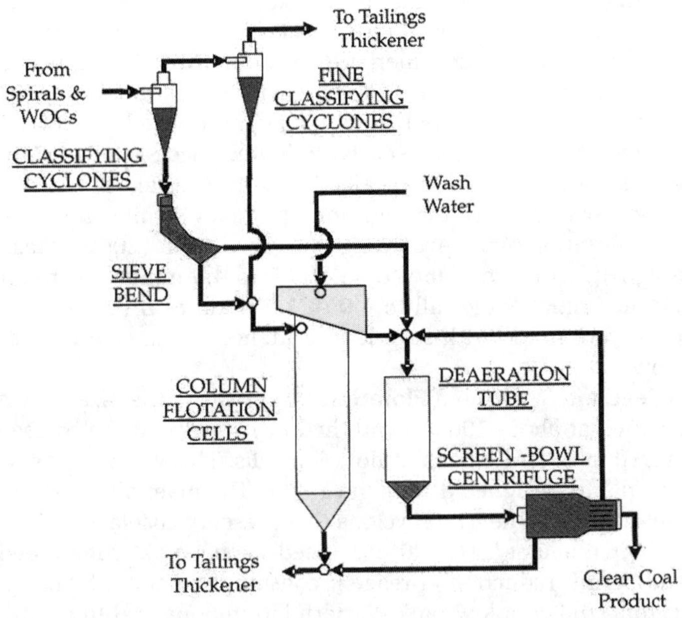

FIGURE 1 Typical deslime-column circuit (Bethell, 2004)

cyclones have also been installed prior to conventional froth flotation cells in two plants in southwestern Pennsylvania. During the life of these two plants, other parts of the cleaning plant were upgraded first, putting a burden on the froth flotation circuit and reducing residence time to the point that coal recovery suffered. By reducing the volume of slurry feeding the flotation banks, the residence time as well as flotation recovery of, especially, the coarser particles has increased.

Bethell (2004) gives some guidelines for determining whether desliming should be used. After careful review of the coal characteristics, coal sales contracts, and the total plant performance, he suggests:

1. Evaluate improvement in total plant yield, not just circuit yield.
2. As long as total moisture is not an issue, 'by zero' flotation should be used for metallurgical coal production. For coals with high clay contents, use column cells. For coals with low clay content, use conventional cells. Use thermal drying as available.
3. For steam coal production, desliming is likely to be the most economical circuitry, unless thermal drying is available. Again, use column cells for high clay content coals and conventional cells (if more economical) for low clay coals.
4. Use filtration with 'by zero' circuits.
5. For steam coal production, evaluate the incremental inerts (ash plus moisture) in each circuit and keep this value essentially the same across circuits.
6. For metallurgical coal production, incremental ash should be the same in each circuit.

Froth Washing. Froth washing or froth sprinkling has been advocated as a way to reduce the carryover of fine ash-forming minerals in the clean coal froth for quite some time (Plaksin and Klassen, 1962; Miller, 1969). Column froth flotation cells routinely use froth washing to clean the froth of these entrained particles. Froth washing is applied in several conventional coal flotation plants (Bethell, 2004). However, this application in conventional circuits meets with limited success unless sufficient capacity or, in other words, residence time is available (Luttrell et al., 2000). The addition of sufficient wash water to remove the entrained particles significantly reduces the residence time in the flotation bank. According to Luttrell et al. (2000), most conventional circuits are designed with a residence time of 3.5 to 4.0 minutes without wash water. The mean residence time would fall to 1.0 to 1.5 minutes if proper wash water rates and froth depths were used. With such low residence times, coal recoveries would be unacceptably low.

Size-Split Feed and Multifeed Flotation Circuits. In the size-split approach, the fine coal is classified at about 100 μm with hydrocyclones and the coarse and fine fractions are recovered in separate bank of flotation cells. This was the preferred circuit for the Bowen Basin plants designed during the 1970s. The main difficulty of this approach lies with the inefficiency of the hydrocyclone that presents misplaced fine particles in the coarse stream. Fuerstenau et al. (2000) discussed the benefits of multifeed flotation circuits that would permit reduction in reagent consumption and a higher product grade. These investigators studied a low-rank Western bituminous coal to compare the performance of single feed and multifeed circuits.

Integrated Hydrocyclone/Flotation Circuit. Meenan (1999) described a flotation circuit design that improves the recovery of the 1.2 mm × 0.3 mm (14 × 48 mesh) particles. The coarser material (greater than 150 μm) is cleaned in recirculating pair hydrocyclones, as shown in Figure 2. The hydrocyclone product is classified in cyclones with fine overflow material reporting to froth flotation. The coarser cyclone underflow material reports to high-frequency vibrating screens to reject the fine pyrite and clays that are carried with the water-split. This circuit design resulted in increased coal recovery of the coarser particles, enhanced fine pyrite rejection, smaller froth cells, and lower reagent consumption when compared to sending all the material to froth flotation.

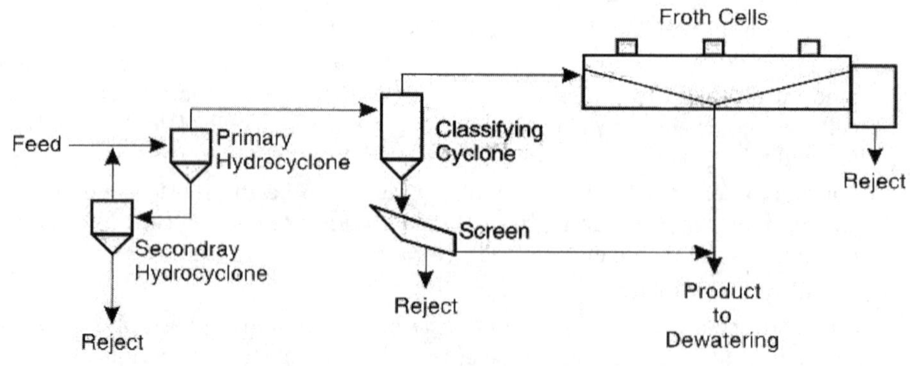

FIGURE 2 Hydrocyclone/flotation circuit (Meenan, 1999)

Equipment Modification and Development

Prior to the 1980s, the conventional, mechanically agitated flotation machines were preferred in coal flotation; but in the last 15 years or so, there has been a marked change and pneumatic column cells that are being used for both new plants and replacement of worn-out mechanical cells. The mechanical cells in use tend to be large, 8.5 to 14 m^3 in size, and are employed in banks of four to six in series.

Canadian researchers in the mid sixties began investigating the flotation process in a countercurrent column. With countercurrent flow patterns of air and slurry, separation can be conducted efficiently and with minimal mixing. Since 1960, many different types of column flotation systems have been developed. This kind of flotation system is more appropriate than the conventional flotation system in many respects. In general, column flotation cells are considered to be advantageous over mechanical flotation cells because they are believed to provide higher concentrate grade, higher recovery, lower maintenance costs and improved process control when designed and operated judiciously. Although there are no moving parts inside the column flotation unit external slurry pumps and air injection systems are required. In many plants, column cells have replaced mechanical cells for improving grade the last cleaning stage. This is done through the use of wash water, which is an important difference between column flotation and conventional flotation cells. The proper design of wash water system is the key to controlling concentrate grade. The wash water is to remove unwanted clay fine particles, which are held in the froth. The performance of column flotation for fine coal recovery from refuse slurry is dependant upon interrelated operating parameters such as wash water rate, feed rate, column height, air flow rate, pulp density and reagent dosage. The effects of frothers on beneficiation of high and low ash Zonguldak coal (80% under 0.5 mm) were investigated by using column flotation (Bayrak et al., 2002). During the experimental study when Dowfroth 250 was used as a frother, the combustible coal recovery increased with increase in frother rate addition. Many researchers agree that froth depth is another important parameter of the column flotation unit. Many studies have shown that froth depth level increases the recovery of fine coal. In column flotation, most of the parameters affecting the results are the same as the general flotation rules but other parameters have an effect because of the shape of the cells. These include the particle size, the coal rank, the type and amount of the collectors, the solid percentage in the pulp and the mixing time.

One of the general advantages of the column cell is the capability of developing deep froth beds and the use of froth washing to remove entrained mineral-matter particles. This allows a fine coal circuit to yield a product much higher grade (low-ash). If flotation feed is deslimed prior to flotation, it is conceivable that froth washing might not be necessary. In such cases, a mechanical cell could give a better performance than a column cell.

Several types of column cells have been developed and the basis for equipment selection is not always clear. Table 1 gives a listing of some recent column cell installations (Luttrell et al., 2004). Examples of commercial column flotation cells include:

Microcel-Column. It has been reported that the Microcel-column installation, developed by the Virginia Polytechnic Institute (Luttrell et al., 1993) has led to an increase in plant yield by 5% at the Peak Down mine in Australia (Brake and Eldridge, 1996). Unlike the conventional flotation column, slurry is taken from the main cell and passed through in-line mixtures into which air is injected and the mixture re-enters the main column. The cell produces deep froths into which large amounts of wash water can

TABLE 1 Recent commercial column installations in the US coal industry (Luttrell et al. 2004)

Company/Location	Brand	No. of Units	Diameter (ft)	Height (ft)	Particle Size
Alpha Natural Resources—White Tail	Eriez/CPT	5	14	28	100 M × 0
Alpha Natural Resources—Middle Fork	Microcel	5	10	25	100 M × 0
Alpha Natural Resources—Holston	Microcel	1	14	30	100 M × 0
Alpha Natural Resources—Roxanna	Microcel	2	13	28	65 M × 0
Alpha Natural Resources—Toms Creek	Microcel	2	14	28	100 M × 0
Alpha Natural Resources—Brooks Run	Microcel	2	15	35	100 M × 0
Carbontronics Fuel—Pawney	Jameson	1	16	12	100 M × 0
Carbontronics Fuel—Gibraltar	Jameson	1	16	12	100 M × 0
Carbontronics Fuel—Linville	Jameson	1	16	12	100 M × 0
Cline Resources—Coal Clean	Eriez/CPT	3	15	28	100 M × 0
CONSOL Energy—Keystone	Jameson	1	16	12	100 M × 0
CQ Inc.—Ginger Hill	Jameson	1	16	12	100 M × 0
CQ Inc.—Pleasant Ridge	Jameson	1	16	12	100 M × 0
Horizon Natural Resources—Marrowbone	Microcel	1	8	26	150 M × 0
Horizon Natural Resources—Lady Dunn	Microcel	3	14	30	100 M × 0
Massey Energy—Power Mountain	Eriez/CPT	2	13	24	100 × 325 M
Massey Energy—Liberty	Eriez/CPT	3	13	24	100 × 325 M
Massey Energy—Bandmill	EIMCO	3	13	24	100 × 325 M
Massey Energy—Marfork	Jameson	4	16	12	100 M × 0
Ohio Valley—Century	Eriez/CPT	2	14	24	100 × 325 M
Peabody—Sugar Camp	Jameson	1	16	12	100 M × 0
Powell Mountain—Mayflower	Ken-Flote	4	8	22	100 M × 0
Sigmon Mining—Sigmon	Eriez/CPT	2	14	28	100 M × 0
TECO Mining—Clintwood Elkhorn	Eriez/CPT	2	10	24	100 × 325 M

be introduced. The full-scale Microcel coal installation consists of 16, 3-m diameter columns and it replaced the old split-feed flotation circuit that used large mechanical cells. Eriez currently markets the Microcel columns in the United States.

CoalPro Column. The *CoalPro*, a joint development between CPT (Canadian Process Technologies) and Eriez, commercializes column flotation cells to meet the special requirements of coal producers. It makes use of patented Slamjet air spargers that are claimed to prevent clogging of nozzles during shutdown periods. Other benefits of the *CoalPro* column include:

- Improved product quality at high product yields
- Low capital and operating cost
- Low maintenance cost
- No large aeration pumps
- No large recirculation pumps
- No downcomers with complex piping

Dual Extraction Flotation Cell. Beneficiation Technologies, LLC, have reportedly installed Dual Extraction Flotation Cells (DEC) in applications such as phosphate, advanced coal flotation and coal-pyrite separation (Kemworks, 2004). Following proper size reduction and desliming, raw feed is introduced as slurry into a feed well then equally distributed through a bottom plate thus entering the upper chamber where the initial separation takes place. Air bubbles are created by introducing fresh water at 30 psi (2068 kPa) into two header pipes with hose attachments leading into sixteen infusers specially designed to create a high volume of finely divided air bubbles. Additionally, low pressure air at 5 psi (34.5 kPa) is provided by a blower to combine in the infusers. As the rising air bubbles move upward in the chamber a physio-chemical separation takes place with desired coal or mineral particles, now coated by a molecular film from the reagentized raw slurry, attaching the bubble and rising as a concentrate to a launder for removal to a dewatering system. The bulk of recoverable material is removed as an upgraded product. Considerable materials in the +28 mesh range along with gangue or waste matter exits the upper chamber via a discharge throat at a controlled rate by means of a dart or plug valve positioned in the opening and connected to a movable shaft capable of up or down movement through a thread bar and adjusting wheel arrangement. This material usually lost to tailings in older flotation machines now has sufficient residence time in the lower chamber where an identical separation as described for the upper chamber takes place. The concentrate from this lower chamber overflows into the launder and combines with concentrate from the upper chamber for dewatering by cyclones or other means. Tailings or waste exits the lower chamber via a tailings discharge pipe to a reject or tailings disposal area.

Jameson Cell. In the Jameson cell the feed is pumped vertically downward through an orifice into a pipe, which is termed as the "downcomer." Air is introduced at this point via the low pressure that is generated through the Venturi action. The apparent countercurrent movement of both the particles and air bubbles is considered to allow for efficient contacting and capture of hydrophobic particles by air bubbles. It is important to note that the counter current movement is apparent because net flow of both the particle and the bubbles is downward in the "downcomer." A 5-m cell will typically have upto 16, 250-mm "downcomers" each with a nominal capacity of 60 m^3/hr slurry (Firth, 1999) requiring considerable amount of piping and feed distribution. Mohanty and Honaker (1999) conducted performance optimization study using laboratory Jameson Cell concluded that a high positive bias factor of about 0.6, which translates to a wash water ratio of 2.5, is required to produce superior coal quality. Gray et al. (2004) presented a case study for a 40,000 tpd copper concentrator in which they compared an all mechanical cell circuit with an all Jameson Cell circuit. They demonstrated that the high selectivity and rapid collection kinetics of the Jameson Cell enabled the separation to be achieved at commercial scale. The power requirement was reduced from 1789 kW in the mechanical cell circuit to 1598 KW for the Jameson Cell circuit. One should be careful in this evaluating the results of their analysis because considerably different residence times were used in the analysis of the two circuits. The residence time was 26 minutes for the rougher for the mechanical cell circuit, whereas it was 4.73 minutes for the Jameson cell circuit. The residence times were 8.4, 12.6, 18, and 12.1 minutes cleaner for the four cleaner stages whereas they were 1.5, 2.0 and 1.6 for the cleaner stages in the Jameson Cell circuit. Furthermore, they omitted the regrind step from their analysis. It is obvious that such a comparison will be very sensitive to hydrophobicity and hence flotation kinetics of the material being floated.

Other Technologies. In addition to column flotation cell equipment modifications, EIMCO developed the SmartCell technology. SmartCells are arranged in a bank of cells similar to a bank of conventional cells. However, they are circular, not rectangular, which reducing manufacturing costs and floor space requirements and eliminates the requirement for baffling in the cell. They are based on the 1+1 mechanism found in WEMCO (now Dorr Oliver-EIMCO) conventional cells. They incorporate an expert control system. CONSOL Energy has installations of SmartCells at both its Bailey and McElroy preparation plants.

Methods of Enhancing the Effectiveness of Reagents

Methods of enhancing the effectiveness of reagents is a complex issue and requires an understanding of interactions between various sub-processes in flotation, which is beyond the scope of this article. Even though many new reagents of been shown to be more effective in promoting grade, recovery or both, they have yet to be put in practice probably because of higher costs and insufficient experience. Non-ionic surfactants and water-soluble polymers have been utilized to modify the coal surface (Harris, 1995). Li et al. (1992) who used a comb-like polymer found that the coal became more hydrophobic with increasing promoter concentration regardless of its original floatability. The PEO/PPO/PEO triblock co-polymers were also found to improve coal flotation and the mechanism of polymer action was a function of the coal rank (Polat et al., 1994, 1997; Polat and Chander, 1995, 1998, 1999). These reagents had double effect on flotation: they modified the coal surface and also they improved the emulsification of the oily collector. For high rank coals, which usually require relatively small amounts of the collector, the surface modifier function of the polymers was dominant over their emulsifier function. The polymer increased ash rejection in flotation primarily because the coal agglomerates, which were observed in the flotation cell, were smaller and considered to be more selective. For medium and low rank coals, where larger oil concentrations were required, the polymer acted both as an emulsifier and a surface modifier. It was suggested based on the surface tension and contact angle studies that adsorption of the block co-polymers at coal/water interface occurred by adsorption of PPO groups by hydrophobic attraction on the most hydrophobic sites, and adsorption of PEO groups by hydrogen bonding on the hydrophilic sites. The coverage of the hydrophilic sites on the surface by the promoter molecules was proposed to be the mechanism by which hydrophobicity increased. For the high rank coals, which contain a relatively small number of hydrophilic sites, the polymer adsorbed on hydrophobic sites, rendering coal less hydrophobic. This was the reason for the observed increase in selectivity for high rank coals since it caused a decrease in the size of the coal agglomerates, hence, in the amounts of ash particles entrapped in the agglomerate structure.

Jia et al., (2000) observed that the addition of oxygenated functional groups to the collector molecule markedly enhances the flotation of lower rank and oxidized coals.

These non-ionic oxygenated promoters (THF series of ester) were more effective collectors than oily collectors for both oxidized and unoxidized coals, attaching to the coal surface through hydrogen bonding for the oxygenated sites or hydrophobic bonding of the hydrocarbon chain for the hydrophobic carbonaceous sites on the coal. They also suggested that nonyl benzene was a better collector than dodecane for the high-sulfur coals, indicating strong interaction of the benzene ring with aromatic sites on the coal surface. Vamvuka and Agridiotis (2001) observed a superior separation when a combination of kerosene and dodecylamine were utilized.

Water Quality

The water chemistry in coal flotation circuit is a product of the chemistry of make up water, soluble salts such as sodium chloride from the raw coal, the soluble products from oxidation of raw coal (both coal and sulfur-bearing minerals), the reagents used in the plant and the adsorption of chemical species on particles as they flow through the plant. The changes in water quality can affect the performance in many ways. Some ions such as dissolved iron and aluminum may adversely affect flotation of coal particles (Somasundaran et al., 2000), whereas dissolved salts and organic substances can alter frothing characteristics. Fulvic and humic acids resulting from oxidation of low rank coals could also have an adverse effect on floatability of coal.

EFFECT OF WATER CHEMISTRY ON COAL FLOTATION

Coal flotation is affected also by precipitation or adsorption of the dissolved mineral species released from the coal itself during grinding and pulping (Somasundaran and Liu, 1998; Somasundaran et al., 2000). The release of mineral ions examined as a function of pH by Liu et al. (1994). The concentrations of dissolved Fe, Al, Ca and Mg decrease as the pH increase, with the mode of alkali addition being irrelevant. This result suggests that if the pH increases during coal processing, there will be precipitation of metal ion species whereas if the pH decreases, there will be dissolution of mineral species. The authors observed a decrease in the flotation recovery to a great extent under the precipitation conditions. They suggested that the hydroxide precipitate from the dissolved mineral species adsorbs on the coal surface and makes surface hydrophilic. It was concluded that the presence of these species could be controlled by manipulating the pH at different stages of processing depending on coal type. In another study of the effects of different hydrolyzable multivalent ions such as Ca and Al, it was observed that the adsorption of Ca increases slightly with pH up to 8 and then sharply above that value, while that of Al exhibits a sharp increase around pH 3–5 (Celik and Somasundaran, 1980). The sharp uptake of these metal ions appears to be governed by the formation of $CaOH^+$ and $AlOH^{2+}$. These results show that the adsorption of multivalent species can drastically affect the hydrophobicity of coal and depress the flotation most probably due to surface adsorption or precipitation.

Fuerstenau et al. (1983), and Fuerstenau and Diao (1992) demonstrated, using film flotation and zeta potential measurements that the maximum flotation response for coal occurs close to its isoelectric point. The work of Harvey et al. (2002), where the effect of electrolyte (NaCl and $MgCl_2$) concentrations on coal flotation was investigated using a modified Hallimond tube, supported this observation. The electrical double layer (EDL) interactions did not appear to control by floatability of coal. The flotation at low electrolyte concentrations decreased due to other effects dominating the EDL interactions, but the recovery increased at high electrolyte concentrations most probably due to compression the EDL. Since the addition of polyvalent cations decrease the zeta potential of coal close to zero, they could be used as a flotation aid within proper concentration range.

Water quality also affects the behavior of clay slimes in flotation. The effect of clay slimes on coal froth flotation has been investigated by many researchers. Some note that the presence of ultrafine clay or other slime particles inhibit flotation (Brown and Smith, 1954; Jowett et al., 1956; Burdon et al., 1976; Mishra, 1978). Others, however, note no loss of recovery (Yancey and Taylor, 1935; Szczypa et al., 1973; Neczaj-Hruzewicz et al., 1974; Firth and Nicol, 1981), and indicate that the major problem with slimes is the dilution of the froth with ash particles. Arnold and Aplan (1986a, 1986b) evaluate both the type of clay and the role of water quality and conclude that both are important in determining the effect of clay slimes on coal flotation. Kaolinite and illite, which do not significantly depress coal flotation, contaminate the floated clean coal largely by carry-over with the froth, though electrostatic attachment of clay to the coal contributes in a lesser degree to higher ash content in the froth. Bentonite greatly depresses all but the most hydrophobic of coals by armour-coating of the bubbles, preventing coarse particle attachment, and increasing slime coatings, all because of its high surface area, charged sites and ion-exchange capacity.

The presence of ions and pH affect the flotation of coal, especially in the presence of clays. Table 2 shows the effect of water quality. At pH 6.5, the addition of 20 percent kaolinite does not reduce coal recovery in tap water. However, coal recovery is reduced from 80.6 to 40.2 percent in distilled water or with no ions present. The addition of ions increases coal recovery as the electrical double layer on both the coal and clay are reduced, keeping the system dispersed. Bentonite is shown to decrease coal recovery to an even greater extent than kaolinite. Also shown in Table 2 is the effect of fuel oil collector addition. Recovery of coal is increased with collector addition even in the presence of clays and without ions present.

The recovery of coal is known to decrease at low and high pH values with the maximum recovery being obtained at or near neutral pH (Zimmerman, 1948). Using kaolinite, Arnold and Aplan (1986b), however, showed little change in coal recovery over a range of pH 3–9. This corresponds with the results of Firth and Nicol (1981). However, bentonite addition depressed coal flotation at low pH while recovery improved somewhat at pH 9.

Many coal preparation plants are now paying close attention to the presence of slimes or ultrafines in a flotation feed and are installing desliming cyclones prior to flotation (Firth, 1999; Bethell, 2004) as described previously.

SERVICE, MAINTENANCE, PLANNED AND UNEXPECTED SHUTDOWNS

In a typical industrial plant, the size range of particles that are presented to the flotation process often depends on the performance of upstream processing units. These include screens, jigs, spirals, cyclones, and heavy media baths. In a typical operation, the coarse coal is deslimed with 0.5 mm wedge wire screens that tend to provide a flotation feed with a top size of 0.7 to 1.00 mm, depending on the amount of wear on the sieve bends (Firth, 1999). The feed may also pass through one or more cyclone stages and inefficiency in those separations may also result in a wide range of particle sizes that report to

TABLE 2 The effect of water quality in a Pittsburgh coal-clay system (Arnold and Aplan, 1986b)

	Tap Water	Distilled Water	40 ppm Ca^{2+} 20 ppm Mg^{2+}	40 ppm Ca^{2+} 200 ppm Mg^{2+}	200 ppm Ca^{2+} 20 ppm Mg^{2+}	1500 ppm $SO_4{}^{2-}$
Coal alone	87.7	80.6	88.8	—	—	76.4
20% kaolinite						
0.1 kg/t MIBC	87.9	40.2	63.6	84.6	81.6	77.3
(+ 0.5 kg/t fuel oil)	(94.3)	(85.0)	(91.6)	—	—	—
5% bentonite						
kg/t MIBC	54.6	12.9	13.0	52.8	50.7	—
(+0.5 kg/t fuel oil)	(89.3)	(52.6)	(48.7)	—	—	—

The numbers in the table correspond to coal recovery, in percent at pH 6.5.

flotation. Large fluctuations in particle size distribution present additional challenges to the flotation plant operator.

While equipment selection is generally based on capital cost and equipment performance guarantees, the modern plant operator increasingly recognizes the importance of maintenance, service and ability to effectively deal with planned and unexpected shutdowns. The demand to increase productivity and efficiency is rather high and these issues play a very important role in both equipment and vendor selection.

Spargers are, perhaps, the area of most column cells that requires the most attention. Selection of spargers that limit the potential for plugging, that are easily replaced, and that are made of wear resistant material limit the maintenance requirement.

It is also important to have a vendor with a real understanding of the flotation process in general so that communications with the reagent supplier are good. A superior knowledge of their particular technology is almost a must and it could be recommended that the vendors be involved in any initial test work for flotation circuit design.

SUMMARY

A brief review of recent advances in coal flotation is presented in this article. The equipment selection and circuit design lacks strong scientific basis and is strongly biased by operator preferences, equipment costs and local availability of service and maintenance vendor. The last factor is especially important since the number of process engineers in the coal cleaning plants, especially in the United States is abnormally low. In many cases, the flotation plants are rarely optimized and loss of coal fines is considered acceptable as they probably decrease the flocculant consumption in blackwater thickener. The advances in coal flotation will continue to be slow as long as alternate, less costly fuels are available and the environmental penalties of throwing combustible matter in tailings ponds are not imposed.

REFERENCES

Aplan, F.F., 1999. The historical development of coal flotation in the United States, in Advances in Flotation Technology, Eds. B.K. Parekh and J.D. Miller, SME, Littleton, Co, 269–287.

Anonymous, 1991. Coal preparation technology–influence and trends, World Mining Equipment, March pp. 20–22.

Arnold, B.J., 2000. The 'Grab and Run' revisited–improving selectivity between organic and inorganic components in conventional coal froth flotation. International J. of Mineral Processing, Vol. 58, pp. 119–128.

Arnold, B.J. and Aplan, F.F., 1986a. The effect of clay slimes on coal flotation, Part I: the nature of the clay, International J. of Mineral Processing, Vol. 17, pp. 225–242.

Arnold, B.J. and Aplan, F.F., 1986b. The effect of clay slimes on coal flotation, Part II: the role of water quality, International J. of Mineral Processing, Vol. 17, pp. 243–260.

Bayrak, N., O'Donnell, J., Toroglu, I., 2002. Recovery of fine coal by column flotation, CHEMECA 2002, Christchurch, NZ, 918, September.

Bethell, P.J., 2004. Froth flotation–to deslime or not to deslime?, CPSA Journal, Vol. 3, No. 1, pp. 12–15.

Brake, I.R. and Eldrich, G., 1996. The development of new Microcel column flotation circuit for BHP Coal's Peak Downs Coal Preparation Plant, Coal Prep 1996, Lexington, Kentucky, pp. 237–252.

Brown, D.J. and Smith, H.G., 1954. Colliery Engineering, Vol. 31, p. 245. (Cited in D.J. Brown, 1962. Coal Flotation, Froth Flotation, D.W. Fuerstenau, ed., AIME, New York, NY, pp. 518–538.

Burdon, R.G., Booth, R.W., and Mishra, S.K., 1976. Factors influencing the selection of processes for the beneficiation of fine coal. Seventh International Coal Preparation Congress, Australia, 25 pp.

Celik, M.S. and Somasundaran, P., 1980. Effect of Pretreatments on Flotation and Electrokinetic Properties of Coal. Colloids and Surfaces Vol 1, No. 1, pp. 121–124.

Firth, B.A., 1999. Australian Coal Flotation Practice, in Advances in Flotation Technology, eds. B.K. Parekh and J.D. Miller, SME, Littleton, Co, pp. 289–307.

Firth, B.A. and Nicol, S.K., 1981. The influence of humic materials on the flotation of coals, International J. of Mineral Processing, Vol. 8, pp. 239–248.

Firth, B.A., Swanson, A.R., and Nicol, S.K. 1979. Flotation circuits for poorly floating coals, International J. of Mineral Processing, Vol. 5, pp. 321–334.

Fuerstenau, D.W. and Williams, M.C., and Hu, W., 2000. Application of multifeed circuits for coal and molybdennite flotation, Proc. XXI International Mineral Processing Congress, Rome, Italy, B9–10.

Fuerstenau, D.W. and Diao., J., 1992. Characterization of coal oxidation and coal wetting behavior by film flotation, Coal Preparation Vol. 10, pp. 1–17.

Fuerstenau, D.W., Rosenbaum, J.M. and Laskowski, J., 1983. Effect of surface functional groups on the flotation of coal, Colloids and Surfaces Vol. 8, pp. 153–173.

Gray, M.P., Harbort, G.J. and Murphy A.S., 2004. Flotation circuit design utilizing the Jameson cell, http://www.xstratatech.com/downloads/jc_flotation.pdf.

Harris, G.H., 1995. Coal flotation with nonionic surfactants, Coal Preparation, Vol. 16, pp. 135–147.

Harvey, A., Nguyen, A.V. and Evans, G.M., 2002. Influence of electrical double layer interaction on coal flotation. J. Colloid and Interface Sci. Vol. 250, pp. 337–343.

Jia, R., Harris, G.H. and Fuerstenau, D.W., 2000. An improved class of universal collectors for the flotation of oxidized and/or low-rank coal, International J. of mineral Processing, Vol. 58, pp. 99–118.

Jowett, A., El-Sinbawy, H., and Smith, H.G., 1956. Slime coating of coal in flotation pulps, Fuel, Vol. 35, pp. 303–309.

Kemworks, 2004. http://www.kemworks.com/DEC.htm

Laskowski, J.S., 2001. Coal Flotation and Fine Coal Utilization, Elsevier, Amsterdam, 386 p.

Li, C., Yu X. and Somasundaran P., 1992. Effect of hydrophobically modified comb-like polymer on interfacial properties of coal, Colloids and Surfaces, Vol. 66, pp. 33–43.

Liu, D., Somasundaran, P., Vasudevan, T. V. and Harris, C. C., 1994. Role of pH and dissolved mineral species in Pittsburgh No.8 coal flotation System: 1. Floatability of coal, International J. Mineral Processing, Vol. 41, pp. 201–214.

Luttrell, G.H., Mankosa, M.J. and Yoon, R.H., 1993. Design and Scale Up Criteria for Column Flotation, Proceedings XVIII International Mineral Processing Congress, Sydney, Vol. 3, pp. 785–791.

Luttrell, G., McKeon, T., and Bethell, P., 2000. An in-plant evaluation of froth washing for conventional coal flotation circuits. SME Preprint # 01–191.

Luttrell, G.H., and Yoon, R.H., 1994. Commercialization of the Microcell column flotation technology, Proc. Eleventh Pittsburgh Coal Conference, Pittsburgh, PA, pp. pp. 1503–1508.

Luttrell, G.H., Stanley, F.L., and Bethell, P.J., 2004. Operating and Design Guidelines for Column Flotation, Proc. 21st Annual International Coal Preparation Exhibition & Conference, Lexington, KY, May 4–6, pp. 227–238.

Meenan, G.F., 1999. Modern Coal Flotation Practices, in Advances in Flotation Technology, eds. B.K. Parekh and J.D. Miller, SME, Littleton, CO, pp. 309–319.

Miller, F.G., 1969. Effect of froth sprinkling on coal flotation efficiency, Trans. AIME, Vol 244, pp. 158–167.

Mishra, S.K., 1978. The slime problem in Australian coal flotation, Mill Operator's Conference, Australasian IMM, Mt. Isa, pp. 159–168.

Mohanty, M.K. and Honaker, R.Q., 1999. Performance optimization of Jameson flotation technology for fine coal cleaning, Minerals Engineering, Vol. 12, pp. 367–381.

Neczaj-Hruzewicz, J., Szczypa, J. and Czarkowski, N., 1974. Influence of flotation reagents on formation of gangue slime coating on coal, Trans. IMM, Vol. 83, pp. C261–263.

Plaskin, I.N. and Klassen, V.I., 1962. Methods of improving the process of coal flotation, Fourth International Coal Preparation Congress, Harrogate, pp. 263–271.

Polat, H., Polat, M. and Chander, S., 1994. Deep cleaning of a hvA bituminous coal using a polyethylene oxide/polypropylene oxide block copolymer, Preprints of paper presented at ACS Meeting, vol. 39, no. 2, pp. 405–410 (1994).

Polat, H. and Chander, S., 1995. Wetting and flotation of coal in the presence of block co-polymers, paper presented at the 1995 Annual Meeting of the SME, Denver, March 6–9.

Polat, H., Polat, M. and Chander, S., 1997. Adsorption behaviour of PEO/PPO triblock copolymers on coals of different rank, Proceedings of the XX International Mineral Processing Congress (Aachen), Vol. 3, Flotation and other Physical Chemical Processes, pp. 725–734.

Polat, H. and Chander, S., 1998. Interaction between physical and chemical variables in flotation of low rank coals, Minerals and Metallurgical Processing, vol. 15, pp. 41–47.

Polat, H. and Chander, S., 1999. Adsorption of PEO/PPO triblock copolymers and wetting of coal, Colloids and Surfaces, Vol. 146, pp. 199–212.

Purcell, R.J., 1982. Circuitry variations for the rejection of pyritic sulfur during coal flotation, M.S. Thesis, The Pennsylvania State University.

Somasundaran, P., Zhang, L., and Fuerstenau, D.W., 2000. The effect of environment, oxidation and dissolved metal species on the chemistry of coal flotation. International J. Mineral Processing, Vol. 58,pp. 85–97.

Somasundaran, P and Liu, D., 1998. Dissolved mineral species precipitation during coal flotation, Fuel and Energy Abstracts, Vol. 39, No. 3, p. 170.

Szczypa, J., Neczaj-Hruzewicz, J. and Sablik, J., 1973. Some properties of slime coatings in coal-gangue systems, Trans. IMM, Vol. 82, pp. C167–169.
Vamvuka, D. and Agridiotis, 2001. The effect of chemical reagents on lignite flotation, International J. Mineral Processing, Vol. 61, pp. 209–224.
Yancey, H.F. and Taylor, J.A., 1935. Froth flotation of coal; sulfur and ash reduction, U.S. Bureau of Mines, RI 3263.
Zimmerman, R.E., 1948. Flotation of bituminous coal, Trans. AIME. Vol. 177, pp. 338–356.

Particle Characterization—Liberation

Practical Optimization of Coal Preparation Plants

Gerald H. Luttrell* **and Rick Q. Honaker†**

A coal preparation plant incorporates a variety of solid-solid separation processes that operate in relatively independent parallel circuits. Studies have shown that total plant performance can be optimized by selecting circuit cutpoints that maximize total clean coal yield irrespective of the performance of individual circuits. Although the optimum cutpoints can be determined using simulation programs, these mathematical routines require detailed washability data and empirical partition information that is not often available to plant operators. In addition, routine fluctuations in the washability characteristics of the plant feed make this approach impractical to implement in industrial facilities. As an alternative, practical optimization criteria can be developed for plant operators through a fundamental understanding of scientific principles involved in coal cleaning operations. This article provides an overview of the most important optimization criteria and presents several industrial case studies that illustrate the financial benefits of this practical approach to plant optimization.

INTRODUCTION

Coal preparation plants make use of a variety of solid-solid separation processes to upgrade run-of-mine coals into high quality products for the utility (steam) and metallurgical (coke) coal markets. The typical range of particle sizes that can be efficiently treated by some of the most common coal cleaning processes are summarized in Figure 1. Because of these limitations, modern plants may incorporate as many as four parallel processing circuits to efficiently treat the coarse (plus 50 mm), small (50×1 mm), fine (1×0.15 mm) and ultrafine (minus 0.15 mm) particles. The clean coal products from

* Virginia Polytechnic Institute and State Unversity, Department of Mining & Minerals Engineering, Blacksburg, Virginia

† University of Kentucky, Department of Mining Engineering, Lexington, Kentucky

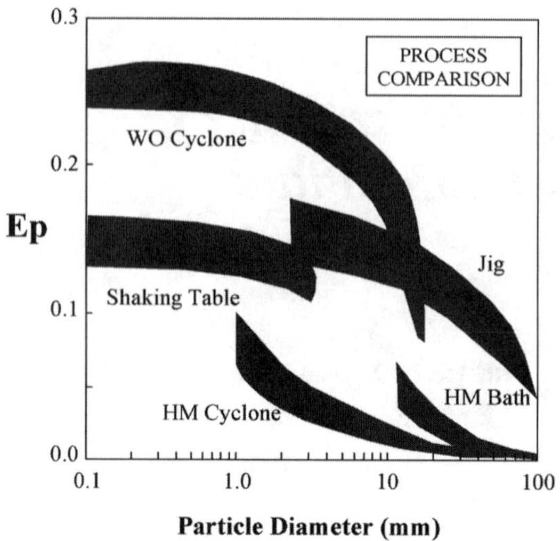

FIGURE 1 Approximate operating efficiencies of coal cleaning equipment (Osborne, 1988)

these circuits are typically blended back together prior to shipment to market. Although some commonalities exist, the selection of what number of circuits to use, which types of unit operations to employ, and how they should be configured, is highly subjective and largely dependent on the inherent characteristics of the coal feedstock (Osborne, 1988). According to Kempnich (2000), the vast majority of mined coal is cleaned in coarse and small coal circuits using dense medium separators and jigs. These processes represent more than two-thirds (68.8%) of the total worldwide installed production capacity. The remaining capacity is treated in fine and ultrafine circuits using water-based separators such as spirals and water-only cyclones (13.1%) and surface-based processes such as froth flotation (13.6%).

The complexities associated with the operation of multiple processing circuits often cause plants to be run under less than optimal conditions. Plant operators must select appropriate cutpoints for each cleaning process to ensure that the overall clean coal products from the plant meet quality criteria dictated by sales contracts. These specifications may include limits on ash, sulfur, moisture, heat content, or other quality indicators. In most cases, plant operators will select cutpoints that provide an identical clean coal quality in every plant circuit. For example, circuits in a given plant may all be set to produce the same clean coal ash of 10% in order to satisfy a contract with a 10% ash limit. While this approach guarantees that the target quality will be met, this method of selecting cutpoints does not generally provide the highest overall yield of clean coal from a global perspective. It may be possible to obtain a higher total yield by operating the coarser circuits at a higher ash and the finer circuits at a lower ash in order to obtain a final blended product containing 10% ash. There is no difference in the quality of the shipped product, but the net increase in yield obtained by optimizing the operating points has a tremendous impact on profitability. In addition, operators are often forced to select cutpoints that are lower than necessary because of contractual limitations that may or may not be justified from economic considerations. This action lowers overall plant yield and reduces the net tonnage of potentially saleable coal. In the present coal

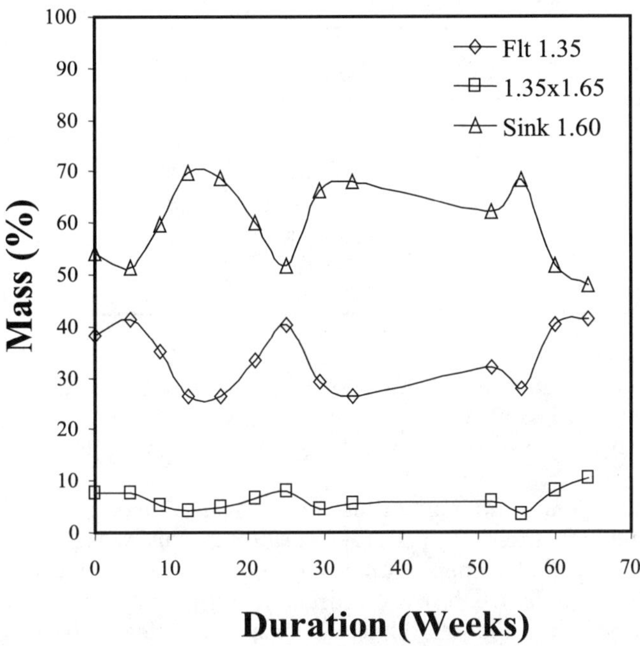

FIGURE 2 Variations in coal washability for a plus 6.35 mm feed coal

market, the loss of just 0.5% of saleable tonnage from a 1,000 tph plant represents a decline in revenues of more than $1 million annually (5 ton/hr × $35/ton × 6,000 hr/yr = $1,050,000).

Many articles have been published in the technical literature that advocate the use of mathematical simulation routines to identify optimum set points for the various coal cleaning processes (Armstrong and Whitmore, 1982; Peng and Luckie, 1991; King, 1999). In this approach, mathematical simulations are performed using experimental float-sink data for different size fractions of the feed coal and partition (Tromp) curves for the different separation processes. The simulations are repeated for all possible cut-points for each circuit and the combination that provides the highest plant yield at the desired target quality is selected as the optimum. Unfortunately, simulation routines require detailed washability data and empirical partition information that is not readily available to most plant operators. This approach is also impractical to implement in modern coal cleaning plants since the feed coals are usually subject to significant variations in terms of size, quality and mineralogical association. For example, Figure 2 shows the variations in mass yield and ash content obtained for the coarse fraction (plus 6.35 mm) of run-of-mine coal from a single underground mine. As shown in Table 1, the percentage of low-ash float 1.35 SG material in the composite feed samples varied substantially from a maximum of 41.4% to a minimum of 26.2% during the 13 month sampling campaign. (The variations are actually larger than these average values since each monthly sample was prepared as a composite of approximately 20–25 individual daily samples taken over each four-week sampling period.) Factors responsible for these variations include natural fluctuations in seam characteristics and changes in the mix of coal

TABLE 1 Analysis of the washability variations for a plus 6.35 mm feed coal

Statistical	Mass (%)			Ash (%)		
Parameter	Flt 1.35	1.35×1.65	Sink 1.60	Flt 1.35	1.35×1.65	Sink 1.60
Maximum	41.43	10.48	69.71	4.63	26.88	89.77
Minimum	26.23	3.64	48.09	3.15	17.76	77.72
Range	15.20	6.84	21.62	1.48	9.12	12.05
Average	33.66	6.30	60.04	3.60	21.94	82.49
Median	33.39	6.07	60.11	3.49	21.98	81.04
Variance	37.67	3.79	62.09	0.18	4.79	13.14
Std. Deviation	6.14	1.95	7.88	0.42	2.19	3.63

entering the plant from multiple mining sections. The routine fluctuations in washability make it essentially impossible to rely on simulation software for plant optimization. This trial-and-error strategy also provides little insight regarding the underlying principles that influence plant optimization.

This article discusses the issue of plant optimization and explains why the lack of insight associated with attempts to improve performance using mathematical simulation routines can actually be detrimental to long-term plant performance. As an alternative, practical operating guidelines based on fundamental scientific principles are provided that plant operators can adopt to effectively maximize clean coal yield. Several case studies are also discussed to illustrate how these practical optimization strategies can be applied to various situations in the coal mining industry.

OPTIMIZATION OF PLANT PERFORMANCE

The effective optimization of coal preparation plants requires the development of a detailed understanding of the underlying principles that govern separation performance. These principles can be generically described using four important guidelines, which are discussed in the following sections.

Guideline 1—Maintain Constant Incremental Quality

The most important optimization concept in coal preparation is the principle of constant incremental quality. This concept, which has long been recognized in coal preparation (Mayer, 1950; Dell, 1956; Abbott, 1982; Rong and Lyman, 1985; Rayner, 1987), states that the clean coal yield for parallel operations is maximum when all circuits are operated at the same incremental quality. This statement is true for any number of parallel circuits and is independent of the particle size and washability (float-sink) characteristics of the feed coal. Mathematical proof of the incremental quality concept is relatively straightforward and has been presented elsewhere (Luttrell et al., 2003). However, the fundamental basis for this concept can be demonstrated using the simple illustrations provided in Figures 3(a) and 3(b). In each diagram, the same two feed streams (A and B) are to be treated. The two streams may represent feed coal to two different cleaning operations in the same plant, the feed streams to two different plants that both service the same customer, or the same cleaning unit at two different points in time. The feed coals are comprised of various amounts of carbonaceous matter (black) and ash-bearing matter (gray). In this particular example, the washability of Stream A makes it relatively easy to upgrade

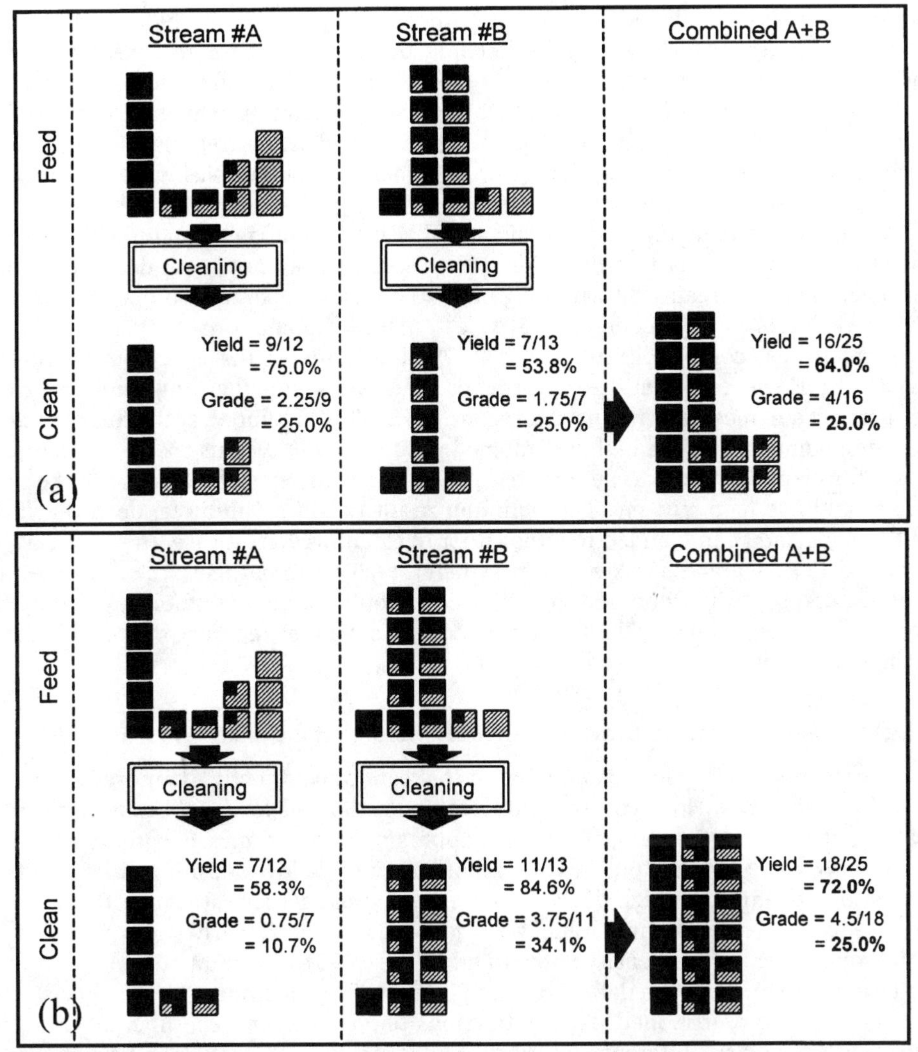

FIGURE 3 Clean coal products obtained by (a) treating parallel streams at constant cumulative ash and (b) treating parallel streams at constant incremental ash

since most of the particles are well liberated. In contrast, Stream B has a difficult washability due to the presence of a large percentage of middlings (composite) particles.

Figure 3(a) shows the result that is obtained when the separation is conducted to provide a target ash content of 25% for both streams. For Stream A, the operating point is raised from left to right until a total of 9 units (blocks) are recovered out of a total of 12 units. This operating point produces a yield of 75% (i.e., 9/12 = 75%) at the desired target ash of 25% (i.e., 2.25/9 = 25%). For Stream B, a yield of only 53.8% (i.e., 7/13 = 53.8%) can be realized before the target ash of 25% (i.e., 1.75/7 = 25%) is exceeded.

The lower yield is due to the poorer liberation of feed Stream B. The clean products from these two streams are blended together to produce a combined yield of 64% at the desired target ash of 25%. At first inspection, this appears to be an acceptable result since the target grade is met. However, a closer examination shows that particles containing 75% ash reported to clean coal when treating Stream A, while at the same time many particles containing just 50% ash were discarded when treating Stream B. This combination is obviously not optimum since higher quality particles were discarded so that poorer quality particles could be taken.

The incremental quality concept dictates that maximum yield can only be obtained when the quality of the last unit recovered from each feed stream is identical. To test this principle, the feed streams shown in Figure 3(b) are each separated so that the last unit recovered in both cases contained 50% ash. This combination of operating points decreased the ash content of Stream A to 10.7% and increased the ash content of Stream B to 34.1%. However, when the two streams are blended together, the combined clean coal product still meets the required target ash of 25%. More importantly, this new set of operating points provides a higher combined yield (i.e., 72% versus 64%). In practice, it is easy for plant operators to be misled and accept the lower yield because of the presence of multiple feed streams. The optimum result would be intuitive, however, if the feed streams had been blended together prior to cleaning. In this case, the pure carbonaceous particles would be recovered first, then the 25% ash particles, then the 50% ash particles, and so on until no additional particles could be taken without exceeding the target ash. This sequence will always provide the maximum recovery of valuable material at a given quality.

Guideline 2—Use Specific Gravity to Control Incremental Quality

Incremental quality is a mathematical term that cannot be directly monitored in operating industrial plants. However, this value can be estimated for ideal separations if the quality parameter of interest is ash. This approximation assumes that the particles in run-of-mine coals contain only two components, i.e., a low-density ash-free carbonaceous component and a high-density pure-ash mineral component. Under this assumption, the ash content of an individual particle must increase linearly with the reciprocal of the composite particle density (Abbot and Miles, 1990). The utility of this expression can be demonstrated by the data shown in Figure 4. This illustration shows the ash contents of narrowly partitioned density fractions obtained from experimental float-sink tests conducted on six different size fractions of a run-of-mine coal. For particles coarser than 28 mesh, the data show that the same individual ash is obtained for any specific gravity regardless of the size fraction treated. The deviation noted for the fractions finer than 28 mesh can normally be attributed to inefficiencies in the experimental float-sink procedures. In any case, plant optimization requires that all parallel circuits be operated at the same incremental ash. Since incremental ash is fixed by particle specific gravity, this implies that total plant yield is maximized when all parallel circuits are operated at the same specific gravity cutpoint (Clarkson, 1992). This statement is true regardless of the size distribution or washability characteristics of the feed coal, provided that very efficient separations are maintained in each circuit.

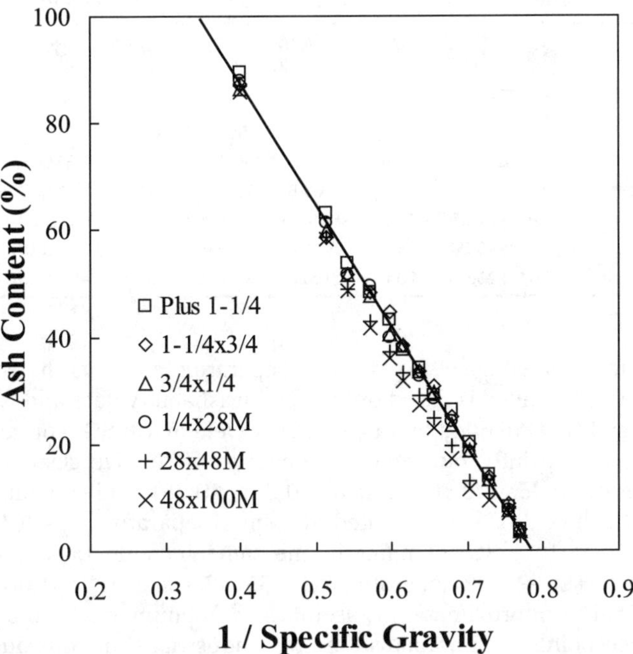

FIGURE 4 Ash content and specific gravity for individual coal particles of different sizes

Guideline 3—Correct Cutpoints to Account for Misplaced Material

According to the previous discussion, total plant yield is maximized when all parallel circuits are operated at the same specific gravity cutpoint. If the separation has a low partition efficiency (which is typical for fine coal separators), then a correction must be made to account of the impact of misplaced solids on incremental quality. This correction generally requires that less efficient units be operated at a slightly higher specific gravity cutpoint (Clarkson, 1992; Luttrell et al., 2003). This phenomenon is due to the fact that a greater proportion of lower density middlings is misplaced into the refuse stream than higher density middlings into the clean coal stream. The shift of higher quality (lower ash) material lowers the effective incremental ash. Therefore, less efficient circuits must be operated at a higher specific gravity cutpoint in order to maintain the same incremental quality. In general, efficiencies of density-based separators tend to decline as the particle size decreases (see Figure 1). Consequently, circuits treating finer particles must use higher specific gravity cutpoints to maintain optimum yield.

The specific gravity correction for misplaced material is particularly important for water-based separators such as spirals and water-only cyclones which tend to operate with significantly lower separation efficiencies. Simulation studies (Luttrell et al., 2003) indicate that the new specific gravity cutpoint (SG*) required to match that of an efficient separator (such as a heavy medium circuit) can be approximated using an empirical expression given by:

$$SG^* = 1.08 + 1.92\ln(SG) - 0.06/C \quad \textbf{(EQ 1)}$$

TABLE 2 Value calculations for various types of coal particles sold on a utility contract

Particle Type									
Specific Gravity	1.30	1.45	1.60	**1.75**	1.90	2.05	2.20	2.35	2.50
Ash Content (%)	0	12.5	25.0	**37.5**	50.0	62.5	75.0	87.5	100
Heat Content (BTU/lb)	15000	13125	11250	**9375**	7500	5625	3750	1875	0
BTU Value Adjustment	$5.00	$1.25	–$2.50	**–$6.25**	–$10.00	–$13.75	–$17.50	–$21.25	–$25.00
Ash Value Adjustment	$5.00	–$1.25	–$7.50	**–$13.75**	–$20.00	–$26.25	–$32.50	–$38.75	–$45.00
Sales Cost	–$5.00	–$5.00	–$5.00	**–$5.00**	–$5.00	–$5.00	–$5.00	–$5.00	–$5.00
Net Value	$30.00	$20.00	$10.00	**$0.00**	–$10.00	–$20.00	–$30.00	–$40.00	–$50.00

in which SG is the desired specific gravity of separation and C is the coal cleanability index. The cleanability index is based on the feed washability data and is defined as the float-sink yield at 1.3 SG divided by the float-sink yield at 1.6 SG. For general use, feed coals may be classified into four general categories of relative cleanability, i.e., easy (C>0.50), moderate (0.35<C<0.50), difficult (0.25<C<0.35), and very difficult (C<0.25). Therefore, a difficult coal (C = 0.3) treated by a spiral separator (Ep = 0.16) would need to operate at about 1.72 SG to maintain the same incremental quality as a heavy medium separator (Ep = 0.02) operating at 1.55 SG (i.e., SG* = 1.08+1.02*ln*(1.55)–0.06/0.3 = 1.72). This approximation is useful since full numerical simulations to determine optimum cutpoints are impractical in most industrial situations due to constantly changing feed blends. However, this is only an estimation procedure and significant errors may occur as a result of unexpected variations in feed coal washability or separator characteristics that may make these predictions less reliable.

Guideline 4—Select Incremental Quality Based on Coal Value

Any discussion of plant optimization must address the issue of coal value. Although coal payments in the steam (utility) market are typically reported on a per ton basis, price adjustments are nearly always applied to compensate for variations in the heat content of the as-received product. For example, consider the tabular analysis given in Table 2. In this simplified example, the pure particles of carbonaceous matter (SG = 1.3) are assumed to contain about 15,000 BTU/lb, while the pure particles of rock (SG = 2.5) contain no heat value. The sales agreement for this example dictates a market price of $25.00 per ton of clean coal with quality specifications of 12,500 BTU/lb and 10% ash. The coal payment schedule is prorated at $0.20 per 100 BTU above/below 12,500 BTU/lb and $0.50 per 1% ash above/below 10% ash. A $5.00 per ton sales related cost has also been applied to account for miscellaneous fees, taxes and other sales expenses associated with the coal production.

The calculations indicate that pure coal particles carry a $5.00 per ton BTU premium (i.e., [15,000–12,500] × $0.20/100 = $5.00) and $5.00 per ton ash premium (i.e., [10–0] × $0.50/1 = $5.00). After subtracting a sales related cost of $5.00 per ton, the net value of these high grade particles is $30.00 per ton (i.e., $25.00 base + $10.00 premium –$5.00 sales = $30.00). In contrast, the pure rock particles carry a $25.00 per ton BTU penalty ([0–12,500]/100 × $0.20 = $25.00) and $45.00 per ton ash penalty ([10–100]/1 × $0.50 = –$45.00). Thus, after subtracting a sales related cost of $5.00 per ton, these low-grade particles have a net negative value of –$50.00 per ton (i.e., $25.00 base–

\$70.00 penalty – \$5.00 sales = –\$50.00). In other words, the producer gains \$30.00 for every ton of pure carbonaceous material shipped to the customer and loses \$50.00 per ton of pure rock shipped. Composite particles containing mixed amounts of coal and rock have market values between these two limits. In order to maximize profitability, the coal producer for this example should obviously ship only those particles that have a positive net value and should never ship particles that have a negative net value. The "zero value" particle, which contains just enough heat content to justify its shipment, occurs at a gravity of 1.75 SG (i.e., at an incremental ash of 37.5%) for this particular case. Therefore, for this simplistic example, the plant operator should set and maintain an effective specific gravity cutpoint of 1.75 SG to ensure that profitability is maximized. This analysis is important since it shows that the specific gravity for optimum plant operation is controlled by the coal sales agreement and is largely independent of the coal washability. In other words, the specific gravity set point which controls what types of particles should be recovered and shipped does not depend on the number of particles in each specific gravity class. Although this example has been greatly oversimplified, the underlying optimization principle illustrated by this analysis can conceptually be applied to any generalized case if the contractual details are available.

To this point, no mention has been made to this point regarding the overall quality (e.g., ash content) of the clean coal product shipped to the customer. In many cases, the optimization methodology described above will provide a cumulative clean coal quality content that is within the upper contractual limits specified by the customer. However, in some situations, the cumulative quality may not be acceptable to the customer. When this occurs, plant operators are generally forced to lower the plant specific gravity and tolerate losses of potentially valuable coal. The fuel chain is obviously less than optimal when operating under this constraint. The resultant mismatch between the coal type and contractual quality specifications provides an opportunity for additional profit by the coal producer and/or customer via renegotiation of the sales agreement.

OPTIMIZATION CASE STUDIES

Case 1—Heavy Medium Circuits

Figure 5 shows a heavy medium circuit at an eastern U.S. coal preparation facility. The circuit consists of a single heavy medium bath which treats approximately 150 tph of plus 10 mm feed coal. Likewise, the 10 × 1 mm fraction of feed is treated using heavy medium cyclones at a rate of 350 tph. The washability data for the feed coals to each circuit are provided in Table 3 (but are not available to the operators). According to the production records at the plant, the operators routinely adjust the SG cutpoints of the bath and cyclones to maintain the same clean coal ash of 7% from both circuits. Because the coarser coal is less liberated, the bath cutpoint must be set to a low value of 1.3 SG in order to meet the 7% ash constraint. In contrast, the coal fed to the heavy medium cyclone circuit is much better liberated and, as such, this circuit must operate at a much higher cutpoint of 1.75 SG (the highest practical value) to maintain the same 7% product ash. Plant sampling programs suggest that the separation efficiency of each of the heavy medium circuits is individually very good. As a result, management believes the plant performance to be very good.

Despite the claims of plant management, it should be readily apparent that the overall performance of the combined circuits cannot be optimum in light of the optimization

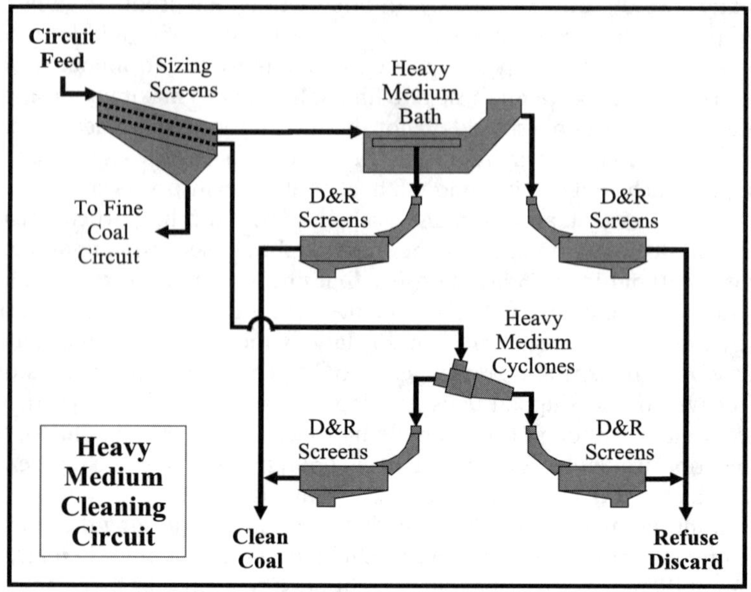

FIGURE 5 Process flowsheet for a typical heavy medium circuit

TABLE 3 Washability data for heavy medium bath and cyclone circuits

	Heavy Medium Bath				Heavy Medium Cyclones			
Float SG	Mass (%)	Ash (%)	Cum. Mass (%)	Cum. Ash (%)	Mass (%)	Ash (%)	Cum. Mass (%)	Cum. Ash (%)
1.30	11.36	3.86	11.36	3.86	59.17	3.54	59.17	3.54
1.35	27.48	8.30	38.84	7.00	6.10	7.58	65.27	3.92
1.40	16.26	14.15	55.10	9.11	3.70	13.09	68.97	4.41
1.45	3.82	20.14	58.92	9.83	2.30	18.53	71.27	4.87
1.50	2.45	25.76	61.37	10.46	1.88	23.79	73.15	5.35
1.55	1.26	29.67	62.63	10.85	1.02	29.03	74.17	5.68
1.60	0.77	33.89	63.40	11.13	0.88	33.19	75.05	6.00
1.65	0.25	37.11	63.65	11.23	0.51	37.21	75.56	6.21
1.70	0.31	40.12	63.96	11.37	0.61	40.23	76.17	6.48
1.80	0.40	44.99	64.36	11.58	1.05	44.49	77.22	7.00
1.90	0.63	52.73	64.99	11.98	0.99	51.58	78.21	7.56
2.00	0.54	62.06	65.53	12.39	0.95	60.75	79.16	8.20
Feed	34.47	88.54	100.00	38.64	20.84	87.22	100.00	24.67

guidelines discussed previously. For efficient units such as heavy medium separators, total plant yield is maximized only when all parallel circuits are operated at the same specific gravity cutpoint. The large different in the SG cutpoint between the bath and cyclone circuits indicates that a better set of operating points is available. In fact, this problem can be observed by comparing the qualities of the last particles recovered in each of the two circuits. When operating at a cumulative ash of 7% (SG = 1.35), the last increment of mass recovered in the bath contained approximately 8.3% ash. To achieve the same cumulative product ash of 7%, the cyclone circuit had to be operated at a much higher cutpoint (SG = 1.75) and the last increment of recovered mass contained about 44.5% ash. The combined plant is obviously not optimum since one circuit is sending particles with more than 8.3% ash to reject, while the other is sending particles with less than 44.5% ash to the clean coal product.

The financial impact of the decision to operate at a constant cumulative ash as opposed to constant incremental ash can be calculated from the washability data given in Table 3. For constant the cumulative ash scenario, the clean coal tonnage and ash can be calculated for the overall plant as:

$$\text{Clean Coal Production} = 150(0.3884) + 350(0.7722) = 329 \text{ tph}$$

$$\text{Clean Coal Ash} = \frac{150(0.3884)(7) + 350(0.7722)(7)}{150(0.3884) + 350(0.7722)} = 7\%$$

Likewise, the results obtained by operating at constant incremental ash can be calculated for cases in which the specific gravity cutpoint in both circuits. In this particular situation, operation of both circuits at a 1.55 SG provides the following clean coal tonnage and ash:

$$\text{Clean Coal Production} = 150(0.6263) + 350(0.7417) = 353 \text{ tph}$$

$$\text{Clean Coal Ash} = \frac{150(0.6263)(10.81) + 350(0.7417)(5.68)}{150(0.6263) + 350(0.7417)} = 7\%$$

These calculations show that the plant can increase overall tonnage by 24 tph without impacting clean coal quality simply by maintaining the same SG cutpoint in both circuits. At a current market price of $35/ton for clean coal, then the net value of this improvement is worth more than $5 million annually (24 tph × $35/ton × 6,000 hr/yr = $5,040,000). Also, operation at a constant SG cutpoint is a much easier condition to maintain since the plant operators are constantly aware of the SG values, but rarely have access to the feed washability data.

Case 2—On-line Control of Plant Quality

In order to maintain a constant clean coal quality, the management of a large preparation plant in the eastern U.S. installed an on-line analyzer to monitor the ash content of the clean coal product generated by the overall plant. The analyzer was deemed necessary because of large fluctuations created by the intermittent addition of lower quality coal from a strip mining operation that frequently entered with the normal plant feed. The control loop was designed to automatically raise the SG cutpoints if the product ash from the plant was below 7.5%, and to drop the SG cutpoints if the ash was above 7.5%. Once the analyzer was properly calibrated, the plant management found the control system to provide a clean coal product with a very consistent quality. Therefore, the capital expenditure of more than $1 million for the on-line analyzer and associated ancillary controls was believed to be well justified.

The problem with the control system described above is that it does not allow a constant incremental quality to be maintained since the SG cutpoints are always changing in response to the feed coal washability. As indicated previously, plant performance can be optimized only by operating all efficient circuits at the same specific gravity cutpoint. This concept is valid not only for a fixed point in time, but also for the entire duration of a given production cycle. As such, a plant that raises and lowers specific gravity cutpoints for quality control purposes will always produce less clean coal than a plant that maintains the same cutpoints. A better approach is to maintain constant SG cutpoints and to use the on-line analyzer to adjust the ratios of the feed coals entering the plant. In this approach, the different feed coals are stored in separate feed stockpiles according to their general washability characteristics. Based on feedback from the on-line analyzer, different ratios of coal from the piles can be automatically fed to the plant as required to maintain a constant clean coal quality for the overall product. The required feed rates can be controlled automatically using ratio feeders and a control loop, or they can be set manually via communication between the plant operator and the stockpile dozer worker who controls how much of each coal is directed to the plant. This approach allows the quality of the clean coal to be adjusted on-line without changing the predetermined cutpoints that optimize plant performance. Precise determination of the washability characteristics of the different coal feeds is not required since the control scheme determines the required mix ratios in real time. The feed coals simply need to be placed into separate piles with generally better and worse washabilities.

The financial benefits of this approach can be illustrated by comparing production data obtained from the plant while operating under (i) SG control and (ii) feed ratio control. To fairly compare these two strategies, a series of plant simulations were performed over a production period of 24 hours (see Figure 6). The simulations were performed using "poor" and "good" plant feeds determined from statistical analyses of the plant feed coals. As expected, the simulation data showed a control system that employs a traditional on-line ash analyzer to maintain a constant product ash of 7.5% via on-line adjustment of SG cutpoints produced 7,787 tons of saleable coal. In contrast, simulation runs performed using the ratio feeding control strategy produced 8,133 tons of saleable coal at the same overall quality of 7.5% ash. The use of the feed ratio control strategy increased the tonnage of saleable coal by 346 tons for this 24 hour period. For a metallurgical coal sold in today's market at $42/ton, this improvement represents nearly $4.8 million in additional revenues annually (i.e., Revenue = 346 tons × $42/ton × 330 days/yr = $4,795,560/yr).

Case 3—Installation of Two-Stage Spirals

The plant described above is currently operating with single-stage spirals with a specific gravity cutpoint of 1.95 SG. In order to maintain a product quality of 7.5% ash, the heavy medium circuits are forced to operate with a cutpoint density of 1.52 SG. Tests conducted by a spiral manufacturer showed that the addition of a second stage of cleaner spirals would reduce the density cutpoint of the overall spiral circuit to 1.75 SG. This upgrade would cost approximately $75,000 for the additional secondary spirals. Unfortunately, the test data showed that this modification would reduce the clean coal tonnage of the spiral circuit from 62 tph for the single-stage spirals down to 56 tph for the two-stage spirals. As a result, the plant management decided to abandon the upgrade because such a move would result in 6 tph less saleable coal from the spiral circuit.

Despite the claims by plant management, the data provided above again suggests that the overall plant production is probably not optimized due to the very large difference in the SG cutpoints between the heavy medium and spiral circuits. For a coal with a

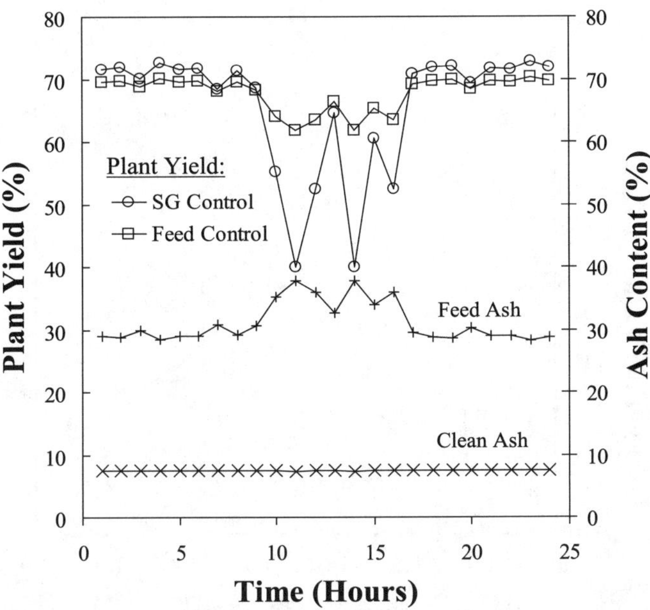

FIGURE 6 Simulation data comparing two online control strategies

relatively difficult washability (C = 0.50), Eq. [1] suggests that the optimum cutpoint for the spirals would be approximately:

$$SG^* = 1.08 + 1.92\ln(SG) - 0.06/C = 1.08 + 1.92\ \ln(1.52) - 0.06/0.50 = 1.76\ SG$$

This value would tend to suggest that a two-stage circuit would be preferred since it provides a cutpoint very close to this value. The oversight by the plant management is that they failed to consider the impacts of reducing spiral SG on the entire plant. A closer examination of the test data showed that the drop in cutpoint from 1.95 to 1.75 SG effectively reduced the incremental ash from about 62% to 38% in the spiral circuit. Although this action reduced the amount of saleable coal by about 6 tph, the tonnage loss was largely due to the rejection of high-ash particles. The removal of high-ash material would allow specific gravity cutpoint in the heavy medium circuits to be raised from 1.52 to 1.57 SG without exceeding the contract ash limit of 7.5%. The experimental test data showed that the increased cutpoint raised the incremental ash in the heavy medium circuits from about 25% to 34% which, in turn, increased the clean coal production from the heavy medium circuit from 430 to 446 tph (a gain of 16 tph). The net gain in saleable tonnage for the overall plant is about 10 tph when using the two-stage spirals. The annual increase in revenue associated with this modification is about $ 2.1 million (i.e., 10 tph × $35/ton × 6,000 hr/yr = $2,100,000). Obviously, the upgrade would provide an extremely attractive payback on the capital investment of $75,000 required to install the two-stage spirals.

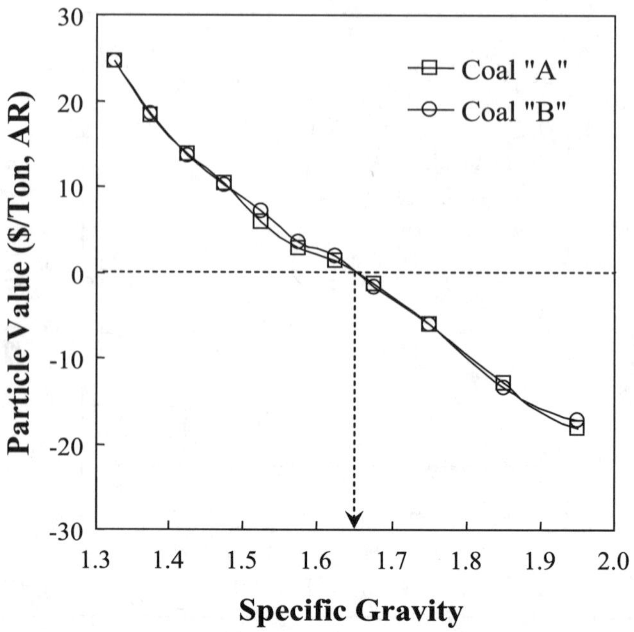

FIGURE 7 Particle value as a function of specific gravity for two processed coals

Case 4—Selection of Appropriate Operating Cutpoints

The marketing group for a major eastern coal operation has been actively selling clean coal from one of the company's preparation facilities. The 1,000 tph facility treats two different run-of-mine coals (A and B) with different washability characteristics. At present, the plant blends these two coals to produce a composite clean coal for a utility customer. In order to further improve profits, an industrial engineer from the marketing group attempted to use linear programming software to identify the best operating points for the facility. Unfortunately, the engineer concluded that the findings from such a study would be unreliable because of uncertainties associated with the separation capabilities of the plant equipment and the large variations in the consistency of the feed washability data. Therefore, attempts to further optimize performance of the plant were abandoned.

While optimization software can be applied very successfully in the coal industry, the engineer in this case failed to understand the underlying factors that provide optimal performance for a coal plant. The incremental analysis shown previously in Table 2 indicates that the optimum operating conditions are actually dictated by the coal sales agreement and are relatively independent of the cleanabilities of the coal feeds. By using the coal contracts for this site, it was possible to calculate the individual worth of each type of particle in the feed as a function of particle specific gravity. At this particular site, the cleaning was performed so as to globally maximize profitability while maintaining a constant fuel cost to the boiler of $1.50/MM BTU. The results of these calculations, which are plotted in Figure 7, show that the point of zero value for both coals occurs at a specific gravity of 1.65 SG. Therefore, if at all possible, the plant cutpoints should be maintained at this value to maximize profitability. The presence of an optimum value can be attributed to a tradeoff between increased clean coal yield and decreased clean coal quality/value that occurs as the plant specific gravity is raised. Particles with specific gravities less than 1.65 SG have sufficient heat content to justify shipment, while those above 1.65 SG do not. Compared to operation at a constant clean coal ash for both coals, this optimization strategy provided additional coal sales revenues of nearly $10 million annually for this particular operation.

SUMMARY

The production of clean coal from a preparation plant can be maximized using an optimization principle commonly referred to as the incremental quality concept. This concept dictates that all plant circuits be operated at the same incremental quality to maximize the net value of delivered clean coal product. For efficient processes such as heavy medium separators, this requirement can be achieved for all practical purposes by operating all parallel circuits at the same effective specific gravity cutpoint. Less efficient water-based separators, such as spirals and water-only cyclones, require slightly higher cutpoints to correct for the misplacement of particles that tend to lower incremental quality. The new cutpoint can typically be estimated from empirical formula developed for each unit operation provided that the approximate cleanability of the coal is known. An incremental analysis of coal value further shows that the optimum specific gravity cutpoints are largely controlled by specifications within the coal sales contract and are generally independent of the washability characteristics of feed coals. This insight makes it possible to identify optimum operating gravities without conducting detailed analysis of the feed coal washabilities. The fundamental understanding provided by these practical optimization principles make it possible for plant operators to effectively improve plant production and to clarify the process of making capital decisions. This strategy is also relatively easy to implement since specific gravity cutpoints can be readily monitored and controlled by plant operators for optimization purposes.

REFERENCES

Abbott, J. 1982. The Optimisation of Process Parameters to Maximise the Profitability from a Three-Component Blend, *Proceedings 1st Australian Coal Preparation Conference*, April 6–10, Newcastle, Australia, 87–105.

Abbott, J. and Miles, N.J. 1990. Smoothing and Interpolation of Float-Sink Data for Coals, *Proceedings International Symposium on Gravity Separation*, September 12–14, Cornwall, England.

Armstrong, M. and Whitmore, R.L, 1982. The Mathematical Modeling of Coal Washability, *Proceedings 1st Australia Coal Preparation Conference*, April 6–10, Newcastle, Australia, 220–239.

Clarkson, C.J. 1992. Optimisation of Coal Production from Mine Face to Customer, *Proceedings 3rd Large Open Pit Mining Conference*, August 30–September 3, Makcay, Australia, 433–440.

Dell, C.C. 1956. The Mayer Curve, *Colliery Guardian*, Vol. 33, pp. 412–414.

Kempnich, R. 2000. Coal Preparation–A World View, *Proceedings 17th International Coal Preparation Exhibition and Conference*, May 2–4, Lexington, Kentucky, USA, 5–48.

King, R. p. 1999. Practical Optimization Strategies for Coal-Washing Plants, *Coal Preparation*, Vol. 20, 13–34.

Luttrell, G.H., Barbee, C.J., and Stanley, F.L. 2003. Optimum Cutpoints for Heavy Medium Separations, *Advances in Gravity Concentration*, Edited by R.Q. Honaker and W.R. Forrest, Society for Mining, Metallurgy, and Exploration, Inc., Littleton, Colorado, 81–91.

Mayer, F.W. 1950. A New Washing Curve. *Gluckauf*, Vol. 86, 498–509.

Osborne, D.G. 1988. *Coal Preparation Technology*, Graham & Trotman, Vol. 1, Norwell, Maine, 600.

Peng, F.F. and Luckie, P.T. 1991. Process Control–Part I: Separation Evaluation, *Coal Preparation*, Edited by J. Leonard, 5th Ed., Society for Mining, Metallurgy, and Exploration, Inc., Littleton, Colorado, 659–716.

Rayner, J.G. 1987. Direct Determination of Washing Parameters to Maximize Yield at a Given Ash, Bull. Proceedings Australia Institute of Mining and Metallurgy, Vol. 292, No. 8, 67–70.

Rong, R.X. and Lyman, G.J. 1985. Computational Techniques for Coal Washery Optimization–Parallel Gravity and Flotation Separation, *Coal Preparation*, Vol. 2, 51–67.

On the Dynamics of Charge Motion in Grinding Mills

Raj. K. Rajamani,* B.K. Mishra,† Sanjeeva Latchireddi,* T.N. Patra,* and Sravan K. Prathy*

The motion of grinding balls has been the subject of study since 1919 when Davis published the first treatise on this subject. Subsequently came the notion that knowing mill power draft, using Bond equation or others, is sufficient to do an analysis of grinding. Next, population balance models provided accurate scale-up from small laboratory size mills to very large plant size mills. However, this success was not repeated when dealing with semiautogenous grinding mill. The study of charge motion with the discrete element method provides support and answers to such issues. Besides providing an accurate picture of charge motion in large mills, the method begs the investigator to study grate and pulp lifter equally critically with fluid dynamic principles. Hence, in recent years, the field of comminution in grinding mills is advancing on a firm footing.

INTRODUCTION

Grinding mills have gone through an evolution both in terms of equipment dimensions and operational sophistication. In the 1940s a typical ball mill had a diameter of just 6 ft. and had steep conical ends. By the 1980s the size had climbed to as much as 18 ft and today ball mills upwards of 24 ft are in operations. In addition, semiautogenous grinding mills came into operation in the mid 70s and today the largest mill of this kind is 40 ft. in diameter. Even though the basic grinding action, tumbling balls breaking particles trapped in the interstices, remains unchanged a number of advances in the design of internal components have taken place. These include, long wear life materials, shell lifters, grate plate design, pulp lifters, discharge trunnion etc. These design evolution was driven by the ever-present challenge of minimizing energy consumption per ton of ore

* University of Utah, Salt Lake City, Utah

† Indian Institute of Technology, Kanpur, India

processed. Comminution is an energy intensive unit operation and hence process evolution is driven by energy minimization.

In the last 80 years much has been understood about operating variables in milling. Many of the findings in laboratory scale research have been translated into plant operation strategies. These include percent solids in the feed, circulating load, proper operation of hydrocyclone bank, ball load, mill speed and recycle crushing. In fact, many of these operating strategies have been imbedded in sophisticated control algorithms. These control system packages, totally automated via computer control systems, are the standards of operation today. In contrast in the early 60s and 70s the grinding circuit operation very much depended on the human operator.

EARLY CHARGE MOTION STUDIES

Quantification of mill charge behavior is essential in understanding size reduction and slurry transport. Work on charge motion dates back to 1919 by Davis who developed a physical and mathematical description of the ball charge motion, although it may be argued that this is based on the earlier work of White (1905). According to Davis, particles move in a locked manner in a circular path until a point is reached where the centrifugal and gravitational forces balance. At this point particles commence free fall in a parabolic path until they impact the mill shell and start their circular path once more. The point of projection of a particle from the body of the charge, under the condition of no interference from other particles in the charge or any sticking effects, was given by a circle, know as the 'Davis Circle', of radius g/ω^2, whose center was vertically above the center of rotation of the mill at a distance of g/ω. The Davis circle idea is presented in Figure 1.

Based on the assumption used for the Davis Circle, the paths of particles do not cross, suggesting that there is no axial mixing of the load, which is an oversimplification. Though this approach provides a mathematical description of particle motion, it does not describe the complete charge shape in tumbling mills. However, it is a fairly accurate approximation of the outer envelop of the charge in an operating mill.

More intense studies on this aspect were initiated in 1970s. Most of this work was related to the prediction of the mill power draw. Hogg and Fuerstenau (1972) considered the widely accepted theory of materials in horizontal rotary kilns and developed an idealized charge motion as shown in Figure 2. They assumed that this shape remains constant with changes to mill speed and load.

PREDICTION OF POWER

Based on the motion of the charge in a ball mill, researchers took different approaches to develop a relationship between the power requirement with the mill dimensions and operating parameters of tumbling mills. The correlation of power input with mill dimensions, speed, and load has been brought about by three principal methods: torque formulae, dimensional analysis, and analysis of the ball charge dynamics.

A number of researchers derived equations, based on the torque principle. The charge profile is considered as a single mass, and the torque necessary to maintain the offset in the center of gravity is calculated as follows:

$$T = M_b r_g \sin\alpha \quad \textbf{(EQ 1)}$$

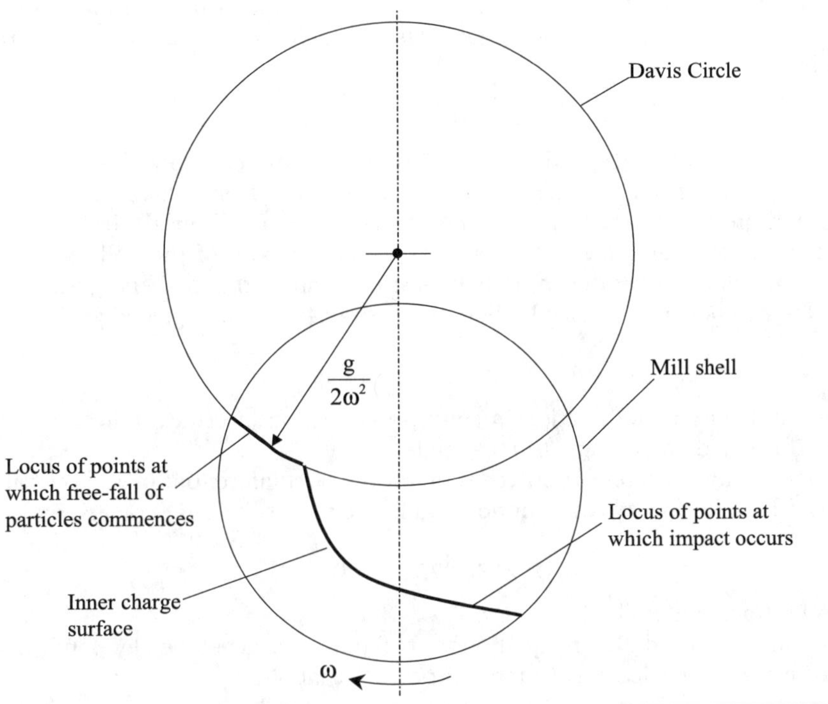

FIGURE 1 The Davis Circle (Redrawn after Davis, 1919)

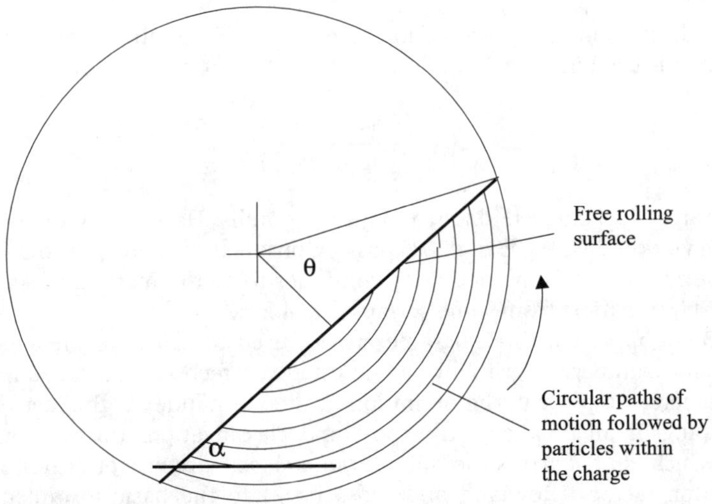

FIGURE 2 Idealised charge motion (Redrawn after Hogg and Fuerstenau, 1972)

where T is the torque, M_b is the mass of the balls, r_g is the distance from the mill center to the center of gravity of the load, and α is the angle of repose. The power draft, P, is given by

$$P = 2\pi TN \quad \textbf{(EQ 2)}$$

where N is the rotational speed of the mill in revolutions per second.

This idea has been exploited by, among others, Hogg and Fuerstenau (1972); they considered an infinitesimal mass in the charge and derived an algebraic equation for potential energy required to lift this mass to the top side of the ball load. Then this energy is integrated over the entire ball mass to compute the total energy supplied from which the mill power is deduced. Their power draft formulae is given by:

$$P = 3.627\rho\phi LD^{2.5}\sin\theta\sin\alpha \quad \textbf{(EQ 3)}$$

where ρ is solid volume fraction in slurry, ϕ is fraction of critical mill speed, *L* is mill length, *D* is mill diameter, and θ is toe angle.

Guerrero and Arbiter (1960) considered the circumferential mass flow rate of the charge to find the energy consumption and hence power:

$$P = 3.627LD^{2.5}f(J) \quad \textbf{(EQ 4)}$$

where J is fractional mill filling.

A commonly used design equation for ball mills was developed by Bond, who incorporated certain empirical results into his power equation:

$$P = 12.262\rho LD^{2.3}J(1-0.937J)(1-0.1/2^{(9-10\phi)}) \quad \textbf{(EQ 5)}$$

BOND MODEL SCALE-UP

In this method, the Bond power equation given in Eq.(5) combined with the following model equation is used in

$$E = 10W_i\left(\frac{1}{\sqrt{P_{80}}} - \frac{1}{\sqrt{F_{80}}}\right) \quad \textbf{(EQ 6)}$$

scaling laboratory locked-cycle data to plant scale mills. Here, E stands for energy per ton, W_i is the work index, P_{80} is the 80% passing product size and F_{80} is the 80% passing feed size. Even though this methodology continues to be the method of scale-up in the mining industry, it suffers from some serious drawback.

In the Bond approach three parameters are used to calculate the specific energy requirement for commercial grinding–a work index, a feed size parameter and a product-size parameter. Implicit in the definition of the work index is the fact that it represents a lumping of all breakage, transport and classification processes into a single parameter, which characterizes the material in a closed circuit wet grinding configuration. Correction factors (Rowland and Kjos, 1997) to the basic formulae (Eq.6) are applied to account for dry grinding, open-circuit grinding, very fine size product and oversize feed. In some cases a diameter efficiency factor is applied to account for capacity and power scale up differences.

POPULATION BALANCE MODEL

In contrast to the Bond model, which has arisen from the correlation of a large amount of data, the population balance model is derived in a physically meaningful way from population balance considerations. This type of model has been shown to provide accurate simulations of the entire size distribution of ground products produced in batch, locked cycle and continues open- and closed circuit grinding in ball-mill scale up design.

The linear size discretized model for breakage kinetics is obtained by dividing the particulate assembly being ground into 'n' narrow size intervals, (x_i, x_{i+1}), i = 1,2, . . ., n. The intervals chosen generally correspond to those of the Tyler $\sqrt{2}$ sieve series. A mass balance for the material in the ith size interval at time t yields for i = 1,2, . . ., n:

$$\frac{d[Hm_i(t)]}{dt} = -s_i Hm_i(t) + \sum_{j=1}^{i-1} b_{ij} s_j Hm_j(t) \quad \textbf{(EQ 7)}$$

Here, $m_i(t)$ is the mass fraction of the material in the ith size interval and H is the total mass of material (hold-up) being ground. In Eq.(7) s_i, the size discretized selection functions for the ith size interval, denotes the fractional rate at which material is broken out of the ith size interval, and b_{ij}, the size discretized breakage function, represents the fraction of the primary breakage product of material in the jth size interval which appears in the ith size interval

To apply Eq.(7) to the problem of design scale up it is apparent that the dependence of the kinetic parameters (s_i, b_{ij}) on mill diameter, mill speed, media load and size, and particle hold up must be known. Considerable experimental effort will have to be devoted to accurately define these relationships. This will entail the estimation of these parameters (by use of non-linear regression algorithms) from mill feed and product size distributions obtained for a wide range of mill diameters, mill speeds, etc., with a variety of feed materials.

The size discretized selection functions, s_i, are proportional to the specific power input the mill

$$ss_i = s_i^E \left[\frac{P}{H}\right] \quad \textbf{(EQ 8)}$$

where the set of proportionality constants, s_i^E (specific selection functions), are independent of mill operating conditions.

The breakage functions is a property of the ore and hence can be taken as invariant with respect to mill diameter. Malghan and Fuerstenau (1976) established the scale-up hypothesis in 5, 10 and 20-inch diameter mills. Later, Herbst and Rajamani (1982) demonstrated the same hypothesis from 10, 15 and 30-inch diameter mills to plant scale mills of 18-ft diameter (Lo et al, 1988). Austin (1997) using a different scale-up methodology yet using the population balance model, demonstrated scale-up from lab mills to plant mills for mineral, cement and coal grinding.

The grand success of the scale up hypothesis raises an intriguing question: How could specific grinding rates measured in mills of diameter as small as 10 inch holds good for the prediction of size distribution in large mills of diameter 16 ft. In other words, specific particle breakage rates do not depend on the milling environment. The huge mass of the ball load in the large mill produces a breakage regime that is more or less identical to the small laboratory scale mill. The reason for this "identical breakage" regime in both the small and large mills is readily answered by the discrete element method simulation of the two mills. Both the mills exhibit only cascading regimes at the same critical speeds. In the cascading regime a few layers of balls are carried upward by mill shell rotation and the balls cascade downward in a smooth fluid-like flow. The relative velocity between adjacent layers exerts a shearing action on the slurry in the interstices. Hence, regardless of the diameter of the mill the shearing action measured as rate of breakage per unit volume of ball load is the same. Thus, when we measure the breakage rate per unit volume of ball load in a laboratory mill, it is identical to the rate that would prevail in a plant scale mill.

In fact, the hypothesis that ball mills operate efficiently in the "cascading regime" is further reinforced by the lifter configuration used in numerous ball mill installations. Most of these mills use a "double wave" or "Noranda-type" lifters. These double waves lifters are approximately two semi-circles in cross section of radius 25 to 50 mm. The trough in the double wave captures a 50 mm diameter ball as it travels upward. These wave type lifters almost invariably produce cascading charge motion, since they release all of the balls at the 3 O'clock position in a counter clockwise rotation mill. Some plant ball mills, however, employ a more vigorous lifter design producing both cataracting and cascading action. The cascading action is much more vigorous in such a situation and hence breakage rates are proportionally higher. However, the cataracting action does not enhance the grinding rates at all. In fact, that portion of the energy is wasted in steel-on-steel collisions within the mill.

SAG MILL SIMULATION

Semiautogenous mills presented a greater challenge to modeling. The overwhelming success of population balance model was quickly extended to modeling semiautogenous mills. First, couple of studies done on 6 ft diameter mills barely established the modeling concepts. Three sets of selection functions and corresponding breakage function sets were needed. In addition to ball-to-particle breakage, there is particle-to-particle breakage and particle self-breakage. It became a challenge to delineate three sets of selection functions from product size distribution data. Nor was it easy to scale these parameters to plant size mills. The difficulty of running 6 ft diameter semiautogenous milling facility in a university setting more or less precluded the modeling advances that ball mill technology went through in the preceding years. Thus semiautogenous mill scale up and design depends on the Bond work index methodology even today.

In the final analysis, a full model incorporating the three modes of breakage rates for semiautogenous milling is lacking as of date. Unlike ball mill grinding circuits, semiautogenous milling circuits exhibit enormous fluctuations in throughput rates. These mills utilize the cataracting motion for self-breakage of larger ore particles and ball-to-ore breakage. In addition, the cascading motion results in fines production. Thus, different ore types require different proportion of cataracting and cascading motion and, more or less retention time in the mill, which contributes to large fluctuations in throughput rates. Despite a clear path way from pilot scale tests to full-scale mill design there were some notable design failures. The semiautogenous mill installed at Freeport, Indonesia's Grasberg operations (Coleman and Veloo, 1996) failed upon start-up due to severe damage to shell lifters. The mill at Alumbrera, Argentina met less than 50% of design capacity (Sherman and Rajamani, 1999) in the first two months of operation.

In a concentrator, all of the auxiliary equipment-pumps, conveyers, screens and hydrocyclones—and two primary resources-steel and electricity-serve primarily to maintain grinding action in the belly of the SAG mill. It is this action that dictates capacity. This being the case, it is understandable to observe this grinding action continuously from the control room and take whatever steps are necessary to keep the grinding field at its highest potential. Unfortunately, the grinding environment within the mill shell is very severe and none of the on-line instrumentation developed so far has survived the impact of large steel balls striking the shell in operation. Since direct observation is impractical today, the next available option is a simulation of the grinding field to understand the intensity of grinding or lack of intensity of grinding.

The inability of the existing models to predict semiautogenous grinding mill power draft as well as inability to explain the severity of the grinding action inside the mill have principally lead to the development of numerical simulation techniques such as the discrete element method.

DISCRETE ELEMENT METHOD (DEM) SIMULATION OF SAG MILLS

The SAG mill is made up of cylindrical shell with two conical shells attached on both ends. Lifter elements are attached in both the cylindrical and conical shell sections. As the SAG mill rotates, typically around 10 rpm, the internal flat walls of the lifter and shell imparts momentum to balls and rocks. The momentum is primarily transferred to particles in direct contact with the plate elements. These particles in turn impart their momentum to adjacent layers of particles. In this manner the motion of the entire charge evolves resulting in what is commonly referred to as cascading and cataracting charge.

The discrete element method (Mishra and Rajamani 1994a, 1994b) replicates the evolution of charge motion as described above. In the simulation the exact dimensions of lifters, plates and balls are used. First, the mill shell is constructed with a series of flat planes joining together to form cylindrical shell and conical shell. Next, a series of planes are constructed on the shell to duplicate shell lifters and end lifters. Next, the grinding balls and rock particles are generated. Usually the rock particles are constructed as spherical particles for ease of computations. The number of spheres so constructed would correspond to 25% filling with 12% filling for grinding balls. In three-dimentional simulation, the number of spheres may be in the range 200,000 to 1,000,000 depending on the size distribution and size of the SAG mill (Rajamani et al, 2002). In two dimensions, the number of spheres is in the 3000–8000 ranges. Now, in the numerical computational point of view, a complex polygonal assembly of rectangular plates encloses a large number of spheres. These spheres have different collisional properties that depend on their size and material make-up.

The critical aspect of DEM is the modeling of the collision between any two spheres or the sphere and a plate element at the contact point a force develops which sends the sphere in a direction away from the point of collision and also there is deformation of the metal or fracture of the rock particle. To account for these two events the collision is modeled by a pair of spring and dashpot. The spring models the opposing force and the dashpot models the dissipation of the force. Force is dissipated as a result of deformation of the material at the point of collision. The spring and dashpots in the model are placed in the normal direction as well as the tangential plane of the collision point. Thus after collision forces are calculated the acceleration and velocity of the pair of spheres are computed with the familiar laws of physics. It is clear then that in this method the forces

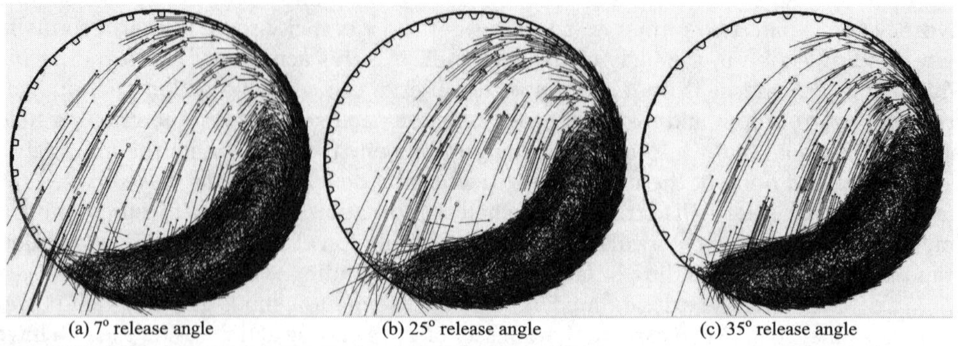

FIGURE 3 Snap shots of charge motion in a 38×24 ft SAG mill

of collision and the dissipation of energy are calculated at a greater level of detail than any other available models. Both two dimensional (Rajamani and Mishra, 1996, 1997b) and three dimensional (Rajamani et al, 2002) models have been developed to study the charge motion in full scale mills.

Hence, one can use forces and dissipated energy to follow lifter wear, shell plate wear and particle breakage. Most importantly, the location and the intensity of impacts on lifter bars can be computationally recorded and a corresponding metal abrasion at that location can be worked out. In a like manner, the energy of impact can be used to fragment the rock particles in the simulation. However, the distribution and number of fragments produced overwhelms the computational task.

SHELL LIFTER DESIGN AND CHARGE MOTION ANALYSIS

A typical example of charge motion analysis with a DEM code known as Millsoft is illustrated in this section. The SAG mill under consideration is a 38×24 ft mill drilled for 60 shell lifters. Therefore, the mill can be fitted with 30 high and 30 low lifter sets or a total of 60 high lifters. The total charge is 27% with 15% balls. The mill is expected to draw 15–18 MW power. The mill speed is set at 76% critical speed. The snapshots of the charge motion at different conditions are shown in Figure 3.

Figure 3a shows the SAG mill fitted with traditional lifters. The high lifters are the top hat type with a 7-degree release angle. The velocity vectors are super imposed on the balls and rock particles. A close look reveals within the cascade the central region where grinding action is minimum. Due to small release angle, considerable amount of rock and ball particles are thrown against the mill shell. The ball-to-liner strike zone extends as high as 9 O'clock mark. The mill may not reach design capacity, especially with hard ore type. Furthermore, considerable damage to liners is imminent within months of operation.

Next, we consider the same mill fitted with 30 high lifters (and 30 low lifters) of 25-degree release angle. Figure 3b shows the snap shot of charge animation. Here, ball strikes on liners are seen nearly up to 8 O'clock position, a considerable improvement over Figure 3a. This lifter is suitable for maintaining a moderately aggressive charge motion at the expense of shorter lifter wear life.

Next, the release angle is increased to 30 degrees with 30 high lifters (and 30 low lifters). The animation is shown in Figure 3c. The cataracting charge lands within the toe of the charge. This type of charge motion is ideal for SAG, for it to preserve the lifters. The mill speed may be increased without fear of damaging liners. Alumbrera mines exploited this concept to increase production. With such lifters, Alumbrera even used 150 mm top ball size.

INDUSTRIAL CASE STUDIES

In recent years, with increasing mill size whose power draw has reached over 20MW, emphasis has been diverted towards the design of shell lifters to increase the grinding efficiency. Lifters are not only consumables for protection against wear, but also critical machine elements. They transfer power to the mill charge and govern the pattern of energy distribution inside the mill, including affecting the grinding kinetics. The performance of liners changes with time, as their shape changes by wear. A good design must take these restrictions into consideration and optimize the performance of liners for the full duration of their useful life. The attainment of expected useful life of a set of lifters depend not only on their initial shape, material and quality of fabrication but a good matching between expected and real operating conditions inside the mill. The main issue is the direct impact of balls on the lifters in the absence of the damping effect of a bed of mineral ore and ball charge. In the following, the capacity increase achieved with newer shell lifter designs using Millsoft or other methods is discussed.

Collahuasi Mines, Chile is a 65,000 tpd operation. It employs two 32 × 15 ft variable speed SAG mills and four 22 × 36 ft ball mills. The original lifters had a face angle of 6 degrees and therefore the SAG mills suffered a throughput restriction. According to Marcelo Villouta (2001), plant manager, “the original SAG mill liner design, Hi-Hi with a 6 deg contact angle, was changed to a 17 deg face angle and then with Millsoft code a new design, Hi-Hi with 30 deg contact angle was installed in SAG mills. A noticeable 11% increase in ore throughput was achieved.

Alumbrera Mines, Argentina (Sherman and Rajamani , 1999) is a copper ore operation treating between 70,000 to 100,000 tpd of ore depending on the rock type presented to mill feed. Two SAG mills of 30x15 ft are used in this circuit. The original design called for 72 rows of 9° lifters. These lifters resulted in drop in the capacity to 50,000 tpd. Within 6 months of start of operations a new lifter design was explored with Millsoft simulations. As a result, the lifter set was changed to 48 rows of 25° lifters and later to 36 rows of 30° lifters. These lifters increased the throughput steadily to 96,000 tpd. Even today, lifter design is constantly modified along with the use of larger size (6" as opposed to 5.5") to increase production while processing hard ores.

Candelaria Mines, Chile is a 65,000 tpd operation. The grinding circuit includes two 36x15 ft SAG mills. The original lifters were 72 rows of rail design with 10 degree release angle. A year later the release angle was changed to 20 degrees. Another year later, the shell lifter design adopted was a 36 rows of 35 degree release angle. This design eliminated packing between lifters, throughput increased by 15%, the wear was slightly better than traditional lifters and the power draw in the first month was much higher compared to previous designs.

Los Pelambres Mines, Chile is 120,000 tpd operation. The SAG mill size is 36x17 ft. the first shell lifter design was 72 rows of 8 degrees release angle. However, severe packing was noticed which was the reason for lower capacity. Subsequently, Millsoft simulation software was used to explore better design. The new lifter design was 36 rows of 30-degree release angle. This design increased throughput by 10,000 tpd.

SLURRY TRANSPORTATION

While the design of shell lifters has advanced considerably due to discrete element models the design and geometry of grate plate and the design of pulp lifters have not advanced in pace with shell lifters. The slurry transport out of the mill into the discharge trunnion very much depends on the grate design and pulp lifter design. The shell lifters may perform perfectly producing ground product slurry, which may or may not be fully discharged out of the grinding chamber. In fact as the slurry advances through the slots in the grate into the pulp lifter, the critical speed of the mill or the centrifugal force due to mill speed opposes the downward flow of slurry through the pulp lifter. Thus, in effect, one might increase mill speed to increase power draft and hence mill throughput, but find that throughput is diminishing simply because the pulp lifter is unable to transport ground slurry out of the mill at the higher mill speed. Hence a theoretical study of slurry transport with computation fluid dynamics would be the next step in advancing semiautogenous grinding mills.

A detailed investigation by Latchireddi (2002), Latchireddi and Morrell (2003a and 2003b) on impact of the discharge grate and pulp lifters on slurry transport in SAG mills revealed the limitations of the conventional designs of pulp lifters (radial and curved) on slurry transport in SAG mills. Significantly inferior performance of the conventional pulp lifters is due to the flow-back phenomena. The flow-back of slurry from pulp lifters back into the grinding chamber through the slots in the grate can be as high as 60% depending on the design and size of the pulp lifter assembly. The fraction of slurry remaining in the pulp lifter gets carried over to the next cycle as carry-over fraction; however, this occurs in mills operating at higher speeds. The overall slurry transportation process in SAG mills is summarized in Figure 4.

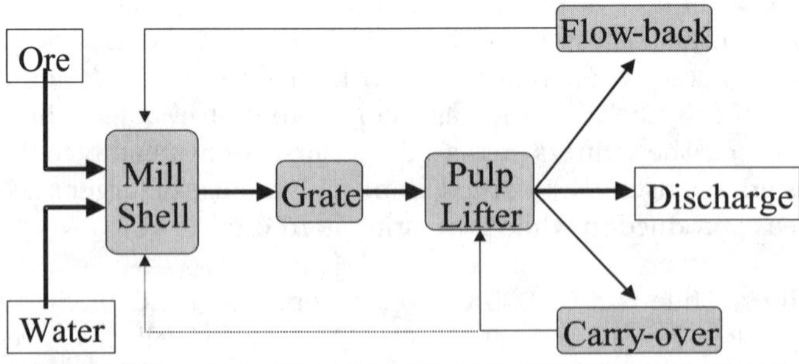

FIGURE 4 Process of slurry transportation in SAG mill

Due to the flow-back process, excess slurry accumulates near the toe region resulting in slurry pool formation. This slurry pool softens the ball strikes at the toe of the charge, which inhibits the impact breakage of particles. It also diminishes the shearing action in the cascading charge, which reduces the fine particle grinding. Both these aspects adversely affect the grinding efficiency thus reducing the mill capacity. Furthermore, the slurry pool applies a counter torque in the mill effectively reducing the power draft. Therefore, the operator either increases the feed rate or increases the critical speed to increases power draft. The power draft increases at first but subsequently decreases, thus giving a false indication to the operator.

CONCLUSIONS

The theory of milling began from the early study of charge motion due to Davis. However, the charge motion was quickly translated into expressions of power draw in the next four decades instead of seriously examining the charge motion itself. In the years when ball mill grinding circuit dominated the grinding scene in concentrators, Bonds method of scale-up became a standard design procedure. Subsequently, population balance model, due to its ability to describe breakage, transport and classification independently, arose as a replacement for the Bonds work index model. Population balance model performed exceedingly well in scaling data from 5,10,15,20 or 30 inch mills to ball mills of diameter as much as 18 ft. In retrospect, we conclude that the success of the scale-up hypothesis is simply due to the fact that breakage in ball mills is due to pure cascading action. As, semiautogenous mills relegated ball mills as secondary-grinding mills the population balance model could not be advanced equally well to the description of these mills. Simply, there were too many breakage mechanisms to cope up with model expressions. Next came discrete element models, which could describe the charge motion of semiautogenous mills more or less accurately since the model is based on physics of collisions. This advancement led to the proper design of shell lifter release angle, which has been put to use in many mine sites around the world. However, the transport of slurry through grate and pulp lifter has not been modeled equally rigorously. Today, one or two empirical models are keeping up with the complexity of this problem. However, a rigorous treatment of this problem through principles of fluid mechanics would greatly advance semiautogenous grinding technology. Thus we have come a full circle, in the last eighty years, from rudimentary charge motion analysis due to Davis to rigorous charge motion analysis with the discrete element method.

ACKNOWLEDGEMENTS

The authors would like to thank the support of the Department of Energy/ Industries of the Future Program for support of the project DE-FC26-03NT41786.

REFERENCES

Austin, L.G., Concepts in process design of Mills, 1997, Comminution Practices, Edit.: S.K. Kawatra, chapter 40, Society for Mining, Metallurgy, and Exploration, Inc., Littleton, Colorado.

Coleman, R and Veloo, C., 1996, "Concentrator Expansion at Freeport Indonesia's Grasberg Operations.", Mining Engineering, 25–33.

Davis, E.W., 1919, Fine crushing in ball mills. A.I.M.E. Transactions, Vol. 61, 250–296.

Guerrero, P.K., and Arbiter, N., 1960, "Tumbling mill power at cataracting speeds". A.I.M.E. Transactions, 217, 73.

Herbst, J.A., and Rajamani, R.K., 1982, "Developing a Simulator for Ball Mill Scale-Up. A Case Study, in Design and Installation of Comminution Circuits, edited by A.L. Mular and G.V. Jergensen II, SME/AIME.

Hogg, R and Fuerstenau, D.W., 1972, Power relationships for tumbling mills, SME-AIME Trans., December, 252, 418–423.

Latchireddi, S.R. and Morrell, S., 2003a, Slurry flow in Mills: Grate-only discharge mechanism (part-1). Accepted by Minerals Engineering.

Latchireddi, S.R. and Morrell, S., 2003b, Slurry flow in Mills: Grate-Pulp lifter discharge mechanism (part-2). Accepted by Minerals Engineering.

Latchireddi, S.R., 2002, Modeling the performance of grates and pulp lifters in autogenous and semiautogenous mills. PhD Thesis, University of Queensland, Australia.

Lo, Y.C., Herbst, J.A, Rajamani, R.K., Arbitter, N., 1988, "Design Considerations for large diameter ball mills, Intnl. J. Miner. Processing, 22 :75–93.

Malghan, S.G., and Fuerestaneu, D.W., 1976, The scale-up of ball mills using population balance models and specific power input. Zerkleinern. DECHEMA-Monograph., 79 (II), No.1586:613–630.

Marcelo Villouta R., 2001, Collahuasi: After two years of operation, SAG 2001, Vancouver, Eds:, Barratt, Allan and Mular, I31–I42.

Mishra B.K., and Rajamani R.K., 1994a, "Simulation of Charge Motion in Ball Mills. Part 1: Experimental Verifications,", International Journal of Mineral Processing, vol. 40, 171–176.

Mishra B.K., and Rajamani R.K., 1994b, "Simulation of Charge Motion in Ball Mills. Part 2: Numerical Simulations,", International Journal of Mineral Processing, vol. 40, 187–197.

Rajamani, R.K., Mishra B.K., and Songfack, P., 1997b, "The Modeling of Rock and Ball Charge Motion in SAG Mills, in Comminution Practices, edited by S.K. Kawatra, Society for Mining, Metallurgy and Exploration, Inc., pp. 195–200.

Rajamani, R.K., Mishra, B.K., 1996, "Dynamics of Ball and Rock Charge in SAG Mills, in International Autogenous and Semiautogenous Grinding Technology, edited by A.L. Mular, D. J. Barratt, and D. A. Knight, Vol. 2, pp. 700–712.

Rajamani, R.K., Mishra, B.K., Joshi, A and Park, J., 2002, Two and Three dimensional simulation of ball and rock charge motion in large tumbling mills, DEM Numerical Modelling of Discontinua, Edit. B.K.Cook and R.P. Jensen, Geotechnical special publication No.117, American Society of Civil Engineers, Reno, Virginia.

Rowland Jr., C.A., and Kjos, D.M., Rod and Ball Mills, 1997, Comminution Practices, Edit. : S.K. Kawatra, chapter 39, Society for Mining, Metallurgy, and Exploration, Inc., Littleton, Colorado.

Sherman, M. and Rajamani, R.K., 1999, "The Effect of Lifter Design on Alumbrera's SAG Mill Performance: Design Expections and Optimization." Proceedings of the Canadian Mineral Processors Conference, Ottawa, Canada, 255–266.

White, H.A., 1905, Theory of tube-mill action. J. Chem. Met. and Min. Soc. of South Africa, Vol. V, 290.

High Fidelity Simulation of the Mineral Liberation Process

J.A. Herbst* and A.V. Potapov*

Mineral liberation modeling can be traced back to the early work of A.M. Gaudin up through more recent population balance modeling by R.p. King, J.A. Herbst and G.T. Adel. A more microscopic look was taken at the University of Utah in the form of the J.E. Sepulveda and J.D. Miller Pargen model which permits the creation of an arbitrary shape, multiphase particle made up of discrete grains. Recent work on image analysis and tomographic analysis by J.D. Miller and C.L. Lin have laid the experimental ground work for model validation, but until now it hasn't been possible to include the necessary physics of the particle stressing process to allow the development of a "first principles" model which does not presuppose a "mode" of liberation (e.g., random or by detachment).

Recent developments in discrete grain breakage (DGB) modeling now make a generalized physics based model feasible. This paper describes current work that is being done to develop a physics based discrete grain liberation (DGL) model. Two-dimensional and three dimensional simulation results exploring the effects of particle size, phase properties and loading type are presented. Relationships between energy based liberation efficiency, degree of random liberation and material properties are identified.

INTRODUCTION

Mineral liberation, the process whereby valuable material is released from waste rock by the application of comminution energy in a mineral processing operation, is an extremely complex process which has resisted rigorous first principles analysis. Early models of liberation by Gaudin (Gaudin 1939) identified extremes of "liberation by

* Metso Minerals Optimization Services, Colorado Springs, Colorado

detachment" (cracks reaching a grain boundary propagate along the boundary "carving out" the valuable grain) and by "random liberation" (cracks propagate across grain boundaries resulting in identical degrees of fragmentation in all phases). The latter extreme, random liberation has been treated by statistical methods at various levels of generality (Gaudin 1939; Weigel 1965) and rigor (King 1979). Alternative phenomenological models have been developed for cases other than random breakage (Andrews and Mika 1975; Herbst et al. 1988; Choi et al. 1988) but their value has been limited by a need to estimate model parameters from data, which is not routinely collected in mineral processing plants (Petersen and Herbst 1985).

In parallel with model development have come increasingly more sophisticated experimental methods for assessing liberation, ranging from manual microscopic counting and specific gravity sorting to 2D and 3D image analysis of data collected from optical images (Lin et al. 1985; King and Schneider 1993), and x-ray micro tomography (Miller and Lin 2004).

Through all of this research it has become known that the degree of liberation achievable from a particular ore is dependent upon several variables including particle/grain size ratio, phase properties and loading type, however, a "first principles" model including all of the variables of this process has been elusive. During the last decade increases in computer hardware speed and improvements in DEM code efficiency have allowed the development of very powerful physics based models (Herbst and Potapov 2004a). Breakage modeling at the microscale has improved very significantly from these developments (Herbst and Potapov 2004b). In the study described here these physics based tools (referred to as High Fidelity Simulation tools) are applied to the prediction of the mineral liberation process using only material properties and applied forces as inputs. The specific model described here is referred to as the discrete grain liberation (DGL) model.

DEVELOPING THE DISCRETE GRAIN LIBERATION MODEL

In order to understand the Discrete Grain Liberation Model (DGL) it is necessary to understand its roots in discrete grain breakage and DEM. The Discrete Grain Breakage (DGB) model is an extension of the Discrete Element Method (DEM) that allows simulation of the particle breakage. Here we will outline the fundamentals of DEM and DGB. Those interested in the details of the DEM can be referred to the paper (Campbell 1997), and those that are interested in the details of DGB for two-dimensional and three-dimensional simulations can be referred to principles (Potapov et al. 1995 and Potapov and Campbell 1996) and applications (Herbst and Potapov 2004a, b) papers.

The Discrete Element Method is a numerical technique designed for the simulation of particle motion in multi-particle systems. In the DEM approach to modeling each particle making up the simulated material is treated as a separate entity, and parameters of motion of every particle (position, velocity, interaction forces etc.) are calculated at each time step of the simulation. While there are two major types of DEM, i.e., soft-particle, and hard-particle techniques, we will describe only the soft-particle technique as it is the most general and widely used of the two.

In the soft-particle DEM approach, the movement of granular particles is tracked by integration of the equation of motion for the solid particles. The particles are generally treated as non-deformable. They are, however, allowed to overlap slightly on the contact with each other. The contact forces and moments are calculated based on the parameters of this overlap. While there are many ways of calculating the contact forces and moments as a function of the overlap parameters, the most widely used is viscous-elastic for the normal force and elastic-frictional for the tangential force (Campbell 1997).

The simulation of particle breakage can be achieved by creating a composite DEM particle that is assembled from a collection of elementary DEM particles. These elementary DEM particles are "glued" together by the contacts that can withstand tensile stresses. These contacts are allowed to break, after the breakage tensile stresses are no longer in place, but compressive stresses can still exist in these contacts.

The elementary particles used to create the breakable composite DEM particles are triangles in two dimensions or tetrahedra in three dimensions. It has been shown that the resulting Discrete Grain Breakage (DGB) model can adequately describe the breakage behavior of brittle-elastic materials with predictable elastic properties (Young's modulus and Poisson's ratio) and crack propagation energy (energy released from the stress particle during the propagation of a unit area crack). It has also been shown through dimensional analysis of single particle impact breakage that there are only three dimensionless parameters relevant to this problem i.e., 1) the ratio of impact speed to the speed of sound in the simulated material, 2) Poisson's ratio and 3) the ratio of the impact energy to the energy released while creating a crack which spans a particles' diameter. Moreover, it has been shown that Poisson's ratio has almost no effect on the results of breakage (as measured by the fragment size distribution), and ratio of the impact speed to the material speed of sound is important only for high-speed collisions (when this ratio is above approximately 0.1). The most important parameter determining the results of the breakage is the ratio of the impact energy to the energy expended in creating a crack of length equal to a particles diameter.

Early computational experiences with this model led to the conclusion that computing times would be much too long for the model to be practical (Potapov et al. 1995). The efficiency of DGB code has recently been increased dramatically (10× or more) through the use of faster computers and shared-memory parallel execution of the code (Herbst and Nordell 2001; Herbst 2002). Prior to a year ago this DGB code has only been used to simulate but only for single-phase particles breakage in a variety of comminution devices (Herbst and Potapov 2003). Here, the extension of the model to the simulation of multi-phase liberation is described.

The process of mineral liberation involves two or more phases. In this case several new dimensionless parameters must be identified to solve the multi-phase breakage problem. For a two-phase system (considered here) one needs three parameters describing energies of the collision; one for each of the two phases and one for the interface. These dimensionless parameters are: 1) the ratio of the impact energy to the impact energy at which breakage is initiated in a particle that consists of only the first material, $E^* = E_{imp}/E_{oA}$, 2) the ratio of the crack energies of the two phases, $\gamma^*_{B/A} = \gamma_B/\gamma_A$, and 3) the ratio of the interfacial crack energy to the crack energy of the first phase, $\gamma^*_I = \gamma_I/\gamma_A$. There is also a dimensionless parameter characterizing the relative elastic properties of the phases $Y^*_{B/A} Y_B/Y_A$. Although Poisson's ratios for the two phases can also be included in this dimensional analysis, we elected not to investigate the effect of this parameter here since it is expected to be unimportant based on results obtained earlier (Potapov and Campbell 1994).

Some additional parameters are required to describe the particle geometry. The first parameter is volumetric content of the second phase B (chosen to be dispersed phase here) in the first phase A (chosen to be the continuous phase), $V^*_B = V_B/V_{A+B}$. In the current study we are going to assume that all the grains of the second material have the same size. Therefore, the geometric parameter describing grain size relative to the particle size is equal to the ratio of the linear sizes of the first phase and the particle size, $d_B^* = d_B/d_p$. Finally, there should be parameter describing the shapes of the grain and particle. These are not easy to define quantitatively for the general case, but will be defined here by irregularity (compared to a circle or a sphere):

$$S_2^* = \frac{\text{perimeter}^2}{\text{area}}/(4\pi) \quad \text{for 2D simulations}$$

$$S_3^* = \frac{\text{surface area}^{\frac{3}{2}}}{\text{volume}}/(6\pi)^{\frac{1}{2}} \quad \text{for 3D simulations}$$

SIMULATION METHODOLOGY

A series of simulations were designed to validate the DGL model and to gain quantitative insight into the effects of key material properties and loading conditions on the liberation process. It is not possible to investigate the full range of all of the dimensionless parameters identified above in detail. To do this one would have to perform thousands of the simulations. Since each two-dimensional simulation currently requires about 20 minutes of CPU time and each three-dimensional simulation requires up to a day of CPU time on a dual processor AMD Athlon computer, this was clearly not possible. Therefore we are going to describe the results for a small subset of these variable values, which are presented as preliminary results of the liberation study in this publication.

A few Preliminary DGL simulations (basically numerical experiments) were conducted to create a basis for analysis. These simulations included two-dimensional (2D) and three-dimensional (3D) simulations involving drop weight breakage of particles in confined and unconfined loading situations. Since energy must be supplied to induce crack propagation and subsequently to liberate valuable grains prior to separation, specific energy has been treated as the principal independent variable in this study. Base case material properties for phase A and phase B were selected and an appropriate range of drop weight energy inputs was identified. The base case material properties (dimensional and dimensionless) are given in Table 1.

The base case particles were chosen as spherical (circular) of diameter 0.1m with 30 included grains of phase B which are also spherical (circular) with equisized diameters of 0.006m, thus the dimensionless size of included grains is $d_B^* = 0.06$ for this case.

TABLE 1 Material properties for base case simulation

	Crack Energy	Young's Modulus	Density
Phase A	$\gamma_A = 192$ Joules/m^2	$Y_A = 10^{10}$ Pascals	$\rho_A = 2600$ kg/m^3
Phase B	$\gamma_B = 192$ Joules/m^2, $\gamma_{B/A}^* = 1.0$	$Y_B = 10^{11}$ Pascals, $Y_{B/A}^* = 1.0$	$\rho_B = 5200$ kg/m^3
Interface	$\gamma_I = 19.2$ Joules/m^2, $\gamma_I^* = 0.1$		

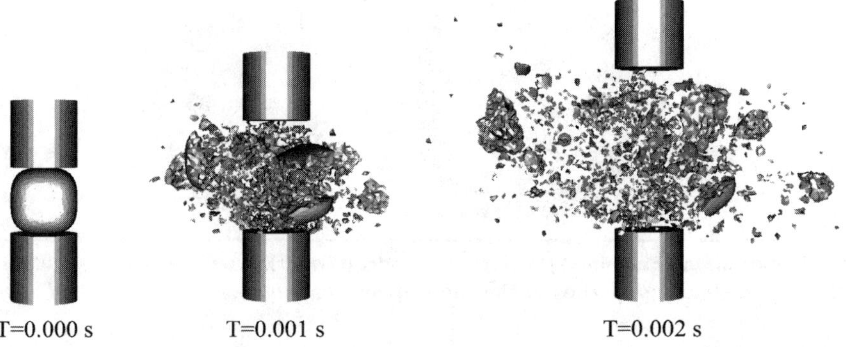

FIGURE 1 Snapshots of 3D simulation of drop weight breakage tests at the beginning, middle and end of an unconfined liberation event

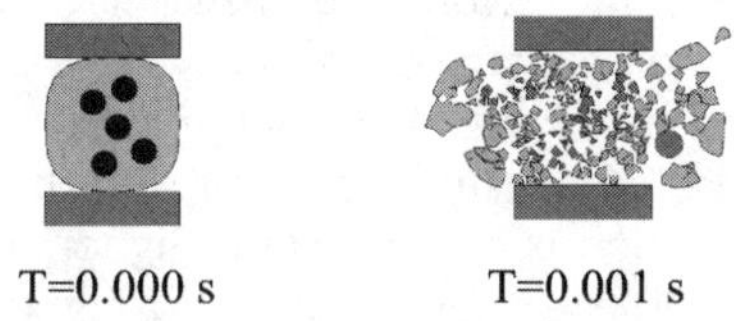

FIGURE 2 Snapshots of 2D simulation of drop weight tests at the beginning and middle of an unconfined liberation event (dark grains are unliberated B, light grains are liberated B)

Drop weight specific energies were chosen from preliminary simulations for the base case which yielded very small (a few %) to very large (approaching 100%) percentages of liberation. The following 12 specific energy values were chosen for the base case and used for each subsequent case run, E_I = 135, 270, 540, 810, 1080, 1350, 1620, 1890, 2160, 2700, 2970 and 3240 joules/kg. Procedures for classifying fragments with respect to degree of liberation were established. Single parameter figures of merit were determined for calculating liberation efficiencies, and a measure of the degree of random/non-random liberation was identified. Simulation conditions were then selected to evaluate the influence of the five dimensionless variables ($\gamma^*_{B/A}$, γ^*_I, $Y^*_{B/A}$, d^*_B, V^*_B, S^*) and loading condition (energy input and degree of shear) on liberation efficiencies.

RESULTS OF SIMULATIONS

Figure 1 shows a series of snapshots of a 3D liberation event in an unconfined state. Here we see that the continuous phase A (light color) is highly rubbleized at the center and near the platens larger "lune" pieces move quickly away from "ground zero". The discrete phase B are largely liberated (light color) throughout the volume of the particle.

The corresponding 2D liberation event for 2D in an unconfined state is shown in Figure 2. Similar behavior to that of Figure 1 is observed.

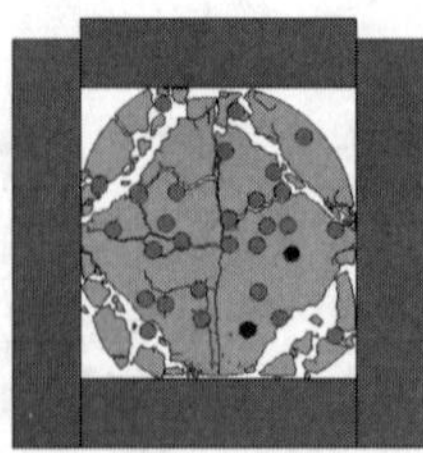
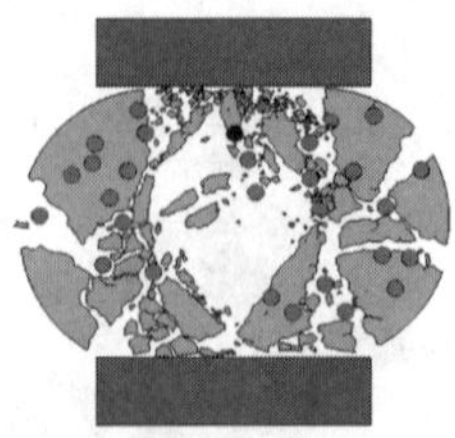

FIGURE 3 Snapshots of confined (left) and unconfined (right) two-dimensional simulations of liberation for base case properties at the same applied energy level

A detailed comparison of the two simulations shows that they yield approximately the same fragment size distributions for each phase and the same degree of liberation. Since the 2D simulations run much faster than the 3D (~75 times) the majority of the work reported here was done in 2D. A second consideration in the preliminary phase of this work is the choice of whether to do simulation in a confined space (with the particles touching rigid walls) or in an unconfined space in which fragments can freely fly away from "ground zero" of the breakage event.

Figure 3 compares snapshots of the liberation process for these two cases when the input energy E_I^* is equal to 4.0. It is visually evident that the fragments are finer for the unconfined case and the liberation is probably also higher. This is probably as expected since the friction losses and the evolution of opposing forces in the confined case undoubtedly reduce the efficiency of crack generation/propagation relative to the unconfined case.

Figure 4 confirms graphically that the fragment size distributions (for phase A and phase B) are indeed finer for $E^* = 4.0$ as well as for $E^* = 1.0$. In addition (as expected) the liberation process for the confined and unconfined states are also significantly different as shown in Figure 5 with the unconfined case producing more liberation at a lower energy than the confined case. Since real world liberation processes all involve some significant level of confinement (particles surrounding the breaking particle or other surfaces such as liners acting to confine) the confined loading situation was adopted as the base case loading for this study.

Because of the stochastic nature of the DGL simulations a procedure had to be developed to underlying trends and to make comparisons between liberation efficiencies for different values of dimensionless parameters. Two key response variables were identified as being important for characterizing liberation efficiency 1) the percentage of phase B grains which are liberated for a given energy input and 2) the percentage of the liberated B grains which are broken during the liberating event.

An example of these key variables versus energy input is shown for base case conditions with particles in two random orientations (Run 1 and Run 2) in Figure 6.

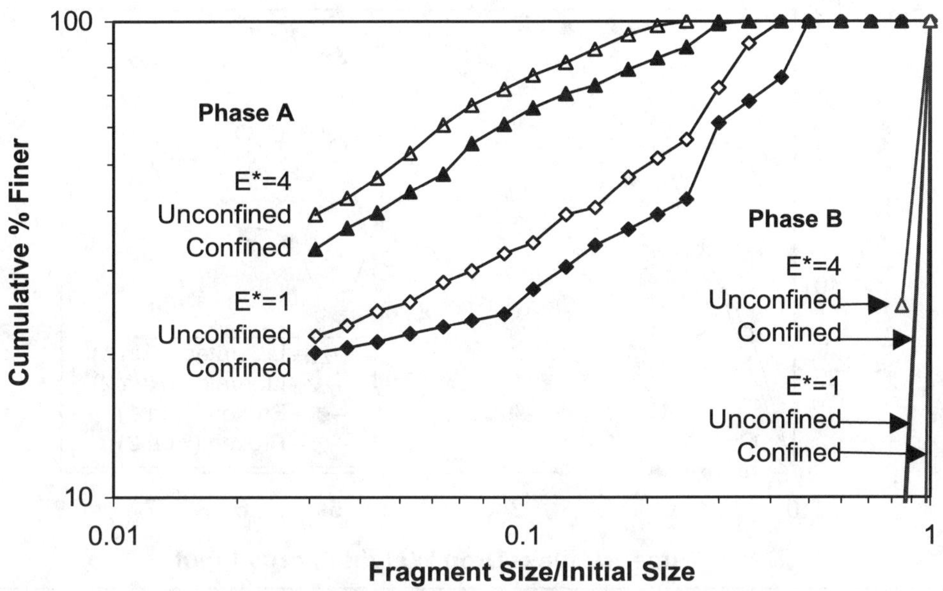

FIGURE 4 Fragment size distributions from confined and unconfined two dimensional simulations at two drop weight energy levels E* = 1.0, E* = 4.0

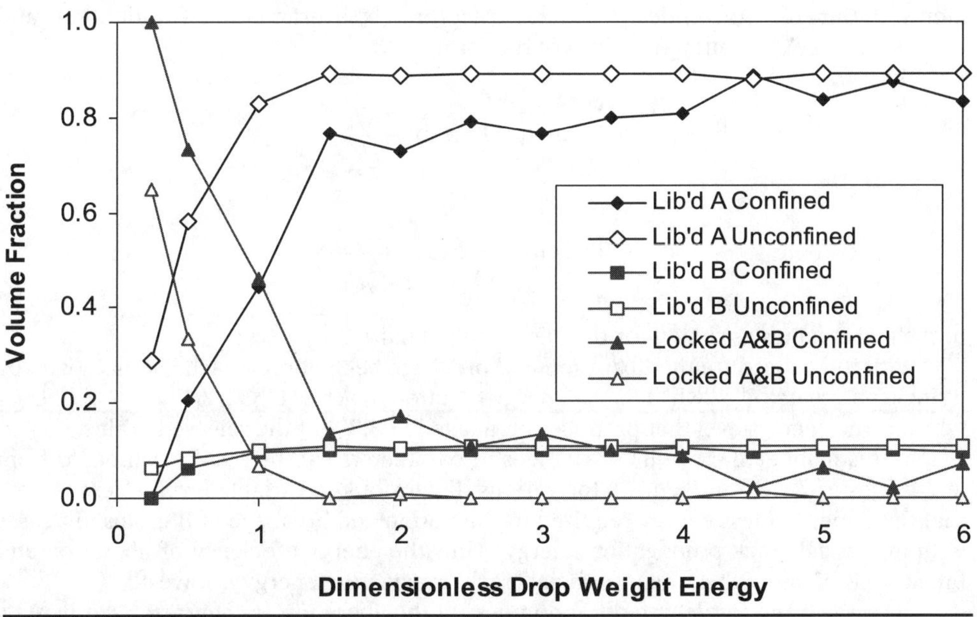

FIGURE 5 Comparison of amounts of liberated A, liberated B and locked A&B as a function of drop weight energy input, E*, for confined and unconfined breakage

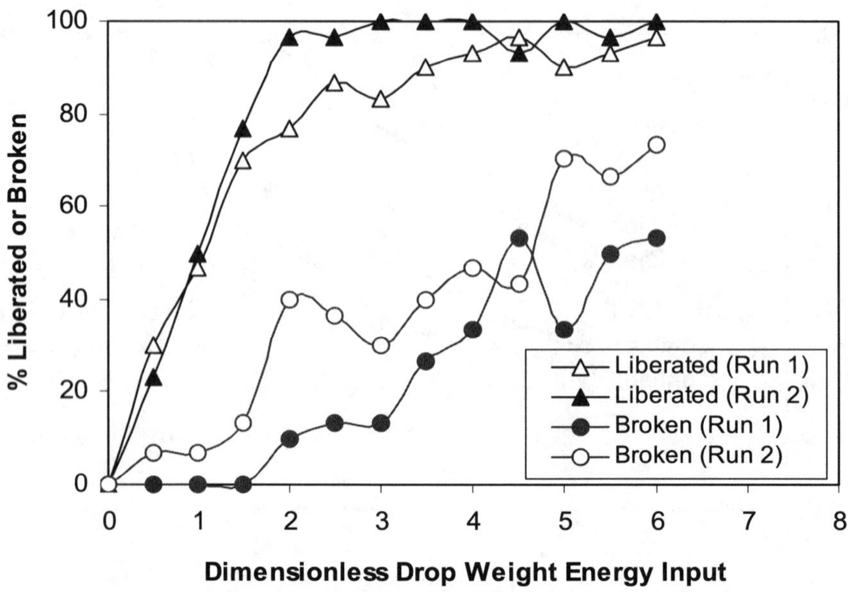

FIGURE 6 Simulated liberation and breakage versus drop weight energy, E*, for base material properties for two particle orientations

Here one sees that the typical statistical fluctuations of these numerical experiments. In this study the underlying trend is captured by averaging the two data sets and then fitting an exponential functions of the form

$$\% \text{ Liberated} = 1 - \exp\left[-\frac{E_I^*}{E_{63.3}^L}\right]$$

$$\% \text{ Broken} = 1 - \exp\left[-\frac{E_I^*}{E_{63.3}^B}\right]$$

to the mean values as shown for the base case in Figure 7.

With this procedure the liberation and breakage behavior are each characterized by a single parameter $E_{63.3}^L$ which is the energy required to achieve 63.3% liberation or $E_{63.3}^B$ which is the energy level that produces breakage in 63.3% of the released grains.

In this manner the results of changes in parameters can be readily estimated from and values as shown in Figure 8 for various dimensionless variable levels. Here we see that the dimensionless energy required for liberation and breakage of B grains increases with interfacial crack propagation energy. Thus the energy efficiency of liberation and breakage is highest when the interfacial crack propagation energy is lowest.

Then the trend for degree of randomness of the liberation (as characterized here by the ratio, $\rho^* = E_{63.3}^L/E_{63.3}^B$) can be easily visualized as shown in Figure 9.

For this case the degree of randomness ρ^* is low when the crack propagation energy of the A/B interface is low, while the point at which the phases and the interface have identical surface energies, i.e., the value of $\gamma_I^* = 1$ yields true random liberation, i.e., $\rho^* \cong 1.0$.

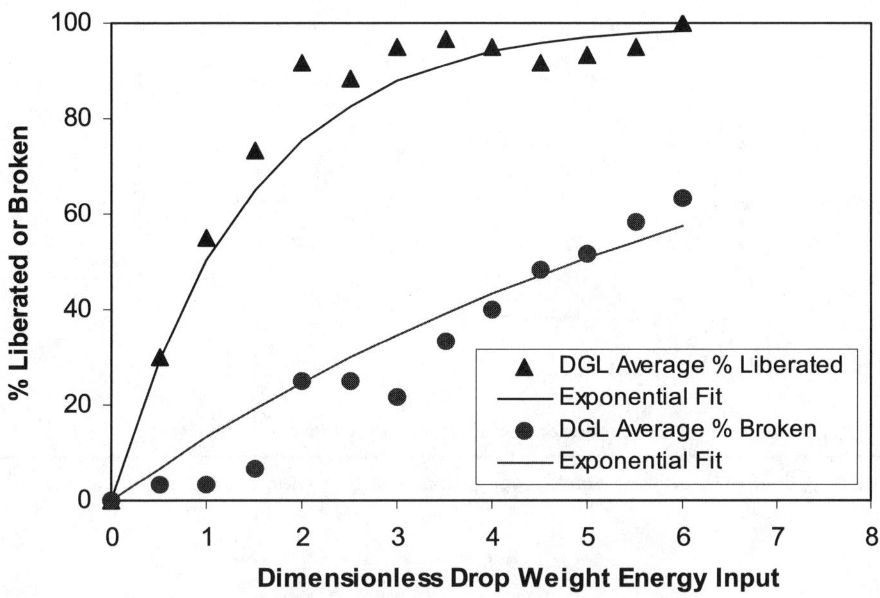

FIGURE 7 Averaging of DGL data from Figure 6 for liberation and breakage as well as resulting exponential fits

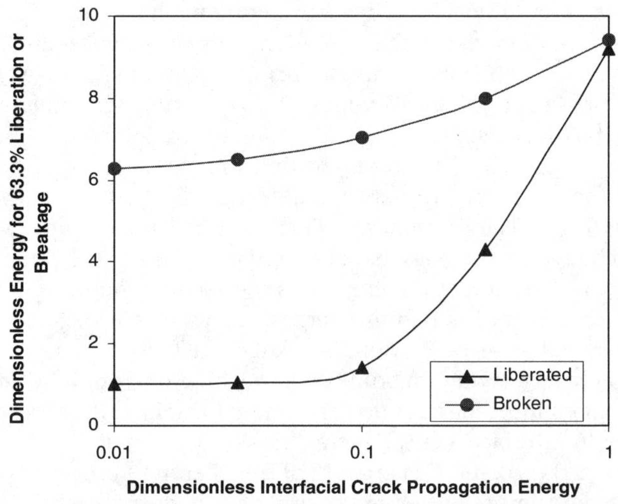

FIGURE 8 Effect of interfacial crack propagation energy, γ_I^*, on $E^L_{63.3}$ and $E^B_{63.3}$

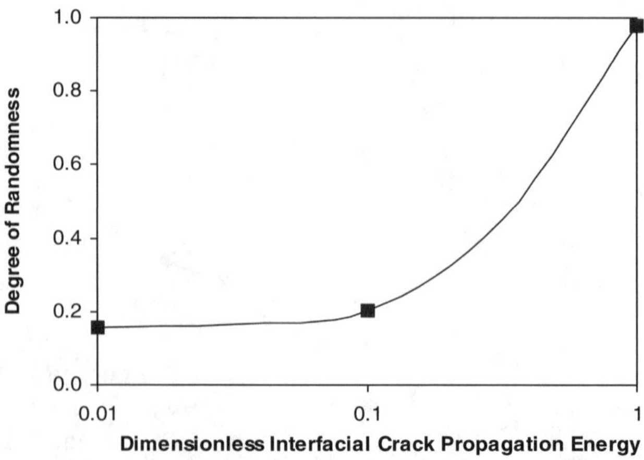

FIGURE 9 Effect of interfacial crack propagation energy, γ_i^*, on degree of randomness of liberation

Continuing this analysis for changes in the crack propagation energy for phase B, Figure 10 shows that for lower crack propagation energies the energy required to liberate and break B grains is low. The energy for breakage of B increases with increasing phase B crack propagation energy. The amount liberated is relatively insensitive to the crack propagation energy of B. These results are consistent with our intuition about what will happen when it becomes more and more difficult to propagate a crack in phase B.

The degree of randomness of the liberation is depicted in Figure 11 where we see that low phase B crack propagation energies result in random fracture (since a crack can pass easily from phase A to phase B) while high crack propagation energies in phase B make it difficult for the crack to cross the boundary therefore the liberation becomes much less random approaching liberation by detachment, i.e., $\rho^* \cong 0$.

The last of the material property variables examined here is the dimensionless Young's modulus $Y^*_{B/A}$. The efficiency of liberation and the amount of grain breakage that results for change in this variable are shown in Figure 12. Here we see that the energy required for liberation of B increases as the Young's modulus increases while the energy required for breakage of B grains decreases slightly with $Y^*_{B/A}$.

In this case the degree of random breakage calculated from Figure 13 shows a decrease with increasing Young's modulus approaching liberation by detachment as the dimensionless value approaches 1000. This can be explained in terms of the increasing strain along the B/A interface as $Y^*_{B/A}$ increases.

Turning now to the geometric variables of the liberation process, we will first examine the effect of the dimensionless grain size of B. Figure 14 shows that the energy required for both liberation and breakage decreases slightly with increasing grain size (at constant volumetric abundance). This is as expected since the grains act as a defect so that the greater the defect the weaker the particle is.

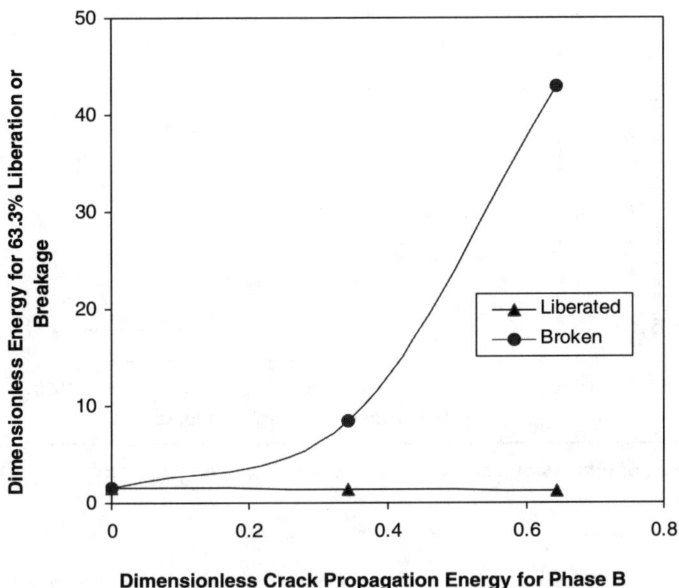

FIGURE 10 Effect of interfacial crack propagation energy for phase B, $\gamma^*_{B/A}$, on $E^L_{63.3}$ and $E^B_{63.3}$

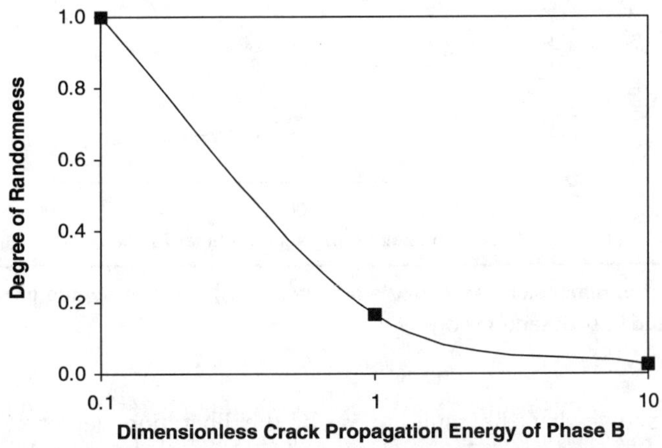

FIGURE 11 Effect of interfacial crack propagation energy of phase B, $\gamma^*_{B/A}$, on degree of randomness of liberation

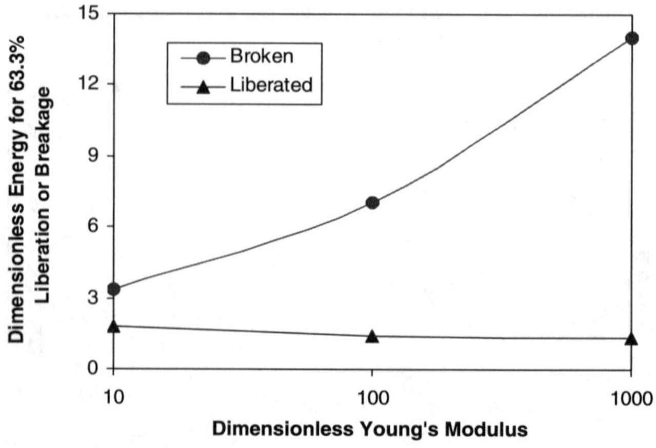

FIGURE 12 Effect of dimensionless Young's modulus $Y^*_{B/A}$, on $E^L_{63.3}$ and $E^B_{63.3}$

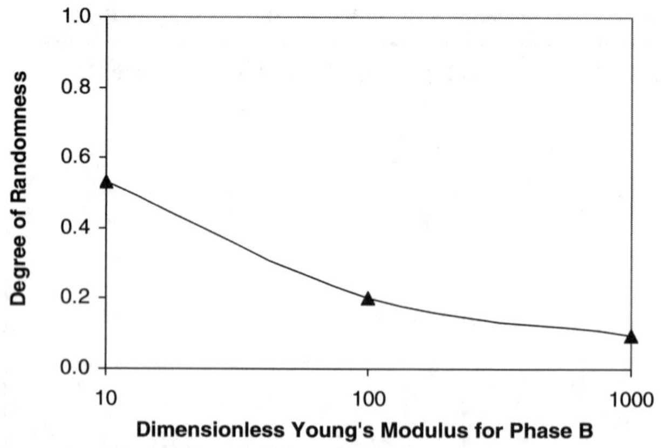

FIGURE 13 Effect of dimensionless Young's modulus, $Y^*_{B/A}$, on the degree of random liberation exhibited in breakage events

Figure 15 shows that the randomness of the liberation decreases slightly with increasing grain size and that it is close to pure detachment. It should also be noted here that this exercise was repeated for different parent sizes with the same d^*_B values and the dimensionless results were virtually identical.

The next geometric variable examined is volumetric abundance. From Figure 16 we see that the energy required to achieve a certain level of liberation and to produce a certain level of breakage of B increases and then decreases with increasing abundance. Sensitivity to this variable is especially low for liberation.

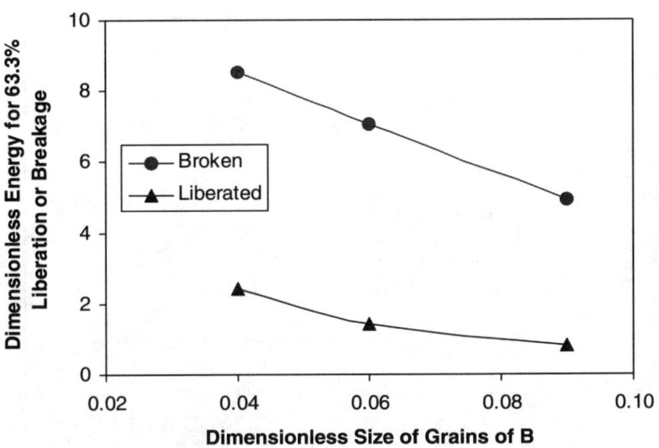

FIGURE 14 Effect of dimensionless grain size, d^*_B, on $E^L_{63.3}$ and $E^B_{63.3}$

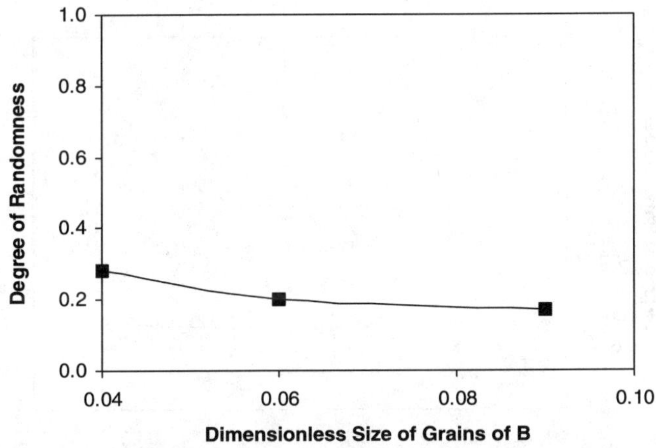

FIGURE 15 Effect of dimensionless grain size, d^*_B, on degree of randomness of liberation

Figure 17 shows that the degree of randomness of the liberation does not change much with the volume of phase B over the range examined.

The final two dimensionless variable effects examined are grain shape, S^*, and ratio of shear energy to total drop weight energy, E^*_S. Figure 18 shows that the energy for liberation is quite insensitive to shape irregularity, while the energy for breakage decreases dramatically for this minor phase with increasing shape irregularity probably due to stress concentrations at sharp corners.

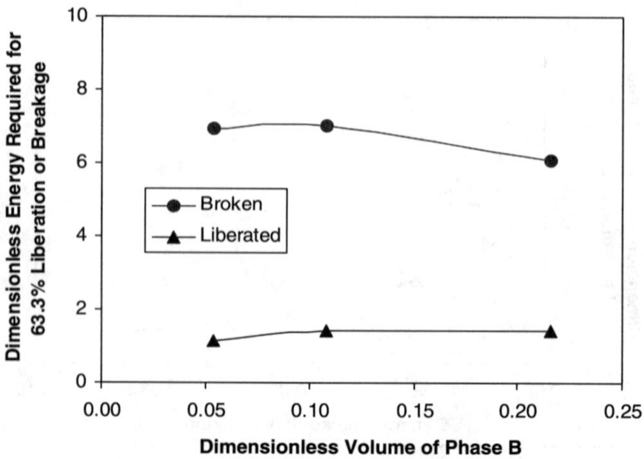

FIGURE 16 Influence of dimensionless volume of phase B, V^*_B on $E^L_{63.3}$ and $E^B_{63.3}$

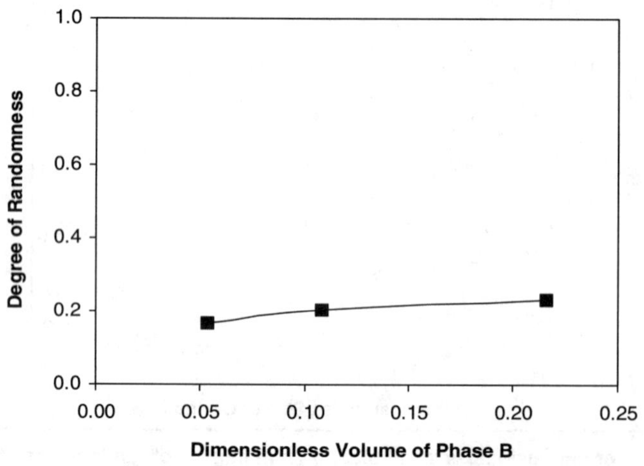

FIGURE 17 Influence of dimensionless volume of phase B, V^*_B, on degree of randomness of liberation

On the other hand, Figure 19 shows that the degree of randomness as the irregularity of shape increases.

To see the influence of adding a shear component to the loading, the drop weight was made to strike the particle at an oblique angle as shown in Figure 20.

The angle was varied from 0° (corresponding to pure impact) to 40° corresponding to 60% shear, 40% impact. The results of these loading experiments are shown in Figures 20 and 21. Figure 20 indicates that the energy for both breakage and liberation decrease slightly with increasing shear component.

Figure 22 shows that the degree of randomness of liberation is quite insensitive to shear.

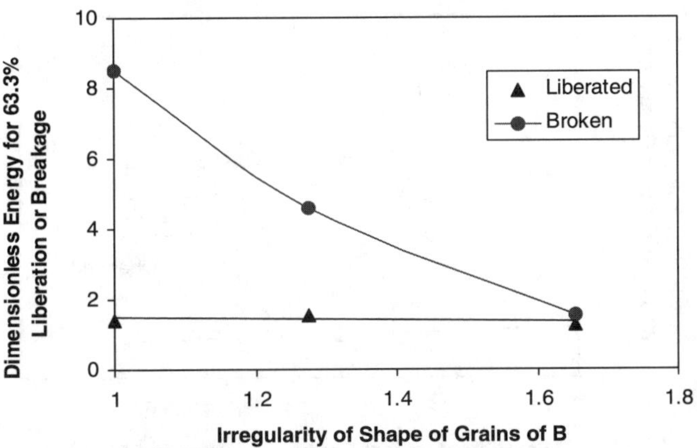

FIGURE 18 Effect of dimensionless grain shape, S* on $E^L_{63.3}$ and $E^B_{63.3}$

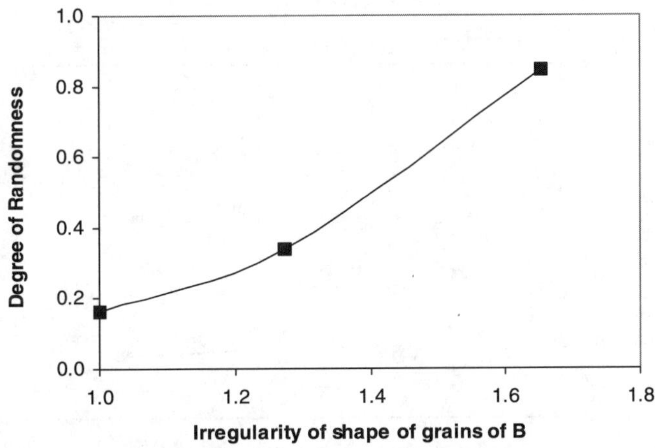

FIGURE 19 Effect of grain shape on degree of randomness of liberation

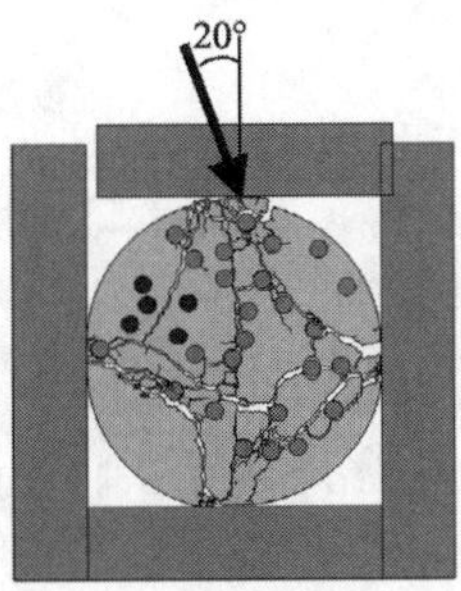

FIGURE 20 Snapshot of breakage simulation from drop weight with a 20° shear component

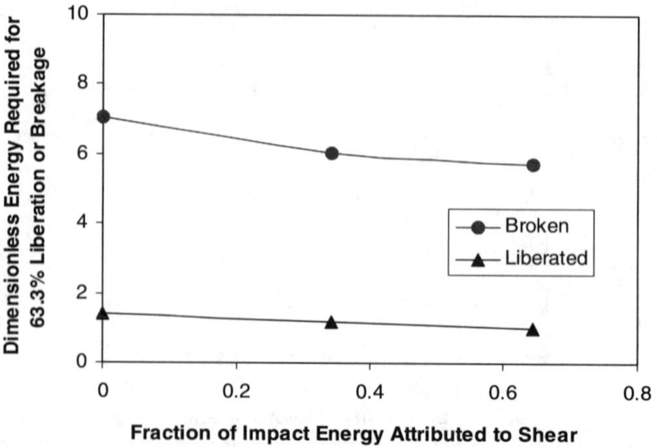

FIGURE 21 Influence of fraction of drop weight energy input applied through shear, E_S^* on $E_{63.3}^L$ and $E_{63.3}^B$

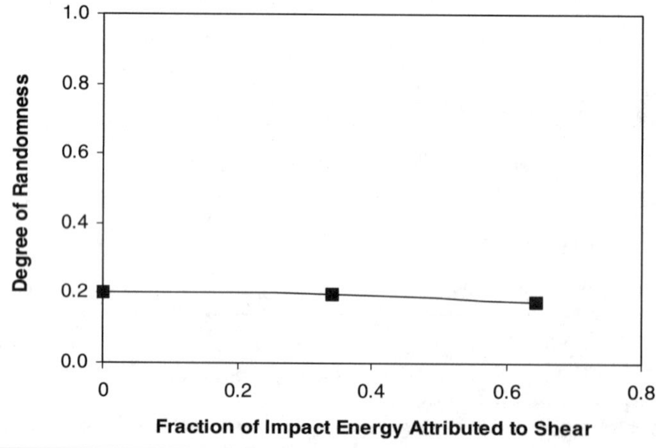

FIGURE 22 Influence of fraction of drop weight energy input applied through shear, E_S^*, on degree of randomness for liberation

CONCLUSIONS

The work described here represents a first attempt at physics based modeling of the liberation process using DGL. The basis of the model was described for both 2D and 3D simulations. A dimensional analysis was completed which showed that there are three dimensionless mechanical properties ($\gamma_{B/A}^*$, γ_I^*, $Y_{B/A}^*$) and three dimensionless geometrical properties of individual particles (d_B^*, V_B^*, S^*), which are important for multiphase systems involving two phases. In addition two dimensionless variables were identified for the loading (E^*, E_S^*) in a liberation event. Two new parameters $E_{63.3}^L$ and $E_{63.3}^B$ were defined to characterize the energy efficiency of a liberation process. The ratio of these parameters, $\rho^* = E^L_{63.3}/E^B_{63.3}$, was found to characterize the degree of randomness of a liberation process. A series of several hundred 2D and 3D simulations were performed with the DGL model and the trends for each of the variables defined above were recorded. Key findings from these simulations are: liberation efficiency is strongly dependent on the dimensionless interfacial crack propagation energy while the degree of random liberation is strongly dependent on 1) the dimensionless Young's modulus 2) the dimensionless crack propagation energy for the minor phase and 3) the dimensionless shape of grains of the minor phase. Because of the dimensionless nature of these results it is expected to apply down to small particle and grain sizes.

The results presented here suggest that the DGL tool has potential not only for predicting the effect of the properties of the individual particles on liberation behavior, but also for prediction of multiparticle liberation behavior in crushing and grinding operation. The authors believe that the current work has merely scratched the surface in identifying the potential fundamental and practical applications of this new tool.

ACKNOWLEDGEMENTS

The authors gratefully acknowledge the assistance of Dr. Ming Song in empirical model fitting and data presentation associated with the work presented here.

REFERENCES

Andrews, J.R.G. and T.S. Mika. 1975. Comminution of a heterogenous material: development of a model for liberation phenomena. Proc. 11th International Mineral Processing Congress, Cagliari, pp. 59–88.

Campbell, C.S. 1997 Computer Simulation of Powder Flows. In: (Gotoh et al eds) Powder Technology Handbook, 2 edn, Dekker, New York, p.777–793.

Choi, W.Z., G.T. Adel and R.H. Yoon 1988. Liberation modeling using automated image analysis. *International Journal of Mineral Processing*. Volume 22. pp. 59–73.

Gaudin, A.M. 1939. Principles of Mineral Dressing. McGraw-Hill: New York, N.Y., p.55

Herbst, J.A. 1988. Development of a multi-component multi-size liberation model. *Minerals Engineering*. Volume 1. Number 2. pp. 97–111.

Herbst, J., and L. Nordell 2001. Optimization of the design of SAG Mill Internals using high fidelity simulation, SAG 2001 Conference, Vancouver, B.C, pp.1–5.

Herbst, J.A. 2002. A microscale look at tumbling mill scale-up using high fidelity simulation. Proceedings of the 10th European Symposium on Comminution. Heidelberg, Germany.

Herbst, J.A. and A.V. Potapov 2004a. Making a discreet grain breakage model practical for comminution equipment performance simulation. *Powder Technology*. Pp. 144–150.

Herbst, J.A. and A.V. Potapov 2004b. Radical innovations in mineral processing simulation. *Mineral and Metallurgical Processing*. Volume 21. Number 2. pp. 57–64.

King, R.P. and C. Schneider 1993. Mineral liberation in continuous milling circuits. Proceedings of the XVIII IMPC. Australian Institute of Metallurgy.

Lin, C.L., J.D. Miller, J.A. Herbst J.E. Sepulveda, and K.A. Prisbrey 1985. Prediction of volumetric abundance from two-dimensional minerals image. *Applied Mineralogy*. Edited by W.C. Park et al. TMS/AIME. p. 157.

Miller, J.D., C.L. Lin 2004. Three-dimensional analysis of particulates in mineral processing systems by cone beam x-ray micro-tomography. *Minerals and Metallurgical Processing*. Volume 21. Number 3. pp. 113–124.

Potapov, A.V., M.A. Hopkins, and C.S. Campbell 1994. A Two-dimensional Dynamic Simulation of Solid Fracture. Part I: Description of the Model, *International Journal of Modern Physics.* Volume C. Number 6. pp.371–398.

Potapov, A.V., C.S. Campbell, and M.A. Hopkins 1995. A Two-dimensional Dynamic Simulation of Solid Fracture. Part II: Examples, *International Journal of Modern Physics.* Volume C. Number 6. pp.399–425.

Potapov, A.V., and C.S. Campbell 1996 A Three-dimensional Simulation of Brittle Solid Fracture, *International Journal of Modern Physics* Volume C. Number 7. pp.717–729.

Sepulveda, J.E., J.D. Miller, and C.L. Lin. 1985. Generation of irregularly shaped multiphase particle for liberation analysis. Proceedings of the XV IMPC. Volume 1. pp. 120–132.

A Simulation Analysis of the Net Effect of Feed Particle Size Distribution on SAG Mill Performance

Jaime E. Sepúlveda*

Semiautogenous grinding (SAG) operators have been realizing the very significant impact of fresh feed particle size distribution on the mill grinding capacity. In actual practice, this impact may be quite difficult to assess because of the natural process variability that may also involve—concurrently—other critical factors such as ore hardness. Recognizing the limitations of the empirical approach, the current publication proposes the use of mathematical models and simulators to explore and assess the net *effect of feed particle size only, independent of any other factors. The simulation analysis here reported provides a basis to claim that the size distribution of the particles being ground (simply characterized by a % – 2" or similar measurement) is one of the most relevant controlling variables of the overall mill performance, regardless of the SAG circuit configuration.*

INTRODUCTION

In mineral ore grinding operations, the size distribution of the particles being ground is—after ore hardness and mill power availability—the most relevant controlling variable of the overall grinding circuit performance, for any given product size specification; particularly in semiautogenous grinding (SAG) applications. In such full scale operations, sudden variations in feed ore particle size distribution may occur, very often in conjunction with changes in other ore properties (like, ore hardness) making it quite difficult for the process analyst to determine to what extent the observed mill performance fluctuations

* Moly-Cop Grinding Systems, Santiago, Chile

are due to feed size changes, ore grindability variations or most typically, both of them concurrently. Changes in feed ore particle size distribution may also be the result of flowsheet modifications, like the incorporation of pebble crushing and precrushing/pre-screening stages to the circuit. In these cases, the process designer would desire to be able to somehow assess the potential benefits associated to the flowsheet modifications, before even materializing the investment. Another frequent cause of systematic feed size differences between two parallel grinding lines is rock segregation in the feed ore stockpile. Upstream the grinding circuit, feed size changes may be the intended result of new operational practices at the mine (e. g., mine-to-mill approach). Again, the concern arises as to how to assess the impact of such improved practices in grinding performance, independent from those frequent and unavoidable ore hardness variations.

As a way to provide reasonable answers to such process analysis concerns, the current publication proposes the systematic use of mathematical models and simulators. The various examples here discussed highlight the clear impact of fresh feed particle size on any given SAG mill performance and so emphasize the need for accurate on-line measurements of such critical variable; area in which the research team led by the University of Utah Prof. Jan D. Miller–being honored at this Special SME Symposium–has made remarkable contributions (Lin and Miller 1993; Lin et al. 1995).

ANALYSIS SCOPE AND METHODOLOGY

Recognizing the practical limitations of obtaining actual, statistically representative operational data - free of the referred ore grindability disturbances–relying on the use of digital process simulators, based on sufficiently detailed models of the unit grinding operations involved, appears to be a very attractive option. By this approach, the various factors affecting process performance can be "decoupled" in order to explore and assess the *net* effect of feed particle size only. The analysis here reported made extensive use of the **Moly-Cop Tools**™ (Sepulveda, 2001a) software package capabilities to simulate fairly complex grinding circuits (Sepulveda, 2002), like those illustrated in Figure 1.

Moly-Cop Tools™ is a set of easy-to-use Microsoft Excel 2000 spreadsheets designed to help engineers characterize and evaluate the operating efficiency of any given grinding circuit, following standardized methodologies and well accepted evaluation criteria. In such computational environment, the main simulation modules applied were the spreadsheets SAGSim_Open and SAGSim_Recycle, both containing a fairly detailed and realistic SAG mill model, previously published elsewhere (Sepulveda 2001b). Such model was derived as an extension of the well known phenomenological models for conventional ball mills. These spreadsheets were properly linked to other complementary spreadsheets developed for screening and crushing applications to simulate the so called complex circuits, like the DSAG, SABC-1 and SABC-2 configurations (see Figure 1).

For the specific purposes of the present analysis, the fresh feed particle size distribution was characterized by a double Weibull expression of the type :

$$F_3(d) = \delta_0\,[1 - \exp\{\ln(0.2)(d/F_{80})^{\delta_1}\}] + (1 - \delta_0)\,[1 - \exp\{\ln(0.2)(d/F_{80})^{\delta_2}\}] \quad \textbf{(EQ 1)}$$

where

$F_3(d)$ = Weight fraction finer than indicated particle size d

F_{80} = 80% passing size of the fresh feed ore

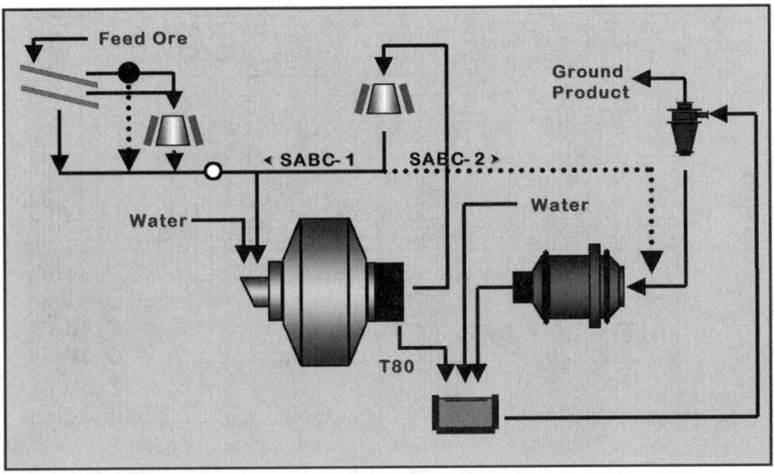

FIGURE 1 Alternative grinding circuit configurations

and δ_0, δ_1 and δ_2 are adjustable parameters allowing to represent a wide variety of alternative size distribution profiles. As an illustration, Figure 2 shows the effect of varying the F_{80} size and Figure 3 shows the corresponding effect of varying the δ_1 parameter.

Then, by loading the appropriate range of parameter values to the corresponding simulation routines, the net impact of the feed size distribution profile on circuit performance may be explored under very extreme feed size conditions, as listed in Table 1.

In this way, the 50% passing size of the fresh feed ore (F_{50}) was varied in the wide range of 25.6 mm to 72.4 mm.

Finally, in terms of processing equipment, the Base Case selected for the various simulation series consisted of a typical 36' $\phi \times$ 17' EGL SAG mill running under the DSAG Configuration (similar to SABC–1 in Figure 1, but without crushers) at 76% of its critical speed, 28% total filling (12% balls), " trommel (or screen), drawing 10,442 kW (gross) and processing 1,300 tonnes/hr of fresh feed ore. The feed size distribution parameters were those labeled "Base Case" in Table 1.

SIMULATION RESULTS

The first series of simulations was designed to explore the response of the traditional DSAG configuration to changes in fresh feed size distribution, at constant ore grindability. As illustrated in Figure 4, there is a noticeable correlation between the grinding capacity (tonnes/hr) developed by the mill and the fineness of the fresh feed; in this case, simply characterized by the % – 2" of "fines" content in the incoming ore. Independent of the feed size distribution profile–which, as indicated above, was varied in a very wide range of values and shapes–the % – 2" seems to be the mill throughput controlling variable. In the example shown in Figure 4, going from say, 45% – 2" up to 60% – 2", would translate into a 15% throughput improvement. As discussed in the following section, SAG mill operations worldwide have been actively searching for alternative ways to

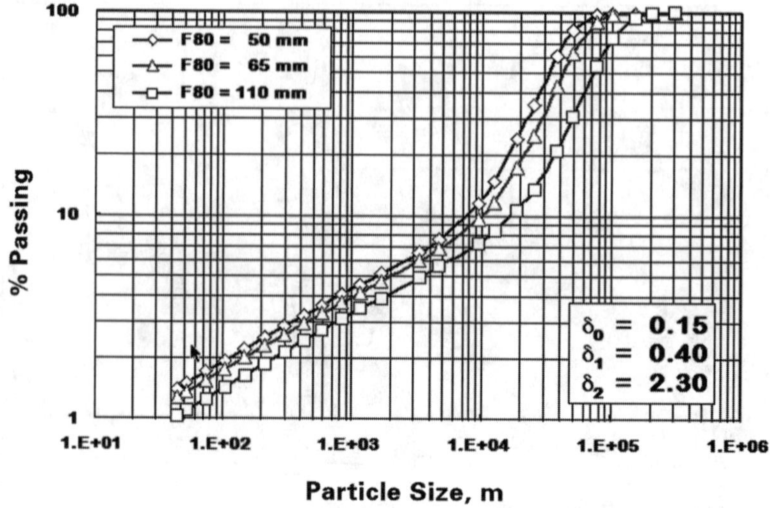

FIGURE 2 Feed particle size distribution profiles for varying F_{80} sizes, as described by the Double Weibull form (Equation 1)

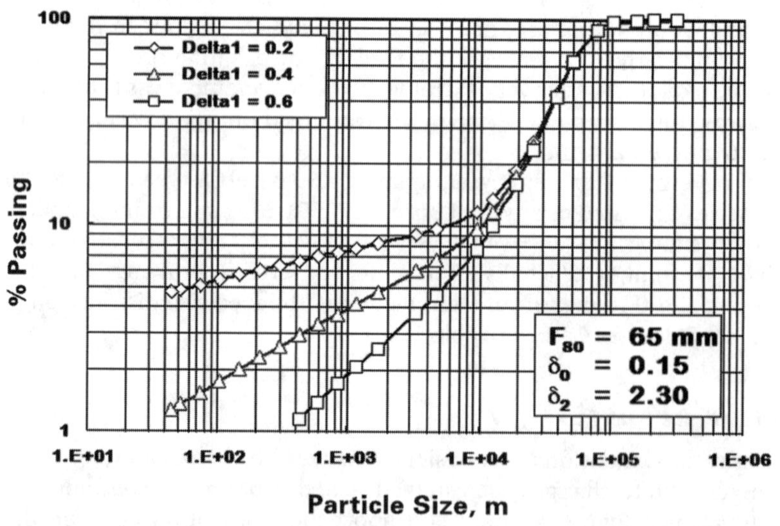

FIGURE 3 Feed particle size distribution profiles for varying d_1 values, as described by the Double Weibull form (Equation 1)

TABLE 1 Feed size distribution parameters (see Equation 1)

	δ_0	F_{80}, µm	δ_1	δ_2	F_{50}, µm
Base Case	**0.15**	**65,000**	**0.40**	**2.30**	**42,880**
D_{80} Effect	0.15	**50,000**	0.40	2.30	33,062
	0.15	**80,000**	0.40	2.30	52,767
	0.15	**95,000**	0.40	2.30	62,962
	0.15	**110,000**	0.40	2.30	72,362
δ_0 Effect	**0.20**	65,000	0.40	2.30	41,860
	0.30	65,000	0.40	2.30	39,528
	0.40	65,000	0.40	2.30	36,646
	0.50	65,000	0.40	2.30	32,964
δ_1 Effect	0.15	65,000	**0.20**	2.30	42,570
	0.15	65,000	**0.30**	2.30	42,725
	0.15	65,000	**0.50**	2.30	43,035
	0.15	65,000	**0.60**	2.30	43,190
δ_2 Effect	0.15	65,000	0.40	**1.00**	25,654
	0.15	65,000	0.40	**1.50**	34,593
	0.15	65,000	0.40	**3.00**	47,107
	0.15	65,000	0.40	**4.00**	50,878

effectively increase the fineness of the feed ore since the significant impact of such variable is now well recognized.

Additional simulation runs were conducted for the same Base Case circuit, converted to the SABC–1 configuration (see Figure 1) by adding an appropriate crusher to remove, crush and recycle pebbles. The SAG mill discharge grate was defined to have 2" slots; therefore, all of the –2" +½" material coming out of the SAG mill was diverted to such crusher. Similar to the previous DSAG case, Figure 5 illustrates that, for the SABC–1 configuration, the % – 2" in the feed ore is also the mill throughput controlling variable, over the wide range of simulated feed size profiles, but even more significant: going from 45% – 2" to 60% – 2" would translate into a 21% throughput improvement, as compared to the 15% improvement highlighted in the previous case.

Finally, Figure 6 further illustrates the critical role of the feed ore fineness for all circuit configurations considered. It is interesting to note that the various feed ore prescreening and precrushing options (see Figure 1), also simulated in the analysis being reported, only expand the range of potential throughput improvements by feeding finer and finer ores, but do not break the correlation with the corresponding % – 2" in such feed ores, as all simulated conditions obey the common trend indicated in Figure 6.

Very importantly, Figure 6 also reveals that the grinding capacity of any given circuit–regardless of the feed size–systematically improves as the circuit evolves from the DSAG to SABC–1 to SABC–2 configurations; that is, as the SAG mill contributes less and less to the overall grinding task !

SUPPORTING EMPIRICAL EVIDENCE

Being this just a simulation analysis, it is interesting to make reference to actual empirical evidence on the same topic here discussed. Palomo (2001) reported actual plant data

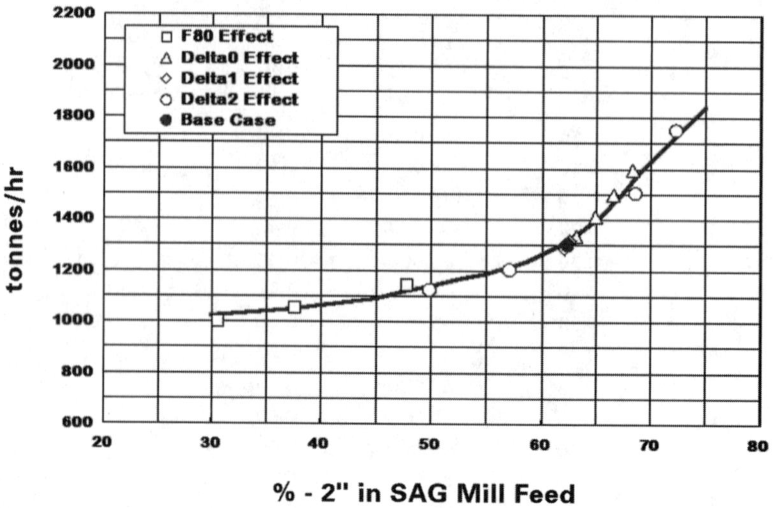

FIGURE 4 Effect of feed fineness on SAG mill throughput, when operating under the DSAG configuration

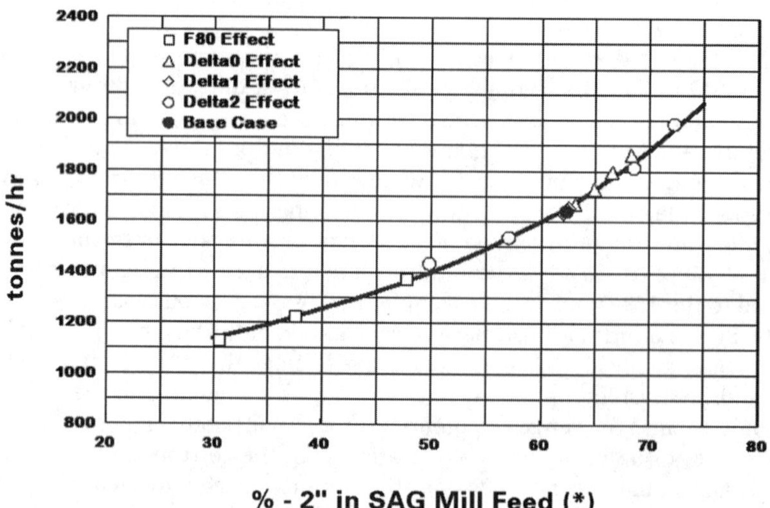

FIGURE 5 Effect of feed fineness on SAG mill throughput, when operating under the SABC-1 configuration. *Not including recycled pebbles.

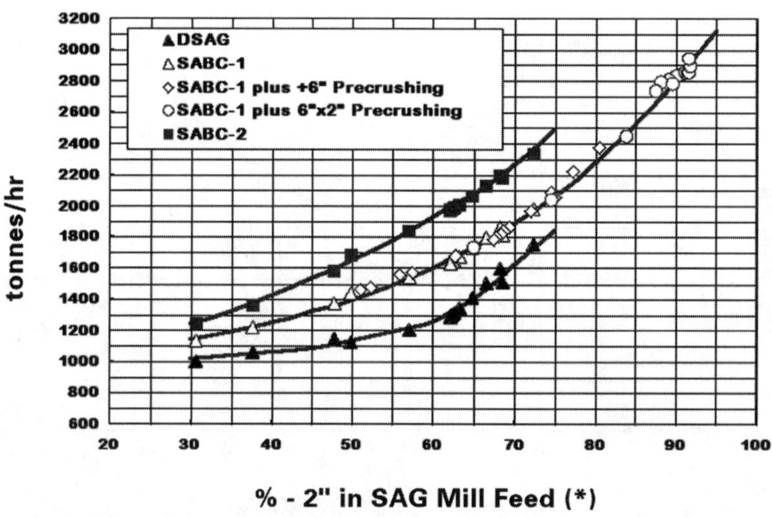

FIGURE 6 Effect of feed fineness on SAG mill throughput, when operating under various circuit configurations. *Not including recycled pebbles.

with reference to the effect of feed size on the performance of the two 36'ϕ × 17' Los Pelambres SAG mills, near Salamanca, Chile, running under the SABC–1 configuration. Figure 7 presents such throughput data as a function of the % – 1¼" in the fresh feed stream, confirming exactly the same trends observed via simulations. The broader variability of the actual plant data is to be expected as ore grindability, mill power draw and balls/rocks ratio changes may have taken place from one sampled condition to another. Quite noticeable, however, the observed improvement in going from 45% to 60% "fines" in the feed is also +21%, same as in the example of Figure 5.

Similar supporting empirical evidence on the critical role of feed particle size may be found elsewhere in the technical literature (Hart et al. 2001, Dance 2001, Morrell and Valery 2001).

CONCLUDING REMARKS

The simulation analysis here reported—to some extent validated by the limited actual data available, added to multiple comments and pieces of evidence collected from several worldwide operations—provides a reliable basis to claim that the size distribution of the particles to be ground (simply characterized by a % – 2" or similar measurement) is one of the most relevant controlling variable of the overall grinding circuit performance in SAG applications.

The design of the most suitable supervisory, expert system for the operational control of the grinding unit should therefore incorporate decision rules related to an on-line feed particle size signal, like those now commercially available.

The observed trends lead to the general conclusion that finer feeds—including precrushing of the coarser rocks—allow for significantly improved SAG mill throughput rates; raising solid objections to the very nature of semiautogenous grinding (i. e., the

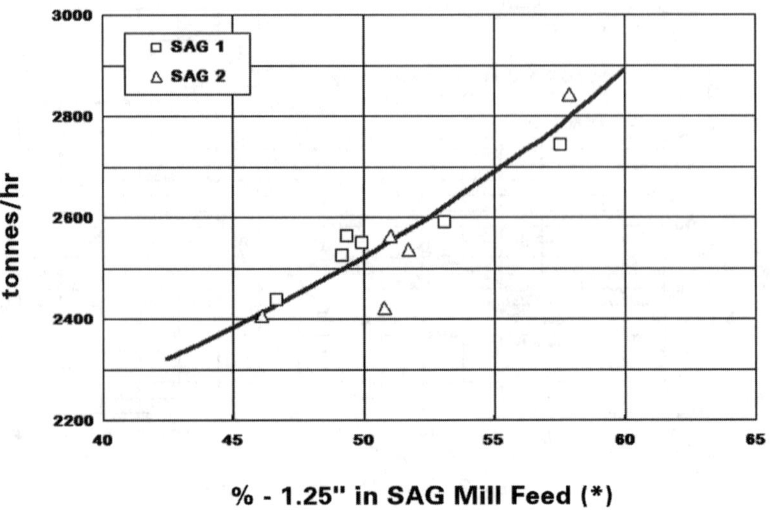

FIGURE 7 Effect of feed fineness on SAG mill throughput, based on actual operational data taken from Los Pelambres, Chile. *Not including recycled pebbles.

use of large rocks as grinding media) and at the same time, reinforcing the evolution of SAG mills back towards conventional ball mills. Therefore, the colloquial statement: "SAG mills are turning into BAG (barely autogenous) mills".

REFERENCES

Dance A., 2001, "The Influence of Primary Crushing in Mill Feed Size Optimization", Proceedings SAG 2001 Conference, Vol. 1, p. 189, Vancouver, B. C., Canada.

Hart S. et al., 2001, "Optimisation of the Cadia Hill SAG Mill Circuit", Proceedings SAG 2001 Conference, Vol. 1, p. 11, Vancouver, B.C., Canada.

Lin, C.L. and J.D. Miller, 1993, "The Development of a PC, Image-based, On-line Particle Size Analyzer", Minerals & Metallurgical Processing, February, 1993.

Lin, C.L., Yen, Y.K. and J.D. Miller, 1995, "On-line Coarse Particle Size Measurement–Industrial Testing", SME Annual Meeting, Denver, Colorado, USA.

Morrell, S. and W. Valery, 2001, "Influence of Feed Size on AG/SAG Mill Performance", Proceedings SAG 2001 Conference, Vol. 1, p. 203, Vancouver, B. C., Canada.

Palomo, R., 2001, "Experiencia Mine-to-Mill en Minera Los Pelambres", IX Moly-Cop Grinding Symposium, Osorno, Chile. (available upon request to molycoptools@molycop.cl)

Sepúlveda, J.E., 2001a, Moly-Cop Tools, Version 1.0, *"Software for the Assessment and Optimization of Grinding Circuit Performance"*, available free of charge, upon request to molycoptools@molycop.cl

Sepúlveda, J.E., 2001b, *"A Phenomenological Model of Semiautogenous Grinding Processes in a Moly-Cop Tools Environment"*, Proceedings SAG 2001 Conference, Vol. 4, pp. 301–315, Vancouver, B.C., Canada.

Sepúlveda, J.E., 2002, *"Desarrollo de un Simulador de Circuitos Complejos de Molienda SAG, en ambiente Moly-Cop Tools"*, IV Simposio Internacional de Mineralurgia, TECSUP, Lima, Perú.

Microscale Characterization and Analysis of Particulate Systems via Cone-Beam X-ray Microtomography (XMT)

C.L. Lin* and C. Garcia†

Characterization and analysis of multiphase particles and/or particulate systems (such as the packed particle bed) are of great technological importance for mineral processing applications. In general, the performance of these applications depends on the statistical characteristics of particle microstructures, such as composition distribution, surface exposure of mineral grains, pore network structure, etc. For continued technological progress in the field of multiphase particulate processes, the need for quantitative spatial analysis of multiphase particles in three dimensions has increased significantly. Such quantitative information must be accurate enough that the measured values can be used as parameters for simulation models, process design procedures, and control strategies.

Modern 3D image-acquisition techniques, such as X-ray microtomography (XMT), offer a unique imaging capability that can produce high-resolution (a few micrometers) three-dimensional images of the internal structure of multiphase particulate samples. These three-dimensional image data present new challenges for image analysis, image processing, and visualization. Furthermore, important issues should focus on how to use the 3D information and derive new insights into the improvement, design, and operation of mineral processing processes. In this paper, utilization of XMT for characterization and analysis of leaching process involving multiphase particulate system is discussed.

* Department of Metallurgical Engineering, University of Utah, Salt Lake City, Utah

† Asistencia de Ingenieria Aplicada, Pedro Aguirre Cerda, Antofagasta, Chile

INTRODUCTION

The ability to obtain accurate 3D geometrical and textural information for individual particles and assemblies of particles is an important tool that can provide information to describe the performance of various processes in the mineral processing and extractive metallurgy industries. In addition to its use in improving existing processes, a 3D tool for particle characterization is of critical importance in developing new production technologies. In this regard, the unique cone-beam X-ray micro-CT facility (Lin and Miller 2002) at the University of Utah can be used to provide 3D information at a resolution of 5 microns regarding particle size, shape, texture, composition, and surface area. Already X-ray microtomography (XMT) has been demonstrated to be an effective diagnostic tool to describe the spatial distribution of mineral phases (Lin and Miller 1996). Recently the application of XMT in mineral processing technologies such as coal washability analysis, mineral liberation analysis, mineral exposure analysis, and analysis of the pore-structure network of packed particle beds was reviewed (Miller and Lin 2004; Lin and Miller 2004). In this paper, the capabilities of XMT are used to quantify leaching reaction progress during column leaching for unsaturated flow conditions.

EXPERIMENTAL METHODS

High-Resolution Three-Dimensional XMT

Application of the principles of cone-beam computed tomography (CT) at the microscale level (microtomography) allows for the quantitative examination of objects in three dimensions at high resolution. Only recently have practical microtomography systems been developed. As the resolution and the techniques for 3D geometric analysis have advanced in the last decade, it has now become possible to map in great detail the mineralogical texture of ore particles in three-dimensional digital space. High-resolution 3D X-ray microtomography (XMT) can be used directly determine the percentage of exposed valuable mineral grain in multiphase particles that vary in size from 100 mm down to a few hundred microns (Miller et al. 2003). Spatial resolution on the order of ten micrometers can be achieved with the use of microfocus X-ray generators.

With cone-beam CT, a whole 3D data set is acquired with only one rotation of the sample. This provides for fast data acquisition and better X-ray utilization. In a cone-beam design, each projection of the object is, in essence, a radiograph. Attenuation measurements are made simultaneously for the entire object rather than for a single slice. Clearly, cone-beam CT has the best prospects for true 3D liberation/exposure analysis and should be able to provide the necessary accuracy to quantitatively describe the micron-sized multiphase systems such as the 3D distribution of mineral phases in multiphase particles.

The University of Utah micro-CT system was designed and assembled to obtain 20482048 pixel reconstruction over a 10-mm diameter, while allowing for the imaging of somewhat larger (40-mm) objects. Specifically, the specimen-positioning stage system can be manually mounted at one of three different locations, providing system magnifications of 1.25, 2, or 5 and spheres of reconstruction with respective diameters of 10, 25, or 40 mm.

The system consists of a microfocus X-ray source, a specimen-positioning stage, and a digital X-ray detection camera. These hardware components are integrated inside an

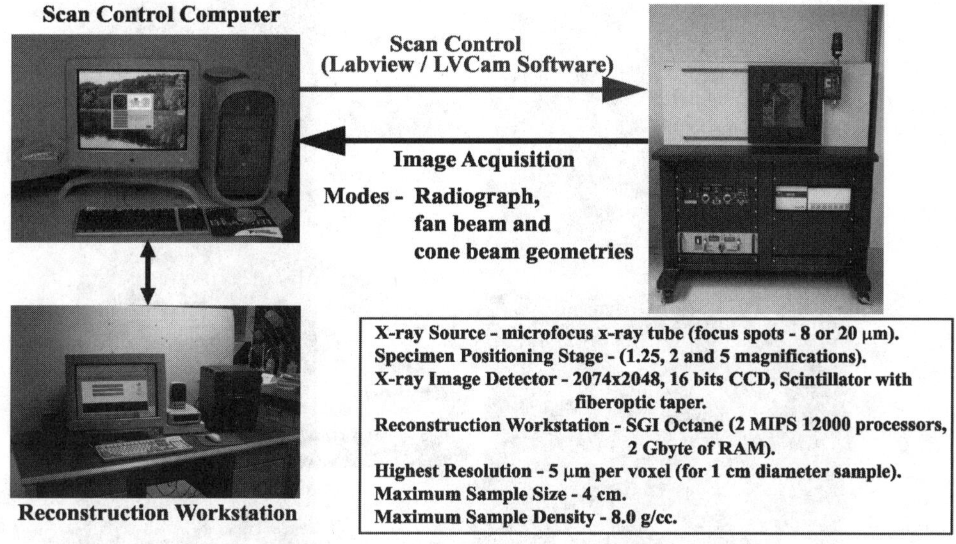

FIGURE 1 The University of Utah cone-beam X-ray micro-CT system

interlocked radiation-safety enclosure. This container is mounted on a vibration-isolated stand, which also supports the power supplies, drivers, and controllers for the X-ray source, CCD camera, and positioning system. A photograph of the University of Utah cone-beam X-ray microtomography system is shown in Figure 1.

Mini-Column Leaching Experiments

Figure 2 illustrates the step-by-step procedures for the mini-column leaching experiments. First, a copper ore sample was loaded into the mini-column reactor, which was about 2.5 cm in diameter and 3 cm in height, to examine the leaching characteristics under unsaturated flow conditions. The particle size distribution (PSD) used for the mini-column test is given in Table 1. Secondly, the column was irrigated at the top with fresh acidic solution, as shown in the right-hand side of Figure 2. The solution was passed through the column once. No solution recycling was done. The inlet solution was maintained at pH 1.5. The flow rate of the solution through the column was kept constant at 8.0 $l/(m^2h)$ using an intravenous (I.V.) system. Solution samples were taken daily for copper assay analysis. Finally, the leaching characteristics of the mini-column test were monitored using XMT with a resolution of 40 microns and associated solution copper assays. Initially, 3D XMT scans were performed for a dry packed mini-column and at different leaching time intervals. A total of 14 sets of XMT data were used for tracking the change of high-density mineral grains during the leaching process.

FIGURE 2 Step-by-step procedures for a mini-column leaching experiment monitored by XMT

TABLE 1 Particle size distribution (PSD) used for mini-column test

Size	Wt. (%)
–6.3 mm + 3.15 mm	50
–3.15 mm + 1.7 mm	50

RESULTS AND DISCUSSIONS

XMT System Calibration

X-ray tomographic reconstruction produces a three-dimensional map of X-ray attenuation coefficients of the irradiated cross-section of the specimen. Differentiation of features within the sample is possible because the linear attenuation coefficient (μ) at each point depends directly on the electron density, the effective atomic number (Z_e) of the material comprising the sample, and the energy of the X-ray beam (E). A simplified equation that illustrates the approximate relationship among these quantities is

$$\mu = \rho\left(a + \frac{bZ_e^{3.8}}{E^{3.2}}\right) \quad \textbf{(EQ 1)}$$

where ρ is the density of the phase, a is a quantity with a relatively small energy dependence, and b is a constant (McCullough 1975; Wellington and Vinegar 1987). When a mixture of atomic species is present, Z_e (the effective atomic number) is defined by

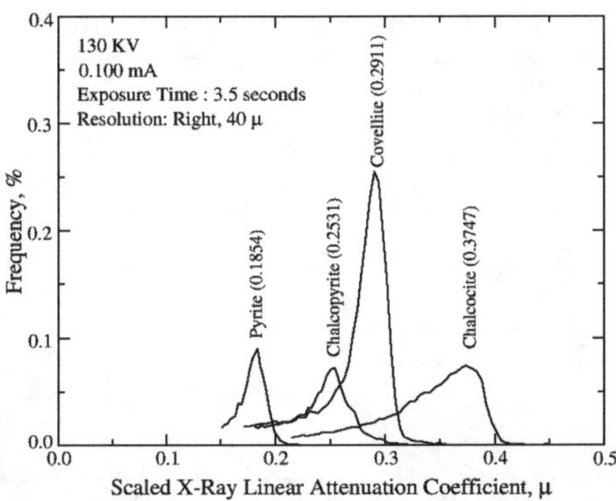

FIGURE 3 Histograms of the X-ray attenuation coefficient (μ) of different natural minerals obtained from XMT scans at 40-μm resolutions. The mean X-ray attenuation coefficient (μ) of each mineral is indicated at the peak of the histograms in the figure.

$$Z_e^{3.8} = \sum_i (f_i[Z_i]^{3.8}) \qquad \textbf{(EQ 2)}$$

where f_i is the fraction of the total number of electrons contributed by element i with atomic number Z_i. Generally, the different mineral phases contained in the ore body can be distinguished using the X-ray CT technique. In this regard, the natural mineral particles were selected as standards for calibration of the CT scans. Figure 3 presents an example of the frequency histogram of X-ray linear attenuation coefficients (μ) obtained for the mineral samples scanned with an energy setting of 130 kV at a resolution of 40 microns and an exposure time of 3.5 seconds. The means of the X-ray linear attenuation coefficients for these natural mineral particles are indicated at the peak positions.

Mini-Column Leaching Experiments

For the mini-column leaching experiments under unsaturated flow conditions, the columns were loaded with coarse ore with a particle size distribution as shown in Table 1. Figure 4 shows the percent of copper recovery versus the leaching time. It is noted that a higher copper recovery rate was observed at the initial stage (about 3 days), which then leveled off and reached ~38% of copper recovery after 67 days of leaching time.

Changes in High-Density Mineral Grains during Mini-Column Leaching Experiments

Beside the solution assays as reported in the previous section, the performance of the mini-column leaching test was monitored using XMT with a resolution of 40 microns. Initially 3D XMT scans were performed for dry packed mini-columns and scanned again at different leaching time intervals. To quantify the roles of surface wetting and diffusion during column leaching of coarse crushed ore, high-density mineral grains were classified and their movements tracked during the leaching periods, based on the mineral types and relative position inside the particle of host rock (liberated, exposed, internal grains).

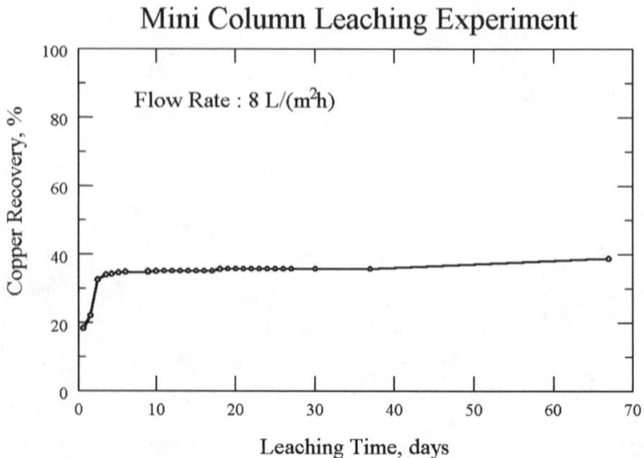

FIGURE 4 Plot of copper recovery from a mini-column test under unsaturated flow conditions

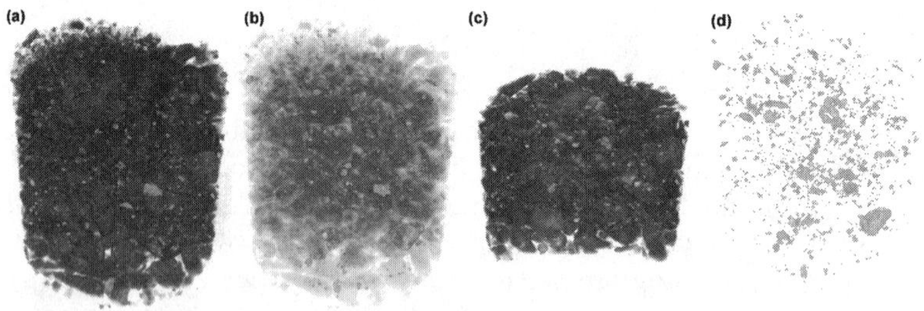

FIGURE 5 (a) Volume rendering image from a packed bed of dry particles for mini-column leaching experiments. (b) Semi-transparent rendering of particles with opaque solids displayed as the high-density mineral grains. (c) Sectional view of (a) by removing part of the volume. (d) High-density mineral grains with the host rock removed digitally.

Detailed information regarding the structure of the packed bed is an important technological matter, since any detailed mathematical model will require this information in order to predict and/or optimize the heap leaching process. Figure 5 shows four kinds of 3D rendering images of the initial dry packed mini-column. Figure 5(a) illustrates a solid packing arrangement of the column. The semi-transparent rendering of particles with the high-density mineral grains displayed as opaque solids is shown in Figure 5(b). A sectional view of the particle bed with part of the volume removed is shown in Figure 5(c). Different mineral phases inside the particles are clearly distinguishable in these images. Finally, the host rock is removed digitally, and only the high-density mineral grains remain unchanged, as shown in Figure 5(d).

Figure 6 shows cross-sectional XMT images taken from the middle of the mini-column at 0 and 14 hours of leaching time, respectively. It is noted that many particles are moved due to compaction during the first day of the leaching process. The sectional

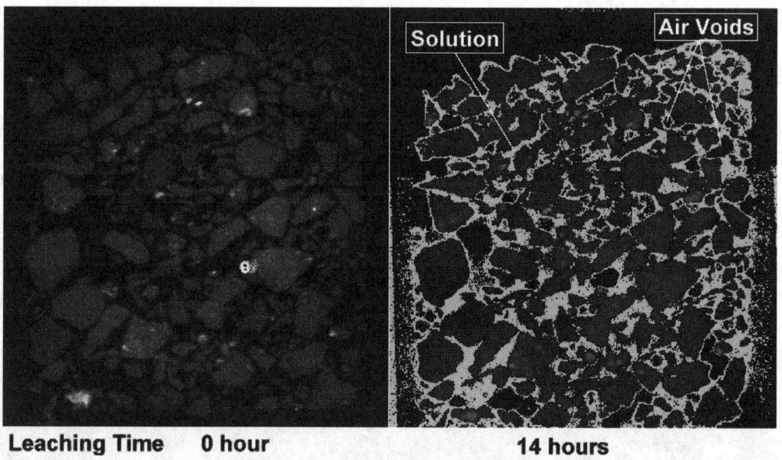

FIGURE 6 Cross-sections of XMT images for different leaching times. The images show air, rocks, and leaching solution found inside the mini-column.

XMT image of the 14-hour leaching time clearly indicates that rocks, leaching solution, and air are in contact inside the column under unsaturated environments.

Several 3D image-processing algorithms were developed to facilitate the tracking of the high-density mineral grains. The steps for tracking the change in location of the high-density mineral grains are summarized as follows:

- Iseparate the high-density grain from the host rock by thresholds,
- perform grain identification using connected-component and provide an index for each grain,
- determine of the centroid, volume and surface area of each grain, ordered by volume,
- measure of the histogram of X-ray linear attenuation coefficients (μ) for each grain, and
- manually associate grain indices for different leaching periods.

Based on these algorithms, individual mineral grains can be identified and tracked during the leaching process. In this regard, the 3D XMT images of high-density mineral grains obtained on the first and 37th days were selected to identify the dissolved grains, as shown in Figure 7. A specific particle containing dissolved high-density mineral grains was identified, as indicated by the circle in the left-hand side of Figure 7. To further study in detail the behavior of grains, this specific particle was extracted from the packed mini-column. As shown previously, the histogram of the X-ray linear attenuation coefficient (μ) can be used to identify and quantify the amount of high-density mineral phase. In this regard, the attenuation histogram of this specific particle was plotted before the leaching process, as shown in Figure 8. The 3D view of this particle is shown as the inset graph in Figure 8. A total of 13 grains were identified, comprising 28.33% by volume of this particle (volume = 4.11 mm^3), as indicated in Figure 8. Most of these grains were identified as chalcocite.

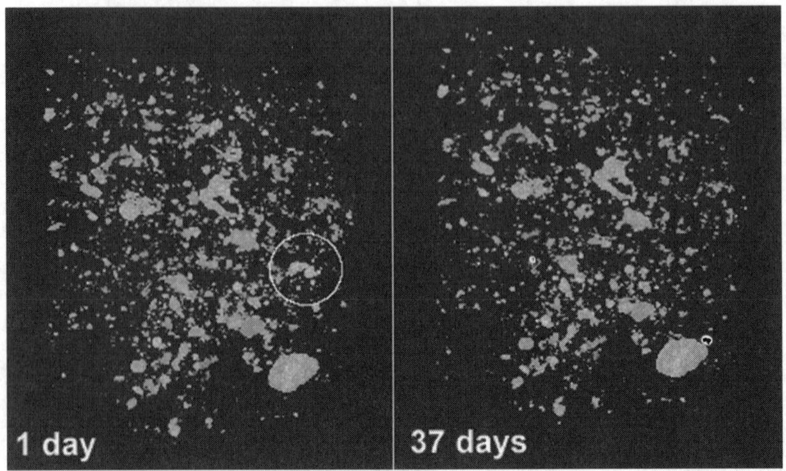

FIGURE 7 Dissolved grains were identified by comparing the XMT images of the high-density grains at leaching periods of 1 and 37 days, respectively. The circle in the left-hand size image indicates such grains.

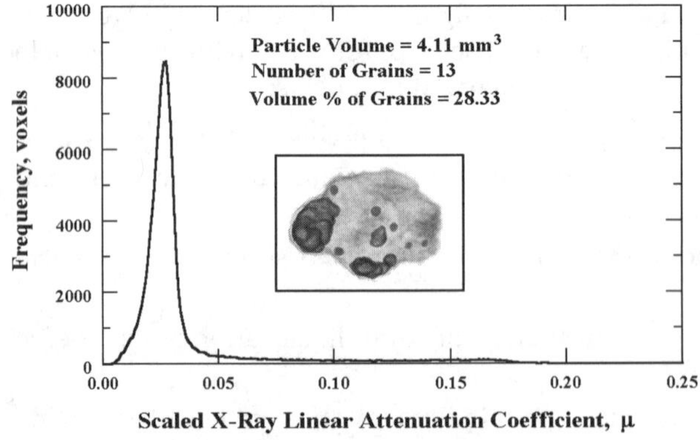

FIGURE 8 Histogram, obtained from XMT scans at 40-micron resolution, of X-ray attenuation coefficient (μ) of a specific particle (4.11 mm^3, 13 grains, 28.33% by volume of high-density grains) before leaching

Figure 9 illustrates the overall attenuation histograms of this particle for different leaching times. The curves related to the high-density mineral grains are clearly shifted to a lower μ, which indicates that some of the chalcocite mineral phase has dissolved. Changes to this particle were observed and characterized in detail during the leaching operation. The 3D views of this particle during the mini-column leaching experiment are shown in Figure 10. As leaching proceeded, most of the chalcocite grains were dissolved, as clearly indicated from these photos.

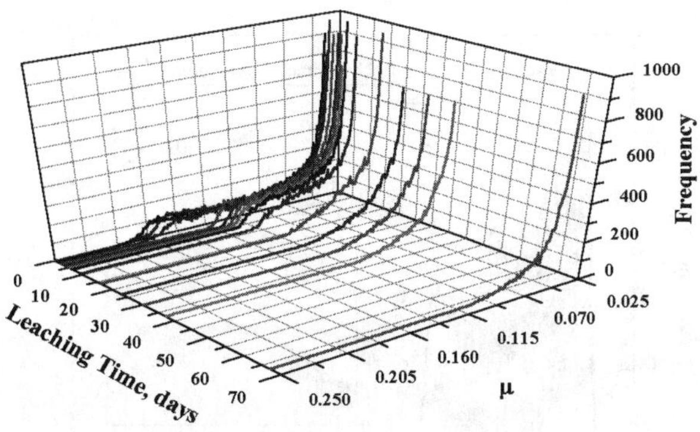

FIGURE 9 Overall histograms of X-ray linear attenuation coefficients (μ) of a specific particle (as shown in Figure 8) after different leaching times. Curves shift to the left due to the dissolution of the chalcocite mineral phase.

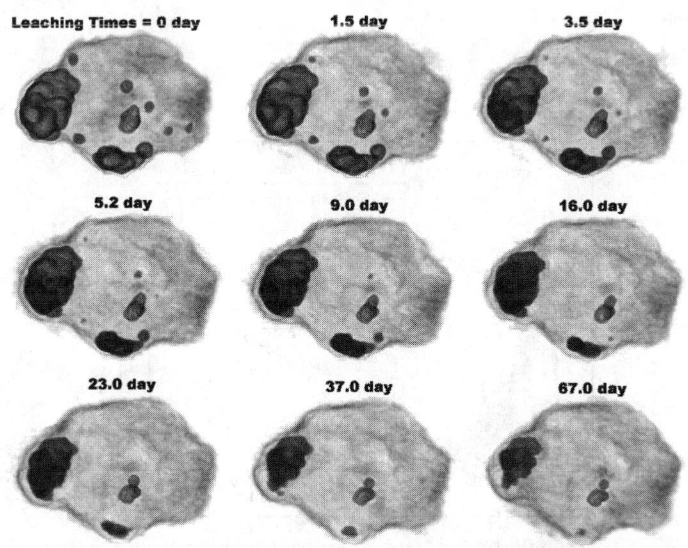

FIGURE 10 Storyboard of the dissolving chalcocite grains during the mini-column leaching experiment

To further study in detail the behavior of grains, the three grains of greatest initial volume were tracked, and the percent volume change after 67 days of the leaching test was classified and is shown for each grain in Figures 11, 12, and 13, respectively. For example, the large grain (grain number 1) with an initial volume of 0.7878 mm^3, located in the left-hand side of the particle, as shown in Figure 10, was tracked for different leaching times. This particular mineral grain was identified as an exposed cluster composed of chalcocite grains. As leaching proceeded, about 75% of the chalcocite

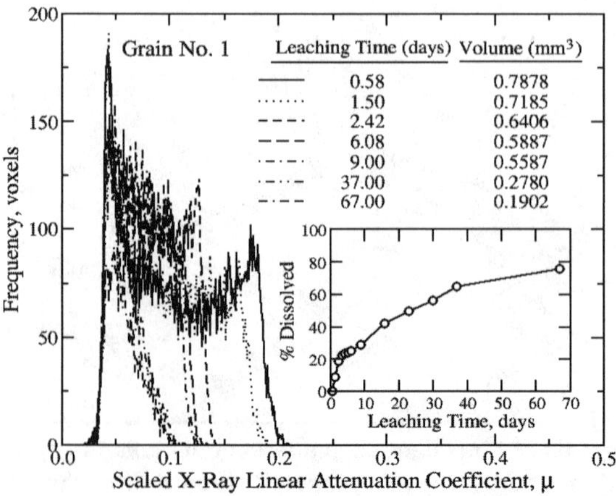

FIGURE 11 Tracking mineral grain number 1 and its attenuation histogram during different leaching times. The initial grain has a volume of 0.7878 mm^3. The rate of dissolution is shown in the inset graph.

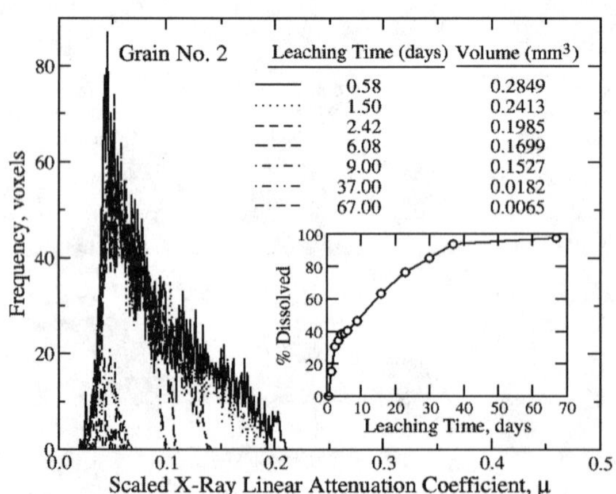

FIGURE 12 Tracking mineral grain number 2 and its attenuation histogram during different leaching times. The initial grain has a volume of 0.2849 mm^3. The rate of dissolution is shown in the inset graph.

phase was dissolved after 67 days of leaching, as clearly indicated from the attenuation histogram of this grain. The rate of dissolution of this grain is shown as the inset graph in Figure 11. In the same fashion, the rate of dissolution for grains number 2 and 3 after 67 days of leaching are about 98% and 45%, respectively. Grains number 2 and 3 are identified as exposed and internal (near the edge) grain-type, respectively. It is evident that exposed grains have a higher percent of volume reduction than internal grains.

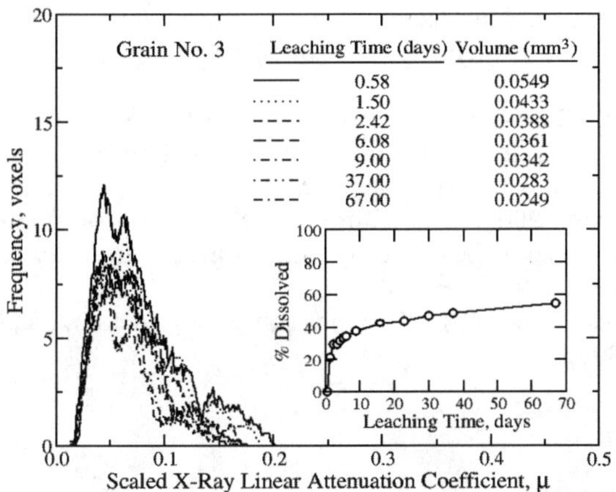

FIGURE 13 Tracking mineral grain number 3 and its attenuation histogram during different leaching times. The initial grain has a volume of 0.0549 mm^3. The rate of dissolution is shown in the inset graph.

SUMMARY AND CONCLUSIONS

X-ray microtomogrpahy (XMT) is an important 3D tool for particle characterization and analysis of particulate systems. Utilization of XMT can provide information to improve the performance of existing processes. Further, this 3D tool is of critical importance in the development of new product technology.

XMT was used to quantify leaching reaction progress during mini-column leaching for unsaturated flow conditions. These preliminary results indicate that XMT is suitable to characterize and analyze in detail the changes of mineral grains during mini-column leaching experiments. The analysis has demonstrated the capability of XMT with respect to the reaction of mineral grains, and this research should be extended to a more complete full-scale study including consideration of fluid flow phenomena.

ACKNOWLEDGEMENTS

The authors would like to thank Christian Roldan for help with mini-column leaching experiments and copper assay analysis.

REFERENCES

Lin, C.L. and J.D. Miller. 1996. Cone beam x-ray microtomography for three-dimensional liberation analysis in the 21st century. *International Journal of Mineral Processing*. 47:61–73.

Lin, C.L. and J.D. Miller.2002. Cone beam x-ray microtomography–a new facility for three-dimensional analysis of multiphase materials. *Minerals & Metallurgical Processing*. 19:65–71.

Lin, C.L. and J.D. Miller. 2004. Pore structure analysis of particle beds for fluid transport simulation during filtration. *International Journal of Mineral Processing*. 73:281–294.

McCullough, E.C. 1975. Photon attenuation in computed tomography. *Medical Physics*, 2:307–320.

Miller, J.D., C.L. Lin, C. Garcia, and H. Arias. 2003. Ultimate recovery in heap leaching operations as established from mineral exposure analysis by X-ray microtomograph. *International Journal of Mineral Processing*. 72:331–340

Miller, J.D. and C.L. Lin. 2004. Three-dimensional analysis of particulates in mineral processing systems by cone beam X-ray microtomography. *Minerals & Metallurgical Processing*. 2:113–124.

Wellington, S.L. and H.J. Vinegar. 1987. X-ray computerized tomography. *J. of Petroleum Technology*. 8:885–898.